T0199668

Agroecology in China

Science, Practice, and Sustainable Management

Advances in Agroecology

Series Editor: Clive A. Edwards

Agroecology in China

Science, Practice, and Sustainable Management

Edited by
Luo Shiming
Professor of Agroecology,
Institue of Tropical and Subtropical Ecology,
South China Agricultural University,
Guangzhou, People's Republic of China

Stephen R. Gliessman
Professor Emeritus of Agroecology,
University of California,
Santa Cruz, USA

CRC Press
Taylor & Francis Group
Boca Raton London New York

CRC Press is an imprint of the
Taylor & Francis Group, an **informa** business

CRC Press
Taylor & Francis Group
6000 Broken Sound Parkway NW, Suite 300
Boca Raton, FL 33487-2742

First issued in paperback 2018

ISBN-13: 978-1-4822-4934-7 (hbk)
ISBN-13: 978-0-367-11251-6 (pbk)

Library of Congress Cataloging-in-Publication Data

Names: Shiming, Luo, 1946- editor. | Gliessman, Stephen R., editor.
Title: Agroecology in China : science, practice, and sustainable management / editors: Luo Shiming and Stephen R. Gliessman.
Other titles: Advances in agroecology.
Description: Boca Raton, FL : Taylor & Francis, 2016. | Series: Advances in agroecology | Includes bibliographical references and index.
Identifiers: LCCN 2015036669 | ISBN 9781482249347
Subjects: LCSH: Agricultural ecology--Research--China. | Agricultural systems--China.
Classification: LCC S471.C6 A366 2016 | DDC 630.951--dc23
LC record available at http://lccn.loc.gov/2015036669

Visit the Taylor & Francis Web site at
http://www.taylorandfrancis.com

and the CRC Press Web site at
http://www.crcpress.com

Contents

Part I
An Overview

Luo Shiming

Part II
Agroecology Research and Practice on a Crop-Field Scale

Li Long, Zhang Weiping, and Zhang Lizhen

Wang Jianwu

Chen Xin

Zhang Jiaen, Quan Guoming, Zhao Benliang, Liang Kaiming, and Qin Zhong

Cao Linkui and Zhang Hanlin

Lin Wenxiong, Fang Changxun, Wu Linkun, and Lin Sheng

Part IV
Perspective

Preface: Toward an Agroecological Future for China

When most Westerners think about agriculture in China, they often recall F.H. King's book *Farmers of Forty Centuries*, published in 1911. King documented a remarkable array of farming practices that today we could probably call agroecological practices: intercropping, crop rotation, use of legumes, integration of crops and livestock, rice–duck–fish farming, the use of night soil, and many other practices that had been applied, as King observed, for at least 40 centuries while maintaining healthy soils and producing food for millions of people on a very limited land base. He also reflected on how in less than 150 years in the United States a very extractive agriculture had already degraded what were initially very rich soils and that we could learn a lot about how to reverse these trends by linking the traditional knowledge of agriculture in China with the modern approaches of the West.

This was my impression of agriculture in China without having ever been to the country and what was on my mind when I was visited by Luo Shiming in 1984 when he stopped to see me in Santa Cruz. I was just a few years removed from my time as a professor at the Colegio Superior de Agricultura Tropical in Tabasco, Mexico, where my own roots in agroecology were strongly influenced by the remarkable farming practices and agroecosystems of the traditional Maya farmers of the region. I remember talking to Luo Shiming about the similarities between the traditional farming systems of both places and how we shared common foundations in agroecology as a result. Both of us were also faced with the challenge of trying to use agroecological knowledge to solve problems that large-scale industrialized agriculture was causing for both the environment and society. I had just recently initiated the Agroecology Program at the University of California at Santa Cruz, and he was just developing the Agroecology Laboratory at the South China Agricultural University in Guangzhou, China.

During my first visit as an agroecologist to China in the spring of 1987, when I traveled there as a fellow in the W.K. Kellogg Foundation Leadership Fellowship program, I marveled at the views of Chinese agriculture that I saw from train and bus windows. I was able to meet with one of the pioneers of agroecology in China, Professor Han Chunru, at the Beijing Agricultural University. There I learned that there was much more support for larger scale, industrial systems of agriculture than there was for small-scale traditional systems, but he impressed upon me how there was a growing ecological agriculture movement beginning to take root around the country. I experienced this movement first-hand in October of the same year when I was invited by Professor Li Zhengfang to participate in an eco-agriculture training course in Nanjing sponsored by the China National Environmental Protection Agency (NEPA). I traveled by boat through the complex and ancient canal network of the Yangtze River delta to a community where I saw traditional rice paddies with associated ducks and fish; biogas digesters powered with pig and human waste; biogas being used for cooking and heating mushroom houses and chicken coops; fish, turtle, and crab ponds being fed with biogas digester sludge; intercropped fruit trees and vegetables; solar hot water and cooking stoves; and much of the production being done organically. Needless to say, I could see many agroecological principles at work in what seemed to be traditional Chinese communities. This visit evolved into a collaborative project funded by the University of California's Pacific Rim Research Program, where we carried out an agroecological study comparing organic and conventional strawberries in three locations (California, Taiwan, and near Nanjing) in support of an emerging organic movement in all three locations.

I finally visited Luo Shiming in Guangzhou in 1996 and was introduced to the many agroecological studies he had been working on in tropical south China. He guided me through a tour of the extensive areas of the famous pond–dike systems that he had spent many years examining from an agroecological perspective. I could see the intensive integration of raised fields with intensive fruit and vegetable production, surrounded by pond-like canals where fish were raised. But, I also could see the beginning of industrialization in these areas as ponds were being filled in and fields covered with an array of buildings, factories, and residential areas. We passed through these areas on our way to see farms in the Pearl River delta where land was being reclaimed from the sea through a massive system of dikes, canals, and raised fields so that any land lost to agriculture would be replaced by reclamation. Even though I was told that it was the government's policy that no land be lost from agriculture, it seemed like there were many fewer crops and many fewer people in these reclaimed lands.

When I returned to China in 2013 to present a talk on agroecology at the 16th National Agroecology Conference in Harbin, northeast China, I was once again overwhelmed by the view from the bus window, but this time by mechanized, monoculture corn as far as I could see. I heard stories of great increases in yields of corn, soybeans, wheat, and rice, but I also heard stories of water pollution, soil erosion, and the movement of rural inhabitants from their farming villages to the newly constructed urban areas around major cities. At the same time, I was impressed by the extensive research that was being done by the Chinese agroecologists from all over the country who gave presentations at the conference. Much research was focused on improving productive output from the land, but there was also a strong focus on correcting industrial agriculture-generated problems through agroecological practices such as reduced tillage for controlling erosion and nutrient leaching, rotations with legumes and grasses to improve soil health, intercropping, agroforestry, and integrated animal–crop systems of many kinds. It occurred to me that most of us outside of China had little knowledge of this agroecological research, not to mention that these agroecology conferences had been held for so long! Discussions I had with Lou Shiming and other agroecologists at this conference convinced us it was time to show the rest of the world what the Chinese agroecology movement looks like. The idea for this book was the result.

As can be seen in this volume, Chinese agriculture, and the work of China's agroecologists, has undergone many shifts and changes over the past 30 years. As one of the major crop and animal domestication centers in the world, China is a country with a history of intensive agriculture that goes back more than 5000 years, but recent industrialization processes have begun to fundamentally change traditional rural life and agricultural practices. Whereas agricultural production and farmers' incomes have increased, industrialization has also brought about serious issues of environmental pollution, resource depletion, and food safety; however, during the same period, agroecology as a discipline has been taught in most of the agricultural universities in China. Eco-agriculture, as the movement and practice of agroecology principles and philosophy, has been recently encouraged by the Chinese government and studied by Chinese scientists with experience from all over China. With contributions from 17 leading experts who have been engaged in agroecology teaching and research, as well as eco-agriculture development, in China for more than 20 years, this book provides a unique look into the way agroecology is being applied.

The book begins with a review of the development of agroecology in China by my co-editor Luo Shiming. As one of the leading agroecologists in the country, he is the ideal person to provide an overview of how agroecologists research and teach agroecology within the context and challenges facing the rural agricultural sector in China.

The second section of the book provides nine examples of field-scale agroecology research and practice from many parts of the country. In Chapter 2, after a brief introduction to the wide variety of intercropping systems that occur throughout China, Li Long and colleagues go into great detail on the research being done on the complex below- and aboveground ecological interactions going on between leguminous and non-leguminous plant species. The understanding of the mechanisms of interaction in the root zone of the mixtures is especially impressive. Intercropping soybeans in sugarcane fields as a way to reduce nitrogen fertilizer use and greenhouse emissions is covered in Chapter 3. Wang Jianwu uses the land equivalent ratio (LER) to demonstrate overyielding in the mixtures, as well as improved nitrogen use efficiency and sugar quality. Chapters 4 to 6 present detailed research results of what perhaps is the classic integrated farming system in China where rice is integrated with water-loving animals. All three chapters go way beyond mere description of the systems and provide remarkable detail on the ecological relationships and mechanisms that make these systems work. After giving a brief description on the general situation of the rice–fish system in China, Chen Xin presents research results on the ecological mechanisms that provide the stability, sustainability, and productivity of this traditional heritage co-culture agroecosystem. Zhang Jiaen and colleagues provide a historical overview and describe the current situation of the mutualistic relationship between rice and ducks, as well as in-depth research results focused on elucidating the agroecological mechanisms of their interaction. Cao Linkui and Zhang Hanlin describe a more recent system design based on rice–frog co-culture, drainage channels planted with aquatic vegetation ("eco-ditches"), and an artificial wetland for reducing nitrogen and phosphorus in the irrigation water effluent while also lowering fertilizer inputs.

The next three chapters in this section delve into how ecological research methods can be applied to improving agroecosystem design and management, especially in ways that improve their environmental performance. In Chapter 7, Lin Wenxiong's team reviews the extensive research done on the role of allelopathy in rice agroecosystems, as well as how to address the issue of the build-up of autotoxicity effects of plant-produced chemicals in the cultivation of Chinese medicinal herbs. In Chapter 8, Weng Boqi and Wang Yixiang present extensive results of research on the structure of grass cover in fruit orchards and how the cover impacts soil properties, biodiversity, and soil carbon sequestration in subtropical China. The effects of no-till rice production on soil properties, greenhouse gas emissions, carbon sequestration, and nitrogen balance in central China are presented by Li Chengfang and Cao Cougui in Chapter 9.

The organic food production situation in China is introduced in Chapter 10 and Chapter 11. In Chapter 10, the researchers headed by Li Ji present long-term results of the conversion of greenhouse vegetable production to organic management, along with impacts on output quality and quantity, soil properties, and nutrition balance. Wu Wenliang gives an historical overview in Chapter 11 of the development stages, labeling standards, management systems, market development, and policy support for organic agriculture in China.

The next five chapters make up the section that reviews research and development of a range of eco-agriculture practices and management experiences in regions with harsh climates and difficult topographic situations in northwest China (Chapters 12 and 13), northeast China (Chapters 15 and 16), and southwest China (Chapter 14). The semi-arid Loess Plateau region of the country is the focus of Chapter 12. Li Fengmin and Guan Yu provide evidence for how eco-agriculture systems can efficiently collect, deliver, and utilize scarce water resources while dramatically increasing crop and animal production in ways that protect local environments. Snowmelt and rainwater runoff from surrounding mountains provide water for isolated arid land valleys, referred to as "oases" by Su Peixi and Xie Tingting in Chapter 13, where eco-agriculture systems provide efficient agroecosystem structure and function for capturing limited water resources. A contrasting arid and hilly region is the site in southwest China for the eco-agriculture experiences presented in Chapter 14 by Ji Zhonghua and Tan Fengxiao. The special adaptations in structure and function of agroecosystems appropriate to arid and degraded landscapes show how important it is to understand ecological function in these regions. Chapter 15, by Wang Hongyan and Wang Daqing, provides an interesting example of how even large-scale mechanized state farms located in the cold temperate zone along the northeast border of China can benefit from an eco-agriculture that emphasizes ecological and environmental quality during farm operation. To conclude this section of the book, in Chapter 16 Wang Shuqing tells a story of the struggle he went through over almost 30 years as a community leader promoting eco-agriculture in his county beginning during the initial development stage of eco-agriculture in the early 1980s. The chapter describes how he had to work to overcome difficulties caused by conflicting goals, the need to locally adapt methods, and working with very limited financial support. His success story is presented as a typical one that reflects the difficult, yet often successful, path of eco-agriculture development in China.

The book closes with a final section and a final chapter that looks at the future prospects for eco-agriculture and agroecology in China. Wu Wenliang and his team end the book in Chapter 17 concluding with an overview of eco-agriculture development as the key practice of agroecology in China. Their review includes major technical approaches, necessary policy support, and possible major development stages that must take place for broader implementation of an agroecological focus for the sustainability of future food systems in China.

What readers will find in this book may be somewhat different from what agroecology has become in the United States, Latin America, or Europe. Under a political and cultural system in many ways unique to China, agroecologists have provided a strong foundation for ecologically sound agroecosystem design and management. They have elucidated the ecological mechanisms of many of China's most important traditional farming systems, as well as pointed out ways to reduce or limit the negative ecological consequences of large-scale industrialized systems. The chapters in this book are models of how to conduct ecological research in a broad array of types of agroecosystems.

Perhaps the biggest challenge faced by Chinese agroecologists, however, is how their research and education activities can also convince decision makers to implement fundamental agricultural policies and legislation that promote, in a country with over 1.4 billion inhabitants, a movement toward the sustainable paths required for future growth and development of their food systems. Ecological sustainability is just one of the components of food systems that must also provide economic opportunity and social well-being for everyone, from the farmers in the field to the consumers gathered around the dinner table. The need to increase yields to feed a growing global population has become a driving force in agricultural research. By carrying out the research outlined in the chapters of this book, Chinese agroecologists are building a strong alternative foundation for the kinds of food and farming systems needed to meet this need, especially when faced with extremely limited land and natural resource availability. By building on the cultural experience in agriculture stretching back many centuries, agroecology can lead the way to an agriculture that not only produces food but also protects water and soil resources, protects biodiversity above and below the ground, sequesters carbon, mitigates climate change, lessens dependence on external purchased farm inputs (especially fossil fuels), and creates opportunity and value for rural communities throughout the country. If they can do this, they will be providing an invaluable model for the rest of us to apply and adapt to food systems elsewhere in the world.

Stephen R. Gliessman

Editors

Luo Shiming completed his undergraduate and graduate study at the South China Agricultural University (SCAU) in Guangzhou, China, and further study in agroecology at the Institute of Ecology, University of Georgia, in the United States. He worked in a rural community as an agricultural technician for 8 years before returning to teach and do research at SCAU, where he ultimately served as president for 11 years. He used his rich experience in farming practice and understanding of the theory and methodology of agronomy and ecology to develop his teaching and research in agroecology. His textbook, *Agroecology*, has had great influence in agricultural universities in China since its first publication in 1987. It is now on its way to a fourth edition. His agroecology research work has included the use of biodiversity in agriculture, energy flow and material circulation in agroecosystems, allelopathy, practical eco-agriculture systems in South China, and computer simulation of rice field systems and artificial wetland systems for wastewater treatment in agriculture. He has served as an expert guide for important agroecology projects launched by the Chinese Ministry of Agriculture, Ministry of Sciences and Technology, and China National Scientific Foundation. His research results have earned him several national and provincial awards in China. Because of his contributions to agricultural research, education, and production, he was awarded an honorary doctoral degree from Pennsylvania State University and was named the China National Outstanding Professor in 2007. He also served as the executive vice president of the World Allelopathy Society, first president of the Asian Allelopathy Society, president of the Agroecology Committee under the China Ecological Society, vice president of the China Ecology Society, and vice president of China Agricultural Society. Since his retirement in 2014, he has stayed very actively involved in various national and local agroecology projects and training programs.

Stephen R. Gliessman, who earned graduate degrees in botany, biology, and plant ecology from the University of California, Santa Barbara, has accumulated more than 40 years of teaching, research, and production experience in the field of agroecology. His international experience in tropical and temperate agriculture, small-farm and large-farm systems, traditional and conventional farm management, hands-on and academic activities, nonprofit and business employment, and organic and synthetic chemical farming approaches has added unique perspectives to his studies as an agroecologist. He has been a W.K. Kellogg Foundation Leadership Fellow and a Fulbright Fellow. He was the founding director of the University of California, Santa Cruz (UCSC) agroecology program, one of the first formal agroecology programs in the world, and was the Alfred and Ruth Heller Professor of Agroecology in the Department of Environmental Studies at UCSC until his retirement in 2012. He is the co-founder of the non-profit Community Agroecology Network (CAN) and currently serves as president of its board of directors. His textbook, *Agroecology: The Ecology of Sustainable Food Systems*, is in its third edition and has been translated into many languages. He is the editor of the international journal *Agroecology and Sustainable Food Systems*. The editor and his family dry farm organic wine grapes and olives in northern Santa Barbara County, California.

Contributors

An Menglong
Economy Institute of Heilongjiang Land
    Reclamation Bureau
Harbin, China

Cao Can
Northeast Agricultural University
Harbin, China

Cao Cougui
Huazhong Agricultural University
Wuhan, China
and
Hubei Collaborative Innovation Center
    for Grain Industry
Jingzhou, China

Cao Linkui
Shanghai Jiao Tong University
Shanghai, China

Chen Xin
College of Life Sciences
Zhejiang University
Zhejiang, China

Dai Lin
Northeast Agricultural University
Harbin, China

Ding Guoying
College of Resource and Environment Science
China Agricultural University
Beijing, China

Du Zhangliu
College of Resource and Environment Science
China Agricultural University
Beijing, China

Fang Changxun
Fujian Agriculture and Forestry University
Fuzhou, China

Guan Yu
Lanzhou University
Lanzhou, China

Guo Yanbin
College of Resource and Environment Science
China Agricultural University
Beijing, China

Hui Han
College of Resource and Environment Science
China Agricultural University
Beijing, China

Ji Zhonghua
Institute of Agro-Environment and Resources
Yunnan Agricultural Academy of Sciences
Kunming, China

Li Chengfang
Huazhong Agricultural University
Wuhan, China
and
Hubei Collaborative Innovation Center
    for Grain Industry
Jingzhou, China

Li Fengmin
Lanzhou University
Lanzhou, China

Li Huafen
College of Resource and Environment Science
China Agricultural University
Beijing, China

Li Ji
College of Resource and Environment Science
China Agricultural University
Beijing, China

Li Long
College of Resource and Environment Science
China Agricultural University
Beijing, China

Li Shengnan
College of Resource and Environment Science
China Agricultural University
Beijing, China

Li Yufei
College of Resource and Environment Science
China Agricultural University
Beijing, China

Liang Kaiming
Institute of Tropical and Subtropical Ecology
South China Agricultural University
Guangdong, China

Liang Long
College of Resource and Environment Science
China Agricultural University
Beijing, China

Lin Sheng
Fujian Agriculture and Forestry University
Fuzhou, China

Lin Wenxiong
Fujian Agriculture and Forestry University
Fuzhou, China

Luo Shiming
Institute of Tropical and Subtropical Ecology
South China Agricultural University
Guangdong, China

Meng Fanqiao
College of Resource and Environment Science
China Agricultural University
Beijing, China

Qiao Yuhui
College of Resource and Environment Science
China Agricultural University
Beijing, China

Qin Zhong
Institute of Tropical and Subtropical Ecology
South China Agricultural University
Guangdong, China

Quan Guoming
Institute of Tropical and Subtropical Ecology
South China Agricultural University
Guangdong, China

Su Peixi
Cold and Arid Regions Environmental and
    Engineering Research Institute
Chinese Academy of Sciences
Lanzhou, China

Tan Fengxiao
Institute of Tropical and Subtropical Ecology
South China Agricultural University
Guangdong, China

Wang Daqing
Northeast Agricultural University
Heilongjiang Land Reclamation Bureau
Heilongjiang, China

Wang Hongyan
Northeast Agricultural University
Heilongjiang Land Reclamation Bureau
Heilongjiang, China

Wang Jian
College of Resource and Environment Science
China Agricultural University
Beijing, China

Wang Jianwu
Institute of Tropical and Subtropical Ecology
South China Agricultural University
Guangdong, China

Wang Shuqing
Bai Quan County Museum of Ecological
    Culture
Heilongjiang, China

Wang Xi
College of Resource and Environment Science
China Agricultural University
Beijing, China

Wang Yixiang
Fujian Academy of Agricultural Sciences
Fujian, China

Weng Boqi
Fujian Academy of Agricultural Sciences
Fujian, China

Wu Linkun
Fujian Agriculture and Forestry University
Fuzhou, China

Wu Wenliang
College of Resource and Environment Science
China Agricultural University
Beijing, China

Xie Tingting
Cold and Arid Regions Environmental and
    Engineering Research Institute
Chinese Academy of Sciences
Lanzhou, China

Xu Maomao
Northeast Agricultural University
Harbin, China

Xu Ting
College of Resource and Environment Science
China Agricultural University
Beijing, China

Yang Hefa
College of Resource and Environment Science
China Agricultural University
Beijing, China

Ying Nie
Economy Institute of Heilongjiang Land
    Reclamation Bureau
Harbin, China

Zhang Hanlin
Shanghai Jiao Tong University
Shanghai, China

Zhang Jiaen
Institute of Tropical and Subtropical Ecology
South China Agricultural University
Guangdong, China

Zhang Lizhen
College of Resource and Environment Science
China Agricultural University
Beijing, China

Zhang Weiping
College of Resource and Environment Science
China Agricultural University
Beijing, China

Zhao Benliang
Institute of Tropical and Subtropical Ecology
South China Agricultural University
Guangdong, China

Zhao Guishen
College of Resource and Environment Science
China Agricultural University
Beijing, China

PART I

An Overview

CHAPTER 1

Agroecology Development in China

Luo Shiming

CONTENTS

This chapter seeks to answer several questions regarding agroecology in China: How did agroecology get started in China? How did it develop as a discipline, a practice, and a movement in China? What were the difficulties encountered in developing agroecology in China? How were they overcome? What is the likely future of agroecology in China? Agroecology in China had its beginnings in the late 1970s. Since then, agroecology has established a solid footing in China and is an area of study that can be found in university curricula, agricultural research, farming practices, and government policy making. Let's go back to 1978 when Deng Xiaoping began to implement the "reform and opening up" policy in China.

1.1 BACKGROUND OF AGROECOLOGY DEVELOPMENT IN CHINA

In the late 1970s, people were allowed and encouraged to liberate their thoughts and to speak out after the 10 years of the Cultural Revolution in China. A lot of ideas and thoughts sprang up like numerous bamboo shoots emerging in a spring forest. Economists became brave enough to say that economic development must follow the rule of economy rather than central planning or a top-down ordering method (although the term "market economy" was not officially recognized until 1992). Dr. Ma Shijun of the China Academy of Science pointed out at the time that social and economic development must follow the rules of ecology. The Ecological Society of China was formally founded in December 1979 and led by Dr. Ma Shijun, although this was long after the British Ecological Society was founded in 1913, the Ecological Society of America in 1915, and the International Association for Ecology in 1967.

China was still a very poor agricultural society in the late 1970s. The gross domestic product (GDP) in 1978 was only 362.4 billion Chinese yuan, with 28% of that coming from the agriculture sector, compared to a GDP of 56,884.5 billion yuan in 2013, with only about 9% coming from agriculture. In the late 1970s, the major focus of ecology in China was on agriculture, and there were concerns about overgrazing; overfishing; deforestation; large-scale reclamation of marginal lands and forest areas; competing uses of crop residue as fuel, fertilizer, and animal feed; and soil erosion and degradation. The ecological concepts of food chains, energy balances, and material flows among ecosystem components triggered the development of many new ideas for a more sustainable way of agricultural development in China.

Professor Shen Hengli, from Shenyang Agricultural University and considered one of the pioneers of agroecology in China, recognized early on that many problems in agriculture could be solved using the concept of the ecosystem (Shen, 1975). Professor Xiong Yi hosted the First Conference of Agroecology at the Nanjing Soil Research Institute, China Academy of Sciences, in 1981 (Zhuang, 1982). The next year, he published an article entitled "The Characteristics and Research Tasks of the Agroecosystem" (Xiong, 1982), in which he stated that agroecology is a science dealing with all the components and processes of agriculture by means of the theory and methods of ecosystem studies. Xu Qi worked with Xiong Yi at the same institute and later became the first director of the Agroecology Committee under the auspices of the Ecological Society of China. The Agroecology Conference sponsored by an Agroecology Committee has been held every two years since then. In 1987, Dr. Ma Shijun, as the first president of the Ecological Society of China, edited *Agroecology Engineering in China*, in which he proposed approaching agriculture within the framework of a nature–social–economic system under the principle of "holistic coordination, regeneration, and recycling" (Ma, 1987).

Universities continued to push forward the development of agroecology in China. The Ministry of Agriculture launched agroecology training programs for teachers from agricultural universities in both 1981 and 1983 at the South China Agricultural University (SCAU). Around 50 to 60 teachers from all over China participated in each of these training programs. They were mainly from the specialties of crop science and farming systems. Ecosystem theory and the book *Fundamentals of Ecology* (Odum, 1971) were introduced at the first training session. In the second training session, prototype textbooks on agroecology that had been written by teachers from SCAU, Huazhong Agricultural University, and Xinjiang Agricultural University were exchanged. Also, the book *Agricultural Ecology* (Cox and Atkins, 1979) was introduced. At that time, I had just come back from studying agroecology at the Institute of Ecology, University of Georgia, in Atlanta. Both Dr. Robert Todd from the University of Georgia and I conducted courses on agroecology and computer data processing methods during this training session.

In conclusion, the initiation of agroecology in China was stimulated by the serious deterioration of its agroecosystems and by the introduction of ecosystem ecology from abroad. The first set of scientists engaged in agroecology had backgrounds primarily in soil science and crop science. By 1984, the development of agroecology had gained a strong foothold in China.

1.2 AGROECOLOGY AS A DISCIPLINE IN CHINA

From 1975, when the first article that included the subject of agroecology was published (Shen, 1975), the quantity of Chinese literature with agroecology content increased steadily—from a few articles a year in the late 1970s to hundreds in the 1980s and 1990s and thousands each year after 2010 (Figure 1.1). Agroecology in China has attracted much attention from academic and business circles, as well as from the public, and has become a well-established discipline and an important field of scientific research in China. Agroecology is providing a set of principles and a methodology to guide agricultural practice in China today.

In the early symposia and conferences on agroecology, arguments arose regarding the boundary, core theory, and methodologies of agroecology (Zhuang, 1982; Lu, 1984), and such disagreements also appeared in the early articles on agroecology discussions (Wu, 1981; Xiong, 1982; Z. Wang, 1985). Eventually, though, agreement was reached regarding three basic concepts of this new discipline. First, as an applied branch of ecology, agroecology should be based on ecosystem theory. Second, this discipline must be able to face the urgent needs of agriculture and rural development in China. Third, agroecology is an interdisciplinary field crossing several disciplines such as crop science, animal science, and soil science. Agroecology pays attention not only to individual components of agroecosystems, as do traditional disciplines, but also to the relationships among components. Based on this consensus, early articles focused on agroecology education, and new textbooks on agroecology for higher education were published in the 1980s (Shen, 1982; X. Zhang, 1985; Han, 1985; Y. Chen, 1985; Wu, 1986; Luo et al., 1987). The textbook *Agroecology* (Luo et al., 1987) had a great influence on subsequent development of agroecology in China. Later versions of the text were released in 2001 and 2009 (Luo, 2009). Since the 1990s, many agroecology textbooks have been published in China by various authors, including L. Wang (1994), Shen (1996), Z. Wang

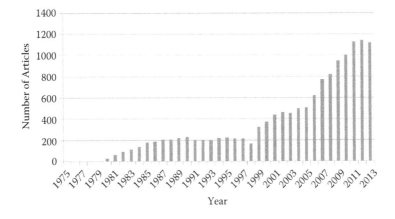

Figure 1.1 Number of articles in Chinese literature with the subject of "agroecology" (www.cnki.net search).

(1997), F. Chen (2002), Zou and Liao (2002), H. Wang and Cao (2008), Cao (2011), and S. Wang and Caldwell (2012). This last textbook was written in English for bilingual teaching in China and is the result of cooperation between Wang Songliang and his Canadian counterpart Dr. Claude Caldwell, among others.

A course in agroecology has been delivered since the late 1980s to undergraduate students in almost all agricultural universities in China offering a major in agronomy and for the major of resources and environmental science. It is also an optional course for undergraduate students in environmental sciences, agriculture economy, weed science, and horticulture. A course in advanced agroecology is a requirement for graduate students majoring in agroecology and agro-environmental study. The content of the study of agroecology in China has experienced some changes. In the early 1980s, the public had little knowledge with regard to ecology, so introducing basic ecology theory and principles to university students was important. About a third to a half of the content of agroecology courses was basic ecology. Today, however, the study of ecology is quite popular in China. Ecology concepts are taught in middle school, and independent ecology courses can be found in university curricula. The basic principles and theories of ecology have been greatly simplified for agroecology courses, which provide only a brief review in order to introduce ecological relationships within agroecosystems. The practical aspects of agroecology are more strongly emphasized. Practical experience accumulated during the first wave of eco-agriculture development in the 1980s and 1990s has proven to be an invaluable educational asset, as has the rich research progress being made in agroecology. Over the past few years, agroecologists in China have begun to acknowledge that social and economic factors can sometimes be even more important than the scientific and technological aspects for sustainable agriculture development. Despite such changes in the field, the main framework of agroecology as a discipline in China has remained quite clear and stable and will be discussed further below.

It is interesting to compare the *Agroecology* series written by leading U.S. agroecologists Stephen Gliessman (2000, 2007, 2015) and Altieri (1983, 1987, 1995) to those written by Chinese authors. Textbooks from both the United States and China share two important common points: (1) the agroecosystem is the main focus of agroecology, and ecological interactions in such systems should be emphasized; and (2) the main task of agroecology is to meet the challenge of the ecological and environmental crisis facing agriculture in both developed and developing countries while exploring alternative approaches to sustainable agriculture development. However, due to differences in natural and socioeconomic conditions between China and the United States, there are also many differences in agroecology practices.

First, the framework, or structure, of agroecology is presented differently. The study of agroecology in the United States is focused more on developing principles and practices for guiding the transition to sustainable agriculture and food systems and encouraging students to adapt these principles to their own personal situations. In China, agroecology textbooks generally have a consistent structure, with the first part focusing on the structure and function of agroecosystems, the second part on how to control and regulate agroecosystems, the third part on why it is important to learn agroecology and on the serious ecological and environmental challenges faced by agriculture, and the fourth part on how to conduct eco-agriculture practice and promote sustainable agriculture development. Also, the examples used in the study of agroecology by U.S. authors come from around the world, with most of them arising from personal study and experience. The examples used in agroecology by Chinese authors come mostly from within China and are usually cited from presentations at conferences and or published Chinese literature. The fact that all of the agroecology books in China have been written by multiple authors from different universities may have limited their diversity and flexibility, but it has also allowed the development of a commonly agreed upon structure and a balanced content that came about thorough exchange and compromise (Luo, 2010b). After more than 35 years of development in China, it is suitable to discuss the common understanding of agroecology as a discipline in China at this stage.

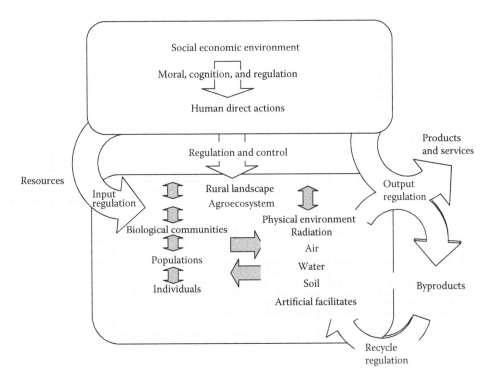

Figure 1.2 Relationships between agriculture and its social and natural environment.

1.2.1 Structure and Function of Agroecosystems

A typical definition of agroecology in China states that "agroecology is a discipline for applying the principles and methodologies of ecology and systems theory to agriculture. It considers agricultural organisms with their surrounding social and natural context as a whole and reveals their interactions, coevolution, regulation, and sustainable development" (Luo, 2009). The object of study within the field of agroecology is the agroecosystem, and the methodology of agroecology was inherited from ecology and systems theory. An understanding of agroecology in China can be expressed by the framework shown in Figure 1.2.

1.2.1.1 Internal Structure and Function of Agriculture

The internal structure of agriculture is similar to natural systems when viewed from an agroecosystem perspective. Organizational levels can be identified—from landscape to ecosystem to community to population to individual. The principles and roles of natural processes that affect agriculture should be respected, carefully observed, and followed. Only in this ecological manner can agricultural activity become more sustainable.

At the landscape level, island biology equilibrium theory, hierarchy theory, scale effect, edge effect, and ecotone phenomena, together with concepts such as patch, corridor, and matrix, are important for optimum landscape design in agricultural regions and rural areas. Depending on the topography, climate, and socioeconomic environment, landscape elements such as windbreak systems, buffering zones along drainage channels, horizontal green fences in hilly areas, biodiversity conservation zones, and patchy arrangements of crops are all examples of landscape design that can be applied to agricultural regions.

At the ecosystem level, the laws governing energy and material flow in nature and the effect of human interference on these flows and its consequences are important to understand. The design of energy and material flow in agriculture includes determining the optimum ratio among producers, consumers, and decomposers; well-established recycling paths; and food chain arrangements.

At the community level, the high efficiency of community structure in nature can be imitated in agriculture. The knowledge of natural community succession can help us to understand the early successional features of field crop production. Successional theory can also be applied to restoration of degraded areas, and the wisdom of natural communities can inform the design of artificial plant and animal communities through the use of intercropping, agroforestry, and cover crops.

At the population level, the dynamic patterns of population growth and the interactions between populations are important considerations in agriculture. Factors affecting populations and their relationships include food supplies and resources, as well as the physical and chemical environment. Chemical ecology, including the concepts of allelopathy and chemical-induced defenses, is important at the population level. The use of repelling and attracting plants for insect control and the use of natural enemies are examples of using population relationships in agriculture.

At the individual level, Liebig's law of the minimum, Shelford's law of the minimum and maximum, and the concepts of ecotype, niche, life form, and biodiversity are important. These concepts and laws are used in agriculture for the arrangement of plant and animal species and varieties and for optimum control of the chemical and physical environments within agricultural production.

1.2.1.2 Input and Output of Agroecosystems

The input and output of an agroecosystem are usually much more intensive than those in a natural ecosystem. The impact of these flows on resources and the environment is also much more significant. For these reasons, the quantity and quality of input and output from agroecosystems should be carefully managed. The concepts of natural resources, socioeconomic resources, renewable resources, non-renewable resources, ecosystem services, ecological footprint, optimum sustainable yield of renewable resources, the law of public resources, the first laws of resources, and Gordon's law of bioeconomic equilibrium are important for developing an understanding of the serious resources issues faced by agriculture in China and many other parts of the world.

1.2.2 Control and Regulation Processes in Agroecosystems

Discussion of control and regulation processes in agroecosystems is totally absent in most general ecology courses. It is even new to many agroecologists who have backgrounds primarily in the natural sciences. However, it is a very important part of agroecology. Only after a period of study and exchange have Chinese agroecologists begun to integrate social and economic concerns into agroecology. A review by Wezel et al. (2009) found that only after the year 2000 did the social dimension of agroecology in the agroecology literature begin to increase rapidly around the world.

1.2.2.1 Agroecosystem Control and Regulation Mechanisms

The three layers of control and regulation mechanisms for an agroecosystem are shown in Figure 1.2. The first layer is inherited from nature and was introduced in the section on the internal structure and function of agroecosystems, above. The second layer of control and regulation comes from direct human actions through agricultural activities, and the third layer comes from those social and economic factors that affect human behavior and hence indirectly affect the structure and function of agroecosystems.

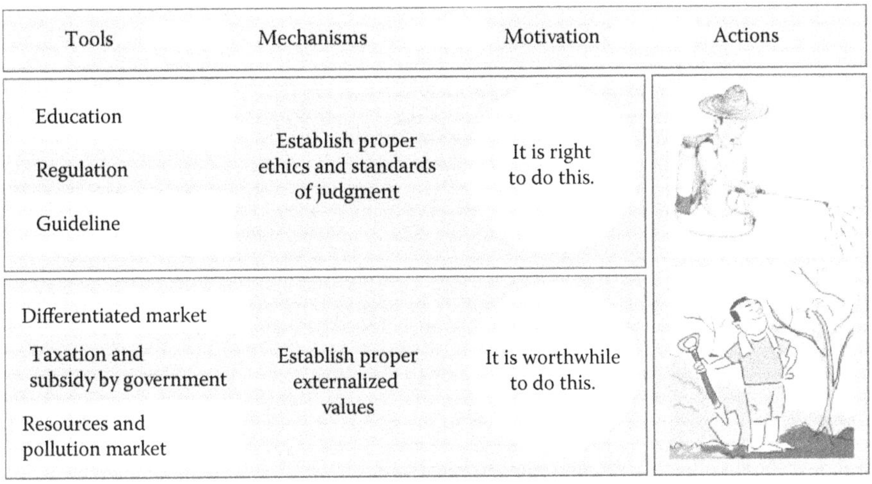

Tools	Mechanisms	Motivation	Actions
Education Regulation Guideline	Establish proper ethics and standards of judgment	It is right to do this.	
Differentiated market Taxation and subsidy by government Resources and pollution market	Establish proper externalized values	It is worthwhile to do this.	

Figure 1.3 Social and economic mechanisms affecting human actions in agroecosystems.

1.2.2.2 Indirect Regulation Mechanism of Agroecosystems

As shown in Figure 1.3, human actions in agroecosystems are mainly driven by the concepts of fairness and benefits. The judgment of whether an action is right or wrong is affected strongly by the moral standards of a culture. The moral standard inside a person is deeply influenced by the education provided by schools, families, and society. So, it is important to introduce greater agroecology knowledge through education and the public media. Laws and government policies usually represent the moral standards of a society and are independent from those of an individual. These laws and policies can regulate the behavior of a person through punishment or reward. In order to promote the importance of ecological and environmental concerns with regard to agriculture development, laws and policies such as ecological zonation policies, ecological subsidy policies, environmental protection laws, clean production laws, circular economy laws, and resource protection laws become essential.

Whether an action is economically beneficial depends heavily on capital flow (Figure 1.4); however, ecological services and environmental quality usually affect many people outside the scope of management and bypass the traditional market system, which could lead to economic externalization and market failure. Methods to overcome this situation include government actions relying on the use of financial and monetary tools, such as adjusting tax, subsidy, and fee levels. According to the Coase theorem, assigning clearer property rights to specific holders and developing new markets for resources or penalties for pollution are also important means to overcome economic externalization (Z. Wang, 1997).

1.2.3 Application of Agroecology Principles and Theory

Applying agroecology principles and theory develops an understanding of how the rules and principles of agroecology exert their impacts on the real world.

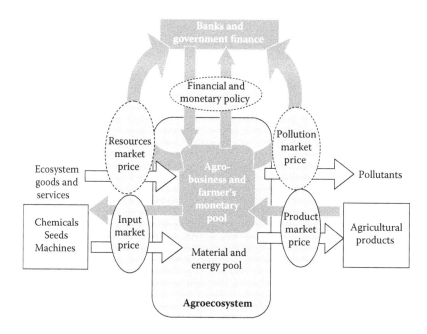

Figure 1.4   Agroecosystem relationships among capital flow (dark arrow), material flow (white arrow), commercial input/output market, new resources/pollution quota market, and government policy.

1.2.3.1 Lessons Learned from the History of Agriculture with an Agroecology Perspective

As described in *Agricultural Ecology* (Cox and Atkins, 1979), agriculture originated independently in many places around the world about 10,000 years ago when the world climate began to warm and dry after the last Ice Age. Many agricultural civilizations that originated in crop and animal domestication centers (e.g., the Babylonian civilization in western Asia) collapsed and disappeared. The common reason for this crisis was the serious damage done to the environment by improper agricultural activities. Agriculture in China also originated around 10,000 years ago in several locations along the Yellow and Yangtze Rivers. Although Chinese civilization has been continuous, its political center was forced to move from the west around Xi'an to the east in Nanjing and then to Beijing mainly because of deterioration of the environment due to human activities along the upper and middle reaches of the Yellow River. The ecological environment was the key factor for the onset of agriculture and also determined the subsequent dramatic shifts of agricultural civilizations.

1.2.3.2 Agroecology View on Agricultural Development

Three distinct stages of agriculture development in China and many parts of the world can be identified: (1) subsistence small-scale farming, (2) industrialized large-scale commercial farming, and (3) sustainable agriculture practice. Examining these stages from an agroecology perspective is important if we are to contemplate our future. Small-scale subsistence farming has been the major form of agriculture around the world for a very long time, and it is still the major type of agriculture in many developing countries. The main tools used are hand tools or draft animals. Social input and product output remain at a low level; however, sound relationships with nature have developed through long-term experience and adaptation. Many ancient traditional agriculture practices in China, such as the use of green manure, intercropping of cereal crops with leguminous crops, pest control by natural enemies, fish–rice co-culture systems, fish–duck co-culture systems, and recycling in fish and silkworm production systems, are still being practiced by farmers today. Our traditional agricultural heritage should be respected, protected, studied, and used today (Luo, 2007). Disadvantages of subsistence agriculture include low labor efficiency and unstable low yields compared to industrialized agriculture.

The problem with industrialized, large-scale agriculture is its lack of sustainability due to its heavy dependence on non-renewable resources and extensive environmental degradation. This sustainability problem is quickly becoming the bottleneck of agricultural development. In some developed countries, an unfair market controlled by a few big companies and the problem of long-distance transport of agricultural products are serious issues. Although such problems have not impacted China during its recent move toward agricultural industrialization and commercialization, they should be noticed and prevented.

The effort to seek sustainable agricultural development dates back to the early stages of agricultural industrialization in Western countries. For example, Dr. F.H. King wrote *Farmers of Forty Centuries, or Permanent Agriculture in China, Korea, and Japan*, in 1911 after he visited China, Korea, and Japan as the chief of the Division of Soil Management of the U.S. Department of Agriculture (King, 1911). Later, such techniques as organic agriculture, biodynamic agriculture, natural agriculture, and eco-agriculture were being utilized and reported on in various industrialized countries and regions throughout Europe, North America, and Japan. In 1991, the Conference on Agriculture and the Environment sponsored by the Food and Agriculture Organization (FAO) of the United Nations and the Netherlands announced the "Den Bosch Declaration and Agenda for Action on Sustainable Agriculture and Rural Development." Since then, more countries have joined the effort to push industrial agriculture to a more sustainable track. Such efforts include multifunctional agriculture in the European Union, environmentally friendly agriculture in South Korea, environment conservation agriculture in Japan, and best management practices in the United States.

1.2.3.3 Eco-Agriculture in China

The serious challenges for agriculture in China include a shortage of resources, overuse of chemicals, food safety issues, and environmental deterioration, which have become motivating factors to push Chinese agriculture practices in a more sustainable direction. A recently applied technique for improving agriculture in China is referred to as *eco-agriculture*. Although many descriptions of eco-agriculture can be found in the Chinese literature, the most significant difference between eco-agriculture and traditional agricultural practices is the awareness of the relationships among components within the agroecosystem. Hence, the basic philosophy and methodology of eco-agriculture in China are related to wise use of the principles and rules of agroecology and caring for nature and the environment. The common understanding of eco-agriculture is that "it is a sustainable pattern of agriculture that considers balancing all ecosystem services provided by agroecosystems through ecologically sound methods." In this statement, "all ecosystem services" include not only goods but also the environment, culture, and spirituality. "Ecologically sound methods" are methods that respect the structure and function of nature, cooperate with natural processes, and learn from the wisdom of nature. Ecologically sound methods are different from normal industrial practices, which usually work against nature by using large amounts of external, non-renewable inputs.

The concept of eco-agriculture in China is not the same as in Europe or in North America such as described by Worthington (1981) or Scherr and McNeely (2008). Ecological agriculture as described by Worthington is confined to small-scale, localized, self-sustaining farms. Eco-agriculture in China has a more flexible meaning, as it can include both big farms and small, local farms and commercial farms, self-sustaining farms and low-input farms. Eco-agriculture as described by Scherr and McNeely refers primarily to landscape planning that coordinates agriculture activity toward achieving conservation goals. In addition to landscape design, eco-agriculture in China also includes design at ecosystem, community, and population levels. The term "sustainable farming" is used quite often around the world. The concept is similar to eco-agriculture practice in China in terms of field production, but China's eco-agriculture approach also considers watershed and landscape levels, as well as important social and economic aspects.

1.3 AGROECOLOGY PRACTICE IN CHINA

The term *agroecology* used for sustainable agricultural practice outside China is quite similar to the term *eco-agriculture* used in China. Eco-agriculture in China was first proposed in 1980 by scientists attending the agricultural eco-economy symposium held in Yinchuan (W. Li, 2003). The first article in the Chinese literature that provided a detailed description of eco-agriculture was published in 1981 (Ye, 1981). In 1983, the number of articles related to eco-agriculture increased to 12. At that time, a common understanding of eco-agriculture in China had been reached among government officials and scientists working on agriculture. It was considered a revolutionary view for agricultural development and replaced the prevailing trend for agricultural modernization through the use of machinery, chemicals, intensive irrigation, and new varieties. A conference in 1983 on eco-agriculture was hosted by the Chinese Academy of Sciences and led by Dr. Ma Shijun. In 1984, a decision by the Chinese State Council on Environmental Protection stated that "eco-agriculture should be actively promoted." A national eco-agriculture symposium hosted by the Ministry of Agriculture was held in 1984 and again in 1987. About 100 counties, 300 townships, and 500 villages were participating in the development of eco-agriculture practice in China by the end of 1990 (Guo, 1988, 1993).

In 1993, 50 counties were chosen by the Ministry of Agriculture in China to be demonstration counties for eco-agriculture development. Five years later, in 1998, the results were encouraging. Agricultural production and farmers' income increased more than the average in China. Soil erosion was reduced by 73.4%, and soil desertification was reduced by 60.5%. The recycling rate of straw reached 49%. In 2000, another 50 counties were chosen as demonstration counties for eco-agriculture development. Among them, seven were named to the Global 500 Honor Roll for Environmental Achievement by the United Nations Environment Programme (UNEP). These demonstration counties directly or indirectly encouraged more than 500 other counties to initiate eco-agriculture practices, accounting for more than 30% of the counties in China. Eco-agriculture practice reached its first peak in the late 1990s and early 2000s. Books about eco-agriculture practices were published, including *Introduction to Multistoried and Multi-Dimensional Agriculture in China* (Lu, 1999) and *Eco-Agriculture: Theory and Practice of Sustainable Agriculture in China* (W. Li, 2003). The number of articles in the Chinese literature that included the subject "eco-agriculture" grew, as shown in Figure 1.5. The jump from one article in 1981 to 3162 articles in 2008 reflects the increasing concern for eco-agriculture practice in China. After that time, though, the number of articles has dropped, to around 2000 in the past 3 years. New eco-agriculture programs supported by the government were launched in 2014.

1.3.1 Eco-Agriculture Practices

"Eco-agriculture pattern" and "eco-pattern" are terms often used to describe a good and stable agroecosystem structure, and "eco-agriculture technique" or "eco-tech" can be used to describe the sound technical package used to operate such an agroecosystem. An eco-pattern can also be referred to as an "eco-agriculture mode," "eco-agriculture system," or "eco-agriculture model."

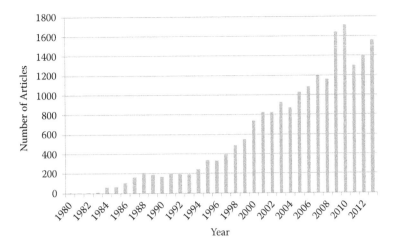

Figure 1.5 Number of articles in Chinese literature from 1980 to 2013 with the subject of "eco-agriculture" (www.cnki.net search).

After more than 30 years' experience with eco-agriculture practice in China, numerous eco-patterns and eco-techs have been generated for different parts of China by farmers, technicians, and agricultural scientists. According to the structure of agroecosystems, those eco-patterns can be divided into (1) landscape planning, (2) agroecosystem cycling design, and (3) biodiversity arrangement (Figure 1.6). The sources of eco-tech include knowledge of traditional agriculture practice, current farming experience, and modern agricultural research. Techniques that can replace and save limited resources, reduce pollutants, or increase output quality are used often in eco-agriculture practice.

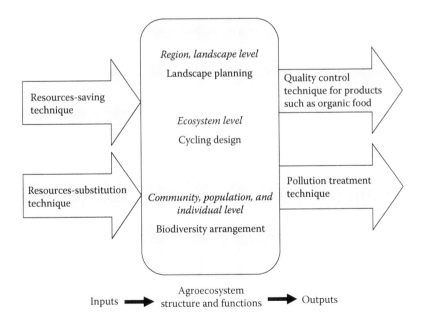

Figure 1.6 Main system design and technical packages used in eco-agriculture.

1.3.2 Eco-Agriculture Patterns

1.3.2.1 Landscape Planning

Landscape planning provides a basic land-use framework for agricultural development. Important aspects of landscape designs for eco-agriculture development include (1) balancing preserved areas and cultivated areas in a watershed; (2) optimum position allocation of land resources for agricultural, industrial, and urban development; (3) vertical production of wood, livestock, fruit, grain, and fish at different altitudes of a landscape; (4) landscape design for soil erosion control; (5) windbreak systems in plains areas; (6) buffer zones along drainage systems; (7) diversified field arrangements for crop production; and (8) visual landscape design for rural residential areas and tourist spots.

Long-term traditional agriculture practice has produced much knowledge at the landscape level in China; for example, a traditional landscape pattern developed in South China in a watershed is shown in Figure 1.7. Forest is preserved on the top and upper part of the mountains and hills. This forest area is important to maintaining the water supply, protecting wildlife, and controlling soil erosion. Economic tree and shrub communities such as chestnut, *Camellia*, *Canarium pimela*, and persimmon, which do not require a lot of labor and provide permanent groundcover, are developed next to the forest. Labor-intensive tea plantations and fruit orchards such as lichi, longan, and orange are located farther down the slope and closer to villages using a contour planting method. On steep mountainous areas, terraces are built on the lower part of the hill for field crop production. Paddy fields prevail on the flat plains area located at the lower reach of the river for the production of various crops such as rice, sugarcane, banana, and vegetables. Fish ponds and crop–dike systems, with internal cycling between pond and dike, or the high bed–deep ditch system, with internal cycling between ditch and bed, are adopted in lowland areas with high water tables (Luo and Han, 1990). Windbreak systems are built along coastal lowlands facing the South China Sea where typhoons land at least twice a year. Houses of villages in mountainous and hilly areas are usually located on southern slopes in order to receive summer tradewinds and to avoid the winter gales from the north. There is usually a large patch of native vegetation called *Fengshui lin* ("geomantic forest")

Figure 1.7 Typical watershed arrangement for agricultural production in South China.

Figure 1.8 (See color insert.) Traditional Yuanyang terrace with large preserved forest above in Yunnan Province.

preserved behind a village. This vegetation actually serves a purpose, beyond spiritual or cultural, as it buffers the micro-environment of the village, provides medicinal herbs, and stabilizes slopes (Cheng et al., 2009). Regardless of whether the village is in a hilly area or on a flat plain, there is usually a fish pond in front of a village that holds and purifies wastewater from the village and stabilizes the temperature and humidity.

Another example of a landscape arrangement is the Yuanyang terraces in the Yunnan Province. Yuanyang terraces are a traditional agroecosystem created by the Hani minority people more than 1300 years ago (Figure 1.8) and are considered one of the Globally Important Agricultural Heritage Systems of the FAO. During a serious winter and spring drought in 2010, Southwest China experienced record low precipitation. Many reservoirs dried up and crops failed. The losses exceeded 35 billion yuan. However, no drought occurred in the Yuanyang traditional terrace areas because of the wise landscape arrangement in the watershed. In Yuanyang, villages are located between the forest and terrace areas. Because terraces are seldom built above the line of the villages, large areas of water-generating forest on the upper parts of the mountains are well preserved. Such a wise landscape arrangement can provide a buffering mechanism to deal with climate variability over a long history (Xu, 2010).

Whenever a reasonable landscape arrangement has not been maintained, serious problems and even disasters have occurred. Large-scale land reclamation in hilly areas has not only destroyed forest but also caused soil erosion and landslides like the one that occurred in Zhouqu, Ganshu Province, in 2010 (J. Wang et al., 2012). Draining lakes for farmland reduces a landscape's water buffering capacity and exacerbates flooding in the lower reaches of a watershed, as happened along the Yangtze River in 1998 (Wu, 1999). Reclamation of land from the grasslands in the north and northwest parts of China resulted in serious wind erosion. A program promoting "farmland retreat for forest and grass" was launched by the government in 1999. The government provided free tree seedlings and grain subsidies for farmers used to farming in the revegetated areas. A total of 14.67 million ha of marginal farmland were returned to tree and grass cover by 2010. Total expenditure for this program reached 430 billion yuan over 10 years. From 2008 to 2011, another 43.6 billion yuan were spent to support renewable rural energy and farmland improvement (http://

www.gov.cn/jrzg/2012-10/07/content_2238558.htm). A program promoting "farmland retreat for lake recovery" was also put into place to recover the buffering capacity of lakes for flood control and water purification.

One example of "farmland retreat for forest and grass" can be found in the Baizhuang watershed in Hebei Province. A total of 3.7 km² in a hilly area suffered from soil erosion, poor soil quality, and drought. In order to improve this situation, a landscape arrangement strategy was devised that put "pine and ash forest on top, fruit orchard in the middle, and terrace located in the lower part." After 4 years of eco-agriculture development, the forest cover increased from 33.3% in 1997 to 75.81% by 2000, mainly on hill tops and steep slopes. All farming operations were removed from this area. In the middle part of the watershed, horizontal green fences were built by planting caragana (*Caragana korshinskii* Kom.) and adsurgens (*Astragalus adsurgens* Pall). Fruits such as apricot, pomegranate, peach, pear, and jujube were interplanted with shrub and grass species such as the leguminous species adsurgens and amorpha (*Amorpha fruticosa* Linn.). This groundcover reduced soil erosion, provided nitrogen to enrich soil fertility, and provided fodder for animals. On the terrace located in the lower part of watershed, apricots were intercropped with soybean and jujube was intercropped with forage grasses. Soil total nitrogen increased from 42.5% to 59.01%, available potassium increased from 13.17% to 43.77%, available phosphorus increased from 8.31% to 28.5%, and soil organic matter content increased from 18.94% to 37.37% in 4 years. Fruit yield and grain yield also increased over the same period. For example, wheat yield increased 1500 kg/hm², corn yield increased 900 kg/hm², and apple yield increased 1620 kg/hm² (J. Chen, 2005).

To protect major farmland areas, reduce the potential for disastrous flooding, and recover the natural balance, four major reforestation projects related to the landscape of agriculture were launched in China. The Three-North (Northeast China, North China, and Northwest China) Shelterbelt Forest Program established 35.6 million ha of shelter forest along the 7000-km ecotone between farmland and desert/erosion sites in northern China. A program aimed at establishing shelter forest along coastal areas resulted in 3.56 million ha of windbreak forest being built along 18,000 km of coastline in the east and the south from 1988 to 2010. Yet another program aimed at establishing shelter forest in the upper and middle reaches of the Yangtze River led to 6.6 million ha of shelter forest being built from 1989 to 2010. Finally, a project to promote forestation of plains areas to protect farmland resulted in 9.3 million ha of shelter belt forest being planted from 1988 to 2000.

Another important policy affecting agricultural landscape arrangements is *key functional zonation*, announced in 2010 by the State Council. Ecologically important zones, ecologically sensitive zones, agricultural production zones, and urban/industry zones have been carefully laid out all over China. All human social and economic activities in China are required to respect this zonation (State Council, 2010). These ecologically important zones are preserved to secure the water supply, to protect wildlife, to buffer environmental fluctuations, to purify air and water, and to give agriculture a stable foundation. Prime farmland should not be less than 120 million ha (equal to 1 Chinese mass unit per person), and major industrial and urban activities are required to concentrate in their respective assigned zones. The economic activities in ecologically sensitive zones are strictly controlled.

Although much experience has been gained and beneficial policies have been put into place for eco-agriculture landscape arrangements in China, greater efforts must be made in the future. For example, because of the high population density and scarcity of land resources, the buffering zone method of forming vegetation strips that can hold soil and absorb nutrients from drainage water while providing habitat for wildlife in or near crop fields has not been well accepted. Because the scale of agriculture production is increasing and non-point-source water pollution is becoming more serious, planting vegetation strips along drainage channels is growing in importance. Similarly, crop diversity layout and species composition of vegetation strips around crop fields are still in the early research stages and have not been put into standard practices so far in China.

1.3.2.2 *Agroecosystem Cycling Design*

Agroecosystem cycling design includes the cycling of (1) crop residues, (2) animal waste, (3) processing waste from factories, (4) organic garbage from villages and towns, and even (5) carbon released to the atmosphere by human activities (Figure 1.9). Crop residues, animal waste, organic waste from food processing, and kitchen waste can be recycled in the form of organic fertilizer for crops; feed for animals; growth media for edible mushrooms, earthworms, and insects; mulch for field crops; and biomass energy (Figure 1.9). Traditional agriculture in China used to be a system without waste. Pigs and chickens usually consumed kitchen and household waste. Through solid composting and liquid fermentation, all organic materials could be returned back to the fields as organic fertilizer. Even returning human waste to crop fields after fermentation is acceptable in traditional farming cultures in many parts of China. A well-known agroecosystem based on recycling is the fish pond–mulberry dike system created in the Pearl River Delta about 600 years ago. Mulberry shrubs were planted on the dike and the mulberry leaves were picked to feed silkworms. The waste from silkworm culture was returned to the fish ponds to feed the fish. The pond mud was dug up and used to fertilize mulberry or other crops such as sugarcane, vegetables, banana, and orange on the dike (Figure 1.10). According to research by Zhong et al. (1993), the organic matter content varied from 2.32 to 2.97% in pond mud samples, compared to 1.30 to 2.01% in soil samples under mulberry from four dike–pond systems. The nitrogen contents ranged from 0.223 to 0.258% in pond mud samples compared to 0.106 to 0.165% in soil samples under mulberry from the same four sampling systems. The recycling rates of nitrogen, phosphorus, and potassium reached 43%, 46%, and 96%, respectively, in this mulberry–silkworm–fish system. The main products from the system were high-market-value silk and freshwater fish.

Mushroom production utilizing straw and animal waste is another traditional way of recycling waste. The growth medium is made by fermenting a mixture of straw/sawdust and cattle/pig dung. After mushroom harvest, the growth medium is usually returned to fields as organic fertilizer. There are more than 60 commercially cultivated mushroom species in China, and, in 2012, 70% of the world's mushroom production took place in China (Su, 2013).

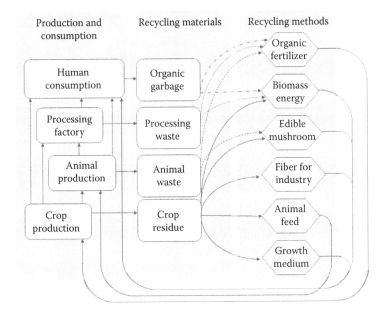

Figure 1.9 Recycling paths related to eco-agriculture design.

Figure 1.10   Farmer on a small boat returning pond mud back to the dike with sugarcane during autumn period in the Pearl River Delta. (Screenshot from *Ecoagriculture in Tropical and Subtropical Regions (3)* video by Luo, S. et al., 1987.)

Two widely used cycling systems known as "four in one" in North China and "three in one" in South China were recommended by the Ministry of Agriculture in the year 2000. The "four in one" system includes pigs, biogas tanks, greenhouses, and vegetables. Residue from vegetable production is used to feed pigs. Pig waste goes to the biogas tank, which is built inside the greenhouse. After fermentation, biogas is used for cooking pig feed and heating the greenhouse. The sludge from the biogas tank is used as organic fertilizer for vegetable production. The "three in one" system includes pigs, biogas tanks, and fruit trees. The biogas tank is built within tropical or subtropical orchards where winters are warm enough to generate biogas all year round. By the end of 2011, 40 million households had biogas tanks, accounting for 23% of rural households in China. Although the number of small biogas tanks (8 to 10 m³) is still growing under the support of government subsidies, mid- and large-size biogas tanks are attracting greater attention. These bigger biogas tanks are usually managed by rural communities or by large animal farms.

Today, however, the traditional ways of recycling are facing challenges due to the increasing scarcity of young workers in rural areas and rapid growth in the scale of production. In 2005, about 84 million tons of crop straw were produced in China. About 81.5% could be collected from fields. Of this collected straw, 35.6% was used as kitchen fuel or new biomass energy, 25.7% was used for animal feed, 6.7% provided raw material for industry, 9.8% was returned to fields as organic fertilizer, 1.5% was used for mushroom production, and 20.7% was wasted by direct burning or discarding (Bi et al., 2010). In order to simplify land preparation and speed up rice stem decomposition after returning crop residues, stem shredding machines, microbial decomposition agents, and chemical agents are quickly being developed and becoming available today. The amount of waste generated from animal production reached 2.5 billion tons in 2011 in China (J. Zhu et al., 2014). The amount is growing quickly because people are consuming more meat in their daily diet, but it is difficult for large-scale animal farms to return their waste directly to nearby farmland. More than half of the animal waste from large animal farms is being discharged into waterways without pretreatment. Fertilizer plants that produce mixed organic and inorganic fertilizers are being installed to solve this problem such that mixed organic and inorganic fertilizer can be transported to cropland farther away from animal farms.

The Circular Economy Promotion Law of the People's Republic of China was announced in 2008 and implemented in 2009. It encourages the use of reduction, reuse, and recycle methods in eco-agriculture. Article 24 states, "Eco-agriculture development is a priority." Article 34 states, "The state government encourages and supports the agricultural producers and related companies to adopt advanced or appropriate technology for the integrative use of crop straw, animal waste, processing by-product, and old plastic sheeting, and to explore the use of biogas and other biomass energy."

1.3.2.3 Biodiversity Arrangement

Biodiversity arrangement includes genetic diversity, species diversity, and ecosystem/landscape diversity. Biodiversity used in agroecosystems not only refers to domestic crops and animals but also includes wild species within or close to agricultural production sites such as natural enemies of crop pests, grass cover in an orchard, phytoplankton and zooplankton in a fish pond, vegetation in windbreak systems, and wild relatives of domestic crops and animals. Long-term evolution of agriculture in China has offered rich experiences in the use of biodiversity in agriculture. New methods using biodiversity in eco-agriculture have undergone continuous innovation in recent years. Biodiversity use at the ecosystem level and landscape level was mentioned in the two previous sections, and the use of genetic diversity and species diversity is introduced here.

A good example of the use of *genetic diversity* is the intercropping of two rice varieties which was shown to eliminate the need for pesticides for the control of rice blast disease in Yunnan Province (Y. Zhu et al., 2000; Y. Zhu, 2004). High-quality traditional glutinous rice varieties (Huang KeNuo, ZiNuo) are susceptible to rice blast disease, but Shanyou 63 and Shanyou 22 are hybrid rice varieties with resistance to rice blast. When one row of a taller, traditional glutinous rice variety was intercropped with four rows of a shorter hybrid rice variety, the incidence of rice blast disease was reduced from 32.43% to only 1.80%. The total rice yield reached 8576 to 8795 kg/ha, and increased 6.5 to 8.7% as compared with hybrid rice monoculture. This method has been widely used in Yunnan, Sichuan, Guizhou, Hunan, and Jianxi provinces.

The same researchers (Y. Zhu et al., 2000; Y. Zhu, 2004) also found that traditional local rice varieties in the Hali terrace area of Yunnan Province can out-compete modern rice varieties because of their disease resistance, stable and reasonable high yield, and good adaptation to the cool climate of high altitudes. It was determined that this outcome is due to the complex genetic background caused by local seed selection processes which are totally different from modern rice breeding programs that result in rice varieties with "pure" genetic backgrounds. Many local traditional crop and animal species can respond well to special environments with good yields. In Guangdong Province, for example, high-quality rice called "Maba Youzhan," which dates back about 4000 years, is well adapted to the high daily temperature fluctuations and high soil phosphorus content in the Maba Township in north Guangdong Province. Another high-quality local rice variety, "Zengcheng Simiao," dating back 500 years in Zengcheng in mid-Guangdong Province, responds well to the sandy soils, high rainfall, and slightly alkaline irrigation water in the area. A tangerine species, "Huazhou Juhong," which is grown in soil with rich biotitic schist and mica carbonate schist in Huazhou County, Guangdong Province, possesses a special quality that allows it to be used as a cough medicine. Modern, high-yielding varieties, however, are quickly replacing traditional varieties. Traditional genetic resources are being lost at an accelerating rate. As one of the major crop and animal production centers in the world, this loss of original genetic diversity is a great threat in China. Research shows that most of the rice varieties used in South China from 1959 to 2005 were comprised of only 14 local varieties and the remainder foreign rice varieties (G. Zhang and Liu, 2009). The 2004 announcement to strengthen the protection and management of biological resources by the Office of State Department represents recognition of this serious situation. Genetic diversity protection both *in situ* and *ex situ* has gained more attention since then.

Species diversity has been widely adopted in traditional agriculture as well as in modern eco-agriculture development in China for its effective use of resources, soil fertility improvement, and pest management. Intercropping, rotation, and relay cropping are common methods of utilizing species diversity in crop production. Rotation between grain crops and leguminous crops has been practiced in China since the Zhanguo Period, around 100 AD, and rotation between paddy rice and upland crops has been in use since the Tang Dynasty, around 700 AD (Hui, 2014). These rotation systems can effectively prevent diseases and maintain soil fertility. Their use has been encouraged in eco-agriculture. For example, milk vetch (*Astragalus sinicus*) as a winter green manure is sown about two weeks before the harvest of second crop rice in south China. Peanut or soybean inter-cropped with sugarcane in the early growth stage of sugarcane can increase soil cover and nitrogen supply. In north China, winter wheat intercropped with summer corn before wheat harvest in May is a common practice. Because young farmers are leaving rural areas today, monocropping methods are becoming more favored. Machinery that can be used for intercropping has become a major development bottleneck for intercropping today.

Another example of using species diversity is fish culture in South China. Traditionally, one fish pond will contain a mix of four to five varieties of carp. Big-headed carp (*Hypophthalmichthys nobilis*) live in the upper layer of the fish pond and feed on phytoplankton. Grass carp (*Ctenopharynodo nidellus*) grow in the middle layer of the pond and use grass as their food source. Common carp (*Cyprinus carpio*) and silver carp (*Carassius auratus*) can tolerate low oxygen levels and live at the bottom of the fish pond. They are omnivorous and help to clean up the pond by consuming debris that drops from the upper layers. A mixture of these fish species in a fish pond can increase feed efficiency and reduce fish diseases. Crop and animal co-culture systems such as the fish–rice co-culture system and the duck–rice co-culture system are also traditional practices in China. These systems are still used by farmers today. New co-culture systems such as rice–crab, rice–frog, and soft-shell turtle (*Trionyx sinensis*) with wild rice (*Zizania latifolia*) have also arisen during development of eco-agriculture in China.

Agroforestry associations have also been widely used. Mixed rubber trees and tea systems have been developed in Hainan Province and Yunnan Province. Not only can the agroforestry system provide shade for the tea to improve its quality but it can also provide better groundcover to reduce soil erosion. The number of spiders that are predators of many insect pests in this mixed rubber–tea community is higher than in a pure rubber community. This system can generate income beginning the third year after planting the tea. This income helps to offset the cost of rubber production, which does not begin until the seventh or eighth year after planting. Another type of agroforestry system is formed by *Paulownia sieb* and crops such as wheat, corn, soybean, or cotton in North China. *Paulownia* is a tall tree species that can be harvested after 10 to 14 years or maintained to form a permanent windbreak system. A *Paulownia* community with 5-m by 40-m spacing is optimum for intercropping. *Paulownia* can stabilize the micro-environment for crops within its borders. Crops can make good use of the land resource and generate more income for the farmer (W. Li, 2003). Species diversity is also useful for biological pest control in eco-agriculture. The production of predator mites (*Neoseiulus cucumeris*) for red mite (*Tetranychus cinnbarinus*) control in citrus orchards and parasitoids (*Trichogramma* spp.) for sugarcane borer control in sugarcane fields has been utilized in large commercial-scale farms in South China.

1.3.3 Techniques Used in Eco-Agriculture

In eco-agriculture, the so-called eco-tech package integrates and modifies techniques from vari-ous resources to form technical packages that can adapt well to operate at the level of the agro-ecosystem as a whole to achieve a balanced result of economic, social, and ecological benefits. Environmentally friendly, resource-saving, quality-improving techniques are often used in the eco-tech package (Table 1.1). Technical packages should be designed according to the target agroeco-system and to the specific resource base and environmental contexts.

Many technologies related to resource replacement and resource saving are available, includ-ing solar water heaters, biomass energy, wind power generators, and hydropower. Techniques often used to replace chemical fertilizer include the use of crop residues, animal waste, and green manure. Biological methods and cultivation methods can be used to replace pesticides. Fertilization based

Table 1.1 Sources of Techniques Often Used in Eco-Agriculture

| Technique | Technique Resources | | |
	Traditional	Current Practice	Modern Technology
Landscape design	Protection of *Fengshui lin* (forest behind and around a village and a pond in front of a village)	Natural preserved areas Ecological corridors, such as wind-break systems	GPS Computer landscape design
Overcoming stressed environment	Pioneer crops such as sesame and cassava used in newly reclaimed land	Soil conservation techniques Salinity soil reform techniques	Identification of stress-resistant genes Selecting new remediation plant species for heavy-metal-contaminated soil
Cycling	Composting Direct straw returning Fish pond–mulberry dike system	Biogas techniques Raising earthworms Growing edible mushrooms	Biomass energy techniques Microbial preparations
Saving resources	Intercropping Crop rotation Saving water from hoeing operation	Water-collecting techniques Drip and sprinkle systems No-tillage methods Soil and plant nutrition testing	Precision agriculture Controlled-release fertilizers Decomposable plastic sheeting
Biodiversity	Natural enemies Green manure Local species Rice–duck co-culture Rice–fish co-culture	Biological pest control techniques Biological remediation methods Living mulch	Functional gene identification and utilization Chemical relationship and allelopathy

Source: Adapted from Luo, S., *Chin. J. Eco-Agric.*, 18(3), 453–457, 2010.

on soil and plant nutrition tests and the use of controlled-release fertilizers are encouraged by the government in China. Drip irrigation, sprinkle irrigation, pivot irrigation, and soil covers are often used to save water resources, especially in arid and semi-arid regions. In addition to the recycling of crop straw and animal waste, artificial wetlands have also been widely adopted for agricultural waste treatment. Although no-tillage or minimum-tillage techniques can help to reduce energy consumption during land preparation, they have not been widely used in China so far.

1.4 AGROECOLOGY AS A RESEARCH FIELD IN CHINA

Since the early 1980s, more and more effort has been put into agroecology research. Research institutions, special academic journals, and long-term agroecology observation stations have been set up. Agroecology research in China has followed the needs of agricultural development and the frontiers of ecology, environmental science, and agricultural science.

1.4.1 Agroecology Research Institutes and Journals

Many research institutes have agroecology as a major research focus, including the Chinese Academy of Sciences, Chinese Academy of Agricultural Sciences, provincial agricultural science academies, and universities (Table 1.2). Research on agroecology at these institutions was initiated mainly after the late 1980s. These various institutions are located all over China in different geological regions, from temperate zones to tropical regions, from wetlands to arid regions, and from plains areas to mountainous and hilly regions. Examples of the agroecology research being conducted at these institutions include agrobiodiversity, biogas, eco-agriculture, cyclical agriculture, farming systems, agro-environment, and crop ecology, among others. Long-term research and observation sites have been set up to observe the changes in soil fertility, soil quality, water balance,

Table 1.2 Major Institutions Involved in Agroecology Research in China

Institution	Year Established	Year Renamed	Main Research Focus and Task Related to Agroecology	Location	Region
Chinese Academy of Sciences					
Institute of Geographic Sciences and Natural Resources Research	1940	1999	Main focus is on regional sustainable development, including agricultural sustainable development. This institute leads the Chinese Ecological Research Network (CERN) and owns three long-term agroecology stations.	Beijing	North China
Shenyang Institute of Applied Ecology	1954	1987	Main focus is on sustainable agriculture and forestry development and on ecological engineering. This institute has seven long-term observation stations for agroecology and forest ecology research.	Shenyang, Liaoning	Northeast China
Northeast Institute of Geography and Agroecology	1978	2002	One focus of research is on agroecology. This institute has five long-term research stations, including three agroecology-related stations.	Changchun, Jilin	Northeast China
Institute of Subtropical Agriculture	1978	2003	Research is focused on integrated agroecosystem ecology in subtropical regions. This institute has four long-term research stations, including two specialized in agroecology.	Changsha, Hunan	Central China
Guangdong Institute of Eco-Environmental and Soil Sciences	1958	1996	Research focus includes soil resources, agricultural environment, and sustainable development.	Guangzhou, Guangdong	South China
Cold and Arid Regions Environmental and Engineering Research Institute	1958	1999	One of the institute's seven research laboratories is an ecology and agriculture laboratory, which includes a long-term Gaolan ecology and agriculture research station.	Lanzhou, Gansu	Northwest China
Xinjiang Institute of Ecology and Geography	1961	1998	A focus of research is on agricultural development in arid oasis regions. This institute has one long-term field observation station for agroecology.	Wulumuqi, Xinjiang	Northwest China
Institute of Mountain Hazards and Environment	1966	1989	The Yanting Purplish Soil Agroecology Research Station was set up in 1991.	Chengdu, Sichuan	Southwest China
Chinese Academy of Agricultural Sciences					
Institute of Agricultural Resources and Regional Planning	1979	2003	Research includes plant nutrition and fertilizers, regional agricultural development and ecology, and environmental issues in agriculture. This institute has nine research and observation stations related to agroecology and agro-environment. Stations are located in Hunan, Inner Mongolia, Shaanxi, Hubei, Shandong, and Beijing.	Beijing	North China

Institution			City	Region
Institute of Environment and Sustainable Development in Agriculture	2001	1953	Research includes agrometeorology, rural water resources, and agroecology and agro-environmental engineering. This institute has agro-environmental research and observation stations in Shanghai, Gansu, Beijing, Hunan, Shanxi, Tibet, Qinghai, Ningxia, Fujian, and Shaanxi.	Beijing	North China
Agro-Environmental Protection Institute	1997	1979	Research is focused on agricultural pollution and its treatment, biodiversity, and eco-agriculture. Monitoring agricultural environments and food safety is an important duty of this institute.	Tianjin	North China
Biogas Research Institute of Ministry of Agriculture	1997	1979	Research focus is on biogas and other biomass energy.	Chengdu, Sichuan	Southwest China
Universities					
Institute of Tropical and Subtropical Ecology, South China Agricultural University	1994	1994	Research topics include agroecology, forest ecology, animal ecology, and environment ecology.	Guangzhou, Guangdong	South China
Agroecology Research Institute, College of Crop Sciences, Fujian Agriculture and Forestry University	1986	1986	Research is conducted on agroecology and allelopathy by using molecular biology methods, among others.	Fuzhou, Fujian	East China
State Key Laboratory of Grassland Agroecosystems, Lanzhou University	2011	1990	Research is focused on sustainable grassland development and agroecology in arid and semi-arid areas.	Lanzhou, Gansu	Northwest China
Circular Agriculture Research Center, College of Crop Sciences, China Agricultural University	2008	2008	Primary focus is on the cycling design of agroecosystems around China.	Beijing	North China
National Engineering Center for Applied Technologies of Agricultural Biodiversity	2007	2002	Primary goals are to promote the use of biodiversity in agriculture and to discover the mechanisms involved in biodiversity and agriculture.	Kunming, Yunnan	Southwest China
Crop Physiology, Crop Ecology and Farming System in Mid-Yangtze River, Key Laboratory of the Ministry of Agriculture, Huazhong Agricultural University	2011	2003	Primary focus is on farming systems and ecology.	Wuhan, Hubei	Central China
Chongming Eco-Agriculture Research and Development Center, College of Agriculture and Biology, Shanghai Jiaotong University	2004	2004	Research is focused on modern eco-agriculture, environmental and biodiversity issues in agriculture, and recycling agricultural wastes.	Shanghai	East China
Provincial Academy of Agriculture Sciences					
Agroecology Institute, Fujian Academy of Agricultural Sciences	2005	1983	Research is conducted on eco-agriculture and Azolla, a nitrogen-fixing fern found in paddy fields.	Fuzhou, Fujian	East China
Tropical Eco-Agriculture Institute, Yunnan Academy of Agricultural Sciences	2005	1987	Research is conducted on eco-agriculture and vegetation recovery in tropical and arid regions in Yunnan Province.	Yuanmu, Yunnan	Southwest China

water quality, crop productivity, and product quality based on how they are influenced by agroecological measures. In addition, other institutes and universities besides those shown in Table 1.2 are also conducting research on agroecology.

The three scientific societies related to agroecology in China are the Agroecology Committee under the Chinese Ecology Society, the Eco-Agriculture Committee under the Chinese Environmental Sciences Society, and the Chinese Association of Agricultural Ecology and Environment under the Ministry of Agriculture. Regular biannual academic conferences have been conducted by the Agroecology Committee and the China Association of Agricultural Ecology and Environment. The number of participants has usually reached 300 to 400 for each of these conferences in recent years. The major journals for publication of agroecology research papers include the *Chinese Journal of Ecoagriculture* (1993–present), *Journal of Agro-Environment Science* (1982–present), *Chinese Journal of Applied Ecology* (1990–present), and *Journal of Ecology and Rural Environment* (1984–present). Other journals, such as *Acta Ecologica Sinica* (1981–present), *Scientia Agricultura Sinica* (1960–present), and *Ecological and Environment Sciences* (1992–present), also publish large numbers of research papers on agroecology.

1.4.2 Overview of Major Research Topics in Agroecology

With the establishment of agroecology as a totally new way of perceiving agriculture in the early 1980s, agroecology research in China began to investigate eco-agriculture practices based on traditional farming. Methods for energy and material flow analysis at the ecosystem level began to appear. In the 1990s, ecological research on climate change, biodiversity, and sustainable development carried over to research on agroecology. A circular economy, low carbon economy, and clean production concepts became hot topics when they were introduced in the late 1990s. The maturation of landscape ecology and molecular biology methods at the beginning of this century enabled agroecology research to stretch in both macro and micro directions. Although more and more research papers by Chinese researchers are being published in international journals, the number of articles with the specific subject of "agroecology" published in China still reflects efforts concentrated within China. The number of papers in Chinese publications with the topics of "agroecology" or "eco-agriculture" has shown a pattern of increasing over the past 30 years (see Figures 1.1 and 1.5). Figure

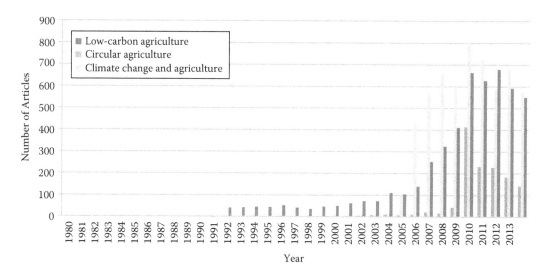

Figure 1.11   Number of articles in Chinese literature from 1984 to 2013 with topics related to energy and material flow (www.cnki.net search).

1.11 provides a breakdown of the articles appearing in Chinese publications that included the topics of "low-carbon agriculture," "circular agriculture," and "climate change and agriculture" as related to material and energy flow in agroecosystems. The topic of "climate change and agriculture" began to attract attention beginning in the late 1980s, but papers addressing the other two topics were very few before 2000. After 2005, however, research on all three topics grew very quickly. It is estimated that 17% of the carbon released in China is caused by agricultural activity. The agricultural-related carbon release increased from about 43.07 million tons per year in 1993 up to 78.43 million tons in 2008 (B. Li et al., 2011). Interest is growing in carbon sequestration by the biochar method in farm fields. Circular agriculture research has been actively supported by the National Scientific Foundation, the Ministry of Sciences and Technology, and various provincial funding sources since 2000.

The number of articles with subjects related to the use of biodiversity in agroecosystems is shown in Figure 1.12. Because intercropping and rotation are traditional Chinese agricultural practices, they became the hot subject of research very early in China, and many articles on these topics were published long before 1980. Today's research on intercropping and rotation focuses not only

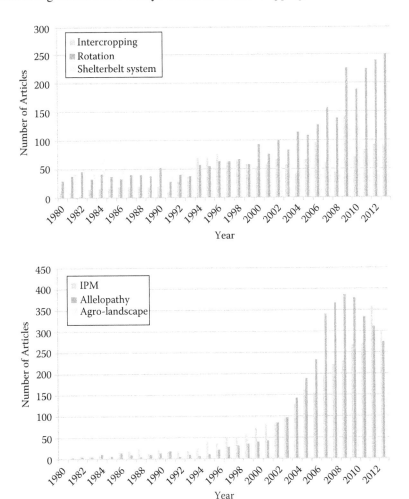

Figure 1.12 Number of articles in Chinese literature from 1980 to 2013 with topics related to biodiversity use in agroecosystem. (Top) Number of articles on intercropping, rotation, and shelterbelt system. (Bottom) Number of articles on integrated pest management, allelopathy, and agro-landscape (www.cnki.net search).

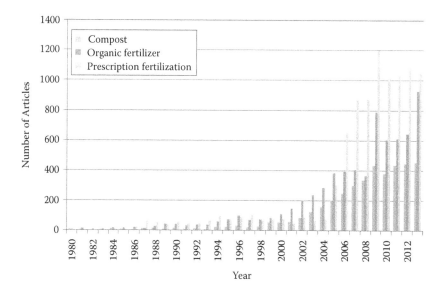

Figure 1.13 Number of articles in Chinese literature from 1980 to 2013 with topics related to plant nutrition inputs (www.cnki.net search).

on their effects on field production but also on the discovery of physical, chemical, and biological mechanisms behind their use. Shelterbelt systems and windbreak systems began development in China from the 1960s to 1970s. Their meteorological effects have been measured, and the biological effects of shelterbelt systems are of great interest today. Although integrated pest management (IPM) was already the subject of considerable research beginning in the early 1980s, it continues to be a hot topic due to public concern for food safety and environmental quality. Also popular is research into allelopathy, which involves the chemical relations between plants and other organisms within agroecosystems and natural ecosystems. This concept was introduced to China in the mid-1980s. Research on agricultural landscapes and buffer zones has been increasing slowly. Only 885 articles including the topic of "agrolandscape" have been published over the past 34 years (Figure 1.12). It can be difficult to conduct research at the landscape level; however, considering the importance of agricultural landscape arrangement and the maturity of landscape ecology, agricultural landscape research should be given more attention in the near future.

Popular topics for agroecology research related to changing inputs include the use of compost and other organic fertilizers to replace chemical fertilizers. To avoid the overuse of chemical fertilizers, prescriptive fertilization methods based on soil and plant nutrition testing have grown in popularity since 2005 and have attracted a significant research effort over the past 10 years (Figure 1.13). Because the use of biogas is very popular in rural areas, it is not surprising that the number of articles related to biogas is high (Figure 1.14). Water-saving techniques such as drip irrigation and plastic sheeting covers to replace open irrigation have been used extensively in arid and semi-arid regions over the past 10 years. The number of articles including the topic of "water-saving agriculture" increased significantly after the year 2000 (Figure 1.14). There has been quite a bit of research on no tillage or minimum tillage in China over the past 20 years despite the low application rate on farms.

Research on output regulation of agroecosystems includes pollution control and food quality control. In order to reduce water pollution in rural areas, research on artificial wetland system structural design and its effects has been growing since the year 2000 (Figure 1.15). The "green food" standard was proposed by the Ministry of Agriculture before the 11th Asian Games held in

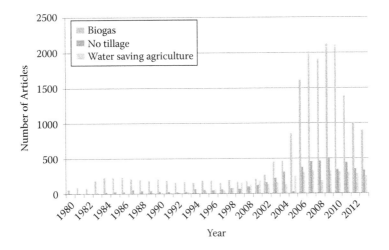

Figure 1.14 Number of articles in Chinese literature from 1980 to 2013 with topics related to energy and water inputs in agriculture (www.cnki.net search).

Beijing in 1990. It strictly controls the use of chemicals in food production and requires good quality soil and irrigation water for food production. The "green food" standard as it applies to various crops has been the focus of research since then (Figure 1.15). A food safety program was proposed by the Ministry of Agriculture in 2001, but the only requirement for earning a "food safety" label in China is that there are no toxic materials in the final products. Articles addressing this subject have increased significantly. A standard for organic food was introduced to China after 1990, and the number of articles related to organic food increased significantly after the year 2000 (Figure 1.15). Dr. Stephen Gliessman, from the University of California, and researcher Li Zhengfang, from Nanjing Institute of Environmental Sciences, made important contributions to the development of organic food and organic agriculture in China in the late 1980s and early 1990s with the support of the Rockefeller Brothers Fund (Z. Li and Zhang, 1988; Z. Li, 1994a,b).

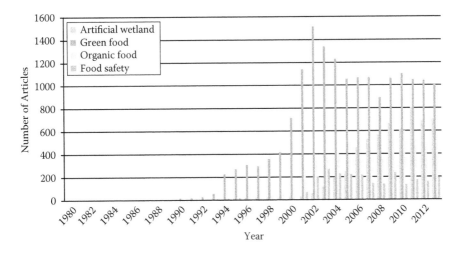

Figure 1.15 Number of articles in Chinese literature from 1980 to 2013 with topics related to agroecosystem outputs (www.cnki.net search).

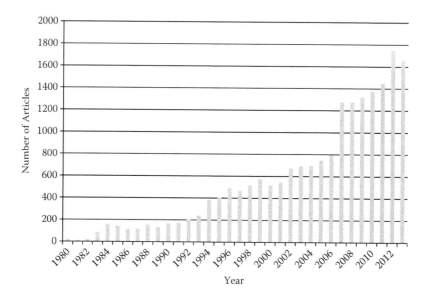

Figure 1.16   Number of articles in Chinese literature from 1980 to 2013 with the topic of traditional agriculture (www.cnki.net search).

1.4.3 Agroecology's Traditional Agriculture Secret

The history of Chinese agriculture can be traced back to about 10,000 years ago. Interest in traditional agriculture research in China has grown since the 1980s, even when industrialization began to speed up. This is apparent from the number of articles shown in Figure 1.16. The focus of agroecology on traditional agriculture is quite different from that of other disciplines such as history, archaeology, crop science, or animal science. Not only does it study interrelationships at the agroecosystem level but it also investigates cooperation and coordination of human activity with natural processes.

Research on the energy and material cycling processes of a traditional fish pond–mulberry dike system in the Pearl River Delta was conducted in the 1980s, as described earlier (Zhong et al., 1993). Research into another traditional cycling system, the high bed–deep ditch system, in lowland areas in the Pearl River Delta was also conducted in the 1980s (Lin and Luo, 1989). The deep ditch can be used to lower the groundwater table, retain nutrients and soil particles washed from the bed, grow rice or taro, and raise fish. It can also help to reduce salinity in newly reclaimed coastal lands. The high bed can be used to grow vegetables, bananas, oranges, or sugarcane. The amount of mud formed by debris from crops and eroded soil each year in the ditch was 51 to 157 t/ha, depending on the crop on the bed. The mud was rich in soil organic matter (3.6 to 5.7%) and nutrients and was returned to the bed each year.

In traditional agriculture in China, soil fertility could be maintained and improved by carefully designed rotation and intercropping systems in addition to the use of organic fertilizer. Research conducted by Li and his team from China Agricultural University (L. Li et al., 2007; Y. Li et al., 2009) discovered the underground mechanism of mutualism in intercropping between leguminous crops such as fava bean and soybean and grain crops such as wheat and corn. The grain crop can stimulate the invasion of rhizobia, help leguminous crops to reduce inhibition of nitrogen fixation, and increase the availability of micro-nutrients such as iron and molybdenum by releasing chelates and altering soil acidity. Leguminous crops provide nitrogen to the nearby grain crop in return. The mechanism to reduce rice blast disease by intercropping one row of the taller traditional rice variety with four rows of a shorter hybrid rice variety has been revealed by Zhu Yongyong (2007). Physical, chemical, and biological mechanisms are involved. The dew on the leaf surface of the tall susceptible rice variety dries earlier because of better air circulation, greater solar radiation received, and higher temperature of the leaf surface. The higher rate of evaporation and transpiration for the local variety also stimulates greater silicon accumulation in rice leaves. These factors reduce the chance of pathogen invasion. Even if invasion occurs, the spread of the disease is greatly reduced by the physical barrier of the four rows of the resistant variety between each row of the susceptible variety.

The history of fish–rice co-culture goes back more than 1200 years in China, and it is still very actively practiced in many hilly and mountainous regions in southern China. A team led by Chen Xin from Zhejiang University discovered the mechanism of mutual facilitation between rice and fish (Xie et al., 2011). Not only do fish feed on weeds, phytoplankton, and debris within the water layer but they also feed on insects and anthers from rice flowers by pushing and shaking rice plants. It has been determined that about one third of rice planthoppers are shaken down by the shaking activity provided by fish. Researchers also found that chemicals released from fish skin could be very effective for inhibiting pathogens such as rice sheath blight disease. The rice plants can reduce fluctuation in water temperature and provide shade for fish. The complementary use of nitrogen between rice and fish results in less nitrogen in the drainage water. Similarly, the duck–rice co-culture system has also been a traditional farming system in China for several hundred years and is still practiced today by farmers in paddy fields. Ducks can help rice to suppress weeds and pests including insects, diseases, and golden apple snail by feeding and paddling around (J.E. Zhang et al., 2002, 2013; J.E. Zhang, 2013). Surprisingly, rice stimulated by the frequent touch provided by duck movement becomes shorter and tougher due to the change of hormone levels of GA and ABA. This morphological change has helped rice to increase lodging resistance.

1.5 CHALLENGES AND FUTURE TRENDS FOR AGROECOLOGY IN CHINA

The challenge for agroecology development in China is significant—as a discipline, as a practice, and for research. Fortunately, new developments in science and technology and the ecological civilization idea proposed in China have provided new momentum for agroecology development in China.

1.5.1 Challenges Faced by Agroecology in China

Traditional agricultural disciplines such as crop science, animal science, soil science, and agricultural engineering deal with very specific subjects and have a narrow focus. Agroecology, however, deals with the multiple relationships among all of these components plus the social and economic aspects of food systems. It can be difficult to define and is a complicated field of study. Not only is it difficult for students to comprehend but it is also confusing for many government officials, farmers, and even scientists. In eco-agriculture practice, government officials and farmers alike often complain that, "Eco-agriculture is just like a basket. You can put anything into it. If it is everything, then it means nothing." It is clear that a simple and solid framework for agroecology as a discipline is necessary to guide practice and research.

When the first wave of eco-agriculture development occurred in the 1980s and 1990s, the prevailing type of agriculture in China was still small-scale traditional farming with surplus farm labor. Most of the traditional practices such as composting and keeping several pigs and chickens in a farmer's household were still widely utilized. Today, many young farmers have left the countryside and are working in urban areas. The urban population in China exceeded the rural population for the first time beginning in 2011 (Figure 1.17). More and more farmland is being transferred to either active local farmers or large companies. The scale of production grows, and labor-saving methods are preferred due to a lack of labor and higher labor costs. Agricultural production has come to rely more heavily on chemical fertilizer, pesticides, and plastic sheeting (Figure 1.18). Various pollution

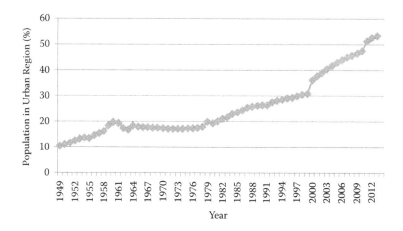

Figure 1.17 Percentage of population in Chinese urban areas from 1949 to 2013 (http://wenku.baidu.com/).

events caused by agricultural activities—such as air pollution due to straw burning, water pollution due to overuse of fertilizer and large-scale animal farms, and "white pollution" caused by covering the soil with plastic film mulch—happen often and are becoming more of a public environmental crisis. Knowledge gained about eco-agriculture during the 1980s and 1990s may not be applicable or sufficient under these new conditions.

1.5.2 Opportunities for Agroecology Development in China

Since the 1980s, agroecology has been recognized as a formal branch of ecology, agricultural science, and environmental science in China. Agroecology can be found in the curricula of many universities and is a formal academic research field in China supported by a variety of funding sources (e.g., central and local governments, various enterprises). This provides a solid base and favorable conditions for the future development of agroecology in China.

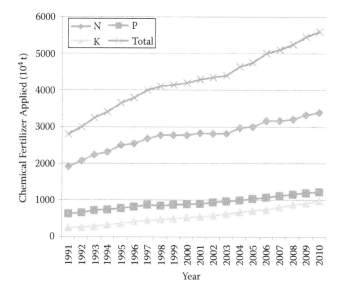

Figure 1.18 Trend for total commercial fertilizer quantity (10^4 t) used from 1991 to 2010 in China. (Adapted from Luan, J. et al., *J. Nat. Resour.*, 28(11), 1869–1878, 2013.)

The environmental and ecological situation in China has deteriorated during the period of rapid industrialization. The news often reports on environmental crises and incidents, such as the heavy smog in North China in 2013, eutrophication incidents in Kunming Lake and Taihu Lake, and the finding by a 2014 soil survey that 16% of soil sampling points all over China were polluted by heavy metals. There are also food safety issues, such as milk powder contaminated by melamine in 2008, rice contaminated by cadmium, and ginger contaminated by the toxic pesticide aldicarb in 2013. This situation has caused much public concern about the ecological and environmental consequences of modern agriculture. People are deeply worried about the quality and safety of the food they consume daily. The central government can no longer ignore this situation and has begun to take steps to reverse it.

The emerging concept of ecological civilization, together with political, economic, social, and cultural issues, was identified as one of the core themes of government work by the 18th National Congress of the Communist Party of China in 2012. "To establish long-term management systems for sustainable development of agriculture" became a key goal for agriculture and rural development in China according to the 2014 No. 1 Document of the State Council, which presented key measures and policies for agricultural development for the coming years. Three main objectives for sustainable agriculture development were named in this document: (1) to promote the development of ecological friendly agriculture, (2) to try a rest-and-recovery approach to agriculture, and (3) to promote the protection and improvement of the ecological base for agriculture. Under the Ministry of Agriculture, the Central Station of Agro-Ecology and Agro-Resources Protection (CSAAP) was established in 2013 with 60 job positions. One of its tasks is to promote eco-agriculture development in China. A new program for eco-agriculture development was established in 2014, and demonstration sites for eco-agriculture development have been selected by CSAAP under this program. This is a strong indication that the second wave of eco-agriculture development in China is coming and along with it strengthening of the agroecology movement.

1.5.3 Next Steps for Agroecology Development in China

The same confusion regarding the boundaries and focus of agroecology is also experienced by the environmental and ecological sciences. Such confusion is caused by philosophies and views that differ significantly from those of traditional disciplines and is quite common during a period of rapid growth for a new discipline. In traditional disciplines, analytical and deductive methods are the foundation of their methodology. The problem of "not seeing the forest for the trees" applies to the unsustainable development situation of a modern society supported by traditional disciplines. A perhaps revolutionary view would be to reintroduce inductive and systematic methods to science and technology in order to see both trees and the forest. It is much easier for people to see a specific object rather than relationships between objects. Fortunately, a holistic view is embedded in the blood of Chinese culture. It is well represented in traditional Chinese medicine, which is still widely accepted in Chinese society today. Another opportunity is the Internet-fueled information revolution. Multidisciplinary or interdisciplinary knowledge exchange is much easier now, and it is very probable that knowledge of agroecology will grow over the next 10 years.

It is necessary to better sort out our knowledge of agroecology as its own discipline. It is clear that there are two types of knowledge related to agroecology: (1) agroecosystem structure and function, topics that are related to the natural sciences and engineering; and (2) the control and regulation of agroecosystems through social and economic systems, which are more related to humanitarian aspects. Although these two types of knowledge are closely linked together, they can still be viewed as two sub-branches of agroecology. For the first, essential knowledge includes the construction of ecologically friendly agroecosystem structures at the biological level, ecosystem level, and landscape level. For the second, the critical knowledge is regulation of human behavior by legislative and economic means. The overall theme of agroecology, though, is the ecological and environmental impacts of agriculture. It is likely that agroecology in China will develop along this line and hence be better understood by government officials, farmers, students, and scientists in their practice and research. What are the key actions to promote the practice of eco-agriculture? According to our understanding of agroecology, based on knowledge accumulated during the first wave of eco-agriculture practice in China and similar practices abroad, we can identify three key actions needed in China: (1) screen and build agroecosystems with ecologically sound structure and function, (2) establish legislative and economic systems that can strongly support the development of eco-agriculture, and (3) switch the dominant top-down mode of action in China to a combination of both top-down and bottom-up.

To build agroecosystems with ecologically sound structure and function, three levels of actions can be identified. The first level can be called *bottom-line* or *red-line actions*, some examples of which include stopping large-scale straw burning in fields, stopping the discharge of animal waste without treatment, reducing fertilizer dosage to a safe level, and using irrigation water within the quotas calculated by water balance budgets. These actions are directly related to regulating resource input and pollution output. The second level can be referred to as *advance actions* or *green actions*, which include the reform of agroecosystems by altering biological structures, material cycling paths, or landscape arrangements. The third level of action can be considered *eco-product* or *eco-farm labeling actions*, which strictly follow the guidelines of green food, organic food, or eco-label production procedures and pass through inspection to be certified and obtain labeling.

The second key action is to establish legislative and economic systems that can support eco-agriculture, based on lessons learned during the first wave of eco-agriculture in China during the 1980s and 1990s. When government support for eco-agriculture ended, more than half of the programs also ended. It is necessary to ensure that eco-agriculture development will receive continuous legal and financial support. A top priority is the passage of legislation that can help to prevent direct pollution and resource exhaustion by agricultural activities and to establish standards for bottom-line actions. Potential economic stimulation policies for green actions include lower taxation rates, easier ways to obtain loans with low interest rates, and government subsidies. For the labeling actions, there are already standards for green food, organic food, and environmental quality. The standard for eco-agriculture products or ecological friendly farms should be developed. Inspection and assessment procedures must be further standardized. The recognition for those labeled products should be introduced to and recognized by more consumers and markets.

The third action requires educating and motivating the various stakeholders. Because China used to have more of a central planning type of government, government and scientists together directed the eco-agriculture movement in the 1980s and 1990s. However, after 36 years of implementation of the "reform and opening up" policy in China, the market system is taking hold. Government also recognizes that the participation of all stakeholders and non-governmental organizations (NGOs) is important to ensure healthy development of the eco-agriculture movement. An environmental law amended this year allows NGOs to represent the public interest in initiating environmental lawsuits. Consumer associations have become quite active and influential recently in China for improving food safety and environment. Government also encourages the activities of farmers' technical associations, agricultural trade associations, and farmer cooperatives. Although these NGOs are still quite weak today compared to similar NGOs in many other countries, they are now enjoying greater opportunities to grow. Only through public awareness and actions rising up from the grassroots will eco-agriculture become a social movement and persist. The combination of top-down guidance and bottom-up participation promises to speed up the development of agroecology in China.

China is experiencing an era of rapid development. The conflict caused by significant supply pressures and ecological pressures in agriculture will last for some time. During this period, it can be predicted that food supply crises, economy fluctuations, environmental issues, and ecological emergencies will occur. Adjustment of the policies and improvement of methodologies to overcome these crises will be a long journey; however, the ship of Chinese agriculture has to and will finally sail its way to a sustainable course.

REFERENCES

Altieri, M.A. 1983. *Agroecology: The Scientific Basis of Alternative Agriculture.* University of California: Berkeley.

Altieri, M.A. 1987. *Agroecology: The Science of Sustainable Agriculture.* Westview Press: Boulder, CO.

Altieri, M.A. 1995. *Agroecology: The Science of Sustainable Agriculture*, 2nd ed. Westview Press: Boulder, CO.

Bi, Y., Wang, Y., and Gao, C. 2010. System constitution and general trend of comprehensive straw resource utilization in China. *J. China Agric Resour. Reg. Plann.*, 31(4): 35–38.

Cao, L. 2011. *Principles of Agroecology.* Shanghai Jiaotong University Press: Shanghai.

Chen, F. 2002. *Agroecology.* China Agriculture University Press: Beijing.

Chen, J. 2005. Ecological agriculture based on soil and water conservation in limestone region of Taihang Mountain. *Bull. Soil Water Conserv.*, 25(1): 82–87.

Chen, Y. 1985. Basic knowledge of agroecology: the first course: systems theory and agroecosystems. *Hebei Agric. Sci.*, 5: 34–37.

Cheng, J., He, F., and Liu, Y. 2009. Progress of research on geomantic forests of Lingnan villages. *Chin. Landscape Archit. J.*, 11: 93–96.

Cox, G.W. and Atkins, M.D. 1979. *Agricultural Ecology: An Analysis of World Food Production Systems.* W.H. Freeman: San Francisco, CA (1987 translation into Chinese by Z. Wang).

Gliessman, S.R. 2000. *Agroecology: Ecological Processes in Sustainable Agriculture.* CRC Press: Boca Raton, FL.

Gliessman, S.R. 2007. *Agroecology: The Ecology of Sustainable Food Systems*, 2nd ed. CRC Press: Boca Raton, FL.

Gliessman, S.R. 2015. *Agroecology: The Ecology of Sustainable Food Systems*, 3rd ed. CRC Press: Boca Raton, FL.

Guo, Sh. 1988. *China Ecological Agriculture.* China Prospects Press: Beijing.

Guo, Sh. 1993. *The Rising of Ecological Agriculture in China.* Hebei Science and Technology Press: Shijiazhuang.

Han, C. 1985. The energy structure and efficiency of agroecosystems. *Rural Ecol. Environ.*, 3: 6–8, 15.

Huang, Z.X., Zhang, J.E., Liang, K.M., Quan, G.M., and Zhao, B.L. 2012. Mechanical stimulation of duck on rice phytomorphology in rice–duck farming system. *Chin. J. Eco-Agric.*, 20(6): 717–722.

Hui, F. 2014. *Traditional Chinese Ecological Culture in Agriculture.* China Agricultural Science and Technology Press: Beijing.

King. F.H. 1911. *Farmers of Forty Centuries, or Permanent Agriculture in China, Korea, and Japan.* Rodale Press: Emmaus, PA.

Li, B., Zhang, J.B., and Li, H.P. 2011. Research on spatial-temporal characteristics and factors affecting decomposition of agricultural carbon emission in China. *China Popul. Resour. Environ.*, 21(8): 80–86.

Li, L., Li, S., Sun, J., Zhou, L., Bao, X., Zhang, H., and Zhang, F. 2007. Diversity enhances agricultural productivity via rhizosphere phosphorus facilitation on phosphorus-deficient soils. *Proc. Natl. Acad. Sci. U.S.A.*, 104(27): 11192–11196.

Li, W. 2003. *Ecoagriculture: Theory and Practice of Sustainable Agriculture in China.* Chemical Engineering Press: Beijing.

Li, Y., Yu, C., Cheng, X., Li, C., Sun, J., Zhang, F., Hans, L., and Li, L. 2009. Intercropping alleviates the inhibitory effect of N fertilization on nodulation and symbiotic N_2 fixation of fava bean. *Plant Soil*, 323: 295–308.

Li, Z. 1994a. Energetic and ecological analysis of farming systems in Jiangsu Province, China. In: *10th International Organic Agriculture IFOAM Conference: Papers, Abstracts, and Conference Programme*, Lincoln University, Lincoln, New Zealand, December 11–16, pp. 142–143.

Li, Z. 1994b. The basic standard of organic agriculture and the development of its products. *Environ. Pollut. Control*, 16(4): 17–21.

Li, Z. and Zhang, G. 1988. Studies of the design and construction of the Nanjing Guquan Rural Ecological Project. *Rural Ecol. Environ.*, 4 (12): 1–4.

Lin, R. and Luo, S. 1989. Structure and function of "high bed and low ditch" farmland ecosystem in Pearl River Delta. *J. Ecol.*, 8(3): 24–28.

Lin, Y., Ma, J., and Qin, F. 2012. The structure, distribution and prospect of China manure resources. *Chin. Agric. Sci. Bull.*, 28(32): 1–5.

Lu, L. 1999. *Introduction to Multistoried and Multi-Dimensional Agriculture in China*. Xichuan Scientific and Technology Press: Chengdu.

Lu, Y. 1984. Agroecology symposium held in Nanjing. *J. Ecol.*, 11(3): back cover.

Luan, J., Qiu, H., Jing, Y., Liao, S., and Han, W. 2013. Determination of factors contributing to the increase of China's chemical fertilizer use and projections for future fertilizer use in China. *J. Nat. Resour.*, 28(11): 1869–1878.

Luo, S. 2007. To discover the secret of traditional agriculture and serve the modern ecoagriculture. *Geogr. Res.*, 26(3): 609–615.

Luo, S., Ed. 2009. *Agroecology*. China Agriculture Press: Beijing.

Luo, S. 2010a. On the technical package for eco-agriculture. *Chin. J. Eco-Agric.*, 18(3): 453–457.

Luo, S. 2010b. Comparisons of agroecology textbooks from China and the United States. *Chin. J. Ecol.*, 29(7): 1458–1462.

Luo, S. and Han, C. 1990. Ecological agriculture in China. In: *Sustainable Agricultural Systems*, Edwards, C.A., Lal, R., Madden, P., Miller, R.H., and House, G., Eds. Soil and Water Conservation Society: Ankeny, IA, pp. 299–322.

Luo, S., Chen, Y., and Yan, F. 1987. *Agroecology*. Hunan Science and Technology Press: Changsha.

Ma, S. 1987. *Agroecology Engineering in China*. Science Press: Beijing.

Odum, E.P. 1971. *Fundamentals of Ecology*. W.B. Saunders: Philadelphia.

Scherr S.J. and McNeely, J.A. 2008. Biodiversity conservation and agricultural sustainability: towards a new paradigm of "ecoagriculture" landscapes. *Philos. Trans. R. Soc. B Biol. Sci.*, 363(1491): 477–494.

Shen, H. 1975. On agroecosystem and land use and land fertility management. *J. Tieling Agric. Coll.*, (2): 65–74.

Shen, H. 1982. The second course: basic principles of agroecosystems (I). *Chin. J. Soil Water Conserv.*, (5): 53–56.

Shen, H. 1996. *Agroecology*. China Agriculture Press: Beijing.

State Council. 2010. *The State Council on the Announcement for National Key Functional Zonation*, Document Series No. 46. State Council: Beijing.

Su, A. 2013. Domestic situation in the mushroom industry and its developmental trend. *Acad. Period. Farm Products Process.*, 334(11): 56–57.

Wang, H. and Cao, Z. 2008. *Agroecology*. Chemical Industry Press: Beijing.

Wang, J. 1997. *The Theory of Pollution Charges*. China Environmental Science Press: Beijing.

Wang, J., Jin, Z., Feng, Y., and Chen, Q. 2012. Zhouqu super-large mud debris flow disaster formation mechanism and management methods. *Gansu Geol.*, 21(2): 67–74.

Wang, L. 1994. *Agroecology*. Shanxi Science and Technology Press: Xi'an.

Wang, S. and Caldwell, C. 2012. *Agroecology*. Science Press: Beijing.

Wang, Z. 1985. On agricultural ecology and ecological agriculture. *J. Ecol. Rural Environ.*, 1(1): 59–64, 72.

Wang, Z. 1997. *Agroecosystem Management*. China Agriculture Press: Beijing.

Wezel, A., Bellon., S., Dore., T., Francis, C., Vallod, D., and David, C. 2009. Agroecology as a science, a movement and a practice: a review. *Agron. Sustain. Devel.*, 29: 503–515.

Worthington, M.K. 1981. Ecological agriculture: what it is and how it works. *Agric. Environ.*, 6(4): 349–381.

Wu, K. 1999. The characteristic and the warnings of the flood in the Changjiang River in 1998. *Progr. Geogr.*, 18(1): 20–25.

Wu, Z. 1981. To guide agricultural production by agroecosystem view. *Agric. Zoning*, (2): 61–67.

Wu, Z. 1986. *The Foundations of Agroecology*. Fujian Science and Technology Press: Fuzhou.

Xie, J., Hu, L., Tang, J., Wu, X., Li, N., Yuan, Y., Yang, H., Zhan, J., Luo, S., and Chen, X. 2011. Ecological mechanisms underlying the sustainability of the agricultural heritage rice–fish co-culture system. *Proc. Natl. Acad. Sci. U.S.A.*, 101(50): E1381–E1387.

Xiong, Y. 1982. The characteristics and research tasks of the agroecosystem. *Sci. Agric. Sinica*, 2: 78–83.

Xu, Y. 2010. Case study on the effect of forest covers on drought resistance and disaster reduction. *Mod. Agric. Sci.*, 22: 214–217.

Ye, X. 1981. Ecoagriculture. *Agric. Econ.*, (4): 3–10.

Zhang, G. and Liu, C. 2009. Genetic Diversity of Modern Rice Varieties in South China and Its Change, PhD thesis. South China Agricultural University: Guangzhou.

Zhang, J.E. 2013. Progress and perspective of research and practice of rice–duck farming in China. *Chin. J. Eco-Agric.*, 21(1): 70–79.

Zhang, J.E., Lu, J.X., Zhang, G.H., and Luo, S.M. 2002. Study on the function and benefit of rice–duck agro-ecosystem. *J. Ecol. Sci.*, 21(1): 6–10.

Zhang, J.E., Quan, G.M., Huang, Z.X., Luo, S.M., and Ouyang, Y. 2013. Evidence of duck activity induced anatomical structure change and lodging resistance of rice plant. *Agroecol. Sustainable Food Syst.*, 37(9): 975–984.

Zhang, X. 1985. Agroecology and ecoagriculture. *Xinjiang Agric. Sci. Technol.*, (3): 31–34.

Zhong, G., Wang, Z., and Wu, H. 1993. *The Interaction between Water and Land in the Pond–Dike System.* Science Press: Beijing.

Zhu, J., Zhang, Z., Fan, Z., and Li, R. 2014. Biogas potential, cropland load and total amount control of animal manure in China. *J. Agro-Environ. Sci.*, 33(3): 435–445.

Zhu, Y. 2004. Using genetic diversity for sustainable rice disease control. In: *Biodiversity for Sustainable Crop Diseases Management: Theory and Technology*, Zhu, Y., Ed. Yunnan Science Press: Kunming.

Zhu, Y. 2007. *Genetic Diversity for Sustainable Management of Crop Diseases.* Science Press: Beijing.

Zhu, Y. et al. 2000. Genetic diversity and disease control in rice. *Nature*, 406: 718–722.

Zhuang, J. 1982. The first conference on agroecology held in Nanjing. *J. Ecol.*, (1): 56–57.

Zou, D. and Liao, G. 2002. *Agroecology.* Hunan Education Press: Changsha.

PART II

Agroecology Research and Practice on a Crop-Field Scale

How Above- and Below-Ground Interspecific Interactions between Intercropped Species Contribute to Overyielding and Efficient Resource Utilization
A Review of Research in China

Li Long, Zhang Weiping, and Zhang Lizhen

CONTENTS

2.1 INTRODUCTION

Intercropping refers to growing two or more crops simultaneously on the same field (Figure 2.1). According to row arrangement and co-growth period, intercropping can be divided into three types: mixed, relay, or strip intercropping. *Mixed intercropping* refers to growing two or more crops simultaneously with no distinct row arrangement. *Relay intercropping* means growing two or more crops simultaneously during part of the life cycle of each; a second crop is planted after the first crop has reached its reproductive stage of growth but before it is ready for harvest. *Strip intercropping* is growing two or more crops simultaneously in adjacent strips that are wide enough to permit independent cultivation but narrow enough for the crops to have agronomic and biological interactions.

In China, multiple cropping accounts for one third of all the cultivated land area and produces half of the total grain yield (Tong, 1994). Intercropping, as a type of multiple cropping system, has been practiced in China for thousands of years. Legume–cereal intercrop systems are productive and sustainable due to effective resource utilization (nutrients, water, and light) and nitrogen input from symbiotic nitrogen fixation within the cropping system (F. Zhang and Li, 2003). In addition, to meet the increasing food requirements of a growing population, a cereal–cereal association has become increasingly popular in China. For example, a wheat (*Triticum aestivum* L.)–maize (*Zea mays* L.) intercropping system has been widely implemented in the irrigated area of the Hexi

Figure 2.1   Various intercropping systems practiced by farmers in northwest China. (Photograph by Li Long.)

Figure 2.2 Wheat–maize strip intercropping in northwest China, with high fertilizer input and productivity, approaching 15 tons per hectare of grain yield annually in the area for one cropping season, such as in Gansu, Ningxia, and Inner Mongolia. The photograph was taken near Zhongwei City, Ningxia. (Photograph by Li Long.)

Corridor in Gansu Province, irrigated areas along the Huanghe River in Ningxia Hui Autonomous Region, northwest China (Figure 2.2), and in the Inner Mongolian Autonomous Regions of China (F. Zhang and Li, 2003). By the 1980s, 25 to 28 million hectares of intercropping were distributed across all of the provinces in China, except for the Tibet and Qinghai provinces. The total intercropped area reached a maximum of 33 million hectares in the 1990s (Liu, 1994; Zou and Li, 2002). It is estimated that 1.5 million hectares of soybean-cultivated area were intercropped with various other crops in southern China by 2008 (Zhou et al., 2010). Based on the WanFang Scholar Database, intercropping can be found in every province throughout China (although data from Taiwan, Hong Kong, and Macao are missing). Generally speaking, the kind of intercropping in eastern China is more diverse than that in the west and is more diverse in the south than in the north.

In 1980, 4.197 million hectares of wheat–maize intercrop were grown in Beijing, Tianjin, Hebei, Shandong, Henan, and Shanxi provinces, accounting for 57.1% of total maize hectares and 74.4% of the double cropping areas of wheat and maize. Wheat–maize–potato intercrops in Sichuan Province totaled 1 million hectares in 1981, accounting for 50% of cultivation area in the province's hilly uplands (Zou and Li, 2002). Agroforested area accounts for more than 70% of the arable land in the Minjiang River dry valley (Bao, 1998). In 1995, 75,100 hectares of wheat–maize intercrops in Ningxia accounted for 43% of the total grain yield for the area. In Gansu, 200,000 hectares have been intercropped annually (L. Li et al., 2001a; F. Zhang and Li, 2003). However, there is a lack of accurate statistical data for the total area intercropped in China in recent years.

2.2 MAIN INTERCROPPING PATTERNS IN DIFFERENT REGIONS OF CHINA

Huang-Huai-Hai Plain in North China has two types of intercropping patterns. The wheat–maize–peanut intercropping system is based on double cropping wheat and maize and is widely practiced in the double cropping area of Huang-Huai-Hai Plain. The second system observed is wheat–maize–soybean intercropping, which is also based on double cropping wheat and maize and widely practiced in the double cropping area of Huang-Huai-Hai Plain. This intercropping system, however, is time and space intensive. Soybean is planted within a wheat–spring maize intercropping system so as to increase both yield and income (Ministry of Agriculture of the People's Republic of China, 2005).

Single cropping systems of rice in the Yangtze River Basin have come to include intercropping rotations. The first type of rotation is a wheat–peanut–green bean intercrop succeeded by hybrid rice. Peanuts are intercropped with wheat, and green beans (*Phaseolus vulgaris*) are then intercropped with peanuts after the wheat harvest. Hybrid rice is transplanted to this more arable land after the peanut and green bean harvest. Hybrid rice is the main grain crop, peanuts and green beans are the main commercial crops, and the peanut and green bean vines can be used as forage. This cropping system is characterized by high land utilization efficiency, high grain yield, and versatile products. In addition, it contains leguminous crops and thus reduces the need for fertilizer. The second type of rotation is a fava bean–spring maize intercrop, succeeded by rice. Fava beans are intercropped with spring maize, and rice is transplanted to this more arable land after the maize and fava bean harvest. In this pattern, rice is the main food crop, beans and fresh maize are the main commercial crops, and maize straw is good forage. This pattern guarantees rice production and is characterized by high land utilization efficiency and significant economic benefits. This pattern can improve farmers' enthusiasm for growing grain crops, as well as for the development of dairy production (Ministry of Agriculture of the People's Republic of China, 2005).

Double cropping of rice in the Yangtze River Basin can include two types of intercropping rotations. The first is a multiple cropping rotation of cover crops (mainly green manure) with a double cropping of rice. Green manure is planted in winter and double-crop rice is planted in summer and autumn. Green manure has a significant effect on the soil fertility of rice paddies, ensuring a high and stable yield of double-crop rice. The second rotation is a peas–Mexican maize intercrop succeeded by rice. Peas are planted in the winter, Mexican maize is intercropped with peas in spring, and rice is planted after the peas and maize are harvested. This system is characterized by high land utilization efficiency and high forage yields and helps to restore soil fertility and ensure rice production (Ministry of Agriculture of the People's Republic of China, 2005).

Southwest China, with its abundant water and greater daylight hours, is a popular area for multiple cropping systems. Wheat–maize–sweet potato and wheat–maize–soybean intercrops (Figure 2.3) are the main cropping systems and account for more than 70% of upland maize in southwestern China (Yong, 2009). Multiple cropping zones in Southeast China include a maize–soybean–potato intercropping system that can produce a high yield with efficient use of resources (Ministry of Agriculture of the People's Republic of China, 2005).

Single cropping areas in Northwest China include spring wheat–spring soybean, spring wheat–spring beans, and peas–maize intercropping systems (Figure 2.4). These systems are very popular in the area's Hexi Corridor region due to thermal conditions that allow for more than one cropping season but not two full cropping seasons. These systems can achieve two harvests in the area, thus enhancing the efficiency of land use and improving the yield per unit of land area. Single cropping areas in Northeast China are some of the main producing areas of maize in China. Maize–soybean systems are among the most important intercropping patterns there.

2.3 YIELD ADVANTAGE, INTERSPECIFIC EXCHANGES OF NUTRIENTS, AND COMPETITION–RECOVERY PRODUCTION PRINCIPLE IN CEREAL–CEREAL INTERCROPPING

Two field experiments in Gansu Province investigated the yield advantage of intercropping systems and compared nutrient uptake by intercropped wheat, maize, and soybeans. At the Baiyun site, the field experiment compared two phosphorus levels (0 and 53 kg P ha^{-1}), two planting densities for

Figure 2.3 Winter wheat–maize–soybean strip intercropping in southwest China. This photograph was taken in August. The wheat was harvested and soybean sown in June; the maize was sown in March and harvested in August. The system features high output and efficient resource utilization. This photograph was taken near the experimental station of Sichuan Agricultural University in Renshou, Sichuan. (Photograph by Li Long.)

Figure 2.4 Pea–maize strip intercropping in northwest China, where pea and maize were sown in March and April, respectively. In contrast to growing maize alone, this system also provides an additional pea harvest. This photograph was taken in Wuwei City, Hexi Corridor, Gansu Province. (Photograph by Li Long.)

wheat and maize, and three cropping treatments (wheat–maize intercropping, wheat monocrop, and maize monocrop). The design for a wheat–soybean intercropping experiment at the Jingtan site was similar, except that it did not include a plant density treatment and phosphorus levels were 0 and 33 kg ha^{-1} (L. Li et al., 2001a).

Yield and nutrient acquisitions by intercropped wheat, maize, and soybeans were all significantly greater compared to those for monocropped wheat, maize, and soybeans, with the exception of potassium acquisition by maize. Intercropping increased yields by 40 to 70% for wheat intercropped with maize and 28 to 30% for wheat intercropped with soybeans. Intercropping also increased nutrient acquisition by wheat as a result of both border- and inner-row effects. Wheat biomass increased two thirds due to the border-row effect and one third due to the inner-row effect. Similar trends were noted for nitrogen, phosphorus, and potassium accumulation. During the co-growth period—about 80 days from maize or soybean emergence to wheat harvesting—yield and nutrient acquisition of intercropped wheat increased significantly while yields and acquisition of maize or soybeans intercropped with wheat decreased significantly. Aggressivity values of wheat relative to either maize (0.26 to 1.63 for A_{wm}) or soybeans (0.35 to 0.95 for A_{ws}) revealed that wheat has a greater competitive ability than either maize or soybeans. The ratio of nutrient competition between crop species (1.09 to 7.54 for wheat relative to maize and 1.2 to 8.3 for wheat relative to soybeans) showed that wheat had a greater ability to acquire nutrients compared to soybeans and maize. Comparison of overall nitrogen and potassium acquisition by intercropped plants with weighted means of the monocropped plants revealed that interspecific facilitation plays a role in nutrient acquisition during co-growth (L. Li et al., 2001a).

The biomass and nutrient accumulation of intercropped soybeans were significantly smaller than in monocropped soybeans before the wheat harvest during a study done at the Jingtan site in 1997. After the wheat harvest, however, soybean biomass and nutrient accumulation increased sharply. The rates of dry-matter accumulation in the intercropped maize (10.0 to 20.1 g^{-2} per day) were significantly lower than those in the monocropped maize (17.1 to 34.8 g^{-2} per day) during the first stage (May 7 to August 3) while mostly intercropped with wheat. After August 3, however, the rates of dry-matter accumulation in intercropped maize (58.9 to 69.9 g^{-2} per day) were significantly greater than in maize alone (22.7 to 51.8 g^{-2} per day) at the same site. Nutrient acquisition showed the same trends during this growth period. At the Jingtan site in 1998, interspecific competition diminished after wheat harvest and disappeared at maize maturity. There appeared to be a recovery growth period after wheat harvesting in wheat–maize and wheat–soybean intercropping; however, the recovery was limited under N_0P_0 treatment. The interspecific competition, facilitation, and recovery effects in different stages together contributed to the observed yield advantage of intercropping (L. Li et al., 2001b).

2.4 IMPORTANCE OF ABOVE-GROUND AND BELOW-GROUND INTERACTIONS BETWEEN CROP SPECIES

Interspecies interactions, including above-ground and below-ground competition and facilitation, play a key role in the structure and dynamics of plant communities in both natural and agricultural ecosystems (L. Li et al., 2001a; W.P. Zhang et al., 2011, 2012; W. Zhang et al., 2013a,b). Interspecific complementary and competitive interactions between maize (*Zea mays* L. cv. Zhongdan No. 2) and fava beans (*Vicia faba* L. cv. Linxia Dacaidou) in maize–fava bean intercropping systems were assessed in two field experiments in Gansu Province in northwestern China. The same study also conducted a micro-plot experiment with three treatments in which root system partitions were used to determine interspecific root interactions. The treatments used included (1) an impermeable plastic sheet partition inserted into the ground between the strips of maize and fava bean to a depth of 0.70 m to prevent interspecific root interactions; (2) a 400-mesh nylon net partition (nominal aperture size 37 μm) inserted into the ground between the two crop species to prevent direct root contact but allow interactions through mass flow and diffusion; and (3) a control treatment with no partition between the two crop species to allow complete intermingling of their root systems. This micro-plot experiment showed a significant positive yield effect on maize when the root systems intermingled freely (no partition) or partly (400-mesh nylon net partition) compared with no interspecific root interaction (plastic sheet partition). Interspecific root interactions between intercropped fava beans and maize played an important role in the yield advantage of the intercropping system. When the roots of the two species intermingled, land equivalent ratio (LER) values based on total yields and grain yields were 1.21 and 1.34, respectively, but when the roots of the two species were separated completely the LER values were reduced to 1.06 and 1.12, greatly diminishing the intercropping advantage. This indicates that the beneficial effect on fava beans intercropped with maize was derived mainly from root intermingling, whereas the beneficial effect on maize resulted mainly from the transfer of some substance or substances from the fava beans to the maize rhizosphere (L. Li et al., 1999).

2.5 LIGHT AND THERMAL RESOURCE UTILIZATION

Though adaptation of crops and cropping systems to climate change has been a high priority in research and agricultural production (Rusinamhodzi et al., 2012), the development of more climate-robust and resource-efficient cropping systems is lacking. Ecological intensification of intercropping systems has occurred across Africa, China, Latin America, and Europe (Hauggaard-Nielsen et al., 2009; Baldé et al., 2011; L. Li et al., 2013). In China, intercropping was developed to enhance land productivity and profitability. In recent years, studies in China of various intercropping systems focused on yield advantages (L. Li et al., 1999; L.Z. Zhang et al., 2008a), resource-use efficiencies (L.Z. Zhang et al., 2008b; Gao et al., 2009), and control of weeds, diseases, and pests (L. Li et al., 2009). Strategies to achieve ecological intensification include developing tools to evaluate crop growth potential under more extreme climatic conditions. To understand the potential of multi- and intercropping systems, we need an agroecosystem approach that integrates knowledge of various disciplines, such as agronomy, crop physiology, crop ecology, and environmental sciences. However, quantification of resource use (e.g., light, heat, water) via experimental evaluation in complex intercropping systems is difficult due to the existence of crop competition and the difficulty of measurement. There is a need to develop crop models that can quantify the complex functional–structural relationships in intercropping systems.

2.5.1 Light Capture and Use Efficiency in Various Intercropping Systems

Light interception (LI) and light use efficiency (LUE) are useful concepts for characterizing the resource capture and use efficiency of cropping systems, including intercropping. Crop light use efficiency is defined as the slope of the often linear relationship between accumulated biomass and cumulative intercepted photosynthetically active radiation (PAR) (Monteith, 1977; Russell et al., 1989). Light interception and its conversion to dry matter by individual components have been quantified in several studies of conventional intercropping systems in China, including wheat–cotton intercrops (L.Z. Zhang et al., 2008a), sunflower–potato intercrops, and jujube–cotton agroforestry (D. Zhang et al., 2014).

Light interception is sometimes increased as a result of mixing two species and growing them together instead of monocropping, either as a result of a lengthening of the period of soil coverage in the intercrop (temporal advantage) or as a result of a more complete soil cover (spatial advantage) in the intercrop (Keating and Carberry, 1993). In a relay intercropping system with a short intercropping period, the negative effect caused by the delayed development and low leaf area index (LAI) of the later crop is more serious than the direct loss caused by reduced light interception during its early stages. Thus, the intercropping systems could be improved by alleviating the delay in crop development by applying a plastic film cover or by increasing plant densities to reduce spatial losses in light interception (L.Z. Zhang et al., 2008a).

As an example, the light intercepted by cotton in a cotton–wheat relay strip intercropping system ranged from 67 to 93% of the amount intercepted in the monoculture. The amount of light intercepted by intercropped cotton was 12 MJ m^{-2} less than by monocropped cotton during the intercropping period. However, light intercepted by cotton was decreased by more than 100 MJ m^{-2} during the period from wheat harvest to open boll stage in both intercropping systems with six rows of wheat and two rows of cotton (6:2) or three rows of wheat and one row of cotton (3:1). Therefore, the direct effect of shading was less than the indirect effect through a lower LAI (incomplete canopy closure). To increase light capture by cotton, the intercropping systems could be improved by alleviating the delay in cotton development by applying a plastic film cover or by increasing plant densities to reduce spatial losses in light interception. The canopy of wheat and cotton intercropping systems captured more light than the monoculture of cotton. Compared to monocropped cotton, a substantial amount of light (420 to 490 MJ m^{-2} from early spring to wheat harvest) in relay intercropping systems was already captured by the wheat crop before cotton plants emerged. On a spatial scale, the wheat in the intercropped system captured about 20% more light than the canopy of monocropped wheat per unit of strip area but not per unit of total land area. Light capture by wheat strips was markedly influenced by planting row density (number of plant rows per unit cropping width, rows per meter, or length of rows [m] per unit ground area [m^2]), which reflects the relative area planted with wheat. Narrow strips (60 cm) are favored over wider strips (120 cm) in terms of increasing light interception by wheat due to more border rows with an increased compensation advantage.

In the jujube tree–cotton intercropping system, light interception by cotton was significantly increased with increasing plant density in association with an increase of LAI (D. Zhang et al., 2014), which is similar to results by Mao et al. (2014). The light use efficiencies (LUEs) of wheat and cotton were not affected by intercropping. Further, the higher productivity of intercrops compared to monocultures can be fully explained by an increase in total light interception. This experiment also shows that light interception and distribution can be modified by strip width and the number of crop rows per strip. The light use efficiency of dominant component crops in intercropping systems are generally not affected, such as millet in a millet–groundnut system (Willey, 1990), sorghum in a sorghum–groundnut system (Matthews et al., 1991), and maize in a maize–cowpea system (Watiki et al., 1993). For the lower, shaded crops, such as groundnut and cowpea in the above systems, it was reported that the LUE was sometimes "slightly" affected. This not only is caused by shading but may also be a result of competition for other resources (e.g., water, nitrogen). A lower LUE of wheat in the grain-filling phase affected by temperature and nitrogen was also found in oilseed rape (Justes et al., 2000).

Light use efficiency of cotton in jujube–cotton intercropping ranged from 1.42 to 1.96 g DM (MJ PAR)$^{-1}$, which is consistent with previous studies in the monoculture and slightly higher than that of a relay intercropping with wheat (L.Z. Zhang et al., 2008a). The shading by the jujube tree during the entire cotton growing season might have caused this increase in LUE due to the more efficient use of diffuse radiation as compared to direct radiation. It might also have been caused by the modification of branch angles of cotton plants, which is typical for strip intercropping (Gu et al., 2014). Light interception and use efficiency of cotton in jujube–cotton agroforestry systems can be improved by optimizing cotton plant density, thereby achieving higher productivity of the system. Cotton at 18 plants m^{-2} gave the highest yield, light capture, and LUE values, compared to 13.5 and 22.5 plants m^{-2} (D. Zhang et al., 2014). Due to the spatial heterogeneity of the jujube–cotton canopy, the light extinction coefficient of intercropped cotton was affected by jujube shading as well as by plant density. A high light extinction coefficient can be the result of the decrease in branching frequency and a flatter canopy in low-light environments (Niinemets, 2010). Indeed, a higher light extinction coefficient was found in the shaded border rows in this study compared to the middle rows. The cotton light extinction coefficient at the highest plant density tested was significantly lower in the middle rows, especially after the cotton flowering stage. This decrease of light extinction at high plant density may have been caused by the increase of branch insertion angles (Gu et al., 2014). The pattern of the extinction coefficient along the width of the cotton strip in our jujube–cotton experiment clearly shows that any study using computer modeling for the light interception in agroforestry systems must use an approach that allows for explicit inclusion of canopy spatial heterogeneity such as row-structured models (L.Z. Zhang et al., 2008a; Talbot and Dupraz, 2012) and functional–structural plant models (Vos et al., 2010; Gu et al., 2014). Approaches like these can process heterogeneous canopies and take into account local shading by trees, such as the jujube trees in the system presented in this study.

2.5.2 Modeling Light Interception in a Heterogeneous Canopy

Simulation models that analyze the behavior of natural and agricultural ecosystems have been developed and used in the past three decades. They have been extremely helpful in integrating knowledge from various disciplines into one framework and improving insight into complex ecosystems. Several models have been developed for the purpose of modeling interactions in intercropping systems, such as INTERCOM (Kropff and Goudriaan, 1994), which simulates crop and weed competition relationships, and GAPS, an object-oriented dynamic simulation model for modeling plant competition (Rossiter and Riha, 1999). Willey (1990) noted that it is a challenge to determine light interception by component crops in intercropping systems. The fractions of the incoming PAR that are absorbed by canopies of component crops in intercropping systems mainly depend on leaf area index and canopy structure (Spitters and Aerts, 1983; Lantinga et al., 1999). Measurement is difficult, especially over an entire growing season, but several modeling approaches have been suggested to calculate light interception in heterogeneous canopies. A simplified approach, based on a block-shaped strip crop structure, was suggested by Goudriaan (1977) and Pronk et al. (2003). This approach was used to calculate light distribution in wheat–cotton intercrops (L.Z. Zhang et al., 2008a; Mao et al., 2014), because it accurately represents the geometry of the mixtures.

Those generic cropping models, however, do not include mechanistic quantification of plant plasticity in occupying available space (J. Zhu et al., 2014). Functional–structural plant (FSP) modeling (Evers et al., 2010; Vos et al., 2010; Gu et al., 2014) is a good tool to address such spatial heterogeneity. This type of modeling combines the quantification of three-dimensional plant structure and the related physiological processes at the plant organ level. By simulating two crop species with their own particular architectural development in a particular strip arrangement, the effects of spatial and temporal heterogeneity and plastic responses to local environmental conditions on crop light absorption and use can be addressed explicitly. Such exploration would well compensate for the difficulty of field measurements in intercropping experiments and provide good parameter values for ecophysiological models.

2.5.3 Crop Development and Quantification in Intercropping

In wheat–cotton intercropping systems, cotton is sown in April in the path assigned during wheat sowing in November of the previous year. The duration of the intercropping period is relatively short, but a fully developed wheat canopy competes for light and nutrients with cotton seedlings. Therefore, the utilization of resources, such as light interception and nitrogen and water uptake, by intercropped cotton during this period is affected by the competitive strength of the wheat crop. The development and growth of cotton in the intercropping systems are suppressed by wheat during an intercropping phase of approximately seven weeks. As a result, the development rate, canopy size, amount of light interception, and total nitrogen uptake of intercropped cotton are decreased. Cotton development in the intercropping systems was delayed by 10 to 15 calendar days. This delay corresponded with 4.7 physiological days (days with optimal temperature conditions) or 115 degree-days expressed as thermal time for the duration from sowing to the first square. The magnitude of the delay was the same in all tested intercrops (L.Z. Zhang et al., 2008b). Beside the development delay, the growth of cotton in the intercropping systems was correspondingly retarded by 6 to 12 days (L.Z. Zhang et al., 2008b). In comparison to the development delay reported for other intercropping systems (Gethi et al., 1993; Bukovinszky et al., 2004), the delay of cotton in wheat–cotton intercropping systems is more serious. It is associated with a 2.7°C decrease in air temperature on a sunny day during the intercropping period as a consequence of shading by wheat. In wheat–soybean intercropping, the development delay of relay intercropped soybean was also observed at the co-growth stage (Wallace et al., 1992). The delay of development in intercropped cotton resulted in a reduced number of fruit branches, nodes, and fruits before the "cut out the top" operation (removal of topmost buds of main stems), thus decreasing harvest index and lint yield, a finding that has also been reported in the literature (Sadras, 1995; Lei and Gaff, 2003). In wheat–soybean intercropping, the development of wheat that is planted later in the intercropped system was also negatively affected at its early growth stage (Wallace et al., 1992).

Shading also modifies other environmental conditions, such as ventilation and soil temperature (Midmore, 1993). The environment can also be modified by management practices, such as soil cover by plastic film or straw (Rana et al., 2004). The decrease in both air and soil temperature was mainly caused by shading of taller crops, because there were no differences in temperature between intercropping and monoculture when the air temperature was low due to the absence of direct radiation (L.Z. Zhang et al., 2008b). A soil cover consisting of plastic film effectively ameliorated soil temperature in intercropping. Licht and Al-Kaisi (2005) found an increased soil temperature as a result of tillage. Thus, tillage may also provide an option to raise soil temperature in intercrops. An improvement of temperature may be also achieved by measures that diminish shading, such as the use of shorter wheat cultivars or planting crops on ridges.

2.6 SYMBIOTIC NITROGEN EXCHANGE IN INTERCROPS OF NON-LEGUME AND LEGUME SPECIES

Intercrops of legumes with non-legumes are widely accepted as productive and sustainable systems all over the world. Most previous studies in this field focused on high utilization efficiency of light, pest and disease suppression (Y.Y. Zhu et al., 2000), and interspecific interactions and nutrient competition (L. Li et al., 2003b). Legumes can assimilate more nitrogen by symbiotic N_2 fixation and also facilitate growth of intercropped cereals by transferring nitrogen (Y. Li et al., 2009a). In order to sustain a higher yield, nitrogen fertilizer is widely used in intensive farming systems, even in a legume–cereal intercropping system; however, nitrogen fertilization may result in a reduction of nodulation and N_2 fixation (Salvagiotti et al., 2008). Researchers at China Agricultural University and Gansu Academy of Agricultural Sciences have studied the effect of nitrogen fertilization on symbiotic nitrogen fixation for many years. They found that nitrogen fixation could be improved by yield maximization in an intercropping system (L. Li et al., 2003b; F. Zhang and Li, 2003; Fan et al., 2006; Y. Li, 2009a,b).

2.6.1 Intercropping Alleviates the Inhibitory Effect of Nitrogen Fertilization on Nodulation and Symbiotic N_2 Fixation in Fava Beans

Symbiotic nitrogen fixation is usually inhibited by nitrogen fertilization in intensive farming systems (Salvagiotti et al., 2008); however, intercropping alleviates the inhibitory effect of nitrogen fertilization on nodulation and N_2 fixation of legumes (Y. Li et al., 2009a). Two years of field experiments (2006 to 2007) with different nitrogen fertilizer rates (0, 75, 150, 225, and 300 kg N ha^{-1}) were carried out in a fava bean (*Vicia faba* L.)–maize (*Zea mays* L.) intercropping system in

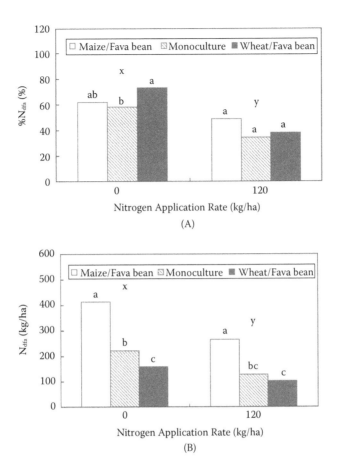

Figure 2.5 (A) Percentage of shoot N derived from the atmosphere (%N_{dfa}) and (B) shoot N derived from the atmosphere (N_{dfa}) of fava bean in three cropping systems. Values labeled with different letters (a, b, and c) indicate significant difference between the three cropping systems within a given N application rate (SAS ANOVA, $p < 0.05$). Values are means ($n = 3$). (From Fan, F. et al., *Plant Soil*, 283(1–2): 275–286, 2006. With permission.)

the northwestern part of China (Y. Li et al., 2009a). The results show that both the nodule biomass and nitrogen derived from the atmosphere (N_{dfa}) in intercropped fava beans were increased by 7 to 58% and 8 to 33%, respectively, at the start of flowering; by 8 to 72% and 54 to 61%, respectively, at peak flowering; by 4 to 73% and 18 to 50%, respectively, at grain filling; and by 7 to 62% and –7 to 72%, respectively, at maturity compared with monocropped fava bean. Intercropping with maize significantly alleviated the inhibitory effect of nitrogen fertilization on nodulation and N_2 fixation and improved the productivity of intercropping (Y. Li et al., 2009a).

Intercropping increased the percentage of nitrogen derived from the atmosphere in the wheat–fava bean system but not that of the fava bean–maize system when no nitrogen fertilizer was applied (Figure 2.5). When receiving 120 kg N ha^{-1}, however, intercropping did not significantly increase the percentage of nitrogen derived from the atmosphere (%N_{dfa}) in either the wheat–fava bean system or the fava bean–maize system in comparison with fava beans in monoculture. The amount of shoot nitrogen derived from the air, however, increased significantly when intercropped with maize, irrespective of N-fertilizer application. Nitrogen derived from the atmosphere decreased when intercropped with wheat, although not significantly at 120 kg N ha^{-1} (Fan et al., 2006).

2.6.2 Effect of Root Contact on Interspecific Competition and Nitrogen Transfer between Wheat and Fava Beans Using Direct and Indirect ^{15}N Techniques

Pot experiments were carried out to investigate the complementary nitrogen (N) use between intercropped fava beans (*Vicia faba* L. cv. Linxia Dacaidou) and wheat (*Triticum aestivum* L. cv. 8354). These pots were separated into two compartments by (1) a solid barrier to prevent root contact and nitrogen movement, (2) a nylon mesh (30 μm) to prevent root contact but allow N exchange, or (3) no root barrier between the compartments to allow root intermingling. Root contact enhanced ^{15}N acquisition by wheat but decreased that by fava beans. The percentages of nitrogen derived from the atmosphere in fava beans were 58%, 80%, and 91% in the treatments with a solid barrier, with a mesh barrier, and without a barrier, respectively. Nitrogen transfer from fava beans to wheat was estimated by the indirect ^{15}N isotope dilution technique and by direct plant labeling via petiole injection of a ^{15}N solution. With the indirect method, the nitrogen transferred from fava beans to the associated wheat was 2 mg with a mesh barrier and 6 mg without a barrier. Using the direct labeling method, nitrogen transferred from fava beans to the companion wheat was 7 mg, equal to 15% of total N in wheat (Xiao et al., 2004).

2.6.3 Nitrogen Movement between Crops

A pot experiment was also conducted to examine the N_2 fixation by peanuts and nitrogen transfer from peanuts to rice at three nitrogen fertilizer application rates (15, 75, and 150 kg N ha^{-1}) using a ^{15}N isotope dilution method. The intercropping advantage was mainly due to the sparing effect for soil inorganic nitrogen contributed by the peanuts. Nitrogen transferred from peanuts accounted for 11.9, 6.4, and 5.5% of the total nitrogen accumulated in the rice plants in intercropping at the same three nitrogen fertilizer application rates, suggesting that the transferred nitrogen from peanuts in the intercropping system made considerable contribution to the nitrogen content of rice, especially in low-nitrogen soil (Chu et al., 2004).

2.6.4 Improved Nitrogen Difference Method for Estimating Biological Nitrogen Fixation in Legume-Based Intercropping Systems

The nitrogen difference method (NDM) is traditionally used for quantifying N_2 fixation. This method is based on the same amount of soil N exploited by N_2-fixing and non-N_2-fixing plants. It may not be suitable for plants with different root traits. It may also be unsuitable for legume-based intercropping, because interspecific competition may lead to a significant difference in nitrogen acquisition from soil between reference plants and intercropped legumes. Thus, we developed an improved nitrogen difference method (INDM) for estimating N_2 fixation in intercropping systems, and we tested NDM and INDM with two field experiments.

In the first experiment, fava beans (*Vicia faba*), peas (*Pisum sativum*), and soybeans (*Glycine max*) were grown in monocrops or were intercropped with maize (*Zea mays*) with two nitrogen application rates (0, 225 kg ha^{-1}). The biomass of fava beans, peas, and maize was significantly increased, whereas the biomass of soybeans was decreased in the intercropping system from the amount in monoculture. Aggressivity analyses revealed a greater nitrogen competition ability of fava beans and peas over maize. Soybeans, however, did not exhibit a greater nitrogen competition ability than maize. When such interspecific nitrogen competition was considered in the estimation, an improved NDM (INDM) could be calculated by

$$N_{fix-int} = \left[N_{leg-int} + \frac{1-x}{x} N_{ref-int} - \frac{N_{ref-sole}}{x} \right] + \left[soilN_{leg-int} + \frac{1-x}{x} soilN_{ref-int} - \frac{soilN_{ref-sole}}{x} \right]$$

where x and $1 - x$ are the planting areas of legumes and non-legumes, respectively, in the intercropping system. Compare this to traditional NDM:

$$N_{fix-int} = \left[N_{leg-int} - N_{ref-sole} \right] + \left[soilN_{leg-int} - soilN_{ref-sole} \right]$$

Percentages of nitrogen derived from the atmosphere (%N_{dfa}) by INDM with two nitrogen application rates (0 and 225 kg ha^{-1}) were decreased by 54.3% and 39.8%, respectively, for fava beans and by 44.7% and 5.0% for peas, but they increased by 113.5% and 191.0% for soybeans. Results indicate that NDM overestimated the %N_{dfa} of fava beans and peas, yet underestimated the %N_{dfa} of soybeans when these legumes were intercropped with maize. INDM is much more suitable for intercropping systems.

The second experiment had fava beans, wheat (*Triticum aestivum* cv. Yongliang 4), maize, and ryegrass (*Lolium perenne* cv. Zhongxu 1) planted in monoculture and fava beans intercropped with either wheat or maize. All treatments received 60 kg N ha^{-1} as urea and 75 kg P$_2$O$_5$ ha^{-1} as triple superphosphate, and the fertilizers were broadcasted to the soil surface before sowing. In the experiment, the %N_{dfa} of fava beans was quantified by NDM, INDM, and the ^{15}N natural abundance method (NA). Only the %N_{dfa} by INDM correlated significantly with that from NA. Both interspecific root interactions and N loss affect %N_{dfa} estimation. Therefore, INDM could be more suitable than NDM for quantifying %N_{dfa} of a N$_2$-fixing plant in intercropping systems (Yu et al., 2010).

2.7 INTERSPECIFIC FACILITATION OF PHOSPHORUS UPTAKE IN INTERCROPS OF NON-LEGUME AND LEGUME SPECIES

2.7.1 Rhizosphere Phosphorus Facilitation in Fava Bean–Maize Intercropping Systems in Phosphorus-Deficient Soils

Legume–grass intercrops are overyielding systems even on phosphorus (P)-deficient soils. Here we show that a new mechanism of overyielding, in which phosphorus mobilized by one crop species increases the growth of a second crop species grown in alternate rows, leads to large yield increases on phosphorus-deficient soils. In our 4 years of field experiments, maize (*Zea mays* L.) overyielded by 43% and fava beans (*Vicia faba* L.) overyielded by 26% when intercropped on a low-phosphorus but high-nitrogen soil. Fava beans had greater capability to acquire insoluble inorganic and organic phosphorus (Figure 2.6). Overyielding of maize was attributable to below-ground interactions between fava beans and maize in another field experiment (L. Li et al., 2007).

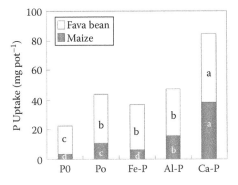

Figure 2.6 Fava bean had a greater ability to acquire insoluble inorganic P and organic P. P0, no P added; Po, organic P; Fe-P, FePO$_4$; Al-P, AlPO$_4$; Ca-P, CaH$_2$PO$_4$. (Adapted from Li, L. et al., *Proc. Natl. Acad. Sci. U.S.A.*, 104(27), 11192–11196, 2007. With permission.)

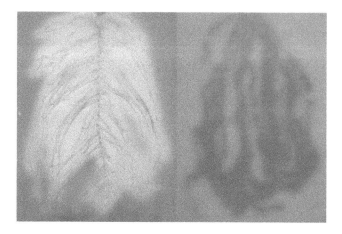

Figure 2.7 (See color insert.) Visualization of rhizosphere acidification of fava bean (left) and maize (right). The roots were imbedded for 6 hr in agar gel containing a pH indicator (bromocresol purple) without P supply. Yellow indicates acidification, and purple indicates alkalization. (From Li, L. et al., *Proc. Natl. Acad. Sci. U.S.A.*, 104(27), 11192–11196, 2007. With permission.)

About a third of terrestrial soils have insufficient available phosphorus for optimum crop production, and many tropical acid soils are highly P deficient. Intercropping of maize and fava beans leads to marked overyielding. Several mechanisms are behind intercropping facilitation on a phosphorus-deficient soil. These factors are related to interspecific rhizospheric effects on phosphorus levels:

1. Rhizosphere acidification by P-efficient species (fava bean) resulted in a pH decrease in the rhizosphere (Figure 2.7), which increased the availability of insoluble inorganic P in soil, such as $FePO_4$ and $AlPO_4$.
2. Carboxylates from root exudation of one species (fava been) chelated calcium (Ca), iron (Fe), and aluminum (Al) and consequently mobilized insoluble soil P, which benefits the species and other species (maize) grown together with it.
3. Greater phosphatase activity in the rhizosphere decomposed soil organic P into an inorganic form that can be used by both species, such as wheat–chickpea and maize–chickpea (L. Li et al., 2007).
4. Fava beans have a greater ability for rhizosphere acidification than maize does and therefore facilitate the utilization of sparingly soluble P in soil (L. Li et al., 2003a, 2007; Zhou et al., 2009).

2.7.2 Chickpea Facilitates Phosphorus Uptake When Intercropped with Wheat or Maize

Organic phosphorus usually comprises 30 to 80% of the total phosphorus in most agricultural soils. Pot experiments were conducted to investigate interspecific complementation in utilization of phytate and $FePO_4$ by plants in a wheat (*Triticum aestivum* L.)–chickpea (*Cicer arietinum* L.) intercropping system under both sterile and non-sterile conditions. It has been shown that chickpeas facilitate phosphorus uptake by intercropped wheat from an organic phosphorus source (L. Li et al., 2003a). In addition, acid phosphatase excreted from chickpea roots was quantified, and the contribution of acid phosphatase to the facilitation of phosphorus uptake by intercropped maize receiving phytate was examined (S. Li et al., 2004). The results showed that average acid phosphatase activity of chickpea roots supplied with phytate was two to three times that of maize. Soil acid phosphatase activity in the rhizosphere of chickpeas was also significantly higher than that of maize regardless of phosphorus sources (Figure 2.8). Chickpeas can mobilize organic phosphorus in both hydroponic and soil cultures, leading to an interspecific facilitation of organic phosphorus utilization in maize–chickpea intercropping (S. Li et al., 2004).

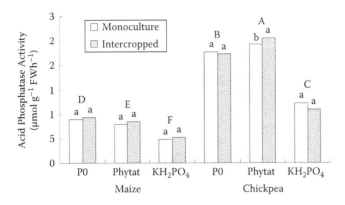

Figure 2.8 Acid phosphatase activity of chickpea roots was greater than that of maize. (From Li, S. et al., *Ann. Bot.*, 94(1), 297–303, 2004. With permission.)

2.7.3 Maize–Fava Bean Intercropping with Rhizobia Inoculation Enhances Productivity and Recovery of Phosphorus Fertilizer

Development of reclaimed desert soils using intercropping systems and rhizobia inoculations is a potentially important strategy for utilizing phosphorus-deficient soils and extending the arable land area. Two years of field experiments with different rates of phosphorus fertilizer (0 and 120 kg P_2O_5 ha^{-1} in 2008 and 0, 60, and 120 kg P_2O_5 ha^{-1} in 2009) were carried out to study the influence of phosphorus fertilizer application on the productivity and phosphorus use of a maize (*Zea mays* L.)–fava bean (*Vicia faba* L.) intercropping system inoculated with *Rhizobium* in a newly reclaimed desert soil. Average grain yields of intercropped fava bean and maize increased by 30 to 197% and by 0 to 31%, respectively (and increased more with zero phosphorus fertilizer application), compared with monocultured crops. Intercropped fava beans showed higher root nodulation and phosphorus accumulation but little response to phosphorus application regardless of cropping system. The apparent phosphorus recovery of the intercropping system was 297.0% greater ($p < 0.001$) than that of monocultured systems (weighted means), and on average was highest at the intermediate phosphorus application rate. Moderate phosphorus fertilizer application enhanced productivity and inoculation of the intercropping system in a reclaimed desert soil, and phosphorus deficiency was ameliorated to some extent as well. The results indicate that *Rhizobium*-inoculated maize–fava bean intercropping could be an efficient cropping system for reclaimed desert soils (Mei et al., 2012).

Only 15 to 20% of the phosphorus applied is recovered by plants during the current growing season, and the remainder accumulates in soil phosphorus pools. Enhanced recovery of phosphorus fertilizer was also observed in a relatively fertile soil that had been cultivated for years under other intercropping systems. Three years of field experiments using different rates of phosphorus fertilizer application (0, 40, and 80 kg ha^{-1}) were conducted with the main intercropped plots of maize with other species and subplots of the respective monocultures. Maize (*Zea mays* L.) was intercropped with oilseed rape (*Brassica napus* L.), turnip (*Brassica campestris* L.), fava bean (*Vicia faba* L.), chickpea (*Cicer arietinum* L.), and soybean (*Glycine max* L.). These studies were carried out to analyze the apparent recovery of phosphorus fertilizer and soil Olsen P in intercrops and monocultures.

Average total grain yields of maize–turnip, maize–fava bean, maize–chickpea, and maize–soybean intercropping were increased by 30.7%, 24.8%, 24.4%, and 25.3%, respectively, compared with the weighted means of the corresponding monocultures. Shoot phosphorus levels of maize–turnip, maize–fava bean, maize–chickpea, and maize–soybean intercropping increased by 44.6%, 30.7%, 39.1%, and 28.6%, respectively, compared with the weighted means of the corresponding monocultures. In addition, average total grain yields and shoot phosphorus levels were highest at 40 kg P ha^{-1} of fertilizer application. Moreover, the average apparent recovery of phosphorus fertilizer of the intercropping systems increased from 6.1% to 30.6% at 40 kg P ha^{-1} of fertilizer application and from 4.8% to 14.5% at 80 kg P ha^{-1} compared with the average of all monoculture systems over 3 years. The results indicated that intercropping combined with a rational phosphorus application rate (e.g., 40 kg P ha^{-1}) resulted in the maximum total grain production and shoot phosphorus levels, maintained a balance of phosphorus inputs/outputs and soil Olsen P at an appropriate level (21.3 mg kg^{-1}), and provided maximum apparent recovery of phosphorus fertilizer (30.6%) through efficient acquisition of phosphorus and other resources by interspecific interactions, thus creating a more sustainable and productive agricultural system (Xia et al., 2013).

2.8 IMPROVEMENT OF IRON NUTRITION IN PEANUTS THROUGH RHIZOSPHERE INTERACTION IN PEANUT–MAIZE INTERCROPS

Peanuts (*Arachis hypogaea* L.), a major oilseed crop in China, account for 30% of cultivated area and 30% of the total oilseed production in the country; however, iron deficiency chlorosis frequently occurs on calcareous soils, especially in north China. Interestingly, iron deficiency chlorosis in peanuts is more severe in peanut monoculture systems than in peanut–maize intercrops on these types of soils; therefore, there is considerable interest in devising practical approaches for the correction or prevention of iron deficiency in crops in Chinese agriculture. Study results have demonstrated that the improved levels of iron of peanuts intercropped with maize were mainly caused by rhizosphere interactions between peanut and maize (Zuo et al., 2000, 2003, 2004).

2.9 ROOT DISTRIBUTION AND INTERACTIONS BETWEEN INTERCROPPING SPECIES

2.9.1 Intercropping Systems with Annual Crops

In wheat–maize intercropping, interspecific interactions lead to an increase for wheat but a decrease for maize, in terms of yield and nutrient acquisition during the co-growth stage (L. Li et al., 2001a). An increase in plant growth and nutrient acquisition of both fava beans and maize has been observed in fava bean–maize intercropping, however, and the interspecific below-ground interactions contributed to yield advantage in this case more than in that of wheat–maize intercropping (L. Li et al., 1999, 2003b). Ecologists and agronomists have considered the spatial root distribution of plants to be important for interspecific interactions in natural and agricultural ecosystems (W.P. Zhang and Wang, 2010; W.P. Zhang et al., 2013c). Yet, few experimental studies have quantified patterns of root distribution dynamics and their impacts on interspecific interactions (L. Li et al., 2006). A field experiment was conducted to investigate the relationship between root distribution and interspecific interactions between intercropped plants. Roots were sampled twice by auger and twice using the monolith method in wheat (*Triticum aestivum* L.)–maize (*Zea mays* L.) and fava bean (*Vicia faba* L.)–maize intercrops and in monocultured wheat, maize, and fava beans up to 100 cm deep in the soil profile (L. Li et al., 2006).

The results showed that the roots of intercropped wheat spread under maize plants (Figure 2.9) and had much greater root length density (RLD) at all soil depths compared to monocultured wheat. The roots of maize intercropped with wheat were limited laterally but had a greater RLD than monocultured maize. The RLD of maize intercropped with fava beans at different soil depths was affected less by intercropping compared to the RLD of maize in wheat–maizet intercrops. Fava beans had a relatively shallow root distribution, and the roots of intercropped maize spread

(A) Roots of maize grown alone

(B) Roots of maize intercropped
with maize

(C) Roots of maize intercropped
with fava bean

Figure 2.9 The roots of maize (A) spread to 40 cm when grown alone; (B) spread only 20 cm when intercropped with wheat away from the maize row; and (C) intermingled with fava bean roots when intercropped with fava bean. (From Li, L. et al., *Oecologia*, 147(2): 280–290, 2006. With permission.

underneath them. The results support the hypotheses that asymmetric interspecific facilitation results in greater lateral deployment of roots, increased RLD, and thus an increased yield. In addition, compatibility of the spatial root distribution of intercropped species contributes to symmetric interspecific facilitation in the fava bean–maize intercropping (L. Li et al., 2006).

Intercropping with wheat leads to greater root weight density and larger below-ground space of irrigated maize at late growth stages (L. Li et al., 2011). In addition, the roots of intercropped maize have a longer life span than the roots of monocultured maize (L. Li et al., 2011). The mechanism involved remains to be elucidated and requires further research.

2.9.2 Agroforestry

A field experiment was conducted to investigate the relationship between root distribution and interspecific interactions between intercropped jujube tree (*Zizyphus jujuba* Mill.) and wheat (*Triticum aestivum* Linn.) in Hetian City, in southern Xinjiang Province in northwest China. Roots were sampled by auger in 2-, 4-, and 6-year-old jujube tree–wheat intercrops; in monocultured wheat; and in 2-, 4- and 6-year-old monocultured jujube down to 100 cm deep in the soil profile. The roots of both intercropped wheat and jujube had less root length density at all soil depths than did the monocropped wheat and jujube trees. The RLD of 6-year-old jujube at different soil depths in jujube–wheat intercrops was influenced by intercropping less than in other jujube–wheat intercropping combinations. Six-year-old jujube exhibited a stronger negative effect on wheat productivity than did 2- or 4-year-old jujube. There was also less negative effect on jujube productivity in the 6-year-old system than in the 2- or 4-year-old jujube trees grown in monoculture. These findings may partially explain the effects of interspecific competition on jujube–wheat agroforestry systems (W. Zhang et al., 2013a).

ACKNOWLEDGMENTS

These research projects were supported financially by the National Basic Research Program of China (973 Program) (Project No. 2011CB100405) and the National Natural Science Foundation of China (Project Nos. 30870406, 31121062, 31210103906, and 31500348).

REFERENCES

Baldé, A.B., Scopel, E., Affolder, F., Corbeels, M., Da Silva, F.A.M., Xavier, J.H.V., and Wery, J. 2011. Agronomic performance of no-tillage relay intercropping with maize under smallholder conditions in Central Brazil. *Field Crops Res.*, 124: 240–251.

Bao, W. 1998. Characteristics of energy input and output in fruit tree–grain crop intercropping ecosystem: a case study from dry valley region in the upper reaches of the Minjiang River. *Eco-Agric. Res.*, 3(6): 50–54.

Bukovinszky, T., Trefas, H., van Lenteren, J.C., Vet, L.E.M., and Fremont, J. 2004. Plant competition in pest-suppressive intercropping systems complicates evaluation of herbivore responses. *Agric. Ecosyst. Environ.*, 102: 185–196.

Chu, G.X., Shen, Q.R., and Cao, J.L. 2004. Nitrogen fixation and N transfer from peanut to rice cultivated in aerobic soil in an intercropping system and its effect on soil N fertility. *Plant Soil*, 263(1–2): 17–27.

Evers, J.B., Vos, J., Yin, X., Romero, P., van der Putten, P.E.L., and Struik, P.C. 2010. Simulation of wheat growth and development based on organ-level photosynthesis and assimilate allocation. *J. Exp. Bot.*, 61: 2203–2216.

Fan, F., Zhang, F., Song, Y., Sun, J., Bao, X., Guo, T., and Li, L. 2006. Nitrogen fixation of faba bean (*Vicia faba* L.) interacting with a non-legume in two contrasting intercropping systems. *Plant Soil*, 283(1–2): 275–286.

Gao, Y., Duan, A., Sun, J., Li, F., Liu, Z., Liu, H., and Liu, Z. 2009. Crop coefficient and water-use efficiency of winter wheat/spring maize strip intercropping. *Field Crops Res.*, 111: 65–73.

Gethi, M., Omolo, E.O., and Mueke, J.M. 1993. The effect of intercropping on relative resistance and susceptibility of cowpea cultivars to *Maruca testulalis* Geyer when in mono and when intercropped with maize. *Insect Sci. Appl.*, 14: 305–313.

Goudriaan, J. 1977. *Crop Micrometerorology: A Simulation Study*. Simulation Monographs, PUDOC: Wageningen.

Gu, S., Evers, J.B., Zhang, L., Mao, L., Zhang, S., Zhao, X., Liu, S., van der Werf, W., and Li, Z. 2014. Modelling the structural response of cotton plants to mepiquat chloride and population density. *Ann. Bot.*, 114(4): 877–887.

Hauggaard-Nielsen, H., Gooding, M., Ambus, P., Corre-Hellou, G., Crozat, Y., Dahlmann, C., Dibet, A., von Fragstein, P., Pristeri, A., Monti, M., and Jensen, E.S. 2009. Pea–barley intercropping for efficient symbiotic N_2-fixation, soil N acquisition and use of other nutrients in European organic cropping systems. *Field Crops Res.*, 113: 64–71.

Justes, E., Denoroy, P., Gabrielle, B., and Gosse, G. 2000. Effect of crop nitrogen status and temperature on the radiation use efficiency of winter oilseed rape. *Eur. J. Agron.*, 13: 165–177.

Keating, B.A. and Carberry, P.S. 1993. Resource capture and use in intercropping: solar radiation. *Field Crops Res.*, 34: 273–301.

Kropff, M.J. and Goudriaan, J. 1994. Competition for resource capture in agricultural crops. In: Kropff, M.J. and Goudriaan, J., Eds., *Resource Capture by Crops*. Nottingham University Press: Nottingham, pp. 233–253.

Lantinga, E.A., Nassiri, M., and Kropff, M.J. 1999. Modelling and measuring vertical light absorption within grass-clover mixtures. *Agric. Forest Meteorol.*, 96: 71–83.

Lei, T.T. and Gaff, N. 2003. Recovery from terminal and fruit damage by dry season cotton crops in tropical Australia. *J. Econ. Entomol.*, 96: 730–736.

Li, L., Yang, S., Li, X., Zhang, F., and Christie, P. 1999. Interspecific complementary and competitive interactions between intercropped maize and faba bean. *Plant Soil*, 212(2): 105–114.

Li, L., Sun, J.H., Zhang, F.S., Li, X., Rengel, Z., and Yang, S. 2001a. Wheat/maize or wheat/soybean strip intercropping. I. Yield advantage and interspecific interactions on nutrients. *Field Crops Res.*, 71(2): 123–137.

Li, L., Sun, J.H., Zhang, F.S., Li, X., Rengel, Z., and Yang, S. 2001b. Wheat/maize or wheat/soybean strip intercropping. II. Recovery or compensation of maize and soybean after wheat harvesting. *Field Crops Res.*, 71(3): 173–181.

Li, L., Tang, C., Rengel, Z., and Zhang, F. 2003a. Chickpea facilitates phosphorus uptake by intercropped wheat from an organic phosphorus source. *Plant Soil*, 248(1–2): 297–303.

Li, L., Zhang, F., Li, X., Christie, P., Sun, J., Yang, S., and Tang, C. 2003b. Interspecific facilitation of nutrient uptakes by intercropped maize and faba bean. *Nutr. Cycl. Agroecosys.*, 65(1): 61–71.

Li, L., Sun, J.H., Zhang, F.S., Guo, T.W., Bao, X.G., Smith, F.A., and Smith, S.E. 2006. Root distribution and interactions between intercropped species. *Oecologia*, 147(2): 280–290.

Li, L., Li, S.M., Sun, J.H., Zhou, L.L., Bao, X.G., Zhang, H.G., and Zhang, F.S. 2007. Diversity enhances agricultural productivity via rhizosphere phosphorus facilitation on phosphorus-deficient soils. *Proc. Natl. Acad. Sci. U.S.A.*, 104(27): 11192–11196.

Li, L., Sun, J.H., and Zhang, F.S. 2011. Intercropping with wheat leads to greater root weight density and larger below-ground space of irrigated maize at late growth stages. *Soil Sci. Plant Nutr.*, 57(1): 61–67.

Li, L., Zhang, L., and Zhang, F. 2013. Crop mixtures and the mechanisms of overyielding. In: Levin, S.A., Ed., *Encyclopedia of Biodiversity*, 2nd ed., Vol. 2. Academic Press: Waltham, MA, pp. 382–395.

Li, S., Li, L., Zhang, F., and Tang, C. 2004. Acid phosphatase role in chickpea/maize intercropping. *Ann. Bot.*, 94(1): 297–303.

Li, W., Li, L., Sun, J.H., Zhang, F.S., and Christie, P. 2003. Effects of nitrogen and phosphorus fertilizers and intercropping on uptake of nitrogen and phosphorus by wheat, maize and faba bean. *J. Plant Nutr.*, 26(3): 629–642.

Li, Y., Sun, J., Li C.J., Li, L., Cheng, X., and Zhang, F. 2009a. Effects of interspecific interactions and nitrogen fertilization rates on the agronomic and nodulation characteristics of intercropped faba bean. *Sci. Agric. Sin.*, 42(10): 3467–3474.

Li, Y., Yu, C.B., Cheng, X., Li, C., Sun, J., Zhang, F., Lambers, H., and Li, L. 2009b. Intercropping alleviates the inhibitory effect of N fertilization on nodulation and symbiotic N_2 fixation of faba bean. *Plant Soil*, 323: 295–308.

Licht, M.A. and Al-Kaisi, M. 2005. Strip-tillage effect on seedbed soil temperature and other soil physical properties. *Soil Till. Res.*, 80: 233–249.

Liu, X. 1994. *Tillage Science*. China Agricultural University: Beijing.

Mao, L., Zhang, L., Zhao, X., Liu, S., van der Werf, W., Zhang, S., Spiertz, J.H.J., and Li, Z. 2014. Crop growth, light utilization and yield of relay intercropped cotton as affected by plant density and a plant growth regulator. *Field Crops Res.*, 155: 67–76.

Matthews, R.B., Azamali, S.N., Saffell, R.A., Peacock, J.M., and Williams, J.H. 1991. Plant growth and development in relation to the microclimate of a sorghum groundnut intercrop. *Agric. Forest Meteorol.*, 53: 285–301.

Mei, P.-P., Gui, L.-G., Wang, P., Huang, J.-C., Long, H.-Y., Christie, P., and Li, L. 2012. Maize/fava bean intercropping with rhizobia inoculation enhances productivity and recovery of fertilizer P in a reclaimed desert soil. *Field Crops Res.*, 130: 19–27.

Midmore, D.J. 1993. Agronomic modification of resource use and intercrop productivity. *Field Crops Res.*, 34: 357–380.

Ministry of Agriculture of the People's Republic of China. 2005. *Main Multiple Cropping Systems in China* [in Chinese]. China Agriculture Press: Beijing.

Monteith, J.L. 1977. Climate and the efficiency of crop production in Britain. *Philos. Trans. R. Soc. Lond. B Biol. Sci.*, 281: 277–294.

Niinemets, Ü. 2010. A review of light interception in plant stands from leaf to canopy in different plant functional types and in species with varying shade tolerance. *Ecol. Res.*, 25: 693–714.

Pronk, A.A., Goudriaan, J., Stilma, E., and Challa, H. 2003. A simple method to estimate radiation interception by nursery stock conifers: a case study of eastern white cedar. *Neth. J. Agric. Sci.*, 51: 279–295.

Rana, G., Katerji, N., Introna, M., and Hammami, A. 2004. Microclimate and plant water relationship of the "overhead" table grape vineyard managed with three different covering techniques. *Sci. Horticult.*, 102: 105–120.

Rossiter, D.G. and Riha, S.J. 1999. Modeling plant competition with the GAPS object-oriented dynamic simulation model. *Agron. J.*, 91: 773–783.

Rusinamhodzi, L., Corbeels, M., Nyamangara, J., and Giller, K.E. 2012. Maize–grain legume intercropping is an attractive option for ecological intensification that reduces climatic risk for smallholder farmers in central Mozambique. *Field Crops Res.*, 136: 12–22.

Russell, G., Jarvis, P.G., and Monteith, J.L. 1989. Absorption of radiation by canopies and stand growth. In: Russell, G. et al., Eds., *Plant Canopies: Their Growth, Form, and Function.* Cambridge University Press: Cambridge, pp. 21–39.

Sadras, V.O. 1995. Compensatory growth in cotton after loss of reproductive organs. *Field Crops Res.*, 40: 1–18.

Salvagiotti F., Cassman, K.G., Specht, J.E., Walters, D.T., Weiss, A., and Dobermann, A. 2008. Nitrogen uptake, fixation and response to fertilizer N in soybeans: a review. *Field Crops Res.*, 108(1): 1–13.

Spitters, C.J.T. and Aerts, R. 1983. Simulation of competition for light and water in crop–weed associations. *Aspects Appl. Biol.*, 4: 467–483.

Talbot, G. and Dupraz, C. 2012. Simple models for light competition within agroforestry discontinuous tree stands: are leaf clumpiness and light interception by woody parts relevant factors? *Agroforest. Syst.*, 84: 101–116.

Tong, P. 1994. Achievements and perspectives of tillage and cropping system in China [in Chinese]. *Cropping Syst. Cultivated Technol.*, 77: 1–5.

Vos, J., Evers, J.B., Buck-Sorlin, G.H., Andrieu, B., Chelle, M., and de Visser, P.H.B. 2010. Functional–structural plant modelling: a new versatile tool in crop science. *J. Exp. Bot.*, 61(8): 2101–2115.

Wallace, S.U., Whitwell, T., Palmer, J.H., Hood, C.E., and Hull., S.A. 1992. Growth of relay intercropped soybean. *Agron. J.*, 84: 968–973.

Watiki, J.M., Fukai, S., Banda, J.A., and Keating, B.A. 1993. Radiation interception and growth of maize/cowpea intercrop as affected by maize plant density and cowpea cultivar. *Field Crops Res.*, 35: 123–133.

Willey, R.W. 1990. Resource use in intercropping systems. *Agric. Water Manage.*, 17: 215–231.

Xia, H.-Y., Wang, Z.-G., Zhao, J.-H., Sun, J.-H., Bao, X.-G., Christie, P., Zhang, F.-S., and Li, L. 2013. Contribution of interspecific interactions and phosphorus application to sustainable and productive intercropping systems. *Field Crops Res.*, 154: 53–64.

Xiao, Y., Li, L., and Zhang, F. 2004. Effect of root contact on interspecific competition and N transfer between wheat and faba bean using direct and indirect N^{15} technique. *Plant Soil*, 262(1–2): 45–54.

Yong, T. 2009. Nitrogen Uptake and Utilization and Rhizosphere Effect in "Wheat/Maize/Bean" Intercropping System, PhD dissertation, Sichuan Agricultural University.

Yu, C., Li, Y., Li, C., Sun, J., He, X.H., Zhang, F., and Li, L. 2010. An improved nitrogen difference method for estimating biological nitrogen fixation in legume-based intercropping systems. *Biol. Fertil. Soils*, 46(3): 227–235.

Zhang, D., Zhang, L., Liu, J., Han, S., Wang, Q., Jochem, E., Liu, J., van der Werf, W., and Li, L. 2014. Plant density affects light interception and yield in cotton grown as companion crop in young jujube plantations. *Field Crops Res.*, 169: 132–139.

Zhang, F. and Li, L. 2003. Using competitive and facilitative interactions in intercropping systems enhances crop productivity and nutrient-use efficiency. *Plant Soil*, 248(1–2): 305–312.

Zhang, F., Li, L., and Sun, J. 2001. Contribution of above- and below-ground interactions to intercropping. In: Horst, W.J., Schenk, M.K., Bürkert, A. et al., Eds., *Plant Nutrition: Food Security and Sustainability of Agro-Ecosystems Through Basic and Applied Research.* Kluwer Academic: Dordrecht, pp. 979–980.

Zhang, L.Z., van der Werf, W., Bastiaans, L., Zhang, S., Li, B., and Spiertz, J.H.J. 2008a. Light interception and radiation use efficiency in relay intercrops of wheat and cotton. *Field Crops Res.*, 107: 29–42.

Zhang, L.Z., van der Werf, W., Zhang, S., Li, B., and Spiertz, J.H.J. 2008b. Temperature mediated developmental delay may limit yield of cotton in relay intercrops with wheat. *Field Crops Res.*, 106: 258–268.

Zhang, W., Ahanbieke, P., Wang, B., Xu, W., Li, L., Christie, P., and Li, L. 2013a. Root distribution and interactions in jujube tree/wheat agroforestry system. *Agroforest. Syst.*, 87: 929–939.

Zhang, W.P. and Wang, G.X. 2010. Positive interactions in plant communities [in Chinese with English abstract]. *Acta Ecol. Sin.*, 30: 5371–5380.

Zhang, W.P., Jia, X., Bai, Y.Y., and Wang, G.X. 2011. The difference between above- and below-ground self-thinning lines in forest communities. *Ecol. Res.*, 26: 819–825.

Zhang, W.P., Jia, X., Morris, E.C., Bai, Y.Y., and Wang, G.X. 2012. Stem, branch and leaf biomass–density relationships in forest communities. *Ecol. Res.*, 27: 819–825.

Zhang, W.P., Jia, X., Damgaard, C., Morris, E.C., Bai, Y.Y., Pan, S., and Wang, G.X. 2013b. The interplay between above- and below-ground plant–plant interactions along an environmental gradient: insights from two-layer zone-of-influence models. *Oikos*, 122: 1147–1156.

Zhang, W.P., Pan, S., Jia, X., Chu, C.J., Xiao, S., Lin, Y., Bai, Y.Y., and Wang, G.X. 2013c. Effects of positive plant interactions on population dynamics and community structures: a review based on individual-based simulation models [in Chinese with English abstract]. *Chinese J. Plant Ecol.*, 37: 571–582.

Zhou, L., Cao, J., Zhang, F., and Li, L. 2009. Rhizosphere acidification of faba bean, soybean and maize. *Sci. Total Environ.*, 407(14): 4356–4362.

Zhou, X., Nian, H., Yang, W., and Han, T.F. 2010. *Status quo* and countermeasures of soybean production and development intercropping with other crops in South China (in Chinese). *Soybean Bull.*, 3: 1–2.

Zhu, J., Vos, J., van der Werf, W., van der Putten, P.E.L., and Evers, J.B. 2014. Early competition shapes maize whole-plant development in mixed stands. *J. Exp. Bot.*, 65: 641–653.

Zhu, Y.Y., Chen, H.R., Fan, J.H., Wang, Y.Y., Li, Y., Chen, J.B., Fan, J.X., Yang, S.S., Hu, L.P., Leung, H., Mew, T.W., Teng, P.S., Wang, Z.H., and Mundtk, C.C. 2000. Genetic diversity and disease control in rice. *Nature*, 406(6797): 718–722.

Zou, C. and Li, Z. 2002. Intercropping and relay intercropping. In: Shi, Y.C., Ed., *Chinese Academic Canon in the 20th Century: Agriculture* [in Chinese]. Fuzhou: Fujian Education Press.

Zuo, Y., Zhang, F., Li, X., and Cao, Y. 2000. Studies on the improvement in iron nutrition of peanut by intercropping with maize on a calcareous soil. *Plant Soil*, 220(1–2): 13–25.

Zuo, Y., Li, X., Cao, Y., Zhang, F., and Christie, P. 2003. Iron nutrition of peanut enhanced by mixed cropping with maize: role of root morphology and rhizosphere microflora. *J. Plant Nutr.*, 26(10–11): 2093–2110.

Zuo, Y., Liu, Y.X., Zhang, F.S., and Christie, P. 2004. A study on the improvement iron nutrition of peanut intercropping with maize on nitrogen fixation at early stages of growth of peanut on a calcareous soil. *Soil Sci. Plant Nutr.*, 50(7): 1071–1078.

CHAPTER **3**

Effects of Reduced Nitrogen Application Rates and Soybean Intercropping on Sugarcane Fields in Southern China

Wang Jianwu

CONTENTS

3.1 INTRODUCTION

Sugarcane (*Saccharum sinensis* Roxb.) is an important industrial crop in southern tropical and subtropical China, including Guangdong, Guangxi, Yunnan, and Fujian provinces, occupying about 1.4 million hectares and providing 90% of China's sugar (Robinson et al., 2011). Excessive nitrogen fertilizer input, however, is a common phenomenon that has become a serious threat to the sustainable development of sugarcane production in this region (Wang and Yang, 1994; Zhou et al., 1998). The average annual nitrogen fertilizer input was 750 kg N ha^{-1}, which is three to five times higher than the rates in Australia and Brazil (Robinson et al., 2011). Consequently, nitrogen use efficiency is less than 20% (Chapman et al., 1994). Excessive nitrogen input results in high residual soil nitrate levels, causing an increase in the nitrate pollution of groundwater in regions close to intensive sugarcane state farms in western Guangdong Province (Z.X. Li et al., 2010, 2011; Ao et al., 2011). Therefore, determining the optimum application of nitrogen has become an urgent issue in recent years, especially for sugarcane production in southern China (Wang and Yang, 1994; Zhou et al., 1998; Yang et al., 2011, 2012, 2013, 2014).

Intercropping of cash crops with sugarcane is widely practiced. Sugarcane has a juvenile period of approximately 100 to 110 days. Both wide row spacing (90 to 150 cm) and a slow growth rate in the initial stage of sugarcane provide space and available natural resources for intercropping in sugarcane fields. Many studies have shown that sugarcane intercropped with species such as watermelon (*Citrullus vulgaris* var. Caliber), peas (*Pisum sativum*), and onions (*Allium cepa*) reduced sugarcane yield yet improved the economic benefits considerably (Nazir et al., 2002; Gana and Busari, 2003; Al Azad and Alam, 2004); however, both sugarcane yield and net income increased in sugarcane–potato (*Solanum tuberosum* cv. Kufri Bahar) intercrops (Imam et al., 1990; S.N. Singh et al., 2010). Control effects on pests, including diseases, insects, and weeds, in sugarcane intercropping systems were also studied (A.K. Singh and Lal, 2008; Berry et al., 2009; C.G. Li et al., 2009; Chen et al., 2011).

Sugarcane depletes soil nutrients considerably, but legume intercropping patterns increase productivity per unit of land and enable crops to more effectively utilize nutrients and improve soil fertility and field conditions (Tang et al., 2005; He et al., 2006). An intercrop partially meets the nitrogen requirement of the companion crop due to the transfer of the symbiotically fixed nitrogen from the legume to the non-leguminous crop (L. Li et al., 2001). The efficiency of nitrogen, phosphorus, and potassium utilization in maize–fava bean intercrops was significantly higher than that of the monoculture system (Z.X. Li et al., 2011), and the organic matter content of sugarcane soil increased due to companion cropping of pulses (Yang et al., 2012). Cereal–legume intercropping systems have several major advantages in increasing yield, land use efficiency (Ghosh, 2004; Dhima et al., 2007), efficiency of natural resources utilization (Harris et al., 1987; F.S. Zhang and L. Li, 2003; Xu et al., 2008), and controlling pests and diseases (Berry et al., 2009; C.G. Li et al., 2009; Chen et al., 2011). Cereal–legume intercrops are becoming a common cropping system around the world (Jensen, 1996; L. Li et al., 2001; Lithourgidis et al., 2011; Eskanddari, 2012).

Sugarcane is generally planted in 80- to 100-cm rows, and soybeans (*Glycine max* L.) can be intercropped between two rows of sugarcane at the same time. Wide-row sugarcane plantations with the preferred row widths of 120 to 140 cm were tested, and it was found that wide rows are conducive to intercropping and benefit both crops in the sugarcane system (Huang, 2002). Sugarcane–soybean intercropping has been recognized as having the potential to reduce nitrogen input and maintain the productivity of sugarcane in southern China (Z.X. Li et al., 2010); however, there is a lack of information on assessing yield, nutrient uptake, and materials cycling in sugarcane intercrops in southern China.

A continuous 5-year field experiment was conducted to explore the yield and quality of sugarcane, as well as interspecific competition, economic benefit, greenhouse gas emissions, and nitrogen and carbon balance in sugarcane–soybean intercrops (Figure 3.1). It was expected that

Figure 3.1 Sugarcane–soybean strip intercropping in Hengli Township in Nansha District, Guangzhou.

intercropping could improve the resource use efficiency, thus increasing yield and even reducing nitrogen input. The objectives of this study were (1) to estimate the intercropping advantage of sugarcane intercropped with soybeans, (2) to assess interspecific competition using different competitiveness indices, (3) to examine the quality of sugarcane in intercropping systems at two rates of nitrogen application, (4) to document greenhouse gas emissions to determine the effect of different rates of nitrogen application on N_2O and CH_4 fluxes from sugarcane soils, and (5) to study the nitrogen and carbon cycling in a soil–plant system and compare the efficiency of nitrogen fertilizer use under different sugarcane–soybean intercropping and management systems.

3.2 LOCATION AND EXPERIMENT DESIGN

The field experiment was conducted from 2009 to 2013 at the farm of the South China Agricultural University, Guangzhou, China (23°8′N, 113°15′E), where an oceanic tropical monsoon climate prevails with an average 1780 hr of annual sunlight. The soil of the experimental field was a latosolic red soil with 21.08 g kg^{-1} organic matter, 75.38 mg kg^{-1} available N, 75.04 mg kg^{-1} Olsen P, and 61.71 mg kg^{-1} K in the upper 30 cm. Meteorological data were collected at the Wushan Weather Station, Guangzhou (Figure 3.2).

A randomized complete block design with three replications was used for the experiment design (Table 3.1). The cropping patterns were arranged in the main plots, and the nitrogen application rate was arranged in the subplots. The cropping patterns included mono-sugarcane (MS), mono-soybean (MB), and sugarcane–soybean intercropping (SB1, SB2). Nitrogen rates were at two levels: N2 was the conventional nitrogen application level (525 kg ha^{-1}) used by local farmers, and N1 was a reduced nitrogen application rate (300 kg ha^{-1}). Plot size was 5.5 m × 4.8 m (Figure 3.3). The planting distance in all treatments was 1.2 m between sugarcane rows and 0.3 m between soybean rows on the same ridge. In sugarcane–soybean intercropping, there were four rows of sugarcane and four (MS1) or eight (MS2) rows of soybeans in each plot. In MS and MB systems, there were 4 rows of sugarcane or 16 rows of soybeans in each plot, respectively.

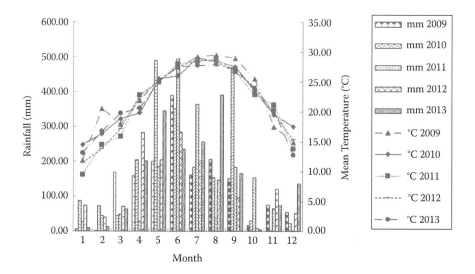

Figure 3.2 Monthly rainfall and mean temperature from 2009 to 2013 in Guangzhou.

According to the weather each year, sugarcane and soybean were sown and harvested on suitable dates (Table 3.2). The sugarcane cultivar was Yuetang 00-236, and the soybean cultivar was Maodou No. 3. Basal fertilizer included potassium chloride (150 kg ha^{-1}), calcium superphosphate (1050 kg ha^{-1}), and compound fertilizer (N:P:K = 15:15:15) (750 kg ha^{-1}), which was applied before sugarcane was planted. The first top dressing fertilizer included 150 kg ha^{-1} potassium chloride and 225 kg ha^{-1} or 113 kg ha^{-1} urea under N2 or N1, respectively, and was applied when sugarcane was in the tillering stage. The second top dressing fertilizer, with 672 kg ha^{-1} or 295 kg ha^{-1} urea under N2 or N1, respectively, was applied at the sugarcane jointing stage. The fertilizer application was the same for the duration of the experiment. Plant and soil samples were collected, greenhouse gas emissions were monitored, and the parameters of crop performance and nutrition were calculated by reported methods (Z.X. Li et al., 2010, 2011; Yang et al., 2011, 2012, 2013, 2014; Y. Zhang et al., 2013).

3.3 LAND EQUIVALENT RATIO AND YIELD

3.3.1 Land Equivalent Ratio of Sugarcane–Soybean Intercropping System

The land equivalent ratio (LER) is used for measuring the advantage of intercropping (Corre-Hellou et al., 2011):

$$\text{LER} = \frac{Y1_{int}}{Y1_{mono}} + \frac{Y2_{int}}{Y2_{mono}} \tag{3.1}$$

where $Y1_{int}$ and $Y2_{int}$ stand for the yields of crops 1 and 2 in the intercropping systems, respectively, and $Y1_{mono}$ and $Y2_{mono}$ stand for the yields of crops 1 and 2 in their monoculture systems, respectively. When LER > 1, the intercropping system is an advantageous one. When LER < 1, the intercropping system is not advantageous.

Based on crop yield, the LER values for all of the intercropped treatments were greater than 1.0 over the 5-year experiment, indicating that the land use efficiency of the sugarcane–soybean intercrops was higher than that of the monoculture systems (Table 3.3). The LER of treatment SB2-N1 was significantly higher than that of SB2-N2 and SB1-N2 in 2009, higher than SB1-N1 in 2010,

Table 3.1   Field Experiment Design of Sugarcane–Soybean Intercropping

Treatment Code	Nitrogen Rate (kg N ha⁻¹)	Cropping Patterns[a]
MS-N1	300	Mono sugarcane
SB1-N1	300	Sugarcane–soybean (1:1)
SB2-N1	300	Sugarcane–soybean (1:2)
MB	0	Mono soybean
MS-N2	525	Mono sugarcane
SB1-N2	525	Sugarcane–soybean (1:1)
SB2-N2	525	Sugarcane–soybean (1:2)

Source: Yang, W.T. et al., *Field Crops Res.*, 146, 44–50, 2013.

[a] 1:1 means one row of sugarcane intercropped with one row of soybean; 1:2 means one row of sugarcane intercropped with two rows of soybean.

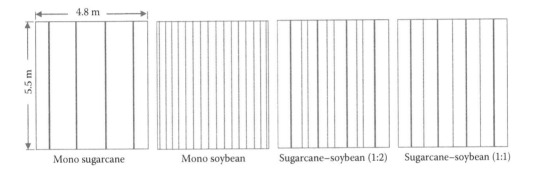

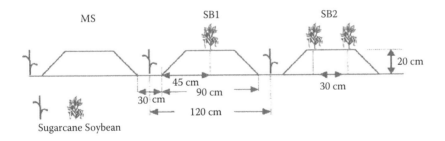

Figure 3.3   Field arrangement of different cropping systems.

Table 3.2   Sugarcane and Soybean Intercropping Dates

Year	Sugarcane Sowing Date	Sugarcane Harvesting Date	Soybean Sowing Date	Soybean Harvesting Date
2009	February 20	January 10	February 21	May 21
2010	March 15	December 26	March 16	June 20
2011	February 26	December 18	March 2	June 2
2012	February 25	December 16	March 10	June 3
2013	March 9	December 8	March 16	June 3

Table 3.3 Yield and Land Equivalent Ratio (LER) of Sugarcane and Soybean under Various Cropping Systems for 5 Years (2009–2013)

Year	N Rate	Cropping Pattern	Sugarcane Yield (t ha⁻¹)		Soybean Yield (t ha⁻¹)		LER
			Intercropping	Monoculture	Intercropping	Monoculture	
2009	N1	SB1	126.06[a]	119.07[a]	4.30[d]	14.85[a]	1.41[ab]
		SB2	135.15[a]		4.97[cd]		1.78[a]
	N2	SB1	113.28[a]	142.53[a]	9.12[b]		1.13[b]
		SB2	115.56[a]		6.85[c]		1.28[b]
2010	N1	SB1	103.91[b]	125.07[a]	1.66[c]	6.41[a]	1.10[b]
		SB2	105.04[b]		1.73[c]		1.42[a]
	N2	SB1	107.83[b]	117.65[ab]	3.72[b]		1.20[ab]
		SB2	114.71[ab]		3.05[b]		1.45[a]
2011	N1	SB1	128.57[a]	127.40[a]	2.33[b]	7.90[a]	1.31[c]
		SB2	126.02[a]		2.44[b]		1.84[a]
	N2	SB1	126.08[a]	112.47[a]	6.46[a]		1.44[bc]
		SB2	119.60[a]		5.08[ab]		1.69[ab]
2012	N1	SB1	97.52[ab]	108.03[a]	3.17[c]	9.31[a]	1.24[b]
		SB2	99.24[ab]		5.73[b]		1.53[a]
	N2	SB1	98.23[ab]	100.10[ab]	3.08[c]		1.31[b]
		SB2	85.70[b]		3.76[c]		1.26[b]
2013	N1	SB1	77.02[a]	89.90[a]	2.74[c]	10.58[a]	1.12[b]
		SB2	85.61[a]		5.74[b]		1.49[a]
	N2	SB1	75.25[a]	80.56[a]	2.64[c]		1.18[b]
		SB2	82.58[a]		4.58[bc]		1.46[a]

Note: Different letters in the same column indicate significant differences at the 5% level by the Duncan multiple range test.

higher than SB1-N1 and SB1-N2 in 2011and 2013, and higher than SB1-N1, SB1-N2, and SB2-N2 in 2012. The field experiment results showed that the LER value of treatment SB2-N1 was significantly higher than other intercropping patterns, and the treatment had an obvious advantage. There were no significant effects on sugarcane yields under different cropping patterns or different nitrogen application rates in 2010, 2012, and 2013 (Yang et al., 2014).

3.3.2 Competition Indices of Sugarcane–Soybean Intercropping System

Aggressivity (AG) measures the interspecies competition in an intercropping system by relating the yield changes of the two component crops (McGilchrist, 1965). This index compares the yields between intercrops and monocultures, as well as their respective land occupancy (L. Li et al., 2001; Williams and McCarthy, 2001; F.S. Zhang and L. Li, 2003; Wahla et al., 2009). In this article, we have employed the aggressivity concept to estimate yield and nitrogen acquisition of sugarcane relative to soybeans in the intercropping system:

$$AG = Y_{is}/(Y_{ms} \times A_s) - Y_{ib}/(Y_{mb} \times A_b) \tag{3.2}$$

where AG is the aggressivity of sugarcane relative to soybeans in the intercropping system. Y_{is} and Y_{ib} are the yields and nitrogen acquisition, respectively, of sugarcane and soybeans in intercrops. Y_{ms} and Y_{mb} are yields and nitrogen acquisition, respectively, of sugarcane and soybeans in monoculture. A_s and A_b are the proportions of the area occupied by sugarcane and soybean in the intercropping system relative to their proportions in monoculture. If AG is greater than 0, then the competitive ability of sugarcane exceeds that of soybeans in intercropping; if it is less than 0, then the soybeans are more competitive.

Table 3.4 Aggressivity Indices of Sugarcane Relative to Soybean in Four Intercropping Treatments

Variable		SB1-N1	SB2-N1	SB1-N2	SB2-N2	Mean	LSD$_{0.05}$
				Treatment			
Crop yield	2009	−0.05^{Aa}	−0.06^{Aa}	−0.55^{Ab}	−0.13^{Aa}	−0.20	0.36
	2010	−0.23^{Aa}	−0.32^{ABa}	−0.17^{Aa}	0.04^{Aa}	−0.17	0.54
	2011	−0.14^{Aa}	−0.70^{Ba}	−0.13^{Aa}	−0.17^{Aa}	−0.29	0.80
	Average	−0.14^{Aa}	−0.36^{ABa}	−0.28^{Aa}	−0.09^{Aa}	−0.22	0.40
Nitrogen acquisition	2009	−0.21^{Aa}	−0.01^{Aa}	−0.79^{Ab}	−0.37^{Bab}	−0.34	0.54
	2010	−0.11^{Aa}	−0.12^{Aa}	−0.36^{Aa}	−0.13^{Aa}	−0.18	0.45
	2011	−0.49^{Aa}	−0.45^{Aa}	−0.15^{Aa}	−0.26^{Aa}	−0.34	0.56
	Average	−0.27^{Aa}	−0.19^{Aa}	−0.43^{Aa}	−0.25^{Aa}	−0.29	0.35

Source: Yang, W.T. et al., *Field Crops Res.*, 146, 44–50, 2013.

Note: Different uppercase letters in a row and different lowercase letters in a column indicate significant differences at the 5% level.

The competitive ratio (CR) is another indicator to assess interspecific competition between species in an intercropping system (Willey and Rao, 1980). It measures the degree to which one crop competes with the other (Dhima et al., 2007; G.G. Zhang et al., 2011). The CR can be calculated as follows:

$$CR = [Y_{is}/(Y_{ms} \times A_s)]/[Y_{ib}/(Y_{mb} \times A_b)] \tag{3.3}$$

where CR is the competitive ratio of sugarcane relative to soybeans. When the CR is greater than 1.0, the competitive ability of sugarcane is higher than soybeans at the co-growth stage in the intercropping system. If the CR is less than 1.0, the competitive ability of sugarcane is less than that of soybeans at the co-growth stage in the intercropping system. The sugarcane and soybean in the intercropping system did not exhibit equal competitive ability based on aggressivity (Table 3.4). All of the aggressivity (AG) indices of sugarcane to soybeans were less than zero, except those of SB2-N2 based on crop yield in 2010. This consistently indicates that soybeans had greater competitiveness in the sugarcane–soybean intercrop. Based on crop yield and nitrogen acquisition, the AG did not differ significantly between the four intercropping systems in 2010 and 2011; however, the AG under SB1-N2 was significantly lower than that under SB1-N1 in 2009. Based on crop yield, the competitive ratios of sugarcane in the sugarcane–soybean intercrops under N1 were always less than 1.0 over the 3 years, but the competitive ratios of SB2-N2 in 2010 and SB1-N2 in 2011 exceeded 1.0. Based on nitrogen acquisition, the competitive ratios of the sugarcane–soybean intercrops were all less than 1.0, indicating that the competition for nitrogen uptake by sugarcane was not as good as soybeans in the field (Table 3.5).

Table 3.5 Competitive Ratio of Sugarcane Relative to Soybean in Sugarcane–Soybean Intercropping Systems

Variable		SB1-N1	SB2-N1	SB1-N2	SB2-N2	Mean	LSD$_{0.05}$
				Treatment			
Crop yield	2009	0.94^{Aa}	0.95^{Aa}	0.59^{Ab}	0.90^{Aab}	0.84	0.34
	2010	0.84^{Aa}	0.73^{Ba}	0.92^{Aa}	1.10^{Aa}	0.90	0.56
	2011	0.91^{Aa}	0.60^{Ba}	1.03^{Aa}	0.92^{Aa}	0.87	0.70
	Average	0.90^{Aa}	0.76^{ABa}	0.85^{Aa}	0.97^{Aa}	0.87	0.37
Nitrogen acquisition	2009	0.85^{Aab}	1.00^{Aa}	0.49^{Bb}	0.69^{Aab}	0.76	0.46
	2010	0.88^{Aa}	0.90^{Aa}	0.72^{ABa}	0.88^{Aa}	0.85	0.37
	2011	0.66^{Aa}	0.65^{Aa}	0.88^{Aa}	0.83^{Aa}	0.75	0.41
	Average	0.80^{Aa}	0.85^{Ba}	0.69^{ABa}	0.80^{Aa}	0.79	0.29

Source: Yang, W.T. et al., *Field Crops Res.*, 146, 44–50, 2013.

Note: Different uppercase letters in a row and different lowercase letters in a column indicate significant differences at the 5% level.

3.4 SUGARCANE JUICE QUALITY AND ECONOMIC BENEFITS

Cropping patterns and nitrogen rates did not significantly affect the quality indices of sugarcane juice in 2009 and 2010 (Table 3.6). Sugar brix, which indicates the percentage of solid material in sugar juice; sugar polarmeter readings, which are positively related to the sugar content of sugarcane juice; and the sucrose content were all not significantly influenced by cropping systems or by

Table 3.6   Sugarcane Juice Quality under Different Cropping Systems in 2009, 2010, and 2011

Treatment	Sugar Brix	Sugar Polarization	Sucrose Content	Apparent Purity	Gravity Purity
		Sugarcane Juice (%)			
Year 2009					
MS-N1	19.83 ± 0.52^a	17.73 ± 0.44^a	17.98 ± 0.42^a	89.40 ± 0.21^a	90.67 ± 0.30^a
SB1-N1	19.57 ± 0.38^a	17.55 ± 0.31^a	17.81 ± 0.30^a	89.69 ± 0.39^a	91.01 ± 0.42^a
SB2-N1	19.40 ± 0.06^a	17.58 ± 0.06^a	17.84 ± 0.06^a	90.65 ± 0.24^a	91.97 ± 0.23^a
MS-N2	19.40 ± 0.80^a	17.35 ± 0.88^a	17.61 ± 0.84^a	89.36 ± 0.80^a	90.73 ± 0.59^a
SB1-N2	19.43 ± 0.30^a	17.56 ± 0.38^a	17.82 ± 0.36^a	90.33 ± 0.63^a	91.66 ± 0.55^a
SB2-N2	19.07 ± 0.20^a	16.98 ± 0.28^a	17.26 ± 0.27^a	89.06 ± 0.52^a	90.51 ± 0.46^a
Results of Two-Way ANOVA Test (F)					
N level	0.68	0.75	0.74	0.65	0.48
Intercrop	0.39	0.22	0.22	0.83	1.17
Intercrop x N	0.06	0.23	0.23	2.51	2.97
Year 2010					
MS-N1	20.33 ± 0.23^a	18.35 ± 0.22^a	18.49 ± 0.28^a	90.22 ± 0.28^a	90.94 ± 0.33^a
SB1-N1	20.17 ± 0.17^a	18.26 ± 0.24^a	18.31 ± 0.26^a	90.55 ± 0.43^a	90.79 ± 0.54^a
SB2-N1	20.07 ± 0.32^a	18.02 ± 0.34^a	18.05 ± 0.34^a	89.79 ± 0.34^a	89.92 ± 0.34^a
MS-N2	19.93 ± 0.49^a	17.97 ± 0.49^a	18.03 ± 0.50^a	90.12 ± 0.45^a	90.46 ± 0.56^a
SB1-N2	19.90 ± 0.32^a	17.83 ± 0.32^a	17.90 ± 0.28^a	89.60 ± 0.27^a	89.94 ± 0.10^a
SB2-N2	20.00 ± 0.15^a	17.97 ± 0.14^a	18.04 ± 0.16^a	89.85 ± 0.05^a	90.22 ± 0.21^a
Results of Two-Way ANOVA Test (F)					
N level	0.98	1.26	1.25	1.52	1.21
Intercrop	0.07	0.14	0.25	0.60	1.37
Intercrop x N	0.15	0.22	0.31	1.33	1.19
Year 2011					
MS-N1	20.00 ± 0.31^a	18.08 ± 0.31^a	18.10 ± 0.31^a	90.39 ± 0.15^a	90.48 ± 0.15^a
SB1-N1	20.03 ± 0.28^a	18.02 ± 0.26^a	18.05 ± 0.25^a	89.96 ± 0.01^{ab}	90.09 ± 0.04^{ab}
SB2-N1	19.50 ± 0.31^a	17.43 ± 0.25^a	17.47 ± 0.24^a	89.41 ± 0.31^{ab}	89.57 ± 0.29^{ab}
MS-N2	20.20 ± 0.00^a	17.97 ± 0.04^a	18.00 ± 0.05^a	88.97 ± 0.22^b	89.12 ± 0.25^b
SB1-N2	19.90 ± 0.03^a	17.66 ± 0.37^a	17.72 ± 0.35^a	88.72 ± 0.63^b	89.00 ± 0.48^b
SB2-N2	20.03 ± 0.47^a	17.90 ± 0.48^a	17.93 ± 0.48^a	89.33 ± 0.63^{ab}	89.46 ± 0.67^{ab}
Results of Two-Way ANOVA Test (F)					
N level	0.62	0.00	0.00	7.88*	7.65*
Intercrop	0.58	0.65	0.66	0.44	0.34
Intercrop x N	0.58	0.91	0.88	1.66	1.51

Source: Yang, W.T. et al., *Field Crops Res.*, 146, 44–50, 2013.

Note: Values are mean ± SE. Different lowercase letters in a column indicate significant differences at the 5% level; * refers to a 5% significance level.

Table 3.7   Economic Benefits of Various Sugarcane Planting Systems

Treatment	Output Income (yuan)		Input Cost (yuan)			Net Return (yuan)
	Sugarcane	Soybean	Pesticide	Fertilizer	Labor	
SB1-N1	44,792	11,883	2256	4695	15,974	33,750
SB2-N1	47,726	25,547	2280	4695	17,174	49,124
SB1-N2	40,514	14,334	2256	5415	14,464	32,714
SB2-N2	41,198	20,248	2280	5415	14,870	38,881
MS-N1	42,561	—	2267	4695	15,021	20,578
MS-N2	50,351	—	2267	5415	17,771	24,898

Source: Li, Z.X. et al., *Chin. J. Appl. Ecol.*, 22, 713–719, 2011.

nitrogen rates, except that the apparent purity (percentage of the sugar polarization reading divided by sugar brix) and the gravity purity (percentage weight of sugar in the solid part of sugar juice) were significantly increased under N1 compared to N2 in 2011. In general, the economic benefits of sugarcane–soybean intercrops were higher than those of sugarcane in monoculture (Table 3.7). The economic benefits of sugarcane in the intercropping treatments with low-nitrogen application treatments (SB1-N1, SB2-N1) were 3.17 to 26.34% higher than those of the high-nitrogen application treatments (SB1-N2, SB2-N2). The net economic returns from sugarcane–soybean intercropped in 1:2 patterns (SB2-N1, SB2-N2) were better than for the 1:1 patterns (SB1-N1, SB1-N2) when nitrogen level was the same.

3.5 GREENHOUSE GAS EMISSIONS

Using the static chamber/gas chromatographic technique, greenhouse gas emissions were monitored during the growing season (Y. Zhang et al., 2013). Results showed that compared with the sugarcane monoculture with low nitrogen dosage (MS-N1), cumulative emissions of CO_2 and N_2O in SB2 under treatments with low nitrogen dosage (SB2-N1) decreased by 35.58% and 56.36%, respectively, but the accumulative CH_4 emissions increased by 7.02% (Table 3.8). Soils in different cropping patterns and nitrogen dosages served as sources of CO_2 and N_2O and as a sink of CH_4

Table 3.8   Accumulated CO_2, N_2O, and CH_4 Emissions from Sugarcane and Soybean Intercropping Treatments and Monocropping Treatments

Treatment	Accumulated CO_2 Emissions (kg ha^{-1})	Accumulated N_2O Emissions (kg ha^{-1})	Accumulated CH_4 Emissions (kg ha^{-1})
MS-N1	5096.89 ± 137.12[ab]	4.61 ± 1.03[a]	−13.68 ± 1.74[ab]
SB1-N1	6422.69 ± 791.95[a]	5.11 ± 2.04[a]	−21.78 ± 8.31[b]
SB2-N1	3283.20 ± 535.71[c]	2.15 ± 0.07[a]	−12.72 ± 3.35[ab]
MS-N2	4103.29 ± 560.66[bc]	3.13 ± 0.92[a]	−5.53 ± 1.10[a]
SB1-N2	4475.84 ± 158.22[bc]	3.72 ± 0.70[a]	−11.36 ± 2.62[ab]
SB2-N2	4775.31 ± 486.46[bc]	5.60 ± 0.82[a]	−4.77 ± 4.31[a]
MB	4780.35 ± 255.29[bc]	3.11 ± 0.27[a]	−9.97 ± 3.36[ab]
Results of Two-Way ANOVA Test (*F*)			
N level	0.84	0.22	3.38
Intercrop	4.54*	0.19	2.13
Intercrop x N	7.01**	3.83	0.05

Source: Zhang, Y. et al., *Chin. J. Eco-Agric.*, 21, 1318–1327, 2013.

Note: Values are mean ± SE. Different lowercase letters in a column indicate significant differences at the 5% level; * and ** refer to a 5% and 1% significance level, respectively.

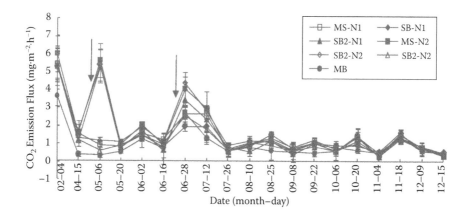

Figure 3.4   Seasonal CO_2 flux from intercropping soils of sugarcane and soybean in 2012.

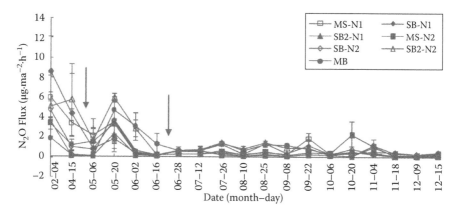

Figure 3.5   Seasonal N_2O flux from intercropping soils of sugarcane and soybean in 2012.

during the growing season. Both cropping pattern and the interaction between cropping pattern and nitrogen level had significant influence on soil CO_2 emissions in 2012 (Figure 3.4). The accumulated soil CO_2 emission of treatment SB2-N1 was significantly lower than treatments SB1-N1 and MS-N1; however, other nitrogen applications and cropping patterns had no significant effects on soil N_2O and CH_4 emission (Figure 3.5). After nitrogen application, the CH_4 absorption rate decreased while CO_2 and N_2O flux rates increased (Figures 3.4 and 3.5).

3.6 NITROGEN UPTAKE AND BALANCE

3.6.1 Efficiency of Soybean Nitrogen Fixation

The efficiency of soybean nitrogen fixation and amount of nitrogen fixation in the monoculture treatment were significantly higher than for the sugarcane–soybean intercrops (Table 3.9 and Table 3.10). Nitrogen input significantly decreased the soybean nitrogen fixation and accumulation (Table 3.10); the two rows of soybean intercrop (SB2) fixed more nitrogen than the one row of intercrop (SB1). The scale of soybean nitrogen fixation was heavily dependent upon the percentage of land covered.

Table 3.9   Soybean Nitrogen Fixation Efficiency (%)* Affected by Intercropping with Sugarcane in 2012 and 2013

Treatment	2012	2013
SB1-N1	44.48 ± 3.15^{c}	21.14 ± 4.67^{b}
SB2-N1	60.06 ± 0.72^{b}	34.91 ± 7.97^{b}
SB1-N2	59.94 ± 2.53^{b}	33.64 ± 2.95^{b}
SB2-N2	63.50 ± 2.54^{b}	26.37 ± 4.85^{b}
MB	71.11 ± 1.21^{a}	67.91 ± 2.13^{a}

Source: Liu, Y. et al., *Chin. J. Appl. Ecol.*, 26(3), 817–825, 2015.

Note: The biological nitrogen fixation of soybean was estimated using the ^{15}N natural abundance technique (Houngnandan et al., 2008). Different letters in the same column indicate significant differences at the 5% level.

* $\%N_{fixed} = (\delta^{15}N_{ref} - \delta^{15}N_{sample})/(\delta^{15}N_{ref} - \delta^{15}N)$.

Table 3.10   Soybean Nitrogen Fixation Rate (kg ha^{-1})* Affected by Intercropping with Sugarcane in 2012 and 2013

Treatment	2012	2013
SB1-N1	12.54 ± 1.66^{c}	10.46 ± 2.97^{c}
SB2-N1	30.24 ± 3.75^{b}	30.97 ± 9.31^{b}
SB1-N2	22.15 ± 3.70^{bc}	15.36 ± 0.81^{bc}
SB2-N2	30.96 ± 3.47^{b}	20.79 ± 2.67^{bc}
MB	60.34 ± 3.41^{a}	78.45 ± 2.18^{a}

Source: Liu, Y. et al., *Chin. J. Appl. Ecol.*, 26(3), 817–825, 2015.

Note: Different letters in the same column indicate significant differences at the 5% level.

*   Nitrogen fixation rate = $\%N_{fixed}$ × Plant N concentration × Dry weight of plant.

3.6.2 Crop Nitrogen Accumulation

Compared with sugarcane in monoculture, sugarcane–soybean intercrops did not significantly affect the sugarcane nitrogen accumulation in 2012 and 2013 (Table 3.11), but the sugarcane nitrogen accumulation of SB2-N2 showed a significant increase compared to both SB1-N2 in 2012 and MS-N1 in 2013 (Table 3.11). Compared with soybeans in monoculture, the sugarcane–soybean intercropping system significantly affected the total levels of soybean nitrogen fixation and accumulation. (Table 3.12). A higher percentage of land planted with soybeans (SB2) resulted in more nitrogen in the soybean pods, but nitrogen levels in soybean stalks were not affected (Table 3.12).

3.6.3 Ammonia Volatilization

Ammonia volatilization was significantly reduced by intercropping in 2012 and 2013 as compared with ammonia volatilization in sugarcane monoculture (MS) (Table 3.13), yet it was not affected by nitrogen input level. The two-year data showed that the reduced nitrogen input (N1) did not significantly reduce the ammonia volatilization compared to the normal nitrogen input (N2) in the two sugarcane–soybean intercropping patterns (SB1, SB2).

Table 3.11 Sugarcane Nitrogen Accumulation (kg ha^{-1})* Affected by Nitrogen Input Level and Intercropping with Soybean in 2012 and 2013

Treatment	2012	2013
MS-N1	176.81 ± 16.21^{ab}	116.33 ± 6.93^{b}
SB1-N1	152.00 ± 11.87^{ab}	121.76 ± 21.02^{ab}
SB2-N1	154.62 ± 12.27^{ab}	117.93 ± 20.7^{ab}
MS-N2	162.47 ± 14.11^{ab}	140.81 ± 15.79^{ab}
SB1-N2	126.83 ± 19.26^{b}	169.09 ± 34.96^{ab}
SB2-N2	184.25 ± 23.86^{a}	195.29 ± 27.13^{a}

Source: Liu, Y. et al., Chin. J. Appl. Ecol., 26(3), 817–825, 2015.

Note: Different letters in the same column indicate significant differences at the 5% level.

* Nitrogen accumulation = Plant N concentration × Dry weight of plant (kg ha^{-1}).

Table 3.12 Soybean Pod and Soybean Stalk Nitrogen Accumulation (kg ha^{-1})* Affected by Nitrogen Input Level and Intercropping with Sugarcane in 2012 and 2013

Treatment	Soybean Pod		Soybean Stalk	
	2012	2013	2012	2013
SB1-N1	13.14 ± 0.82^{d}	12.97 ± 0.34^{d}	19.2 ± 2.7^{b}	17.45 ± 2.20^{b}
SB2-N1	27.11 ± 1.75^{b}	30.53 ± 1.33^{b}	26.33 ± 6.38^{b}	25.92 ± 5.83^{b}
SB1-N2	14.77 ± 0.33^{d}	15.35 ± 0.44^{d}	16.66 ± 2.46^{b}	17.26 ± 1.86^{b}
SB2-N2	20.28 ± 1.18^{c}	22.53 ± 0.72^{c}	28.31 ± 3.90^{b}	29.49 ± 3.93^{b}
MB	50.37 ± 0.45^{a}	45.02 ± 4.55^{a}	41.92 ± 0.62^{a}	50.39 ± 3.13^{a}

Source: Liu, Y. et al., Chin. J. Appl. Ecol., 26(3), 817–825, 2015.

Note: Different letters in the same column indicate significant differences at the 5% level.

* Soybean pod nitrogen accumulation = Soybean pod N concentration × Dry weight of soybean pod (kg ha^{-1}). Soybean stalk nitrogen accumulation = Soybean stalk N concentration × Dry weight of soybean stalk (kg ha^{-1}).

Table 3.13 Effect of Reduced Nitrogen Rates and Sugarcane–Soybean Intercropping on Ammonium Volatilization (kg ha^{-1}) in 2012 and 2013

Treatment	2012	2013
MS-N1	0.57 ± 0.02^{b}	0.96 ± 0.05^{a}
SB1-N1	0.56 ± 0.01^{b}	0.75 ± 0.04^{b}
SB2-N1	0.56 ± 0.01^{b}	0.74 ± 0.03^{b}
MS-N2	1.11 ± 0.13^{a}	0.97 ± 0.02^{a}
SB1-N2	0.57 ± 0.02^{b}	0.83 ± 0.02^{b}
SB2-N2	0.6 ± 0.03^{b}	0.83 ± 0.03^{b}

Source: Liu, Y. et al., Chin. J. Appl. Ecol., 26(3), 817–825, 2015.

Note: Ammonia volatilization from field soil was determined by the venting method (Z.H. Wang et al., 2002). Different letters in the same column indicate significant differences at the 5% level.

Table 3.14   Effect of Reduced Nitrogen Rates and Sugarcane–Soybean Intercropping on Nitrogen Leaching (kg ha^{-1}) in 2012 and 2013

Treatment	2012	2013
MS-N1	60.71 ± 2.89[a]	90.04 ± 1.06[a]
SB1-N1	53.29 ± 2.35[a]	91.57 ± 3.12[a]
SB2-N1	60.81 ± 2.92[a]	83.25 ± 7.51[a]
MS-N2	60.24 ± 2.23[a]	78.2 ± 1.15[a]
SB1-N2	63.14 ± 1.15[a]	91.39 ± 2.30[a]
SB2-N2	59.46 ± 3.01[a]	78.92 ± 2.95[a]

Source: Liu, Y. et al., *Chin. J. Appl. Ecol.*, 26(3), 817–825, 2015.

Note: Nitrogen leaching (NO_3^--N) was assessed by the ion-exchange resin bag method (Fang et al., 2005). Different letters in the same column indicate significant differences at the 5% level.

3.6.4 Nitrogen Leaching

Compared to monoculture and normal nitrogen inputs, sugarcane–soybean intercropping patterns and reduced nitrogen inputs did not significantly affect nitrogen leaching (Table 3.14). Heavy rainfall in the subtropical region could be the dominant factor affecting nitrogen leaching in the field.

3.6.5 Nitrogen Balance

The annual nitrogen balance results indicated a surplus (the amount of N inputs are more than crop uptake and all N outputs) in all treatments (Table 3.15). The surplus amounts in treatments with a conventional nitrogen application rate (N2) were significantly higher than those for treatments with a reduced nitrogen application rate (N1). Even the reduced nitrogen input (N1) in sugarcane–soybean intercropping showed surplus results. Surplus nitrogen application could increase nitrogen loss in sugarcane production. Whether sugarcane is in monoculture or in an intercropping system with soybean, continuing sugarcane production could significantly reduce soil total nitrogen content. The total soil nitrogen content was more than 1.0 g kg^{-1} in 2009 and 2010 but was reduced to less than 1.0 g kg^{-1} in all treatments with sugarcane in 2011; soil fertility levels dropped from medium to low in that time period (Yang et al., 2013).

3.7 CARBON CYCLING AND BALANCE

The carbon balances of all treatments were calculated in order to judge carbon sink intensity under different fertilization and intercropping situations.

3.7.1 Soil Carbon Sequestration

The change in the soil carbon pool in 2012 is shown in Table 3.16. The soil carbon sequestration (SCS) of sugarcane–soybean intercrops was significantly greater than that of monoculture. The SCS of SB2-N1 and SB1-N1 increased by 166.71% and 139.28%, respectively, compared with that of MS-N1 (Table 3.16). In the treatment with conventional nitrogen input levels (N2), the SCS of SB2-N2 was 133.21% higher than that of MS-N2. Meanwhile, the SCS of monoculture soybeans was the highest among all the treatments due to the nitrogen fixation ability of soybeans.

Table 3.15   Nitrogen Balance (kg ha^{-1}) in Sugarcane–Soybean Intercropping Systems in 2012 and 2013

Item		2012							2013						
		MS-N1	SB1-N1	SB2-N1	MS-N2	SB1-N2	SB2-N2	MB	MS-N1	SB1-N1	SB2-N1	MS-N2	SB1-N2	SB2-N2	MB
N input	Fertilizer N	300	300	300	525	525	525	—	300	300	300	525	525	525	—
	Fixed N	—	12.54	30.24	—	22.15	30.97	60.34	—	10.46	30.98	—	15.36	20.8	78.45
	Seed N	20.7	22.67	24.63	20.7	22.67	24.63	7.86	20.7	22.67	24.63	20.7	22.67	24.63	7.86
	Straw N	—	19.2	26.33	—	16.66	28.31	45.18	—	17.45	25.92	—	17.26	29.49	50.39
N output	Sugarcane N	176.81	152	154.65	162.47	126.84	184.2	—	116.33	121.76	117.93	140.81	169.09	195.29	—
	Bean pod N	—	13.14	27.1	—	14.77	20.28	40.28	—	12.97	30.53	—	15.35	22.53	45.02
	Ammonium volatilization	0.57	0.56	1.12	0.56	0.57	0.6	—	0.96	0.75	0.74	0.97	0.57	0.83	—
	Leaching N	60.71	53.29	60.81	60.24	63.14	59.46	—	90.04	91.57	83.25	78.2	91.39	78.92	—
Balance		82.61[c]	135.42[bc]	137.52[bc]	322.43[a]	381.16[a]	344.37[a]	73.1[c]	113.37[b]	123.53[b]	149.08[b]	325.72[a]	303.89[a]	302.35[a]	91.68[b]

Source: Liu, Y. et al., Chin. J. Appl. Ecol., 26(3), 817–825, 2015.

Note: The overall field N balance was calculated based on the input and output balance method. The N input includes fertilizers, biological fixed N, and seeds and straw returned N. The loss of N was mainly caused by N removed in harvested sugarcanes and beans and by leached N and ammonium. The balance is equal to the quantity of N input minus the N output. A positive number indicates a nitrogen gain; a negative number indicates a nitrogen deficiency. Different letters in the same row indicate significant differences at the 5% level.

Table 3.16   Change of Soil Carbon Pool in Sugarcane–Soybean Intercropping System in 2012

Treatment	Before Planting		After Harvest		Soil C Sequestration (kg ha⁻¹)
	Total C (g kg⁻¹)	Soil C (kg ha⁻¹)	Total C (g kg⁻¹)	Soil C (kg ha⁻¹)	
MS-N1	7.27	2509.83	7.86	2682.02	172.19[b]
SB1-N1	7.94	2934.40	8.94	3346.42	412.02[a]
SB2-N1	7.53	2641.75	8.88	3101.00	459.25[a]
MS-N2	8.06	3053.85	8.49	3257.19	203.34[b]
SB1-N2	7.91	3092.31	8.49	3309.74	217.43[b]
SB2-N2	7.51	2768.21	8.77	3242.42	474.21[a]
MB	8.44	3191.98	9.70	3683.24	491.25[a]

Source: Zhang, Y. et al., *Chin. J. Eco-Agric.*, 21, 1318–1327, 2013.

Note: Soil C sequestration is equal to soil C after harvest minus soil C before planting. Different letters in the same column indicate significant differences at the 5% level.

3.7.2 Carbon Cycling of Sugarcane Fields

The amounts of carbon input, output, and rates of return in different treatments were significantly different (Table 3.17). The main field carbon inputs were in the form of fertilizer, sugarcane seedlings, soybean seeds, and compost from animal wastes. The total carbon input of SB2-N1 was 25.11% and was 11.25% greater than that of MS-N2 and SB1-N2. The main field carbon outputs were from the harvest of sugarcane and soybeans, as well as CO_2 and CH_4 emissions. The carbon output of SB2-N1 was 4859.46 kg ha⁻¹, which represented a decrease of 4.9% and 28.87% compared with MS-N1 and SB1-N1, respectively.

Table 3.17   Carbon Cycling of Sugarcane–Soybean Intercropping Systems in 2012

Item		MS-N1 (kg ha⁻¹)	SB1-N1 (kg ha⁻¹)	SB2-N1 (kg ha⁻¹)	MS-N2 (kg ha⁻¹)	SB1-N2 (kg ha⁻¹)	SB2-N2 (kg ha⁻¹)	MB (kg ha⁻¹)
C input	Fertilizer C	268.68	268.68	268.68	470.19	470.19	470.19	—
	Sugarcane seeding C	2659.80	2659.80	2659.80	2659.80	2659.80	2659.80	—
	Soybean seed C	—	651.67	1303.34	—	651.67	1303.34	2606.67
	Litter C	2607.04	2644.78	2693.55	2673.32	2648.99	2671.47	—
	Sum	5535.52[f]	6224.93[d]	6925.37[b]	5803.31[e]	6430.65[c]	7104.80[a]	2606.68[g]
C output	Bean pod harvesting C	—	181.86	368.63	—	209.77	277.73	550.26
	Sugarcane harvesting C	1748.15	1726.45	1220.36	1617.10	1494.43	1659.10	—
	CH_4	−13.68[ab]	−21.78[b]	−12.73[ab]	−5.53[a]	−11.36[ab]	−4.77[a]	−9.97[ab]
	CO_2	5096.89[ab]	6422.69[a]	3283.20[c]	4103.29[bc]	4475.84[bc]	4775.31[bc]	4780.35[bc]
	Sum	6831.36[ab]	8309.22[a]	4859.46[c]	5714.86[bc]	6186.69[bc]	6707.37[b]	5320.63[bc]
Return	Soybean stalk	—	294.29	431.19	—	271.90	459.64	800.01
Return/input		—	0.05	0.06	—	0.04	0.06	0.31
Output/input		1.24	1.34	0.70	0.99	0.96	0.94	2.04

Source: Zhang, Y. et al., *Chin. J. Eco-Agric.*, 21, 1318–1327, 2013.

Note: The field carbon budget was calculated based on the input and output balance method. Different letters in the same row indicate significant differences at the 5% level.

Table 3.18   Carbon Budget (kg ha^{-1}) of Sugarcane–Soybean Intercropping Systems in 2012

Item	MS-N1	SB1-N1	SB2-N1	MS-N2	SB1-N2	SB2-N2	MB
Input	5535.52	6224.93	6925.37	5803.31	6430.65	7104.80	2606.68
Soil sequestration	172.19	412.02	459.25	203.34	217.43	474.21	491.25
Return	—	294.29	431.19	—	271.90	459.64	800.01
Output	6831.36	8309.22	4859.46	5714.86	6186.69	6707.37	5320.63
Budget balance	−1123.65	−1377.98	2956.35	291.79	733.29	1331.28	−1422.69

Source: Zhang, Y. et al., *Chin. J. Eco-Agric.*, 21, 1318–1327, 2013.

3.7.3 Carbon Budget of Sugarcane Fields

The net carbon sequestration rates of various treatments are listed in Table 3.18. The carbon balances of treatments SB2-N1, MS-N2, SB1-N2, and SB2-N2 were positive and acted as carbon sinks, but treatments MS-N1, SB1-N1, and MB became carbon sources. The potential carbon sequestration capacity of SB2-N1 was 2956.35 kg ha^{-1}, which was the highest among all of the treatments.

3.8 CONCLUSIONS AND DISCUSSION

3.8.1 Conclusions

This study demonstrated that sugarcane–soybean intercropping systems improved land use efficiency compared to the efficiency of sugarcane monoculture systems. LERs of sugarcane–soybean intercrops based on crop yield were around 1.08 to 1.84, and sugarcane–soybean (1:2) intercropping patterns under reduced nitrogen rates had the highest LER values in all 5 years. Soybeans had higher competition ability than sugarcane in the intercropping systems, as indicated by the AG and CR values. This means that soybeans close to sugarcane rows would result in a reduced sugarcane yield and nitrogen acquisition. On the other hand, soybeans could grow better in an intercropping system than in monoculture. The quality of sugarcane juice was not affected by intercropping with soybeans. Only excessive nitrogen application (N2) in 2011 reduced sugar content in sugarcane juice by measuring the apparent purity and gravity purity from both monoculture (MS-N2) and intercropping (SB2-N2) treatments.

After nitrogen application, the CH_4 absorption rate decreased while CO_2 and N_2O flux rates increased. Because CH_4 emissions fluctuated strongly due to multiple factors, soil could actually become either a source or a sink of CH_4. Cropping patterns, and the interaction of cropping patterns with nitrogen levels, had a significant influence on soil CO_2 emissions. The accumulated soil CO_2 emissions of treatment SB2-N1 were significantly lower than those of SB1-N1 and MS-N1. Nitrogen application rates and cropping patterns had no significant effect on soil N_2O and CH_4 emission.

Compared with monoculture and normal nitrogen rates, sugarcane–soybean intercrops and reduced nitrogen rates did not significantly affect soil nitrate, ammonium, available nitrogen content, or soil microbial biomass nitrogen. The 3-year data showed that the nitrogen surplus under the reduced nitrogen rate was distinctly lower than under the normal nitrogen rate. SB2-N1 had the largest carbon sequestration potential, with 2956.35 kg C ha^{-1} in 2012 and 872.59 kg C ha^{-1} in 2013.

In conclusion, sugarcane–soybean intercropping (1:2) under N1 is the optimum practice in terms of land use efficiency, nitrogen use efficiency, crop yield, production cost, and environmental protection in south China. Sugarcane–soybean intercropping (1:2) with reduced nitrogen application is feasible in practice when considering the lowered costs and sustained sugarcane yield.

3.8.2 Discussion

3.8.2.1 Advantages of Sugarcane–Soybean Intercrops

Many previous studies have reported that cereal–legume intercropping systems had higher LER values compared to monoculture systems (Ghosh, 2004; Lithourgidis et al., 2011; Eskanddari, 2012; Mei et al., 2012). Some studies demonstrated that sugarcane intercropped with maize or potato also had higher LER values than did monoculture systems (Govinden, 1990; C.G. Li et al., 2009). Our results are consistent with these studies. In our study, all of the LER values based on crop yield and nitrogen acquisition in the 5-year experiment were greater than 1.0. This indicates that sugarcane–soybean intercropping systems have an advantage in crop yield and nitrogen acquisition and could improve land use efficiency compared to monoculture systems. The LER values of sugarcane–soybean (1:2) intercropping were higher than those of sugarcane–soybean (1:1) in our study. Sugarcane–soybean (1:2) intercropping with N1 (300 kg ha^{-1}) was the optimal intercropping pattern in terms of nitrogen efficiency and land use efficiency. The advantage of sugarcane legume intercropping systems in this study could be attributed to the significant complementarities in the utilization of natural resources including nutrients, water, and light in terms of both space and time (Martin and Snaydon, 1982; F.S. Zhang and L. Li, 2003; Hauggaard-Nielsen and Jensen, 2005). Sugarcane took 2 or 3 months to mature from planting to full canopy cover. During this period, the efficiency of light, water, and nutrient use by sugarcane was very low (Mendoza, 1986; Parsons, 1999). Soybeans intercropped between rows of sugarcane could make good use of those natural resources, resulting in increased nutrient use efficiency and crop yield.

3.8.2.2 Aggressivity and Competitive Ratios for Sugarcane–Soybean Intercrops

Interspecific competition obviously plays an important role in determining crop yield in an intercropping system (L. Li et al., 2001; L.H. Zhang et al., 2007). In general, sugarcane is considered the dominant crop in sugarcane–legume intercropping systems (L. Li et al., 2001; Dhima et al., 2007; Z.X. Li et al., 2011). However, different results were observed in our study, as indicated by the competitive indicators of AG and CR. The AG values based on crop yield and nitrogen acquisition over 3 years for each intercrop pattern were negative, except for sugarcane–soybean (1:2) intercropping under N2 (525 kg ha^{-1}) in 2010. Similarly, most of the CR values based on crop yield and all CR values based on nitrogen acquisition in sugarcane–soybean intercropping were less than 1.0, suggesting that sugarcane yield and nitrogen acquisition have been negatively affected by the interspecific competition caused by introducing soybeans into the system. Previous sugarcane intercropping experiments have also documented that sugarcane yield was reduced by intercropping (Nazir et al., 2002; Gana and Busari, 2003). The sugarcane tillering in intercropping was less than that in mono-sugarcane, resulting in a decrease in the number of millable stalks in the intercropping system. The sugarcane tiller densities tend to decrease with the decreasing light interception (Singels and Smit, 2009). In our study, soybean planting reduced the sugarcane tillering at the co-growth stage. Interspecific competition between sugarcane and cabbage or between sugarcane and maize was actually less than the intraspecific competition within the companion crops (cabbage or maize) in a monocropping system (Parsons, 1999). Similarly, soybeans were able to acquire more natural resources (water, light, and nutrients) in an intercropping pattern than in a monoculture in this study. When soybeans were in the flowering stage, sugarcane was still in the seedling and tillering stage. The soybean canopy reached and fully covered the space in sugarcane rows. It decreased the radiation intercepted by sugarcane, thus reducing the number of total tillers and number of millable stalks of sugarcane, in addition to reducing the sugarcane yield in the intercropping system.

3.8.2.3 Sugarcane Juice Quality

Sugarcane quality, particularly sucrose content, is very important for sugarcane production. The sucrose content of the juice is one of the key factors in maximizing sucrose yield (Lingle and Wiegand, 1997). Sugarcane juice purity is an important index in estimating sugarcane quality and is one of the key factors in predicting sugar yield. Our study demonstrated that sugarcane–soybean intercropping did not significantly affect sugarcane juice quality. Other quality indices, including sugar brix, sugar polarization, and apparent and gravity purity, were also not significantly affected by intercropping. Similar results have been reported by other sugarcane intercropping experiments (Mahadevaswamy and Martin, 2002; Nazir et al., 2002). Nitrogen is the primary nutrient limiting the quality of sugarcane production in most production environments. Some Indian research reported that high nitrogen rates did not affect sugarcane juice quality (Madhuri et al., 2011; A.K. Singh et al., 2011). These results were consistent with our results in 2009 and 2010. Similar to our study in 2011, Wiedenfeld (1995) found that increased nitrogen application had a slight positive effect on sugarcane juice purity and sugar content in the cane at first, and then sugarcane juice purity declined with increasing nitrogen application rate in the first or second ratoon of sugarcane. Through a pot experiment in China, Dao et al. (2011) found that sugarcane gravity purity decreased with increasing nitrogen fertilizer. These results suggest that higher nitrogen fertilizer application rates might not result in better sugarcane quality.

3.8.2.4 Pests, Diseases, and Biodiversity of Sugarcane Farming System

The performance of the whole ecosystem, including pests, changes when a monoculture farming system shifts to an intercropping system. The situation of pests is worthy of investigation in our future research. Soil biodiversity also needs to be measured in order to understand the mechanism of nitrogen and carbon balance. Another team from South China Agricultural University reported that the soil microbe numbers varied greatly under both intercropping and monoculture treatments (X.P. Li et al., 2013). When compared to sole plantations of sugarcane or soybeans, the numbers of bacteria, fungi, and actinomyces in intercroppng systems increased significantly. The differences in soil microorganisms might be attributed to combined factors such as root biomass, root exudates, and the microclimatic environment within the intercropping community.

ACKNOWLEDGMENTS

This study was financed by the State Key Development Program for Basic Research of China (Grant No. 2011CB100400), the National Science and Technology Support Program of China (Project No. 2012BAD14B16-04), and the Science and Technology Development Program of Guangdong (Project No.2012A020100003).

REFERENCES

Al Azad, M.A.K. and Alam, M.J. 2004. Popularizing of sugarcane based intercropping systems in non-military zone. *J. Agron.*, 3: 159–161.

Ao, J., Jiang, Y., Huang, Z., Lu, Y., Huang, Y., Zhou, W., Chen, D., and Li, Q. 2011. Strengthen the sugarcane nutrient management and reduce the sugarcane production cost [in Chinese]. *Guangdong Agric. Sci.*, 38: 31–34.

Berry, S., Dana, P., Spaull, V., and Cadet, P. 2009. Effect of intercropping on nematodes in two small-scale sugarcane farming systems in South Africa. *Nematropica*, 39: 11–34.

Chapman, L.S., Haysom, M.B.C., and Saffigna, P.G. 1994. The recovery of N-15 labelled urea fertilizer in crop components of sugarcane and in soil profiles. *Aust. J. Agric. Res.*, 45: 1577–1585.

Chen, B., Wang, J., Zhang, L., Li, Z., and Xiao, G. 2011. Effect of intercropping pepper with sugarcane on populations of *Liriomyza huidobrensis* (Diptera: Agromyzidae) and its parasitoids. *Crop Prot.*, 30: 253–258.

Corre-Hellou, G., Dibet, A., Hauggaard-Nielsen, H., Crozat, Y., Gooding, M., Ambus, P., Dahlmann, C., von Fragstein, P., Pristeri, A., and Monti, M. 2011. The competitive ability of pea–barley intercrops against weeds and the interactions with crop productivity and soil N availability. *Field Crops Res.*, 122: 264–272.

Dao, J.M., Guo, J.W., Cui, X.W., Fan, X., Liu, S.C., and Zhang, Y.B. 2011. Effects of different nitrogen application on yield and quality of sugarcane. *Sugar Crops China*, 33: 22–23.

Dhima, K., Lithourgidis, A., Vasilakoglou, I., and Dordas, C. 2007. Competition indices of common vetch and cereal intercrops in two seeding ratios. *Field Crops Res.*, 100: 249–256.

Eskanddari, H. 2012. Yield and quality of forage produced in intercropping of maize (*Zea mays*) with cowpea (*Vigna sinensis*) and mungbean (*Vingna radiate*) as double cropped. *J. Basic Appl. Sci. Res.*, 2: 93–97.

Fang, Y.T., Mo, J.M., and Zhou, G. 2005. Ion-exchange resin bags method research of forest soil NO_3-N and its response to nitrogen deposition [in Chinese]. *Ecol. Environ.*, 14: 483–487.

Gana, A. and Busari, L. 2003. Intercropping study in sugarcane. *Sugar Technol.*, 5: 193–196.

Ghosh, P.K. 2004. Growth, yield, competition and economics of groundnut/cereal fodder intercropping systems in the semi-arid tropics of India. *Field Crops Res.*, 88: 227–237.

Govinden, N. 1990. Intercropping of sugarcane with potato in Mauritius: a successful cropping system. *Field Crops Res.*, 25: 99–110.

Harris, D., Natarajan, M., and Willey, R. 1987. Physiological basis for yield advantage in a sorghum/groundnut intercrop exposed to drought. 1. Dry-matter production, yield, and light interception. *Field Crops Res.*, 17: 259–272.

Hauggaard-Nielsen, H. and Jensen, E.S. 2005. Facilitative root interactions in intercrops. *Plant Soil.*, 274(1–2): 237–250.

He, J.F., Huang, G.Q., Liao, P., Liu, X.Y., and Su, Y.H. 2006. Effects on disaster reduction of maize/soybean intercropping ecological system on upland red soil. *Meteorol. Disaster Reduct. Res.*, 29: 31–35.

Houngnandan, P., Yemadje, R.G.H., Oikeh, S. et al. 2008. Improved estimation of biological nitrogen fixation of soybean cultivars (*Glycine max* L. Merril) using ^{15}N natural abundance technique. *Biol. Fertil. Soils*, 45: 175–183.

Huang, R.H. 2002. The primary report of comparative experiments of wide-rows cultivation for sugarcane [in Chinese]. *Guangxi Sugarcane*, 4: 16–17.

Imam, S.A., Delwar Hossain, A.H.M., Sikka, L.C., and Midmore, D.J. 1990. Agronomic management of potato/sugarcane intercropping and its economic implications. *Field Crops Res.*, 25: 111–122.

Jensen, E.S. 1996. Grain yield, symbiotic N_2 fixation and interspecific competition for inorganic N in pea–barley intercrops. *Plant Soil*, 182: 25–38.

Li, C.G., He, X.H., Zhu, S.H., Zhou, H.P., Wang, Y.Y. et al. 2009. Crop diversity for yield increase. *PLoS ONE*, 4(11): 1–6.

Li, L., Sun, J.H., Zhang, F.S., Li, X., Yang, S., and Rengel, Z. 2001. Wheat/maize or wheat/soybean strip intercropping: I. Yield advantage and interspecific interactions on nutrients. *Field Crops Res.*, 71: 123–137.

Li, X.P., Mu, Y.H., Cheng, Y.B., Liu, X.G., and Nian, H. 2013. Effects of intercropping sugarcane and soybean on growth, rhizosphere soil microbes, nitrogen and phosphorus availability. *Acta Physiol. Plant.*, 35: 1113–1119.

Li, Z.X., Feng, Y.J., Yang, W.T., and Wang, J.W. 2010. The progress of research on sugarcane intercropping [in Chinese]. *Chin. J. Eco-Agric.*, 18: 884–888.

Li, Z.X., Wang, J.W., Yang, W.T., Shu, Y.H., Du, Q., Liu, L.L., and Shu, L. 2011. Effects of reduced nitrogen application on the yield, quality and economic benefit of sugarcane intercropping with soybean [in Chinese]. *Chin. J. Appl. Ecol.*, 22: 713–719.

Lingle, S.E. and Wiegand, C.L. 1997. Soil salinity and sugarcane juice quality. *Field Crops Res.*, 54: 259–268.

Lithourgidis, A.S., Vlachostergios, D.N., Dordas, C.A., and Damalas, C.A. 2011. Dry matter yield, nitrogen content, and competition in pea–cereal intercropping systems. *Eur. J. Agron.*, 34: 287–294.

Liu, Y., Zhang, Y., Yang, W.T., Li, Z.X., and Guan, A.M. 2015. Effects of reduced nitrogen application and soybean intercropping on nitrogen balance of sugarcane field [in Chinese]. *Chin. J. Appl. Ecol.*, 26(3): 817–825.

Madhuri, K.V.N., Hemanth Kumar, M., and Sarala, N. 2011. Influence of higher doses of nitrogen on yield and quality of early maturing sugarcane varieties. *Sugar Technol.*, 13: 96–98.

Mahadevaswamy, M. and Martin, G.J. 2002. Production potential of wide row sugarcane intercropped with aggregatum onion (*Allium cepa*) under different row ratios, fertilizer levels and population densities. *Indian J. Agron.*, 47: 361–366.

Martin, M. and Snaydon, R. 1982. Root and shoot interactions between barley and field beans when intercropped. *J. Appl. Ecol.*, 19: 263–272.

McGilchrist, C.A. 1965. Analysis of competition experiments. *Biometrics*, 21: 975–985.

Mei, P.P., Gui, L.G., Wang, P., Huang, J.C., Long, H.Y., Christie, P., and Li, L. 2012. Maize/fava bean intercropping with rhizobia inoculation enhances productivity and recovery of fertilizer P in a reclaimed desert soil. *Field Crops Res.*, 130: 19–27.

Mendoza, T.C. 1986. Light interception and total biomass productivity in sugarcane intercropping. *Philipp. J. Crop Sci.*, 11: 181–187.

Nazir, M.S., Jabbar, A., Ahmad, I., Nawaz, S., and Bhatti, I.H. 2002. Production potential and economics of intercropping in autumn-planted sugarcane. *Int. J. Agric. Biol.*, 4: 140–142.

Parsons, M.J. 1999. Intercropping with sugar cane. *Proc. S. Afr. Sug. Technol. Assoc.*, 73: 108–115.

Robinson, N., Brackin, R., Vinall, K., Soper, F., Holst, J., Gamage, H., Paungfoo-Lonhienne, C., Rennenberg, H., Lakshmanan, P., and Schmidt, S. 2011. Nitrate paradigm does not hold up for sugarcane. *PloS ONE*, 6: e19045.

Singels, A. and Smit, M. 2009. Sugarcane response to row spacing-induced competition for light. *Field Crops Res.*, 113: 149–155.

Singh, A.K. and Lal, M. 2008. Weed management in spring planted sugarcane (*Saccharum* spp. hybrid)-based intercropping systems. *Indian J. Agric. Sci.*, 78: 35–39.

Singh, A.K., Lal, M., and Singh, S.N. 2011. Agronomic performance of new sugarcane genotypes under different planting geometries and N levels. *Indian J. Sug. Cane Technol.*, 26: 6–9.

Singh, S.N., Yadav, R.L., Yadav, D.V., Singhand, P.R., and Singh, I. 2010. Introducing autumn sugarcane as a relay intercrop in skipped row planted rice–potato cropping system for enhanced productivity and profitability in the Indian sub-tropics. *Exp. Agric.*, 46: 519–530.

Tang, J.C., Mboreha, I.A. She, L.N., Liao, H., Chen, H.Z., Sun, Z.D., and Yan, X.L. 2005. Nutritional effects of soybean root architecture in a maize/soybean intercropping system [in Chinese]. *Sci. Agric. Sin.*, 38: 1196–1203.

Wahla, I.H., Ahmad, R., Ehsanullah, A.A., and Jabbar, A. 2009. Competitive functions of components crops in some barley based intercropping systems. *Int. J. Agric. Biol. (Pakistan)*, 11: 69–71.

Wang, X.L. and Yang, D.T. 1994. Absorption and distribution of nitrogen phosphorus and potassium in different growth stages of sugarcane [in Chinese]. *Chin. J. Soil Sci.*, 25: 224–226.

Wang, Z.H., Liu, X.J., Ju, X.T. et al. 2002. Field *in situ* determination of ammonia volatilization from soil: venting method [in Chinese]. *Plant Nutr. Fertil. Sci.*, 8: 205–209.

Wiedenfeld, R.P. 1995. Effects of irrigation and N fertilizer application on sugarcane yield and quality. *Field Crops Res.*, 43: 101–108.

Willey, R.W. and Rao, M.R. 1980. A competitive ratio for quantifying competition between intercrops. *Exp. Agric.*, 16: 117–125.

Williams, A.C. and McCarthy, B.C. 2001. A new index of interspecific competition for replacement and additive designs. *Ecol. Res.*, 16: 29–40.

Xu, B.C., Li, F.M., and Shan, L. 2008. Switchgrass and milkvetch intercropping under 2:1 row-replacement in semiarid region, northwest China: aboveground biomass and water use efficiency. *Eur. J. Agron.*, 28: 485–492.

Yang, W.T., Li, Z.X., Shu, L., and Wang, J.W. 2011. Effect of sugarcane/soybean intercropping and reduced nitrogen rates on sugarcane yield, plant and soil nitrogen [in Chinese]. *Acta Ecol. Sin.*, 31: 6108–6115.

Yang, W.T., Li, Z.X., Feng, Y.J., Shu, L., and Wang, J.W. 2012. Effects of sugarcane–soybean intercropping on soybean fresh pod yield and agronomic traits [in Chinese]. *Chin. J. Ecol.*, 31: 577–582.

Yang, W.T., Li, Z.X., Wang, J.W., Wu, P., and Zhang, Y. 2013. Crop yield, nitrogen acquisition and sugarcane quality as affected by interspecific competition and nitrogen application. *Field Crops Res.*, 146: 44–50.

Yang, W.T., Li, Z.X., Lai, J.N., Wu, P., Zhang, Y., and Wang, J.W. 2014. Effects of sugarcane–soybean intercropping and reduced nitrogen application on yield and major agronomic traits of sugarcane [in Chinese]. *Acta Agron. Sin.*, 40: 556–562.

Zhang, F.S. and Li, L. 2003. Using competitive and facilitative interactions in intercropping systems enhances crop productivity and nutrient-use efficiency. *Plant Soil*, 248: 305–312.

Zhang, G.G., Yang, Z.B., and Dong, S.T. 2011. Interspecific competitiveness affects the total biomass yield in an alfalfa and corn intercropping system. *Field Crops Res.*, 124: 66–73.

Zhang, L.H., van der Werf, W., Zhang, S.P., Li, B.G., and Spiertz, J.H.J. 2007. Growth, yield and quality of wheat and cotton in relay strip intercropping systems. *Field Crops Res.*, 103: 178–188.

Zhang, Y., Wang, J.W., Wang, L., Yang, W.T., Wu, P., Liu, Y., and Tang, Y.L. 2013. Effect of low nitrogen application and soybean intercrop on soil greenhouse gas emission of sugarcane field [in Chinese]. *Chin. J. Eco-Agric.*, 21: 1318–1327.

Zhou, X.C., Liu, G.J., Portch, S., Zeng, Q.P., Yao, J.W., and Xu, P. 1998. Effect of fertilizer K, S, Mg and nutrient characteristics of high yield sugarcane [in Chinese]. *Soils Fertil.*, (3): 26–28, 32–33.

CHAPTER 4

Integrated Rice–Fish Agroecosystems in China

Chen Xin

CONTENTS

4.1 IMPORTANCE OF INTEGRATED RICE–FISH SYSTEMS IN CHINA

Natural resources (e.g., arable land, freshwater) are very important to ensure sustainable food production for the world (Foley, 2011), especially for countries such as China that have large populations and relatively limited land and water. China has almost one quarter of the world's population but only one tenth of the global arable land and one quarter of the per-capita global renewable water resources (Liu and Yang, 2012; FAOSTAT, 2013). China currently faces great challenges in feeding its population with such limited natural resources.

Rice is the main ingredient of daily diets in China, providing roughly half of the grain consumed by Chinese people (FAOSTAT, 2013). Rice plants are adapted to periodic submergence under water, and rice production requires ample water. Freshwater aquaculture provides one third of aquatic-grown protein for Chinese people and is nearly the only source of aquatic-grown protein in some inland areas. Freshwater aquaculture also requires large quantities of freshwater, which is also needed for irrigation, drinking, and industrial use, as well (Frei and Becker, 2005a). In addition, new land suitable for aquaculture is limited, and intensive freshwater aquaculture can cause environmental problems (e.g., water pollution, spread of disease). For these reasons, efforts have been made to find an efficient approach to aquaculture in order to meet the increasing needs for aquatic-grown protein in China.

Rice fields offer a suitable environment for raising fish (note that "fish" in this chapter refers to a wide range of aquatic animals, including fish, freshwater prawn, marine shrimp, crabs, turtles, frogs, etc.). Actually, culturing fish with rice in paddy fields (also referred to as the rice–fish system, or RFS) has a long history in China and many other Asian countries (Halwart and Gupta, 2004; You, 2006). Rice–fish farming systems have tremendous potential for increasing food security and alleviating poverty in rural areas. RFS is also an efficient way of using the same land resource to produce both carbohydrates and animal protein, either concurrently or serially. In rice–fish farming systems, water is used to simultaneously produce the two basic foods.

Rice–fish farming systems also hold potential for environmental and resource conservation. On the one hand, raising fish in paddy fields can reduce or eliminate pesticide and insecticide use, because fish are able to eradicate weeds by eating or uprooting them, and they devour some insect pests. Raising fish in rice fields can also reduce fertilizer requirements for rice, because rice plants can use the unconsumed fish feed and fish excretions as organic fertilizers (Oehme et al., 2007). Thus, integrating rice cultivation with aquaculture results in efficient resource use. This system in turn leads to a cleaner and healthier rural environment. On the other hand, rice–fish farming can help to solve some problems generated by freshwater aquaculture; for example, nutrients in effluents from raising fish are absorbed by rice plants rather than becoming a pollution source.

4.2 DEVELOPMENT OF RICE–FISH SYSTEMS IN CHINA

China has 27.4 million hectares of irrigated rice fields, next only to India in the amount of land under rice production with similar irrigation. China is the world's largest freshwater aquaculture producer. Integrating rice culture with aquaculture has a long history; however, it was not until after the founding of the People's Republic of China in 1949 that rice–fish farming developed quickly throughout the country. Over the last 60 years, Chinese RFS has experienced an uneven and discontinuous development, which can be divided into three main phases (Ministry of Agriculture of the People's Republic of China, 1980–2012, 2000–2012). Each of these phases is related to policy.

4.2.1 1949 to 1978

After the founding of the People's Republic of China in 1949, the Ministry of Agriculture (MOA) of the Chinese government tried hard to develop RFS to meet the urgent need for aquatic-grown food, especially for people living in mountainous areas. In 1954, the fourth National Aquaculture Meeting by the MOA proposed a plan to develop RFS nationwide, which greatly promoted the development of traditional RFS. By 1959, the area under RFS production had been expanded to 666,000 hectares, although RFS continued to be managed in traditional ways with low fish and rice yields. At that time, carp species (big-head carp, silver carp, black carp, and grass carp) were the main aquaculture species in RFS. Since the middle of the 1960s, however, interest had gradually waned, mainly due to the use of chemical pesticides and herbicides in the early attempts to boost rice productivity. RFS was officially discouraged. By the end of the 1970s, the area of RFS had declined to 103,200 hectares.

4.2.2 1980 to 2005

Interest in RFS was renewed in the early 1980s, when the land-contract system for farmers was implemented in rural areas and individual families became the main farming units. In addition, rapid development of aquaculture was needed to provide animal protein. In the context of aquaculture development, RFS was added to the framework of the National Aquaculture Development Plan and was identified as an approach to improve rural economies. RFS was greatly promoted nationwide through research, policies, and official activities. The area under RFS production more than doubled between 1985 (648,660 ha) and 2000 (1,532,381 ha). The fish yield in RFS also increased from 125 kg ha^{-1} in 1985 to 487 kg ha^{-1} in 2000. The higher income from fish in RFS, however, led farmers and extension campaigns to overemphasize fish yield and its income and ignore rice in RFS. For example, some local governments suggested that the farmers reduce rice production to expand fish production. In some areas, rice fields were even totally converted to fishponds for intensive aquaculture. These inappropriate RFS practices contributed to a decline of RFS from 2001 to 2005. As the coastal regions experienced rapid economic development, small-scale RFS could not generate enough economic return, which may have been preventing further development of RFS in these regions.

4.2.3 2007 to 2015

After the 17th National Congress of the Communist Party of China in 2007, the rice–fish system was again promoted in order to increase food security (e.g., grain, vegetables, animal protein) and to save scarce land and water resources. In 2011, the MOA listed RFS within the national fishery development plan in the 15th Five Year Plan (2011 to 2015). In addition, the MOA set up a long-term (2012 to 2016) nationwide program called "Technology and Models of Rice–Aquaculture Combination" to improve traditional RFS and to develop diverse RFS based on local socio-environmental conditions. Diverse, large-scale, intensive, high-efficiency, and technology-based RFS is also currently being demonstrated in the main rice production regions led by the National Aquaculture Technical Extension Station of MOA. The area under RFS cultivation has reached 150,000 hectares in recent years.

Table 4.1 Production and Economic Income in Rice Monoculture and Rice–Fish Systems

Factor	Traditional Rice–Fish Co-Culture (TRF)	Improved Rice–Fish Co-Culture (IRF)	Rice Monoculture (RM)
Rice yield (ton ha^{-1})	5.98 ± 0.14[a]	6.03 ± 0.26[b]	6.05 ± 0.13[b]
Fish yield (kg ha^{-1})	374.63 ± 14.54[a]	1012.93 ± 88.05[b]	—
Rice income (RMB ha^{-1})	11,960.27 ± 274.93[a]	12,095.96 ± 256.78[b]	12,095.06 ± 512.96[b]
Fish income (RMB ha^{-1})	14,985.24 ± 569.08[a]	40,517.27 ± 3522.72[b]	—
Total income (RMB ha^{-1})	26,945.51 ± 844.01[a]	52,613.23 ± 3779.50[b]	12,095.06 ± 512.96[c]
Net income (RMB ha^{-1})	24,883.77 ± 581.19[a]	49,581.95 ± 3374.61[b]	9632.60 ± 237.91[c]

Note: Means with different letters are significantly different at the 5% level.

4.3 ECOLOGICAL FUNCTIONS OF RICE–FISH FARMING SYSTEMS

Rice–fish farming systems should be evaluated to determine what ecosystem functions they generate and whether they are worthwhile endeavors for farmers. Here, we focus on rice and fish yields, chemical inputs, resource use, and environmental impacts of rice–fish systems.

4.3.1 Productivity and Economic Benefits

Rice–fish systems generally increase land productivity (Tsuruta et al., 2011). Through sampling and comparative analysis of the productivity and economic benefits of rice monoculture (RM), traditional RFS (TRF, without formula feed input and with low-density fish stocking), and improved RFS (IRF, with formula feed input and with higher density fish stocking), a study found that there was no significant difference in rice yields among RM, TRF, and IRF. However, fish yields in IRF were significantly higher than in TRF (Table 4.1). The average fish yields were 374.63 ± 14.54 kg ha^{-1} and 1012.93 ± 88.05 kg ha^{-1} for TRF and IRF, respectively (Table 4.1). Net income per hectare of the rice field was significantly lower in RM than in TRF and IRF. Both total and net incomes per hectare of the rice field were higher in IRF than in TRF. As Table 4.1 shows, the increase in economic income from RFS mainly came from an increase in fish yields.

A field experiment was conducted to further survey the productivity and economic benefits among TRF with low fish density (targeted yield 750 kg ha^{-2}) and IRF with higher fish densities. Results showed that the productivity of IRFs was significantly higher than that of TRFs, and, with increased fish density, system productivity and economic benefits were obviously improved (Table 4.2). This was mainly because of the extra economic income from fish.

A recent meta-analysis of 12 research papers published within the past 20 years indicates that the rice–fish system has a positive effect on rice yield, regardless of fish culture type (i.e., single fish species or mixed fish species) (Ren et al., 2014).

4.3.2 Chemical Input

Generally, the use of pesticides is greatly reduced or entirely eliminated in RFS, and fertilizer use is also significantly reduced. Through a survey of 120 farmers in Vietnam, Berg (2002) showed that pesticide use in RFS was reduced by 43.8% compared with monocultures. In Indonesia, there was no herbicide use in RFS, and pesticide use was only 23.24% of that in monoculture systems (Dwiyana and Mendoza, 2008). In a field survey, we randomly selected 31 different areas (natural villages, 25 in hilly areas and 6 in plains areas) throughout the study site located in China's southeast Zhejiang Province (Chen et al., 2011). In each area, three to five pairs of rice monoculture (RM) fields and rice–fish co-culture (RF) fields were selected. Records were kept of farming details such as plant

Table 4.2 Production and Economic Income of Traditional and Fish-Density-Increased Rice–Fish Co-Culture Systems

	Rice Yield (kg ha^{-1})	Fish Yield (kg ha^{-1})	Cost* (yuan ha^{-1})	Total Income (yuan ha^{-1})	Net Income** (yuan ha^{-1})
Treatment					
Rice monoculture (RM)	6171 ± 299.68[a]	—	2462 ± 184.80[ab]	20,981 ± 1018.9[a]	18,519 ± 834.1[a]
Rice–fish co-culture (RF)	5898 ± 415.74[a]	484 ± 70.9[a]	2062 ± 262.82[a]	49,093 ± 5667.5[b]	47,031 ± 5404.7[b]
Fish monoculture (FM)	—	414 ± 37.53[a]	441 ± 101.74[c]	24,840 ± 2251.8[c]	24,399 ± 2150.1[c]
Fish Density in Rice–Fish Systems					
Target yield 750 kg ha^{-2}	5044 ± 308.52[a]	686 ± 54.38[a]	2955 ± 403.01[b]	42,572 ± 3100.8[bd]	39,617 ± 2697.8[d]
Target yield 1500 kg ha^{-2}	5126 ± 508.86[a]	1190 ± 104.43[b]	4091 ± 643.59[d]	62,978 ± 5703.8[e]	58,887 ± 5060.2[e]
Target yield 2250 kg ha^{-2}	5150 ± 465.66[a]	1578 ± 243.85[b]	5227 ± 884.17[de]	78,570 ± 11,151[f]	73,343 ± 10267[f]
Target yield 3000 kg ha^{-2}	4911 ± 297.21[a]	2159 ± 199.54[c]	6363 ± 1124.75[e]	101,093 ± 8873[g]	94,730 ± 7748.5[g]

Note: Means with different letters are significantly different at the 5% level.

* Cost includes fish seeds, rice seeds, fertilizers, feeds, and pesticides.

** Net income = Total income – cost.

variety, sowing, transplanting, harvest, pest control, fertilization, fish stocking density, fish feeding, and outputs during the rice-growing season at each farm. The farmers were approached and surveyed by the research team members in both the fields and their homes. It was assumed that the survey did not influence the farming activities conducted by the farmers themselves. Total pesticide use was expressed as active ingredient per hectare, and organic and chemical fertilizers were calculated as nitrogen (N), phosphorus (P), and potassium (K) per hectare. The results showed that inputs of chemical fertilizers and pesticides were significantly higher in RM than that in TRF and IRF (Table 4.3). No herbicides were applied during the rice-growing season in TRF and IRF, which suggested that fish could effectively control the weed population in paddy fields. Organic fertilizers were dominant in TRF, and chemical fertilizers were dominant in IRF; however, inputs of chemical fertilizers were significantly higher in IRF than in TRF (Table 4.3). It could be the reason that rice yields in IRF were slightly higher than those in TRF. Pesticide inputs were slightly higher in TRF than in IRF (Table 4.3), mainly due to the slightly increased levels of pests and diseases in TRF.

Table 4.3 Rice Production and Inputs in Various Farming Systems

		Traditional Rice–Fish Co-Culture (TRF)	Improved Rice–Fish Co-Culture (IRF)	Rice Monoculture (RM)
Rice yield (ton ha^{-1})		5.98 ± 0.14[a]	6.03 ± 0.26[b]	6.05 ± 0.13[b]
Chemical fertilizers (kg ha^{-1})	N	43.80 ± 5.41[a]	101.32 ± 6.96[b]	166.35 ± 8.46[c]
	P_2O_5	12.24 ± 2.22[a]	26.77 ± 2.35[b]	40.67 ± 3.44[c]
	K	14.17 ± 2.90[a]	30.29 ± 3.37[b]	59.27 ± 5.99[c]
	Total	70.21 ± 5.23[a]	158.39 ± 16.97[b]	266.28 ± 18.15[c]
Organic fertilizers (kg ha^{-1})	N	102.21 ± 12.63[a]	70.15 ± 4.82[b]	18.48 ± 0.94[c]
	P_2O_5	33.66 ± 5.17[a]	18.54 ± 1.63[b]	4.52 ± 0.38[c]
	K	33.06 ± 6.76[a]	20.97 ± 2.31[b]	6.59 ± 0.55[c]
	Total	163.83 ± 17.21[a]	109.56 ± 11.75[b]	29.59 ± 2.02[c]
Pesticides (kg ha^{-1})	Insecticides	1.53 ± 0.17[a]	1.54 ± 0.14[a]	3.26 ± 0.86[b]
	Fungicides	0.23 ± 0.02[a]	0.27 ± 0.03[a]	0.58 ± 0.13[b]
	Herbicides	0.00 ± 0.00[a]	0.00 ± 0.00[a]	0.38 ± 0.09[b]
	Total	1.76 ± 0.18[a]	1.81 ± 0.15[a]	4.22 ± 0.36[b]

Note: Means with different letters are significantly different at the 5% level.

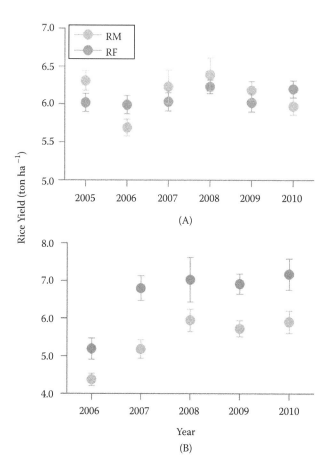

Figure 4.1   Rice yield in rice monoculture (RM) and rice–fish co-culture (RF). (A) Rice yield in field survey, and (B) rice yield in field experiment. Error bars are SE.

Achieving higher rice yields in rice monoculture generally requires higher inputs of chemical fertilizers and pesticides. To test whether RFS can help maintain high rice yields with lower chemical inputs, we conducted a 6-year study. The stability of rice yields in rice monoculture (RM) and rice–fish co-culture (RF) was compared through both a farmers' field survey and a field experiment. The temporal stability of rice yields was measured as the degree of consistency of yields around its mean over the same time interval (Tilman et al., 2006). In the survey of farmers' fields, the temporal stability of rice yields was determined using data collected annually from 31 sampling units (each unit containing three to five subsamples) over a period of 6 years. In the field experiment, the temporal stability of rice yields was also determined using data from each experimental plot over a period of 5 years. Survey results from farmers' fields showed that, although rice yields did not differ between RM and RF over the 6 years (Figure 4.1A), the temporal stability of rice yields was higher in RF than in RM. There was 68% more pesticide and 24% more fertilizer (Table 4.4) applied in RM than in RF. Pesticide application was also more variable in RM than in RF (Table 4.4). In our field experiment, however, where we did not apply pesticides, rice yields were significantly higher in RF than in RM (Figure 4.1B). The temporal stability of rice yields was also higher in RF than in RM (Xie et al., 2011a).

Table 4.4   Fish Yield and Utilization of Pesticides, Fertilizers, and Fish Feed in Rice Monoculture and Rice–Fish Co-Culture As Determined by Farmer Surveys

Variables		Rice Monoculture (RM)	Rice–Fish Co-Culture (RF)
Fish yield	Average (kg ha^{-1})	—	522.66 ± 80.54
	Coefficient of variation (%)	—	3.66 ± 0.41
Fish feed input	Average (kg ha^{-1})	—	1108.13 ± 145.79
	Coefficient of variation (%)	—	1.99 ± 0.13
Pesticides input	Average (kg a.i. ha^{-1})	14.85 ± 1.50	4.75 ± 0.75
	Coefficient of variation (%)	18.80 ± 1.48	11.92 ± 1.49
Total fertilizer input	Average (kg NPK ha^{-1})	282.27 ± 26.53	215.20 ± 19.30
	Coefficient of variation (%)	3.14 ± 0.23	3.72 ± 0.28
Nitrogen-fertilizer input	Average (kg N ha^{-1})	183.54 ± 15.84	149.49 ± 14.51
	Coefficient of variation (%)	1.90 ± 0.33	1.67 ± 0.10

Note: Values are means ± SE for 2005 to 2010. N, nitrogen; P, phosphorus; K, potassium.

To test whether the temporal stability of rice yields in rice monoculture (RM) is more dependent on pesticides than in rice–fish co-culture (RF), a correlation analysis between the temporal stability of rice yields and pesticide use was conducted. It was found that the temporal stability of rice yields was positively correlated with the quantity of pesticides applied in RM ($R^2 = 0.526$, $p = 0.001$, $n = 31$), but not in RF ($R^2 = 0.104$, $p = 0.077$, $n = 31$). Many factors can cause temporal variation in rice yields, including year-to-year changes in climate, pest incidence, use of new rice varieties, and levels of fertilizer application. Throughout the study, paired farmers' fields in the survey and plots in the field experiment did not change in rice variety, irrigation scheme, or levels of applied fertilizer. In addition, weather (temperature, precipitation, and relative humidity) during the rice growing seasons did not vary widely over the 6 years. In the survey, the factor that varied most was pesticide quantity; therefore, the results indicate that the temporal stability of rice yield in RM may largely depend on pesticides and that the higher stability in RF compared to RM was at least partly dependent on the presence of fish (Xie et al., 2011a).

4.3.3 Nitrogen Use Efficiency

Compared to intensive aquaculture farms, traditional rice–fish co-culture farms are smaller and produce lower fish yields (Halwart and Gupta, 2004). Although fish yields in the co-culture system can be greatly increased by increasing fish stocking density and by applying more fish feed, these methods could increase environmental risks (Ding et al., 2013). Thus, the question arises: Do increases in fish yield threaten the environment in the traditional rice–fish co-culture system? We conducted several experiments in China to answer this question (Hu et al., 2013).

In the first experiment, we compared the impact of added nitrogen (N) and the efficiency of N use in three systems: traditional rice–fish co-culture, fish monoculture, and rice monoculture. In the second experiment, we tested how the impact of added N and the efficiency of N use in rice–fish co-culture affected fish by stocking density and the rate of fish feed application. In the third experiment with rice–fish co-culture, we examined whether fish yield can be increased without decreasing rice yield and without increasing the total N application rate or the risk of N pollution. This experiment was accomplished by managing the ratio of N supplied in fish feed vs. N supplied in rice fertilizer. We found that fish yields increased while rice yields did not change as fish stocking density and fish feed application increased in rice–fish co-culture. As fish yields increased,

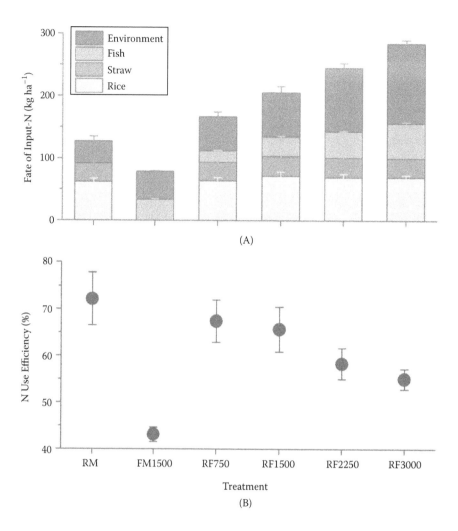

Figure 4.2 Effect of fish stocking rate in rice–fish co-culture on fate of (A) added N and (B) N use efficiency. Values are means ± SE. Bars with different letters refer to the difference of the same fractions across treatments based on LSD ($p < 0.05$) by LSD multiple comparison at the 95% level. RM, rice monoculture; FM1500, fish monoculture with target fish yield of 1500 kg ha^{-1}; RF750, RF1500, RF2250, and RF3000, rice–fish co-culture with target fish yields of 750, 1500, 2250, and 3000 kg ha^{-1}, respectively.

however, nitrogen lost to the environment through air in the form of NH_4 and N_2O and through drainage water also increased (Figure 4.2A) and the efficiency of N use in the system decreased (Figure 4.2B). Thus, a trade-off between fish yield and N use should be considered when researchers and farmers attempt to increase fish yield by increasing fish density in rice–fish co-culture. In the second experiment, fish yields were similar, with stocking densities of 10,000 and 15,000 fish per hectare. Significantly less N remained in the environment with 10,000 rather than 15,000 fish per hectare. In the third experiment, a stocking density of 10,000 fish per hectare was used to determine whether the quantity of N remaining in the environment at harvest can be reduced and whether the efficiency of N use can be enhanced by regulating the relative percentages of feed-N and fertilizer-N while holding constant the total amount of N added. The results indicated that different percentages of feed-N and fertilizer-N did not affect rice yields or the efficiency of nitrogen use, but they did affect fish yields. Fish yields were highest at 63% feed-N and 37% fertilizer-N

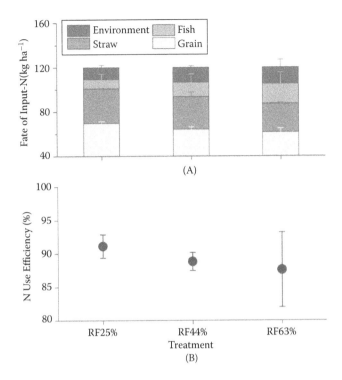

Figure 4.3 Effects of different percentages of feed-N vs. fertilizer-N in rice–fish co-culture (RF) on (A) the fate of added N, and (B) N use efficiency. Values are means ± SE. Bars with different letters refer to the difference of the same fractions across treatments based on LSD ($p < 0.05$). RF25%, RF with 75% fertilizer N and 25% fish-feed N; RF44%, RF with 56% fertilizer N and 44% fish-feed N; RF63%, RF with 37% fertilizer N and 63% fish-feed N.

(Figure 4.3). These results suggest that fish yields can be increased and fertilizer-N can be reduced in traditional rice–fish co-culture without increasing the quantity of N lost to the environment through air and water during crop production.

4.3.4 Field Aquatic Environment

The contents of nitrogen, phosphorus, and chemical oxygen demand (COD) in rice field water can reflect the influence of RFS on aquatic environments. These conditions are described below.

4.3.4.1 Total Nitrogen Content in Field Water

Traditional RFS does not use additional feed and thus may not contribute to environmental problems. As the fish stocking density increases and more fish feed is applied, however, high-yielding RFS may experience environmental problems just as intensive aquaculture does. Experiments showed that total N levels in field water did not vary significantly among rice monoculture (RM), rice–fish co-culture (RF), and fish monoculture (FM) systems when fish stocking density was low ($F_{2,9} = 0.304$; $p = 0.745$) (Figure 4.4). Total N levels in field water in all treatments also had no significant change with rice growth; however, total N content in field water tended to increase as fish density increased. Total N content in field water also gradually increased during the rice growth period, probably due to continuous fish-feed input. Unused fish feed led to the increase of total nitrogen content in field water.

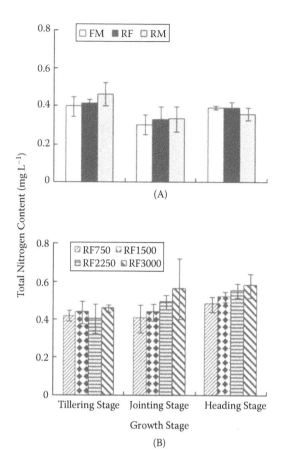

Figure 4.4 Total nitrogen in field water in (A) traditional and (B) fish-density-increased rice–fish co-culture systems. Values are means ± SE. RM, rice monoculture; RF, rice–fish system; FM, fish monoculture; RF750, RF1500, RF2250, and RF3000, rice–fish co-culture with target fish yields of 750, 1500, 2250, and 3000 kg ha⁻¹, respectively.

4.3.4.2 Total Phosphorus Content in Field Water

Like total nitrogen content in field water, there was no significant difference in total phosphorus (P) among rice monoculture (RM), rice–fish co-culture (RF), and fish monoculture (FM) systems when fish stocking density was low ($F_{2,9} = 0.376$; $p = 0.697$) (Figure 4.5A). Total P content in field water gradually decreased as the rice grew ($p < 0.05$) (Figure 4.5A). The high content of total P at the early stage of rice growth was likely caused by phosphorus used as basal fertilization. The gradual decrease over time was due to plant absorption of P. When fish density increased, total P content in the field water had significant differences in different months ($p < 0.05$). Total P content in field water under higher fish stocking density treatments increased significantly with the growth of rice, probably due to high feeding rates. The results of multiple comparisons showed that the total P content in field water had no significant difference among all treatments in either the tillering stage or the jointing stage ($p < 0.05$), while it increased with the increase of the fish target yields of different treatments at the heading stage. The total P content in field water significantly increased as the fish stocking density increased at the heading stage ($p > 0.05$) (Figure 4.5B). The experiment also indicated that when the target fish yield of rice–fish system reached 3000 kg ha⁻¹, the risk of water pollution increased significantly (Figure 4.5B).

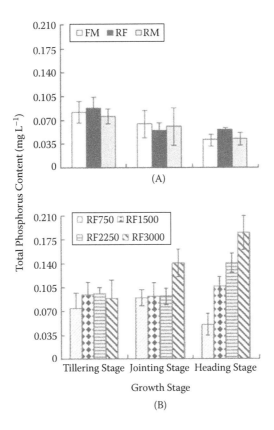

Figure 4.5 Total phosphorus in field water in (A) traditional and (B) fish-density-increased rice–fish co-culture systems. Values are means ± SE. RM, rice monoculture; RF, rice–fish system; FM, fish monoculture; RF750, RF1500, RF2250, and RF3000, rice–fish co-culture with target fish yields of 750, 1500, 2250, and 3000 kg ha^{-1}, respectively.

4.3.4.3 NH$_4^+$-N Content in Field Water

In traditional RFS, the decrease of NH$_4^+$-N in field water was similar to the trends of total P and total N in field water over time. There was no significant difference between any of the treatments ($F_{2,9} = 0.341$; $p = 0.720$) (Figure 4.6A). In RFS with increased fish densities, the increasing trend of NH$_4^+$-N content in field water was similar to the total P content in field water. There were significant changes in NH$_4^+$-N content in field water over time ($p < 0.05$). The NH$_4^+$-N content in field water was significantly higher at the heading stage of rice than at the tillering stage and jointing stage in rice–fish fields with high fish stocking densities (Figure 4.6B). Multiple comparisons showed that there was no significant difference in NH$_4^+$-N content in field water between any of the treatments in the tillering stage or jointing stage ($p > 0.05$). In the heading stage, NH$_4^+$-N content in field water in the low fish stocking density treatment (RF750) was significantly lower than that in high fish stocking density treatments (RF1500, RF2250, and RF3000) ($p < 0.05$), and there was no significant difference between RF1500 and RF2250 ($p > 0.05$), although both were significantly lower than RF3000 ($p < 0.05$) (Figure 4.6B). These experimental results may be attributed to two reasons. On the one hand, through chemical processes, unused feed could increase NH$_4^+$-N content in field water. On the other hand, when fish density was high, fish feces could also be the main source of the increase of NH$_4^+$-N content in field water.

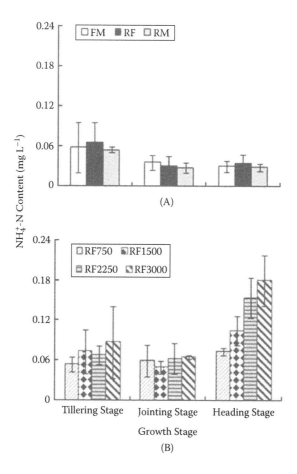

Figure 4.6　Ammonium-N in field water in (A) traditional and (B) fish-density-increased rice–fish co-culture systems. Values are means ± SE. RM, rice monoculture; RF, rice–fish system; FM, fish monoculture; RF750, RF1500, RF2250, and RF3000, rice–fish co-culture with target fish yields of 750, 1500, 2250, and 3000 kg ha^{-1}, respectively.

4.3.4.4 Chemical Oxygen Demand in Field Water

Chemical oxygen demand (COD) is the index for measuring the organic matter content of water. The results of multiple comparisons showed that the COD of field water did not vary significantly among fish monoculture (FM), rice–fish co-culture (RF), and rice monoculture (RM) in either the tillering stage or jointing stage ($p > 0.05$), and the COD in field water in FM was significantly higher than in RF and RM at the heading stage ($p < 0.05$) (Figure 4.7A), whereas there was no significant difference between RF and RM ($p > 0.05$). In RFS with increased fish densities, there was no significant difference between any of the treatments ($p > 0.05$). The COD in field water increased gradually in each treatment with the growth of rice ($p < 0.05$) (Figure 4.7B).

4.3.5 Greenhouse Gases

Rice–fish farming systems are thought to be globally important in terms of three environmental issues: climate change, shared water, and biodiversity. CH_4 and N_2O are the major greenhouse gases emitted from rice fields. Experiments showed that integration of fish in rice cultivation had impacts on the emission of these two greenhouse gases. For example, Frei et al. (2007) reported an increase in

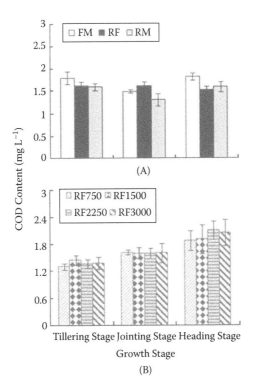

Figure 4.7 Chemical oxygen demand (COD) in field water in (A) traditional and (B) fish-density-increased rice–fish co-culture system. Values are means ± SE. RM, rice monoculture; RF, rice–fish system; FM, fish monoculture; RF750, RF1500, RF2250, and RF3000, rice–fish co-culture with target fish yields of 750, 1500, 2250, and 3000 kg ha^{-1}, respectively.

CH_4 emissions in rice–fish treatments that resulted from what can be called a biological perturbation effect created by the movement of fish. However, in a field experiment, N. Li (2013) found that mean CH_4 flux in RFS without fish feed input decreased by 26% compared to rice monoculture; yet, feed input increased CH_4 emissions. For N_2O emissions, Datta et al. (2009) found that although fish rearing increased CH_4 emissions fish stocking reduced N_2O emission from field plots. C.F. Li et al. (2009) also reported that N_2O emissions were reduced in rice–fish systems compared to rice monoculture systems.

4.4 ECOLOGICAL MECHANISMS UNDERLYING THE SUSTAINABILITY OF RICE–FISH SYSTEMS

Unlike pond aquaculture, fish share a habitat with rice in rice–fish farming systems. Studies showed that positive interactions and complementary nutrient use between rice and fish are the foundation of RFS and help explain why this system can remain productive over time with low levels of pesticide and chemical fertilizer inputs (Figure 4.8).

4.4.1 Facilitation between Rice and Fish

In a well-designed RFS, rice and fish can help each other. For example, rice plants protect fish from predation and from strong direct solar radiation by providing refuge and shade. Rice plants also create better aquatic environments for fish through nutrition absorption. Meanwhile, fish remove rice pests (rice planthoppers, rice sheath blight with possible pathogens and weeds) from

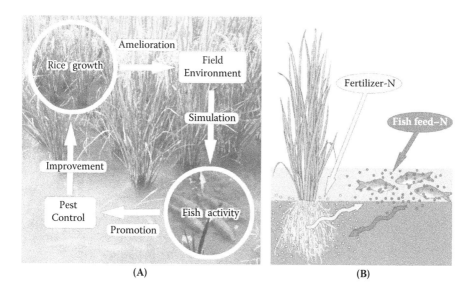

Figure 4.8   (See color insert.) Positive interactions and complementary use of nitrogen (N) between rice and fish explain why the rice–fish co-culture system can remain productive for long periods with low input of chemicals. (A) *Positive interactions between rice and fish:* Fish remove pests from rice through feeding activity, and rice plants moderate the field environment for fish which in turn promotes fish activity and pest removal. (B) *Complementary use of N by rice and fish:* Unused fertilizer N promotes plankton in paddy fields that is consumed by fish. The unconsumed fish feed acts as an organic fertilizer, and the N in the unconsumed feed can be gradually used by rice. Thus, rice and fish use different forms of N, resulting in a high efficiency of N utilization in RF.

the rice fields. The facilitation between rice and fish has multifunctional effects within the rice field ecosystem by reducing pests, stabilizing rice yields, and increasing the efficiency of nutrient use (Figure 4.8A).

4.4.1.1 Rice Provides a Better Environment for Fish

Rice plants can provide a better environment for fish by cleaning field water and providing shade during the hot summer season. For example, a 5-year field experiment showed that field water temperature was significantly lower in rice–fish fields than in fish monoculture fields. Compared to fish monoculture, the temperature and light intensity of surface water in RFS decreased 2.56°C ($F_{1,6}$ = 437.587, p = 0.000) and 1456.29 µmol m^{-2} s^{-1} ($F_{1,6}$ = 254.531, p = 0.000), respectively, from 12:00 p.m. to 2:00 p.m. (Xie et al., 2011a).

The NH$_4^+$-N levels in water were also significantly lower in rice–fish fields than in fish monoculture fields during the rice-growing season ($F_{1,30}$ = 10.620, p = 0.000) (Figure 4.9A). Total soil nitrogen tended to increase in fish monoculture but did not substantially change in rice–fish farming systems during the 5-year experiment (Figure 4.9B). Total soil N accumulated over time in fish monoculture was greater than in rice–fish farming systems at the end of the experiment ($F_{1,30}$ = 2.783, p = 0.044) (Figure 4.9B). RFS with low ammonia-N accumulating in water and low total soil N accumulation compared to fish monoculture suggests that rice–fish co-culture can reduce the risk of environment pollution.

The micro-environment ameliorated by rice plants was favorable for fish, which contributed to an increase in fish activity, especially during the hot season. Compared to fish monoculture, fish in rice fields were more active. The frequency of fish activity was 57.6% higher in RF than in FM (p < 0.05) when the air temperature was around 35°C (Xie et al., 2011a).

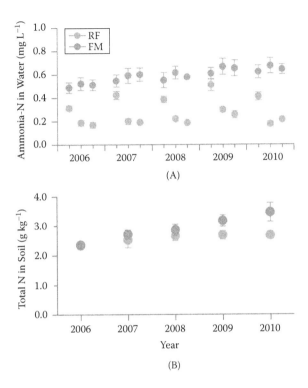

Figure 4.9   (A) Ammonium-N in water, and (B) total N in soil. RF, rice–fish co-culture; FM, fish monoculture. Error bars are SE.

4.4.1.2 *Fish Remove Pests of Rice*

Insect pests, diseases, and weeds are important factors inhibiting rice growth. Many experiments have shown that fish can encourage rice growth by partially removing these pests from the rice–fish system. Xiao et al. (2001) reported that fish can disrupt or eat the mycelia of the pathogenic fungus *Rhizoctonia solani* Kühn and thus inhibit the development of rice sheath blight. Wang et al. (2007) showed that, besides sheath blight, the incidence of rice blast disease and *Ustilaginoidea virens* (Cooke) Tak. in rice–fish fields was lower than in conventional rice fields with rice monoculture.

A number of studies showed that stocking fish in rice fields can help to remove insect pests. Vromant et al. (1998) demonstrated a pest controlling effect in the case of rice caseworm, but they found no significant effect in a further study on leafhoppers and planthoppers (Vromant et al., 2002). However, Sinhababu and Majumdar (1981) found that the common carp feeds on brown planthopper (*Nilaparvata lugens*). Vromant et al. (2003) found fish reduced populations of *Nymphula depunctalis*, and Frie et al. (2007) found that fish significantly reduced the number of dipteran insects. Recently, more and more experimental evidence has suggested that fish could reduce pest populations in a rice–fish culture. Experiments in China indicated a reduction of 30 to 60% in planthopper populations (Xie et al., 2010, 2011a). Further experiments indicated that the reduction in rice planthoppers was partly due to fish bumping rice plant stems. This bumping caused rice planthoppers to fall into the water where they were possibly consumed by the fish. The removal rate of rice planthoppers by fish was 26% (Xie et al., 2011a).

The effectiveness of fish as a biocontrol agent, however, depends on how well they are distributed within a rice field. If fish stay mostly in the pond refuge, then they cannot be effective in controlling rice pests. Halwart et al. (1996) found that in rice fields provided with a pond refuge that made up 10% of the field and stocked with either *Cyprinus carpio* or *Oreochromis niloticus*, more fish were present among the rice plants than in the pond. Because feeding is a major impulse for the diurnal activity of fish, the distribution pattern supports the hypothesis that fish are potentially important in pest control.

Fish in rice fields are capable of removing weeds by eating or uprooting them. A field trial by Patra and Sinhababu (1995) in India confirmed a reduction of weed biomass by 39% due to the presence of the common carp fry. Rothuis et al. (1999) reported a reduction of up to 100% of submerged and floating weeds, which are readily available as fish feed. Frei and Becker (2005a) noted a complete elimination of filamentous algae in rice plots due to consumption by Nile tilapia and common carp. In a 5-year field experiment, fish substantially reduced weeds by 91.25% (Xie et al., 2011a).

4.4.2 Complementary Use of Resources between Rice and Fish

Living together in RFS, rice and fish can use resources in complementary ways, thus reducing both fertilizer input and fish feed input. On the one hand, rice plants can use the nutrients from fish waste or unconsumed fish feed. For example, Chakraborty and Chakraborty (1998) reported that 75 to 85% of the nitrogen in the fish waste was composed of ammonium ions that can be absorbed directly by rice plants. Experiments also indicated fish can use bioresources (e.g., weeds, algae, phytoplankton) and some agricultural by-products as food. These results suggest that fish can transform N inaccessible to rice into a form that is easily absorbed by rice. Another field experiment that we conducted can help to explain why RFS can reduce the use of N fertilizers. In this experiment, the impact of N inputs on rice monoculture (RM), rice–fish co-culture (RF), and fish monoculture (FM) systems was examined. The experiment found that rice yield was significantly higher in RF plots with fish feed input than in RF plots without feed input. Rice yield also tended to be higher in RF plots with feed than in RM plots, even though N input was 36.5% higher in the RM plots than in the RF plots with fish feed. Fish feed input significantly increased fish yields in both FM and RF plots. Only 11.1% and 14.2% of the N in fish feed was assimilated into fish bodies in RF and FM, respectively. In RF, however, rice plants used the unconsumed N in fish feed and reduced fish feed N in the environment (i.e., in soil and water). A comparison of RF with and without fish feed indicated that 31.8% of the N contained in rice grain and straw was from fish feed (Xie et al., 2011a).

In turn, fish can use the bioresources (e.g., weeds, algae, phytoplankton) that increased due to fertilizer use. Fish can also use some agricultural by-products. In the experiment described above, subtraction of the fish N in RF from the fish N in FM indicated that 2.1% of the fertilizer N was assimilated into fish bodies in the RF. Thus, integrating rice with fish results in more efficient nutrient use through both complementary nutrient use and recycling. Many of the by-products in rice fields can serve as fertilizer and feed inputs to either rice or fish. This in turn leads to a cleaner and healthier rural environment.

4.4.3 Fish and Rice Together Improve the Soil Environment

Aquatic organisms in rice fields have an obvious influence on the soil environment (temperature, dissolved oxygen, and nutrients). For example, the dissolved oxygen content in rice–fish fields is significantly higher than in rice monoculture fields, which is not only conducive to the growth of fish but also advantageous to rice root growth and development by improving the field soil aeration. In addition, fish activities stir up surface soil, thus increasing water and soil temperatures, a result that is beneficial to rice growth in late spring when air temperatures are usually higher than water and soil temperatures. Cagauan (1995) documented that fish can increase soil porosity by stirring up the soil, making nutrients in the soil more likely to contact and be efficiently absorbed by the fibrous roots of rice.

Numerous studies have found that raising fish in rice fields can increase available soil nitrogen and phosphorus content significantly. Soil organic matter and other nutrient contents also tend to increase (Panda et al., 1987; Datta et al., 2009), but NO_3-N and total N in soil tended to decrease (C.F. Li et al., 2009). Oehme et al. (2007) found that soil nutrients increased with fish feed input in the rice–fish system. Huang et al. (2001) also found that soil organic matter, total N, and total P content increased by 15.6 to 38.5% in rice–fish systems compared to rice monoculture systems. The reasons why fish improve the nutrient content in rice–fish fields may include the following:

1. Fish waste directly increases the amount of available nutrients in the paddy fields (Vromant et al., 2001).
2. Fish disturb the soil layer to the extent that the nutrients fixed in the soil are released (Cagauan, 1995). At the same time, the dissolved oxygen concentration increases and improves soil reduction/oxidation (REDOX) conditions, which may promote N mineralization and nitrification (C.F. Li et al., 2009).
3. Fish waste contains abundant and easily degradable organic carbon that is favorable for microbial growth, which would promote the nutrient cycle and the activation of soil nutrients (Rochette et al., 2000).
4. The feeding behavior of fish directly inhibits the growth of weeds and plankton and reduces nutrients used by these organisms, leading to more nutrients remaining in the water and soil (Oehme et al., 2007); however, Vromant and Chau (2005) found that fish reduce phosphate content, although the NO_3^--N and NH_4^+-N levels in the water were unchanged.

4.5 TECHNOLOGY FOR IMPROVING RICE–FISH SYSTEMS

Achieving acceptable and sustainable levels of rice and fish production in RFS requires a technology package that is much more sophisticated than that required for either rice or fish monoculture. This technical package should include the following components: (1) installation of temporary physical structures such as trenches and pits to protect the fish during field operations and to prevent them from escaping; (2) rice and fish varieties more adapted to RFS, including rice varieties that are adapted to water that is deeper than in rice monoculture and fish varieties that are adapted to water that is shallower than in fish monoculture; and (3) daily field management procedures, including the coordination of irrigation, fertilization, pest control, and fish feeding.

4.5.1 Modification of Rice Fields for Rice–Fish Co-Culture

Raising fish in a rice field requires several physical modifications of the field (e.g., embankments, fish refuges) for fish habitation. These field modifications reduce the area for rice cultivation. Thus, a proper field configuration for rice culture and fish should be considered and designed. Rice field embankments are typically low and narrow because the usual rice varieties do not require deep water. To make the rice field more suitable for fish, the height of the embankment needs to be increased in most cases. Usually, embankments with a height of 40 to 50 cm (measured from ground level to crown) are satisfactory because, although the water level for rice does not normally exceed 20 cm, these taller embankments are sufficient to prevent most fish from jumping over.

A fish refuge (e.g., trench, pit) is a deeper area provided for the fish within a rice field. This can be in the form of a trench or several trenches, a pond, or even just a sump or a pit. The purpose of the refuge is to provide a place for the fish in the event water in the field dries up or is not deep enough. It also serves to facilitate fish harvest at the end of the rice season or to contain fish for further culture during rice harvest (Halwart et al., 1998). In conjunction with the refuge, trenches are often made to provide the fish with better access to the rice field for feeding (Halwart and Gupta, 2004).

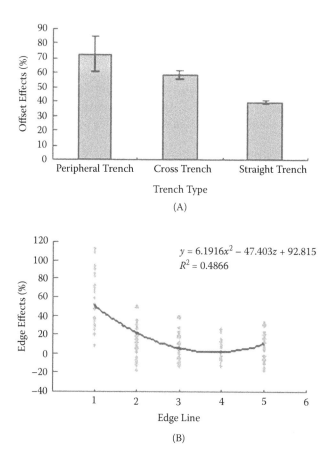

Figure 4.10 (A) Edge effect on rice yield, and (B) compensation of edge effects on rice yield under different types of trenches. Edge lines indicate rice rows.

Whether or not constructing fish refuges reduces rice yields has been a concern in the development of RFS. Several studies indicate that refuges definitely reduce the field area available for rice cultivation, but a proper size refuge will not reduce rice yields (Wu et al., 2010). For example, when the trenches are just wide enough and deep enough to safely accommodate all fish during drainage, transplanting, or harvesting, the trenches usually require only the removal of two rows of rice seedlings. In this manner, the trenches do not significantly affect the production of the rice crop. The trenches should be approximately 40 to 50 cm wide, with a suggested minimum depth of 50 cm (Koesoemadinata and Costa-Pierce, 1992; Halwart and Gupta, 2004). A proper refuge may not reduce rice yields due to the "edge effect." An experiment showed that a positive edge effect of rice yields occurred in the trenches (Figure 4.10A). The edge effect can somewhat compensate for the loss in yield caused by the reduction of planting space occupied by refuges (Figure 4.10B).

There are several ways in which the trenches could be developed. The simplest way involves digging a central trench longitudinally in the field. China has practiced digging trenches in the shape of a cross and even a "double-cross," a pair of parallel trenches intersecting with another pair. An experiment in the Philippines reported a design with a 1-m-wide central trench with water from a screened inlet flowing directly into it from a narrow peripheral trench. Another experimental design in the Philippines used an "L-trench" with a width of 3.5 m, involving two sides of the rice field, and occupying 30% of the rice field area (Halwart and Gupta, 2004).

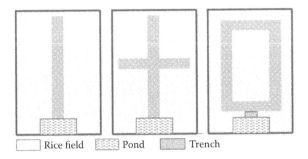

Rice field   Pond   Trench

Figure 4.11 Examples of various trench–pond layouts in rice–fish farming.

Different trench designs can result in different edge effects on the rice yields of trenches. A field experiment was conducted to evaluate how three trenches (central trench, cross trench, and circle trench) affected the rice yield (Figure 4.11) (Wu et al., 2010). The results showed that average compensation rate of the edge effect on rice yields was above 80%. Compensation rates for central trenches, cross trenches, and circle trenches were 58.02%, 85.58%, and 95.89%, respectively (Figure 4.10B). There were no significant differences between the fish yields among the three trenches (Figure 4.12).

4.5.2 Rice Varietals

Nutrient cycling and the biocommunity may change when raising fish in rice paddies; thus, rice cultivation (selection of rice varietal, propagation, irrigation, and fertilization) in RFS will differ from that in a rice monoculture. A compatible rice variety should be selected based on the following: (1) tolerance to lodging in order to prevent the negative impacts of possible lodging caused by fish activity and a high water table; (2) tolerance to high soil fertility that often occurs because of unconsumed fish feed and fish excrement; and (3) tolerance to diseases, insect pests, and weeds so that pesticides can be reduced or eliminated.

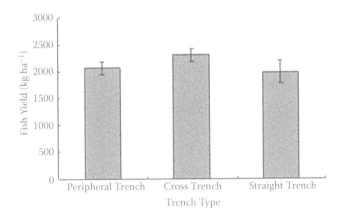

Figure 4.12 Fish yields under various types of trenches.

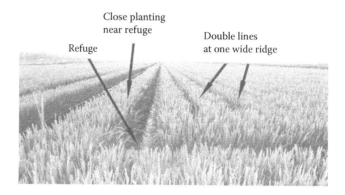

Figure 4.13   Diagrammatic pattern of rice planting for rice–fish systems.

4.5.3 Planting Patterns

Rice planting patterns and rice density are important for both rice yields and fish yields in RFS. For example, a planting pattern that incorporates alternating narrow and broad row spacing (Figure 4.13) will generate a better rice plant population structure that also allows fish to easily move around the rice field. When raising fish in rice fields, rice seedling density is reduced so that fish can easily move among rice plants. An experiment showed that lower seedling density (30 cm × 35 cm) did not reduce rice yields but did increase fish yields by 50% (Figure 4.14) compared to regular seedling density (20 cm × 25 cm).

4.5.4 Water Management

It is important to adopt a proper water management scheme in RFS because rice and fish have different water requirements. In rice cultivation, intermittent irrigation is better for the growth of rice plants, but it may not be the best option for rice–fish co-culture, as it requires concentrating the fish in trenches or pits every time the rice field is dried out. Continuous submergence is preferable in rice–fish co-culture, so the rice field is kept flooded from transplanting to about 2 weeks before harvest.

Rice is sensitive to water at the tillering and reproduction stages. Generally, shallow water is suitable for tillering, while the reproduction stage requires deeper water. Thus, the depth of water should be adjusted according to rice growth stages and fish needs. For example, a water depth of

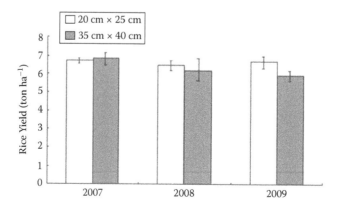

Figure 4.14   Effects of rice plant density on rice yield.

Table 4.5 Yields of Rice and Fish in Three Experiments

	Rice Yield (t ha⁻¹)		Fish Yield (kg ha⁻¹)	
Treatment	Grain	Straw	Gross*	Net**
RF25%	5.67 ± 0.09^a	3.84 ± 0.10^a	655.62 ± 75.08^b	338.37 ± 70.71^b
RF44%	5.36 ± 0.12^a	3.49 ± 0.24^a	829.01 ± 34.84^{ab}	516.26 ± 37.22^{ab}
RF63%	5.37 ± 0.29^a	3.21 ± 0.07^a	1024.20 ± 73.82^a	689.65 ± 79.07^a

Note: Values are means ± SE. Numbers with different letters are statistically different based on LSD ($p \leq 0.05$). RF25%, RF with 75% fertilizer N and 25% fish-feed N; RF44%, RF with 56% fertilizer N and 44% fish-feed N; RF63%, RF with 37% fertilizer N and 63% fish-feed N.

* Gross yield was the total fresh fish weight per ha at harvest.
** Net yield was calculated by subtracting fresh fish weight before stocking from gross weight at harvest.

6 to 8 cm is reasonable between transplanting and late tillering, while the fish are still small. The water depth should be increased to 15 to 20 cm at the early reproduction stage, when the fish also need deep water.

One experiment suggests a suitable standing water depth of 15 to 20 cm for RFS (Halwart and Gupta, 2004; Yang et al., 2010). At that depth, and assuming the fish refuge has a depth of 50 cm below field level, there is an effective water depth of 65 to 70 cm available to the fish in the refuge. This is sufficient to provide the fish with a cooler area when shallow water over the rice field warms up to as high as 40°C during the summer. Field experiments also showed that this depth (15 to 20 cm) did not affect rice growth and did not reduce rice yield (Yang et al., 2010).

4.5.5 Fertilizer Management

Application of fertilizers, organic or inorganic, is beneficial to both rice and fish. It has been thought that rice–fish farming might consume 50 to 100% more fertilizer than rice monoculture. Additional fertilizer is necessary to support phytoplankton production as the base of the food chain for fish culture; however, experiments indicate that the presence of fish in the rice field may actually boost fertility in the field and lower the need for fertilizer, because part of the unconsumed fish feed can be used by the rice. Thus, fertilizers for rice should be integrated into fish feed input. One experiment compared the ratios of fertilizer N and fish-feed N under a constant total N that meet both rice and fish needs. With a total N input at the same level of 250 kg ha⁻¹, the ratio of fertilizer N to fish-feed N was optimum (37% and 63%) and could increase fish yields without reducing rice yields (Table 4.5) or the efficiency of N use and without increasing the release of N into the environment (Figure 4.3).

4.5.6 Pest Management

Although raising fish in rice paddies can reduce the occurrence of diseases, insect pests, and weeds, pest control cannot be neglected for rice in RFS. The general methods for pest control include light traps, biocontrols, and chemical controls with low-toxicity pesticides.

4.5.7 Fish Varieties

Fish species for stocking in rice fields should be capable of tolerating a harsh environment. Unlike fish ponds, field environments are characterized by shallow water, high temperatures (up to 40°C in summer) and variable temperatures (range of 10°C in one day), low oxygen levels, and high turbidity (Halwart and Gupta, 2004). Only highly tolerant species can survive in this kind of environment. Fast growth is also an important characteristic, so the fish can attain marketable size by the time the rice is ready for harvest. Although there are very few commercially valuable species suitable for such adverse environmental conditions, recent rice–fish farming practices around the world have revealed that practically all of the major freshwater species now being farmed, including a salmonid and even a few brackish water species, have been successfully raised in rice field ecosystems along with several crustacean species. In China, cyprinids and tilapias are the two groups of fish growing in traditional rice–fish farming systems. Carp family species (cyprinids), particularly the common carp and *Carassius*, have the longest documented history for rice–fish cultivation in China. In recent years, however, carp, crab, shrimp, soft-shell turtle, loach, and other valuable species have been grown successfully in rice fields throughout China.

4.5.8 Stocking Pattern and Density

Like aquaculture using fishponds, rice–fish co-culture may involve the stocking of young fry for the production of fingerlings (nursery operation) or the growing of fingerlings into marketable fish (grow-out operation). Rice–fish farming may be either the culture of only one species (monoculture) or a combination of two or more species of fish and crustaceans (polyculture). Thus, the stocking density varies depending on the type of culture as well as the number of species used. A final factor determining the stocking is the type of modification in the rice field and the area for fish activity. The stocking density negatively affected the survival rate and average body weight (ABW) of small fingerlings, although final fish yields increased; however, high stocking density of fingerlings around 10 g each did not decrease the survival rate and did increase final fish yields (Table 4.6). Another experiment showed that increased fish density did not reduce rice yields and did increase fish yields (Figure 4.15).

4.5.9 Fish Feed

Fish graze and feed on a wide range of plant and animal organisms; however, preferences vary between species as well as with developmental stages within species. For example, among cyprinids the common carp has the widest range in food options and can feed on a variety of plant and animal matter. Another important factor is the presence and abundance of food organisms; for example, juveniles of the rice-consuming apple snail (*Pomacea canaliculata*) may become a major food item of common carp in rice fields (Halwart et al., 1998).

Supplemental feeds often consist of what is locally available; consequently, rice bran is a common supplemental feed in practically all rice-producing countries (Halwart and Gupta, 2004). In Bangladesh, wheat bran and oil cake are used, as well; in the Philippines, where coconut is an

Table 4.6 Effects of Stocking Density of Fish on Fish Yield and Survival Rate

Stocking Density (number ha⁻¹)		Fish Yield (kg ha⁻¹)	Fish Survival Rate (%)	Rice Yield (kg ha⁻¹)
Large fry (50 g each)	3000	577.50 ± 129.15[a]	89.00 ± 4.25[a]	5875.05 ± 124.95[a]
	4500	1009.05 ± 128.40[ab]	89.97 ± 2.72[a]	7430.55 ± 367.50[b]
	6000	1159.5 ± 91.50[b]	52.47 ± 4.08[a]	7249.95 ± 348.00[b]
Small fry (10 g each)	22,500	848.25 ± 58.80[a]	55.93 ± 1.86[a]	7354.05 ± 352.95[a]
	33,750	1286.70 ± 98.25[b]	59.20 ± 4.50[a]	8011.35 ± 411.30[a]
	45,000	1159.50 ± 91.50[b]	52.47 ± 4.08[a]	8481.30 ± 714.90[a]

Note: Means with different letters are significantly different at the 5% level.

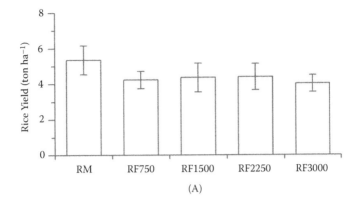

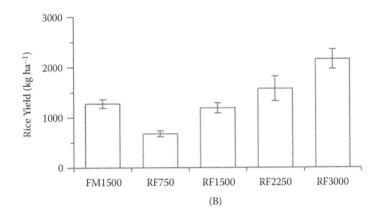

Figure 4.15 Effects of fish density on rice and fish yield. FM1500, fish monoculture with target fish yield of 1500 kg ha^{-1}; RF750, rice–fish co-culture with target fish yield of 750 kg ha^{-1}; RF1500, rice–fish co-culture with target fish yield of 1500 kg ha^{-1}; RF2250, rice–fish co-culture with target fish yield of 2250 kg ha^{-1}; RF3000, rice–fish co-culture with target fish yield of 3000 kg ha^{-1}.

important product, copra meal is employed. In China, feed may consist of wheat bran, wheat flour, oilseed cakes (e.g., rapeseed, peanuts, soybeans), grasses, and green fodder. In Malawi, maize bran and Napier grass are a few examples. Studies have shown that the use of supplemental feeding results in higher survival rates of 67% vs. 56.1% without supplemental feeding, and with a corresponding yield of 337.5 kg ha^{-1} compared to 249 kg ha^{-1} without supplemental feeding (Halwart and Gupta, 2004).

Recently, commercial feeds have been used to increase fish yields. A field experiment was conducted to compare how local feed, commercial feed, and the combination of local and commercial feeds affect fish yields. The results showed that fish yields increased as the ratio of commercial feed to local feed increased; the experiment also showed that feed application patterns affected fish yields (Figure 4.16). Applying commercial floating feed for the entire growth duration of the fish can significantly increase fish yields (Figure 4.16); however, fish quality would be improved by applying commercial feed at the early growth stage of the fish and then applying local feed (wheat bran, wheat flour, and oilseed cakes) at the late growth stage of the fish. In addition, water quality may become an important issue when employing supplemental feeds, as it can deteriorate rather quickly if the field is "overfed."

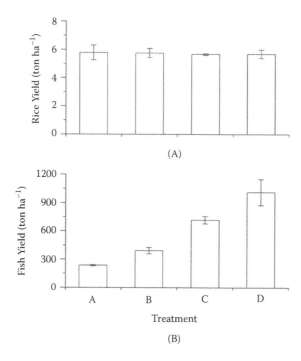

Figure 4.16　Effects of feed types on rice and fish yield. (A) Local feed for the entire growth duration (rice and wheat bran, 4.86 kg plot^{-1}). (B) 50% local feed for the early growth duration (2.43 kg plot^{-1}) and 50% formulated feed for the late growth duration (2.43 kg plot^{-1}). (C) Formulated feed for the entire growth duration (86 kg plot^{-1}). (D) Floating formulated feed for the entire growth duration (4.86 kg plot^{-1}).

4.6 PROSPECTS FOR RICE–FISH SYSTEMS IN THE FUTURE

Rice–fish systems could help increase food security, conserve resources, and protect the environment. The potential to increase the area under RFS cultivation worldwide is enormous, if RFS is adopted in the millions of hectares of rice fields with suitable conditions. Rice and aquatic food are essential components of local diets in many countries, including Egypt, India, Indonesia, Thailand, Vietnam, the Philippines, Bangladesh, and Malaysia. Integrated rice–fish farming has been practiced in these countries for some time but has failed to become a common practice. Now is an opportune time to promote rice–fish farming. First of all, the growth of fisheries has become constrained by limited resources in many areas. Increasing aquaculture production is one obvious solution to meet the growing demand for fish, and the world's rice fields represent millions of hectares of potential aquaculture areas. Second, there is a growing recognition of the need to work "with" rather than "against" nature. Integrated pest management (IPM) is being promoted as an alternative to extensive pesticide use, and fish have been found to be effective pest control agents. Chemical pesticide is a double-edged sword that can be as injurious to human health and the environment as it is to its targeted pests.

Third, freshwater is a limited resource and the integration of fish with rice is one way of using water more efficiently by producing both aquatic animals and rice in the same volume of water. In addition, new land suitable for aquaculture is limited and the culture of fish together with rice is an effective way of utilizing scarce land resources. Fourth, rice is not a purely economic crop. In many countries and especially in China, it is a "political crop," as well, because rice is one of the most important crops for national food security. The price of rice is not completely based on market demand, but is partially regulated by government policy, which does not always secure a just

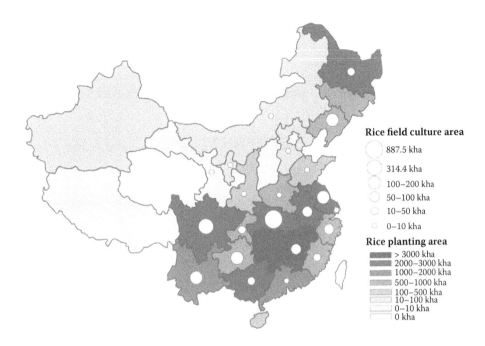

Rice field culture area

- 887.5 kha
- 314.4 kha
- 100–200 kha
- 50–100 kha
- 10–50 kha
- 0–10 kha

Rice planting area

- > 3000 kha
- 2000–3000 kha
- 1000–2000 kha
- 500–1000 kha
- 100–500 kha
- 10–100 kha
- 0–10 kha
- 0 kha

Figure 4.17 Rice planting areas and integrated rice–fish farming areas in provinces of China.

economic return to farmers. The price of fish, however, is usually based on market demand; hence, growing fish in a rice field can greatly increase farmers' income which can also encourage rice farmers to continue to cultivate rice, thus stabilizing rice production. Fifth, consumers' recognition of organic foods or "green" foods is increasing. Rice or fish products produced by RFS are welcome because these products experience little or no chemical inputs. All of these developments serve as an impetus for promoting rice–fish farming. Together, these trends address various concerns of all of the sectors involved in rice farming. Despite this, only 4.48% of the 27.4 million hectares of well-irrigated rice paddies are currently co-cultured with fish in China (Figure 4.17). There are several challenges to the development of rice–fish farming:

1. The greater water depth required in rice–fish farming than in traditional rice cultivation may be a limiting factor if the water supply is inadequate.
2. A small percentage of the cultivatable area is lost through the construction of drains and shelter holes, resulting in a reduction of the rice growing area and a decrease in rice yields.
3. The farmer requires more sophisticated technologies for both planting rice and culturing fish.
4. The farmer has to make a greater initial investment for installation, such as higher bunds, borders, drains, and shelter holes in the rice field.
5. Small-scale household farming makes it difficult to develop high-yielding, commercialized rice–fish systems.
6. Specialized machinery is still lacking for refuge construction, feed broadcasting, rice transplanting, and harvesting once a large-scale commercialized rice–fish system is developed.
7. Products (rice, grain, and fish) from rice–fish systems are not yet widely recognized and accepted by consumers.

For these reasons, greater effort is necessary to ensure the sustainable development of rice–fish farming, including (1) assessment of rice paddies as potential rice–fish systems (including factors such as water supply and water quality for the rice field); (2) evaluation of new fish species for rice field cultures and rice varietals for rice–fish co-culture; (3) development of technology packages for culturing rice and fish, including optimum fertilization rates for rice and optimum feeding rates for fish; (4) development of a field configuration for rice culture and aquaculture; (5) determining optimal rice planting patterns for rice–fish farming, as well as the carrying capacity and optimum fish stocking densities; (6) development of unique machines for rice–fish systems; and (7) helping farmers create product identities and brands that will garner consumer and societal acceptance.

REFERENCES

Berg, H. 2002. Rice monoculture and integrated rice–fish farming in the Mekong Delta, Vietnam: economic and ecological considerations. *Ecol. Econ.*, 41: 95–107.

Cagauan, A.G. 1995. Overview of the potential roles of pisciculture on pest and disease control and nutrient management in rice fields. In: Symoens, J.J. and Micha, J.C., Eds., *The Management of Integrated Freshwater Agro-Piscicultural Ecosystems in Tropical Areas.* Technical Centre for Agricultural and Rural Cooperation: Wageningen; Royal Academy of Overseas Science: Brussels, pp. 203–244.

Chakraborty, S.C. and Chakraborty, S. 1998. Effect of dietary protein level on excretion of ammonia in Indian major carp, *Labeo rohita*, fingerlings. *Aquacult. Nutr.*, 4: 47–52.

Chen, X., Wu, X., Li, N., Ren, W.Z., Hu, L.L., Xie, J., Wang, H., and Tang, J.J. 2011. Globally Important Agricultural Heritage System (GIAHS) rice–fish systems in China: an ecological and economic analysis. In: Li, P.P., Ed., *Advances in Ecological Research.* Jiangsu University Press: Jiangsu.

Datta, A., Nayak, D.R., Sinhababu, D.P., and Adhya, T.K. 2009. Methane and nitrous oxide emissions from an integrated rainfed rice–fish farming system of Eastern India. *Agric. Ecosyst. Environ.*, 129: 228–237.

Ding, W.H., Li, N., Ren, W.Z., Hu, L.L., Chen, X., and Tang, J.J. 2013. Effects of productivity improvement of traditional rice–fish system on field water environment [in Chinese]. *Chin. J. Eco-Agric.*, 21: 308–314.

Dwiyana, E. and Mendoza, T.C. 2008. Determinants of productivity and profitability of rice–fish farming systems. *Asia Life Sci.*, 17: 21–42.

FAOSTAT. 2013. Food and Agriculture Organization of the United Nations, Statistics Division, http://faostat3.fao.org/home/index.html#HOME.

Foley, J.A., Ramankutty, N., Brauman, K.A., Cassidy, E.S., Gerber, J.S., Johnston, M., Mueller, N.D., O'Connell, C., Ray, D.K., and West, P.C. 2011 Solution for a cultivated planet. *Nature*, 478: 337–342.

Frei, M. and Becker, K. 2005a. Integrated rice–fish culture: coupled production saves resources. *Nat. Resour. Forum*, 29: 135–143.

Frei, M. and Becker, K. 2005b. A greenhouse experiment on growth and yield effects in integrated rice–fish culture. *Aquaculture*, 244: 119–128.

Frei, M. and Becker, K. 2005c. Integrated rice–fish production and methane emission under greenhouse conditions. *Agric. Ecosyst. Environ.*, 107: 51–56.

Frei, M., Khan, M.A.M., Razzak, M.A., Hossain, M.M., Dewan, S., and Becker, K. 2007. Effects of a mixed culture of common carp, *Cyprinus carpio* L., and Nile tilapia, *Oreochromis niloticus* (L.), on terrestrial arthropod population, benthic fauna, and weed biomass in rice fields in Bangladesh. *Biol. Contr.*, 41: 207–213.

Halwart, M. and Gupta, M.V. 2004. *Culture of Fish in Rice Fields.* Food and Agriculture Organization and World Fish Center: Rome.

Halwart, M., Borlinghaus, M., and Kaule, G. 1996. Activity pattern of fish in rice fields. *Aquaculture*, 145: 159–170.

Halwart, M., Viray, M.C., and Kaule, G. 1998. *Cyprinus carpio* and *Oreochromis niloticus* as biological control agents of the golden apple snail *Pomacea canaliculata*: effects of predator size, prey size. *Asian Fish. Sci. (Philipp.)*, 11: 31–42.

Hu, L.L., Ren, W.Z., Tang, J.J., Li, N., Zhang, J., and Chen, X. 2013. The productivity of traditional rice–fish co-culture can be increased without increasing nitrogen loss to the environment. *Agric. Ecosyst. Environ.*, 177: 28–34.

Huang, Y.B., Wong, B.Q., Tang, J.Y., and Liu, Z.Z. 2001. Effects of rice–*Azolla*–fish symbiotic system on soil environment. *Chin. J. Eco-Agric.*, 9(1): 74–76.

Koesoemadinata, S. and Costa-Pierce, B.A. 1992. Development of fish–rice farming in Indonesia: past, present, and future. In: De la Cruz, C.R. et al., Eds., *Rice–Fish Research and Development in Asia*: *ICLARM Conference Proceedings*. ICLARM: Manila, pp. 45–62.

Li, C.F., Cao, C.G., Wang, J.P., Zhan, M., and Yuan, W.L. 2009. Nitrous oxide emissions from wetland rice–duck cultivation systems in southern China. *Arch. Environ. Contam. Toxicol.*, 56: 21–29.

Li, N. 2013. Ecological Analysis of Leading Rice–Fish Ecosystems in China, PhD thesis, Zhejiang University, Hangzhou.

Liu, J.H. and Yang, W. 2012. Water sustainability for China and beyond. *Science*, 337: 649–650.

Ministry of Agriculture of the People's Republic of China. 1980–2012. *China Agriculture Yearbooks*. Beijing: China Agriculture Press.

Ministry of Agriculture of the People's Republic of China. 2000–2012. *China Fishery Yearbooks*. Beijing: China Agriculture Press.

Oehme, M., Frei, M., Razzak, M.A., Dewan, S., and Becker, K. 2007. Studies on nitrogen cycling under different nitrogen inputs in integrated rice–fish culture in Bangladesh. *Nutr. Cycl. Agroecosys.*, 79: 181–191.

Panda, M.M., Ghosh, B.C., and Sinhababu, D.P. 1987. Uptake of nutrients by rice under rice-cum-fish culture in intermediate deep-water situation (up to 50-cm water depth). *Plant Soil*, 102: 131–132.

Patra, B.C. and Sinhababu, D.P. 1995. Weeds in rainfed lowland rice–fish systems. *Oryza*, 32: 121–124.

Ren, W.Z., Hu, L.L., Zhang, J., Tang, J.J., Sun, C.P., Yuan, Y.G., and Chen, X. 2014. Can positive interactions between cultivated species help to sustain modern agriculture? *Front. Ecol. Environ.*, 12: 507–514.

Rochette, P., Bochove, E., Prevost, D., Angers, D.A., Cote, D., and Bertrand, N. 2000. Soil carbon and nitrogen dynamics following application of pig slurry for the 19th consecutive year. II. Nitrous oxide fluxes and mineral nitrogen. *Soil Sci. Soc. Am. J.*, 64: 1396–1403.

Rothuis, A.J., Vromant, N., Xuan, V.T., Richter, C.J.J., and Ollevier, F. 1999. The effect of rice seeding rate on rice and fish production, and weed abundance in direct-seeded rice–fish culture. *Aquaculture*, 172: 255–274.

Sinhababu, D.P. and Majumdar, N. 1981. Evidence of feeding on brown plant hopper, *Nilaparvata lugens* (Stall) by common carp, *Cyprinus carpio* var. *communis* L. *J. Inland Fish. Soc. India*, 13(2): 16–21.

Tilman, D., Reich, P.B., and Knops, J.M.H. 2006. Biodiversity and ecosystem stability in a decade-long grassland experiment. *Nature*, 441: 629–632.

Tsuruta, T., Yamaguchi, M., Abe, S., and Iguchi, K. 2011. Effect of fish in rice–fish culture on the rice yield. *Fish. Sci.*, 77: 95–106.

Vromant, N. and Chau, N.T.H. 2005. Overall effect of rice biomass and fish on the aquatic ecology of experimental rice plots. *Agric., Ecosyst. Environ.*, 111: 153–165.

Vromant, N., Rothuis, A.J., Cuc, N.T.T., and Ollevier, F. 1998. The effect of fish on the abundance of the rice caseworm *Nymphula depunctalis* (Guenee) (Lepidoptera: Pyralidae) in direct seeded, concurrent rice–fish fields. *Biocontr. Sci. Technol.*, 8(4): 539–546.

Vromant, N., Chau, N.T.H., and Ollevier, F. 2001. The effect of rice-seeding rate and fish stocking on the floodwater ecology of the rice field in direct-seeded, concurrent rice–fish system. *Hydrobiology*, 445: 151–164.

Vromant, N., Nam, C.Q., and Ollevier, F. 2002. Growth performance and use of natural food by *Oreochromis niloticus* (L.) in polyculture systems with *Barbodes gonionotus* (Bleeker) and *Cyprinus carpio* (L.) in intensively cultivated rice fields. *Aquacult. Res.*, 33: 969–978.

Vromant, N., Nhan, D.K., Chau, N.T.H., and Ollevier, F. 2003. Effect of stocked fish on rice leaffolder *Cnaphalocrocis medinalis* and rice caseworm *Nymphula depunctalis* populations in intensive rice culture. *Biocontrol Sci. Technol.*, 13: 285–297.

Wang, H., Tang, J.J., Xie, J., and Chen, X. 2007. Effects of species co-existence on diseases, insect pests and weeds in rice field ecosystem. *Chin. J. Appl. Ecol.*, 18(5): 1132–1136.

Wu, X. 2012. The Utilization of Nutrients in Traditional Rice–Fish Co-Culture System, PhD thesis, Zhejiang University, Hangzhou.

Wu, X., Xie, J., Chen, X., Chen, J., Yang, X.X., Hong, X.K., Chen, Z.J., Chen, Y., and Tang, J.J. 2010. Edge effect of trench–pond pattern on rice grain and economic benefit in rice–fish co-culture. *Chin. J. Eco-Agric.*, 18(5): 995–999.

Xiao, X.C., Kan, X.L., Liu, Y.H., and Zou, Z.H. 2001. Raising fish in rice fields to control diseases, insect pests and weeds. *Jiangxi J. Agric. Sci. Technol.*, (4): 45–46.

Xie, J., Wu, X., Tang, J.J., Zhang, J., and Chen, X. 2010. Chemical fertilizer reduction and soil fertility maintenance in rice–fish co-culture system. *Front. Agric. China*, 4(4): 422–429.

Xie, J., Hu, L.L., Tang, J.J., Wu, X., Li, N., Yuan, Y., Yang, H.S., Zhang, J., Luo, S.M., and Chen, X. 2011a. Ecological mechanisms underlying the sustainability of the agricultural heritage rice–fish co-culture system. *Proc. Natl. Acad. Sci. U.S.A.*, 108: E1381–E1387.

Xie, J., Wu, X., Tang, J.J., Zhang, J., Luo, S.M., and Chen, X. 2011b. Conservation of traditional rice varieties in a Globally Important Agricultural Heritage System (GIAHS): rice–fish co-culture. *Agric. Sci. China*, 10(5): 754–761.

Yang, X.X., Xie, J., Chen, X., Chen, J., Wu, X., Hong, X.K., and Tang, J.J. 2010. Effects of different irrigation depths on yields of rice and fish in rice–fish system. *Guizhoug J. Agric. Sci.*, 38(2): 73–74.

You, X.L. 2006. Rice–fish culture: a typical model of sustainable traditional agriculture [in Chinese]. *Agric. Archaeol.*, 4: 222–224.

CHAPTER 5

Rice–Duck Co-Culture in China and Its Ecological Relationships and Functions

Zhang Jiaen, Quan Guoming, Zhao Benliang, Liang Kaiming, and Qin Zhong

CONTENTS

5.1 A REVIEW OF THE HISTORY AND CURRENT STATUS OF RICE–DUCK CO-CULTURE IN CHINA

5.1.1 Brief Introduction to Rice–Duck Co-Culture in China

China is a large country with a long agricultural history. Rice has been cultivated in China for more than 7000 years, and ducks have been domesticated there for about 3000 years (Xiong and Zhu, 2003). Raising ducks in paddy fields (known as rice–duck co-culture or integrated rice–duck farming) is a typical farming technique in traditional Chinese agriculture. Records of this technique, originally an attempt to use ducks to control locusts, date back to the Ming Dynasty (Zheng et al., 2005). Since then, this farming practice has become quite popular on rice farms in China.

Generally, the development of rice–duck co-culture in China can be divided into several stages, from its beginning to present day (H. Huang et al., 2003; Zheng et al., 2005). The first stage is usually described as one of *moving ducks to different paddy fields and various regions* and took place from the Ming Dynasty to the 1970s. In this stage, ducks were driven from one paddy field to another, as well as from one location to another over a long distance. Ducks were continually moved among different rice fields to graze during either the rice growing seasons or after the rice harvest. The main purpose of this practice was to give ducks access to natural feed such as weeds, pests, algae, phytoplankton, and micro-animals in the paddy fields, in addition to saving grain fodder and lowering the cost of duck production. The most significant characteristic of this co-culture system is that the rice planting and duck breeding processes were usually separate and managed by different farmers.

The second stage can be considered to be the *daytime rice–duck co-culture* that took place from the 1970s to the mid-1980s. This practice involved grazing ducks in several fixed paddy fields during the day and bringing them home after dark. In this farming system, ducks can continually eat weeds and prey on pests and insects during the day. The main goal of this method is to improve rice yields by raising ducks to control rice pests and weeds, and duck breeding is usually complementary. Although rice planting and raising ducks were combined and closely overlapped in this method, the two processes were still partially separate.

The third stage, *day and night rice–duck co-culture*, has been practiced from the late 1980s to the present. Here, a certain number of ducks are raised in a specific area of a fixed paddy field day and night. The rice and ducks grow together until the rice heading stage. The paddy field provides a good habitat and various kinds of natural foods for the ducks; in turn, ducks support rice growth in such ways as weeding, pest and disease control, muddying, and supplying nutrients. The complementary nature of this co-culture system allows the elimination of chemical pesticide and fertilizer use either partially or fully. Thus, the system can produce two kinds of high-quality foods simultaneously: rice and ducks.

5.1.2 Current Status of Rice–Duck Co-Culture Practices

It is well known that rice–duck co-culture is an ideal farming practice; however, due to rapid population growth, large food demands, and industrialization during the last century, pursuing high grain yields became a dominant goal. Rice–duck farming fell out of popularity, giving way to intensive agriculture that depends heavily on large inputs of chemical pesticides and fertilizers. Unfortunately—due to the long-term development of intensive chemical and industrial agriculture—environmental pollution, biodiversity loss, and food security are becoming increasingly greater threats to humankind today. In order to overcome these challenges, rice–duck co-culture has been gradually brought back and improved upon in many countries and regions since the 1990s. Over the last two decades, rice–duck integrated farming systems have been widely used in Asian countries including China, Japan, South Korea, Philippines, Malaysia, Thailand, and Vietnam, as well as in other countries and regions of the world. This farming system is therefore regarded as one of the common agricultural techniques of Asia (Shen, 2003). In present-day China, in order to meet increasing demands for pollution-free foods, rice–duck co-culture has become popular in almost all of the southern provinces where rice is the main crop, including Jiangsu, Hunan, Hubei, Guangdong, Jiangxi, Zhejiang, Guangxi, Fujian, Sichuan, Yunnan, Guizhou, and Chongqing. It can be also found in some northern parts of China, including Liaoning, Jilin, Heilongjiang, Xinjiang, and Ningxia provinces (J.E. Zhang, 2013).

5.2 STRUCTURE OF RICE–DUCK CO-CULTURE SYSTEMS

5.2.1 Basic Composition of Rice–Duck Co-Culture System

Various rice–duck co-culture practices can be found in different countries and regions. These differences are usually found in the choice of rice varieties, duck species, number of ducks per hectare of paddy field, application of chemical fertilizers, field management methods used for ducks, water and soil management methods for rice, and other factors. Regardless of these regional differences, there are still some basic commonalities in each system's composition and components (Figure 5.1).

Figure 5.1 (See color insert.) Rice–duck co-culture in the field.

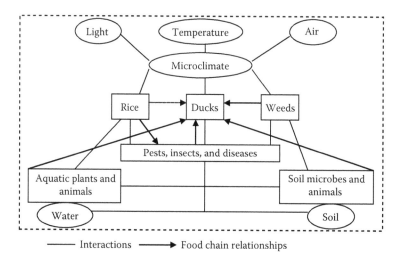

Figure 5.2 Composition of and relationships in integrated rice–duck co-culture systems.

From an ecological perspective, the rice–duck co-culture system can basically be divided into two kinds of components: environmental factors and biological factors. The environmental factors involve the soil, water, air, light, and microclimates present in the paddy field. The biological factors usually include rice, ducks, weeds, rice pests and their natural enemies, rice diseases, aquatic plants and animals, soil microbes, and soil animals. These environmental and biological factors make up an integrated ecosystem that can sustain different kinds of matter cycling, energy flow, and information exchange in the paddy field (Figure 5.2) (J.E. Zhang et al., 2002, 2006; Quan et al., 2005a).

To improve and strengthen ecological functions of the rice–duck farming system, other organisms are often artificially introduced into the paddy field together. For example, fish are usually combined with ducks in the paddy fields, creating what is known as a rice–duck–fish farming system. The addition of *Azolla* aquatic ferns results in a rice–duck–*Azolla* system or even a rice–duck–*Azolla*–fish farming system.

5.2.2 Structure and Practice of Rice–Duck Co-Culture

In practice, paddy fields are usually divided into different plots over a certain area; each plot is enclosed or fenced using plastic netting, bamboo, or other available materials. This enclosure is used to protect ducks from predators and to prevent them from escaping. The area of each enclosed plot is often around 0.267 to 0.333 ha (Figure 5.3). After the plots are fenced, ducks are released into each plot. The number of ducks used in the plot depends on a comprehensive analysis of several key factors, including duck species, weed density, pest levels, rice planting density, soil nutrient supply, and soil structure in a specific paddy field. Usually, the number of the released ducks in a paddy field varies from 150 to 375 individuals per hectare (J.E. Zhang et al., 2002, 2005).

In the integrated rice–duck farming system, different organisms occupy different niches. The vertical structure of the farming system can be divided into five layers from upper to lower: the upper rice canopy layer, middle canopy layer, lower canopy layer, water layer, and soil layer. Usually, most pests and their biocontrols are concentrated from the upper layer to the lower layer of the rice canopy; weeds exist in the lower or middle layers; aquatic plants and animals mainly concentrate in the water layer; and soil microbes and soil animals live in the soil layer. Ducks can touch almost all of the vertical layers in such a farming system.

Figure 5.3 Enclosure plots preventing ducks from escaping.

There is a general production procedure for the rice–duck co-culture (Figure 5.4). At the beginning, the duck eggs are hatched or brooded for ducklings while rice seeds are sown. About 7 to 10 days after transplanting rice seedlings, small ducklings that are about 1 week old are released into the paddy field at a concentration of approximately 225 to 375 ducklings per hectare. After that, the ducklings are raised in the paddy fields day and night until the stage of rice ear heading. This period usually lasts about 60 days and is referred to as the rice–duck co-culture phase. When rice growth is at the heading stage, the ducks should be taken out of the paddy fields in order to prevent them from eating rice ears, which concludes their role in supporting rice growth in paddy fields. About a month after duck removal, the rice is harvested (J.E. Zhang et al., 2002).

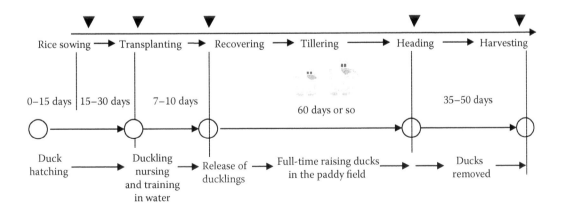

Figure 5.4 Basic production procedure of the rice–duck co-culture system.

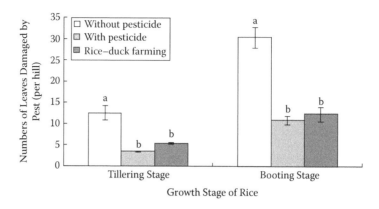

Figure 5.5   Rice leaves damaged by chewing pests under various treatments.

5.3 EFFECTS OF RICE–DUCK CO-CULTURE ON PESTS

It is well known that the duck is an omnivorous animal and a very strong predator that preys on insects, weeds, and small animals such as golden apple snails. Ducks eat some rice pests directly; they also press pests into the water with their wings and chase pests out of paddy fields. Finally, simply by their presence in the rice paddy ducks disturb the activities of pests (Takao, 2001).

5.3.1 Effect of Ducks on Rice Pests in Paddy Fields

Many related studies have reported that ducks have good control effects on rice planthopper, leafhopper, and stink bug populations, and they have certain impacts on rice borer, leafroller, and other pests. Generally, the total control effect of ducks on rice pests is no lower than the control effect of chemical pesticides (Lin and Jin, 2002; Quan et al., 2005b). Based on our experimental results, the numbers of leaves damaged by chewing pests in the rice–duck treatment were significantly lower than those in the pesticide-free treatment ($p < 0.05$). No significant difference was found between the rice–duck and pesticide treatments ($p > 0.05$) (Figure 5.5). The average numbers of leaves damaged at the tillering and booting stages of rice growth were 12.50 and 30.40 per m^2 sample, respectively, in the pesticide-free treatment, but only 5.40 and 12.30, respectively, in the rice–duck treatment (J.E. Zhang et al., 2009c).

The numbers of leaves rolled by rice leafrollers in the rice–duck co-culture treatment tended to be quite stable during the entire experimental period, but two peak periods of damage appeared in the other two treatments. The peaks of the damage in the pesticide-free treatment appeared at about 34 days and 59 days after transplanting, and at 39 and 55 days after transplanting in the pesticide treatment. At both the tillering and booting stages, the numbers of rolled leaves were significantly higher in the pesticide-free treatment than in those stages in the other two treatments ($p < 0.05$); however, there was no difference in the numbers of rolled leaves between the pesticide treatment and the rice–duck co-culture treatment at those stages ($p > 0.05$) (Figure 5.6) (J.E. Zhang et al., 2009c).

The numbers of rice plants killed by rice stem borers (*Chilo suppressalis* Walker) were significantly lower in the rice–duck farming treatment than in the pesticide-free treatment; yet, there was no significant difference between the rice–duck treatment and the pesticide treatment at the rice tillering stage ($p > 0.05$) (Figure 5.7). At the rice booting stage, the numbers of rice plants killed were significantly lower in the rice–duck treatment than in both the pesticide and pesticide-free treatments ($p < 0.05$) (Figure 5.7) (J.E. Zhang et al., 2009c).

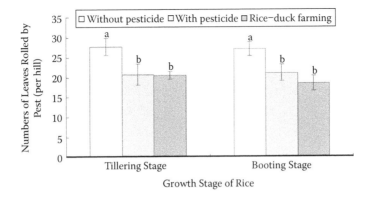

Figure 5.6   Rice leaves rolled by chewing pests under various treatments.

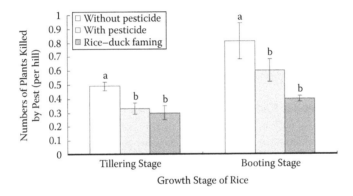

Figure 5.7   Plants killed by rice stem borer under various treatments.

No significant difference in panicle numbers per hill was found among the three treatments ($p > 0.05$) (Table 5.1). The numbers of filled grains per panicle, grain weights, and yield were significantly lower in the pesticide-free treatment than in the other two treatments ($p < 0.05$), but there were no significant differences in those indices between the pesticide and rice–duck treatments ($p > 0.05$) (J.E. Zhang et al., 2009c).

In conclusion, damage to rice plants by brown planthoppers, rice stem borers, leafhoppers, and leafrollers can be significantly reduced in rice–duck co-culture systems without pesticide application. Additionally, no reduction in rice yields occurred in rice–duck farming systems compared to conventional farming with chemical pest control.

Table 5.1   Rice Yield Components and Rice Grain Yield

Treatment	Panicles per Hill	Filled Grains per Panicle	Grain Weight (g/1000 grains)	Yield (kg ha⁻¹)
Without pesticide	14.6 ± 1.9^a	67.8 ± 7.4^b	15.88 ± 0.18^b	3036.95 ± 220.57^b
With pesticide	14.2 ± 1.5^a	115.3 ± 6.9^a	17.36 ± 0.24^a	4973.85 ± 582.22^a
Rice–duck farming	13.4 ± 0.9^a	124.9 ± 10.8^a	16.83 ± 0.17^a	4894.11 ± 325.21^a

Note: Numbers with different letters are significantly different at $p \leq 0.05$ (least significant difference test).

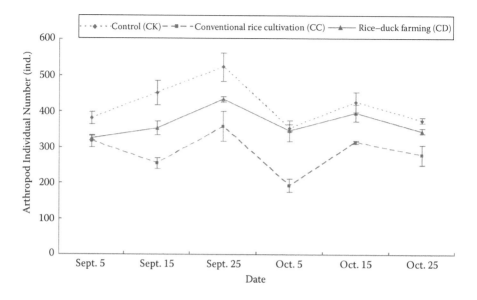

Figure 5.8 Variations in the number of arthropods in the three treatments.

5.3.2 Effect of Ducks on Paddy Field Arthropods

Whether or not ducks in rice fields could eliminate the entire arthropod community is important to the natural balance of the local insect community. Past research on the direct effects of ducks on the abundance and diversity of arthropod communities has generally been confined to the observations of interactions between specific pests and ducks. In order to further evaluate the impact of duck foraging on the overall diversity of arthropod communities in the rice–duck co-culture system, a study was conducted to investigate that impact in the field, using three treatments: integrated rice–duck cultivation (RD), conventional rice cultivation (CC), and a control treatment (CK) (Qin et al., 2011). The results showed that during the entire observation period individual numbers of arthropod species in the three treatment groups were found in the following order: CK treatment > RD treatment > CC treatment. Significant differences between RD and CK numbers were detected during the period of September 5 to 25. Differences in individual numbers of arthropod species between RD and CC were significant during the entire study period, except September 5 (Figure 5.8).

The data in Table 5.2 show that arthropod community biodiversity in all three treatments experienced a moderate decrease when rice growth moved from the tillering stage to the booting stage. No significant differences in arthropod community biodiversity were found between RD and CK treatments during the tillering and booting stage. Arthropod community biodiversity was significantly higher in the RD treatment than in the CC treatment by 11.69% in the tillering stage and 17.36% in

Table 5.2 Richness, Evenness, and Shannon Indices of Arthropod Community for Two Critical Growth Periods

| Treatment | Richness Index (*R*) | | Evenness Index (*J*) | | Shannon Index (*H′*) | |
	Tillering Stage	Booting Stage	Tillering Stage	Booting Stage	Tillering Stage	Booting Stage
CK	1.56 ± 0.04[a]	1.48 ± 0.05[a]	0.60 ± 0.04[a]	0.62 ± 0.03[a]	1.84 ± 0.01[a]	1.74 ± 0.02[a]
CC	1.36 ± 0.05[b]	1.19 ± 0.02[b]	0.52 ± 0.03[b]	0.54 ± 0.02[b]	1.62 ± 0.03[b]	1.47 ± 0.05[b]
RD	1.54 ± 0.02[a]	1.44 ± 0.02[a]	0.59 ± 0.01[a]	0.56 ± 0.02[b]	1.79 ± 0.03[a]	1.74 ± 0.03[a]

Note: Numbers with different letters are significantly different at $p \leq 0.05$ (least significant difference test).

the booting stage. Duck activity may have a minor effect on arthropod community biodiversity in late-growth rice fields when compared to the pesticide application treatment. There were no significant differences in the evenness index values between RD and CK during the tillering stage of late-growth rice. The evenness index value for RD was 11.86% higher than the CC treatment during the tillering stage, which contributed to significant differences between these two treatments. No significant differences were found between RD and CC during the booting stage (Table 5.2). No significant differences in Shannon index values were found between RD and CK during the tillering and booting phases of the late-growth rice. Shannon index values for RD were 11.47% and 15.52% higher than control, which were significantly higher than those in CC during the two stages (Table 5.2).

In summary, the numbers of arthropod individuals in both RD and CC treatments showed varying levels of reduction when compared with the CK treatment. However, the arthropod richness index (R), evenness index (J), and Shannon index (H') in the RD treatment were similar to those in CK and significantly higher than for the CC treatment in both tillering and booting periods of the late-growth rice. These findings indicate that rice–duck co-culture was favorable for maintaining arthropod diversity in the paddy field and may help to lessen the decline in abundance and diversity of rice farmland arthropods caused by the utilization of chemical pesticides and fertilizers.

5.4 EFFECTS OF RICE–DUCK CO-CULTURE ON PADDY FIELD WEEDS

The weeding effect is one of the most important and easily understood aspects of the power of ducks. Ducks can eat weeds and weed seeds directly from the mud, disturb the germination of weed seeds by moving and swimming, and reduce sunlight and oxygen levels in the water through constant plowing and muddying, thus preventing weed growth and seed germination (Takao, 2001). Many researchers have reported that ducks can effectively and sustainably control most weeds, except for barnyard grass. They observed that ducks have weed control rates of 95% or more, which is obviously better than that of chemical herbicide control (Liu et al., 2004; Quan et al., 2005b). In order to further investigate the impacts of duck activities on the local weed community, an experiment was conducted with three treatments: a rice–duck co-culture with duck grazing and disturbance effects, a rice–duck co-culture with duck disturbance effects only, and a control (rice monoculture, no duck effects). The main experimental results are described below (J.E. Zhang et al., 2009a).

5.4.1 Effect of Ducks on the Biodiversity of the Weed Community

The total number of weed species differed significantly ($p \leq 0.05$) among the treatments at both sampling stages (Table 5.3). At the rice heading stage, fewer species were found in the treatment with duck grazing and disturbance (2.33) and in the disturbance only (7.00), while 11.75 species were observed in the monoculture. At the harvesting stage, the number of weed species in the monoculture treatment was significantly higher ($p \leq 0.05$) than those in the treatments with duck

Table 5.3   Species Richness and Total Density of the Weed Community Under Different Treatments

Factor	Species Richness (N)		Total Density (no. of plants m^{-2})	
	Heading Stage	Harvesting Stage	Heading Stage	Harvesting Stage
Grazing and disturbance	2.33 ± 1.20[c]	6.00 ± 0.41[b]	1.5 ± 0.76[c]	13.63 ± 5.16[b]
Disturbance	7.00 ± 0.58[b]	6.25 ± 1.18[b]	8.67 ± 2.17[b]	14.67 ± 4.18[b]
Control	11.75 ± 1.65[a]	11.00 ± 0.71[a]	64.33 ± 25.04[a]	48.25 ± 16.50[a]

Note: Numbers with different letters are significantly different at $p \leq 0.05$ (least significant difference test).

Table 5.4   Total Biomass of the Weed Community

Treatment	Heading Stage (dry weight, g m^{-2})	Harvesting Stage (dry weight, g m^{-2})
Grazing and disturbance	3.32 ± 1.67[c]	1.16 ± 0.21[b]
Disturbance	22.96 ± 3.82[b]	2.20 ± 0.39[b]
Control	56.01 ± 10.07[a]	9.14 ± 2.02[a]

Note: Numbers with different letters are significantly different at $p \leq 0.05$ (least significant difference test).

grazing and disturbance and with disturbance only. No significant difference was found between the treatments involving ducks. The total density of the weed community differed significantly among the treatments at these two sampling stages (Table 5.3). At the heading stage, the lowest total density was found in the treatment with duck grazing and disturbance. A significantly higher density of the weed community was found in the monoculture control. At the harvesting stage, there was no significant difference in the density of the weed community between the treatment with duck grazing and disturbance and with disturbance only; however, the total density in the monoculture control was significantly higher than that in the other two treatments.

5.4.2   Effect of Ducks on Weed Biomass

The weed biomass differed significantly ($p \leq 0.05$) among treatments at both sampling dates (Table 5.4). The weed biomass in the rice–duck system with grazing and disturbance was significantly lower than that in the other treatments at both sampling stages. No significant difference in the weed biomass was found between the treatments with grazing and disturbance and with disturbance only at the harvest stage. The above experiments provide evidence that the weed community diverged after ducks were introduced to the rice field. At the rice heading stage, the control and duck disturbance-only plots were dominated by broadleaf weeds, whereas the plot with duck grazing and disturbance was dominated by grass. Water and soil disturbances caused by ducks walking in the rice field could inhibit weed germination and the growth of seedlings, thus impacting weed development indirectly in a rice–duck farming system (Q.S. Wang et al., 2004). Full-time walking, swimming, and plowing by ducks created a murky and muddy field, which resulted in inhibited germination and growth of weeds due to reduced light penetration of the water (Wei et al., 2006). In our experiment, in addition to the treatment with duck grazing and disturbance, the weed density and biomass were also reduced in the plot with duck disturbance only. Compared with the monoculture rice, the weed biomass was reduced by 98% in the treatment with grazing and disturbance on soil and water, and by 84% in the treatment without grazing but with soil and water disturbance only. This result implies that, in the rice–duck field, only 14% of the weeds were controlled by duck predation and 84% of the weeds were inhibited by duck activities (e.g., walking, plowing, mudding). Moreover, the weed seeds in the soil profile changed because of the constant duck movement in the field (Wei et al., 2006). The results suggest that duck activities, such as grazing weeds and disturbance of the soil and water, had a significant effect on the weed community structure and that weeds could be well controlled without herbicide application in a rice–duck farming system.

5.5   EFFECTS OF RICE–DUCK CO-CULTURE ON RICE DISEASES

5.5.1   Prevention and Control Mechanisms of Duck Activities on Rice Diseases

Generally, duck activities can effectively reduce incidences of rice diseases, either directly or indirectly, and usually via the following pathways. First, the activities of ducks can change the microclimate of the rice canopy in paddy fields as they move around together. Duck activities can improve levels of aeration, wind, sunlight, humidity, and temperature in the rice canopy, all of which are very closely related to the occurrence of rice diseases. An improved microclimate can disturb and change the conditions leading to outbreaks of rice diseases. This means that duck activities can indirectly result in mitigating the outbreak of some rice diseases. Second, ducks can reduce the spread of paths or sources of some anthophilous diseases by eating certain pests and septic or dead rice leaves. For example, because ducks can effectively suppress the brown rice planthopper and green leafhopper, which are causative pests, the incidence of rice stripe disease, rice black streaked dwarf virus, yellow dwarf disease, and verticillium wilt disease also decline to a certain extent synchronously. Third, duck activities can also stimulate rice growth and increase its lodging resistance (J.E. Zhang et al., 2013), which would indirectly enhance its resistance to some diseases of the rice plant. A lot of studies have already shown that activities of ducks can effectively decrease the outbreak of rice sheath blight, as well as demonstrate a good controlling effect on rice blast, bacterial leaf blight, and stripe virus disease (Tong, 2002; Quan et al., 2005b).

5.5.2 Effects of Different Rice–Duck Co-Culture Systems on Major Rice Diseases

Studies have shown that rice–duck co-culture is one of the methods of organic rice farming in which weeds, diseases, and insects can be effectively controlled with minimal or no pesticide and herbicide application; however, in conventional rice–duck farming systems, the controlling effects on diseases, insects, and weeds slowly decrease when ducks are taken out of the paddy field at the rice heading stage. In order to avoid a period where ducks are absent from the paddy field, we designed two new rice–duck farming systems that improve upon the conventional rice–duck farming system. In these new systems, two batches of ducks were raised with rice within one rice-planting season (Figure 5.9) (Liang et al., 2012).

To further examine the effects of different rice–duck co-culture systems on major rice diseases, pests, and weeds, a comparison experiment was conducted with four treatments: (1) conventional rice cultivation treatment (CR treatment), in which no ducks were reared; (2) conventional rice–duck pasturing (CD treatment), in which five 7-day-old ducklings were released into each plot at the beginning of the rice early growth stage after transplanting and were removed at the heading stage;

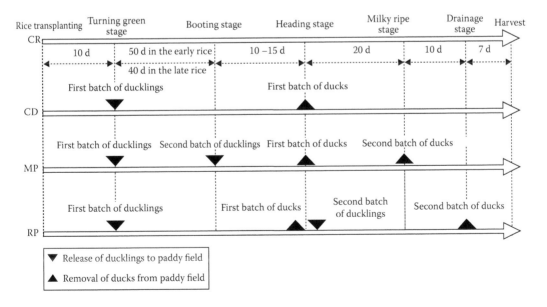

Figure 5.9 Flow diagram of various rice–duck farming systems.

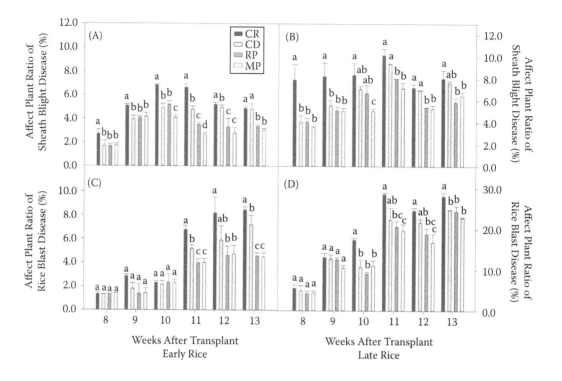

Figure 5.10 Dynamic changes of rice diseases in the different treatments described in the text.

(3) rotational pasturing treatment (RP treatment), in which two batches of ducklings were pastured in sequence during the entire rice season and the treatment of the first batch was consistent with CD, while the second batch (seven ducklings at 7 days old) was released into each plot at the rice heading stage (after the first batch was removed from the field) and removed at the paddy drainage stage; and (4) mixed pasturing treatment (MP treatment), in which the treatment of the first batch of ducks was consistent with CD and the second batch (six ducklings at 5 days old) was released into each plot 10 to 15 days before removal of the first batch of ducks and was driven out of the paddy field 10 days before the drainage stage (Figure 5.9).

The experimental results indicated that CD treatment was efficient at controlling sheath blight disease in both the early and late rice stages (Figure 5.10A,B). The ratios of disease-infected plants declined significantly ($p < 0.05$) by 22.1 to 37.8% in the early rice and by 15.7 to 49.2% in the late rice ($p < 0.05$), respectively, compared with the CR treatment. The effect was insignificant, however, after the heading stage when ducks were removed from the plots, although, compared with the CD treatment, the diseased plant ratios of RP and MP in the early rice decreased by 15.9% and 26.1%, respectively. For the late rice, the ratio decreased by 13.5% and 19.8% after the heading stage, indicating that both RP and MP treatments had a stable and long controlling effect on rice sheath blight.

As shown in Figure 5.10C,D, there was no significant difference between the number of plants infected by blast disease in the CR and CD treatments for most periods; however, after the rice heading stage, the incidence of blast disease in RP and MP was significantly ($p < 0.05$) lower than in CR treatments in both early and late rice stages. The ratio of disease-infected plants was significantly ($p < 0.05$) lower in RP than in the CD treatment in the early rice stage, and the ratio of diseased plants in MP was significantly ($p < 0.05$) lower than in CD in both rice stages. In summary, these three types of rice–duck co-culture systems all exhibited a certain influence over controlling rice sheath blight and blast in the paddy field. Moreover, impacts of the two-batch rice–duck systems on major rice diseases were significantly ($p < 0.05$) higher than those of the conventional rice–duck system. This improvement may mainly be due to the lengthening of pasture time and increasing effects of duck activities. These two new rice–duck co-culture systems may have potential application as bio-control agents in organic rice farming systems (Liang et al., 2012).

5.6 EFFECTS OF RICE–DUCK CO-CULTURE ON PADDY SOIL AND AQUATIC ENVIRONMENT

5.6.1 Effects of Rice–Duck Co-Culture on Soil Fertility and Soil Microbes

Duck activities can directly influence soil structure and fertility. First of all, ducks excrete feces directly onto the soil daily. During a 2-month period of co-culture, it is estimated that the total excreted feces per duck can reach 10 kg, which contains 47 g N, 70 g P, and 31 g K (Xiong et al., 2003). This is a good continuous supply of organic fertilizer deposited onto the soil. Second, duck stirring and intertillage can improve elements of the soil environment such as soil air, texture, and soil structure (Takao, 2001). Finally, duck activities can enhance the decomposition of soil organic matter and nutrient transformation, which benefits the growth of rice plants.

Most researchers report that rice–duck co-culture could improve most soil properties to some extent; for example, a case study was conducted regarding the effects of rice–duck co-culture on soil fertility (Yang et al., 2004). The results showed that compared with conventional paddy fields the soil organic matter, total N, available N, available P, and available K increased by 11.3 to 29.3%, 3.0 to 15.1%, 5.9 to 9.6%, 3.8 to 9.7%, and 23.4 to 27.3%, respectively, in the rice–duck farming system (Table 5.5). Physical soil characteristics were also improved. Bulk soil density decreased by 0.01 g cm^{-3}. The >0.25-mm soil aggregate and soil structure coefficient increased by 2.65 to 3.12% and 2.56 to 6.63%, respectively (Table 5.6). Meanwhile, the redox status of the soil was also markedly ameliorated.

Table 5.5   Chemical Characteristics of Soils in Rice–Duck Complex Ecosystem

Treatment	Organic Matter (g kg⁻¹)	Total N (g kg⁻¹)	Available N (mg kg⁻¹)	Available P (mg kg⁻¹)	Available P (mg kg⁻¹)
		Early Rice			
Rice–duck	39.38	2.189	198.54	68.69	131.46
Control	30.46	1.902	181.23	66.16	103.24
		Late Rice			
Rice–duck	40.92	2.189	191.15	42.40	106.84
Control	36.76	2.126	180.44	38.65	86.61

Note: Sampling time was the harvesting season of the early rice and late rice in 2001.

Table 5.6   Physical Characteristics of Soils in Rice–Duck Complex Ecosystem

Treatment	Bulk Density (g cm⁻³)	Aggregate >0.25 mm (%)	Aggregate <0.001 mm (%)	Dispersion Coefficient (%)	Structure Coefficient (%)
		Early Rice			
Rice–duck	0.926	19.36	1.38	5.36	94.64
Control	0.936	16.71	3.13	11.99	88.01
		Late Rice			
Rice–duck	0.980	15.14	2.23	7.84	92.16
Control	0.990	12.02	2.66	10.40	89.60

Note: Sampling time was the harvesting season of the early rice and late rice in 2001.

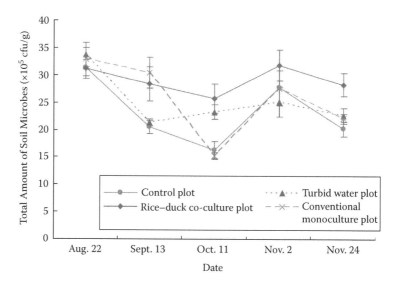

Figure 5.11   Effects of rice–duck farming on quantity of soil microbes.

It should be noted that when using rice–duck co-culture to produce organic or green foods that are required to be free of chemical fertilizers soil fertility should be maintained by an external supply of nutrition in the form of organic fertilizers, crop straw, or duck feed. Rice–duck co-culture not only can improve soil fertility but can also have positive effects on soil organisms. To further understand the effects of rice–duck farming on the population dynamics of microbial flora in paddy soil, a field experiment was conducted to investigate the changes of soil microbial communities by using an agar plate bioassay. The experiment was composed of four treatments: a rice–duck co-culture plot, a turbid water plot being muddied by ducks, a clear water plot without any duck activities or agrochemical materials input, and a conventional rice farming plot with chemical fertilizer and pesticide application. The results showed that the total amounts of culturable bacteria, actinomycetes, and fungi in paddy soil were significantly higher under the rice–duck farming treatment at the heading stage (October 11) and ripening stage (November 24) than those of conventional rice farming, the turbid water plot, and the clear water plot treatments (Figure 5.11) (J.E. Zhang et al., 2009b).

5.6.2 Effect of Rice–Duck Co-Culture on Aquatic Environment in Paddy Fields

Usually, ducks stir up soil and make water muddy using their feet and bills both day and night in the paddy fields. This muddying process can obviously change aspects of the aquatic environment including sunlight, temperature, dissolved oxygen, and nutrients, which can make the distribution of water and soil nutrients in the field more even. This muddying-water phenomenon can be mostly found in the rice–duck co-culture field. To investigate the effects of rice–duck farming on the aquatic paddy environment, a field experiment was conducted with two treatments: conventional rice monoculture and rice–duck co-culture. The results showed that under rice–duck farming, the temperature and pH value of the surface water in the paddy fields decreased. In the same treatment, electrical conductivity, oxidation–reduction potential, turbidity, and the contents of nitrogen (N), phosphorus (P), and potassium (K) increased. The total N, P, and K increased by 1.85 to 5.06 times, 2.01 to 8.70 times, and 42.79 to 109.18%, respectively, as compared to conventional rice monoculture (Figure 5.12). All of these results illustrate that the rice–duck co-culture improved the aquatic environment and nutrient supply of the paddy field, making this treatment very beneficial for rice growth (Quan et al., 2008).

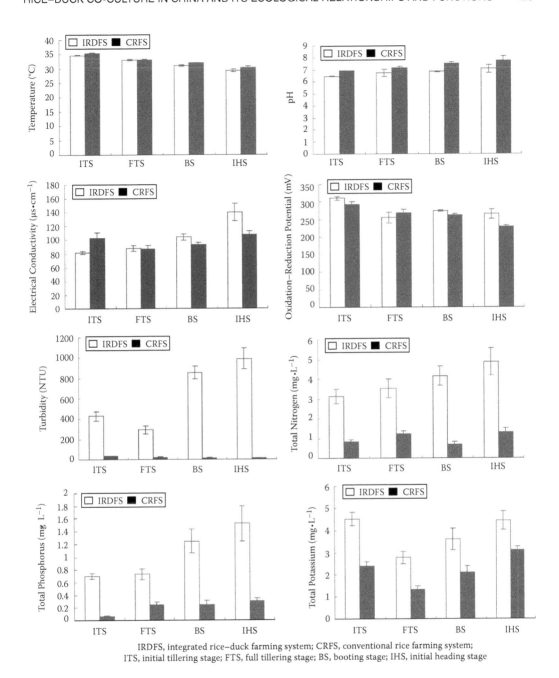

IRDFS, integrated rice–duck farming system; CRFS, conventional rice farming system;
ITS, initial tillering stage; FTS, full tillering stage; BS, booting stage; IHS, initial heading stage

Figure 5.12 Dynamics of temperature, pH, electrical conductivity, oxidation–reduction potential, turbidity, total nitrogen, total phosphorus, and total potassium of water in paddy field.

Furthermore, duck activities not only change the aquatic environment in the paddy field but can also alter the aquatic algal community. A related study showed that 6 algal phyla, 57 algal genera, and 108 species could be found in all experimental paddy water during sampling periods. The phyla found were Cyanophyta, Chlorophyta, Bacillariophyta, Euglenophyta, Cryptophyta, and Pyrrophyta. The difference of the algal numbers in the two treatments was not significant, but the algal density and biomass in the rice–duck (RD) complex ecosystem were lower than in the conventional (CK) paddy ecosystem. The algal biomass increased in paddy water with the growth of rice. The predominant algal species during the early periods of rice growth were *Merismopedia glanca*, *Sceaedsmus quadricauda*, *Euglenav iridis*, and *Cryptomonas erosa*, but in the late periods (45th day after putting ducks in the paddy field) the predominant algal species in CK were *Closterium acerosum*, *Cosmarium quadrum*, and *Cosmarium formosulum* (belonging to the Chlorophyta phylum), and the predominant algal species in RD switched to *Euglena iridis*, *Euglena oxyuris*, and *Phacus longicauda* (belonging to the Euglenophyta phylum). Compared with CK, the algal density and biomass in RD decreased significantly within 15 days after putting ducks into the paddy field. Moreover the genus diversity was reduced during the late periods of rice growth (45th day after putting ducks in the paddy field) (J.P. Wang et al., 2009).

5.7 STIMULATING EFFECTS OF DUCK ACTIVITIES ON RICE PLANTS

The ducks constantly stimulate most parts of the rice plant by touching and rubbing against rice plants with their whole body, striking the rice leaves by flapping their wings while swimming and running, poking their heads inside the rice stalks, rubbing against the roots and stalks of rice plants, and removing the mud at the base of the stalks with their bills and feet. This physical stimulation can cause some changes of shape, height, stalk thickness, effective tillering, and other growth characteristics of rice plants (Takao, 2001; Shen, 2003; J.E. Zhang et al., 2007, 2011b; Wang et al., 2008; Z.X. Huang et al., 2012; J. Zhang et al., 2012). In order to investigate the impacts of duck activities—including touching, rubbing, pecking, flapping, and diving—on the growth characteristics of rice plants, field and laboratory experiments were conducted to investigate the mechanical stimulations of duck activities on rice physiological characteristics, and hence on rice morphology and anatomy (J.E. Zhang et al., 2013).

5.7.1 Effect of Mechanical Stimulation by Duck Activities on Rice Morphology

The internode of the rice culm is composed of a coat or epidermis, mechanical tissue, thin-wall tissue, and vascular bundle. There are four internodes for each culm (N1 to N4). The culm wall thicknesses of all of the four internodes of individual rice plants were checked for both treatments. There was a distinct difference in the culm wall thickness between the RD (rice with duck) and RND (rice with no ducks) treatments, except at internode N3. The RD treatment was 35% thicker than RND treatment for internode N4. The resistance strengths to lodging or breakage of internodes N2 and N3 were 101 and 119 g higher, respectively, in the RD treatment than in the RND treatment, while the lodging indexes of internodes N2 and N3 were 21 and 15 smaller in the RD treatment than in the RND treatment, respectively (Table 5.7).

5.7.2 Effect of Mechanical Stimulation by Duck Activities on Microstructures of Internodes

Comparison of the light electron microstructures of the N1, N3, and N4 internodes for the rice culm between the RND and RD treatments showed that the vascular bundles were much better organized and structured in the RD treatment than in the RND treatment (Figure 5.13). There are two vascular bundle rings inside the epidermal layer of the rice culm: the outer ring and the inner ring. Figure 5.14 shows the outer ring of vascular bundles for N4, N3, and N1 internodes between the two treatments. The sizes and areas of the vascular bundles for the internodes were larger in the RD treatment than in the RND treatment. The outermost part of the rice culm is the epidermis,

Table 5.7   Stem Wall Thickness, Breaking Resistance Strength, and Lodging Index of Internodes N1, N2, N3, and N4 of the Rice Culm

Treatment	N1	N2	N3	N4
Stem Wall Thickness (mm)				
Rice and ducks (RD)	0.44 ± 0.02[a]	0.65 ± 0.04[a]	0.79 ± 0.05[a]	0.91 ± 0.06[a]
Rice with no ducks (RND)	0.31 ± 0.02[b]	0.54 ± 0.02[b]	0.66 ± 0.03[a]	0.67 ± 0.03[b]
Breaking Resistant Strength (g)				
Rice and ducks (RD)	—	560.18 ± 16.38[a]	725.09 ± 12.40[a]	—
Rice with no ducks (RND)	—	458.61 ± 37.62[b]	605.47 ± 39.32[b]	—
Lodging Index				
Rice and ducks (RD)	—	116.34 ± 6.05[b]	91.91 ± 3.45[b]	—
Rice with no ducks (RND)	—	137.41 ± 7.84[a]	107.22 ± 4.37[a]	—
Internode Length (cm)				
Rice and ducks (RD)	31.25 ± 0.76[a]	15.78 ± 0.36[a]	10.42 ± 0.21[a]	5.65 ± 0.10[b]
Rice with no ducks (RND)	30.85 ± 1.05[a]	15.52 ± 0.29[a]	10.93 ± 0.27[a]	6.85 ± 0.56[a]

Note: Same letters after the values in the same column in the table indicate no statistical significance.

which is composed of a thin but dense layer made up of various thick-walled cells in a remarkably uniform pattern. Figure 5.15 shows a cross section of the epidermis of the rice culm. This photograph demonstrates that the epidermis of the rice culm in the RD treatment was significantly denser (or more compact) than the RND treatments, especially for the N4 and N3 internodes. Within the epidermis is the small and thick-walled sclerenchyma that forms the mechanical tissue and provides major mechanical support for non-elongating regions of the rice plant. Figure 5.15 further reveals that the mechanical tissue of the three internodes was thicker in the RD treatment than in the RND treatment.

5.7.3 Effect of Mechanical Stimulation by Duck Activities on Rice Plant Hormone ABA

It is common to observe rice plants maturing at shorter heights in rice–duck fields than in adjacent rice monoculture fields. In our mechanical touching study, we found that artificial mechanical stimulation or touch on the rice stalks for 5, 10, or 15 seconds per day increased the content of abscisic acid (ABA) in the leaves of the rice plant compared with the control, especially for the last measurement of touching treatments (Table 5.8). The increase in ABA content inhibited stem elongation and resulted in thicker culm walls, which improved lodging resistance. Additionally, duck activities may improve soil potassium and silicon influxes into the rice plant through the plant's larger vascular bundles. These two factors contributed to an increase in the stiffness of the rice culm. The denser layer and thicker mechanical tissue of the epidermis of the rice culm in the N4 and N3 internodes in the RD treatment suggested that duck activities promoted stronger internodes and higher resistance strength, especially near the base of the rice plants, thereby enhancing the lodging resistance of the rice plant. This study showed that duck activities induced the anatomical structure changes in the rice culm and enhanced the lodging resistance of the rice plant. This evidence sheds light on the role of mechanic stimulation of animal (e.g., duck, fish) activities on plant growth and should encourage researchers to develop a novel, organic, and substantial rice farming pattern for economic, environmental, and ecological benefits.

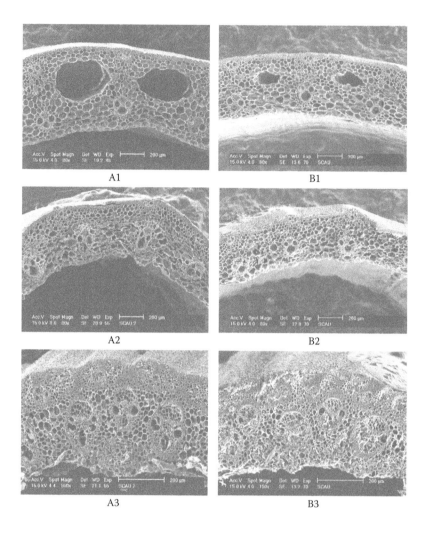

Figure 5.13 Microstructures of internodes N4 (A1), N3 (A2), and N1 (A3) of the rice culm for the RD treatment. Microstructures of internodes N4 (B1), N3 (B2), and N1 (B3) of rice culm for the RND treatment. Amplifying factor of the microscope was 80 for A1, A2, B1, and B2 and 150 for A3 and A3.

5.8 MECHANISM OF RICE–DUCK CO-CULTURE AND ITS EFFECTS ON PADDY CH$_4$ EMISSIONS

5.8.1 Effect of Rice–Duck Co-Culture on CH$_4$ Emissions

Global warming, resulting from the elevated concentrations of greenhouse gases in the atmosphere, has emerged as one of the most prominent global environmental issues. Methane (CH$_4$) is an important greenhouse gas and has approximately 25 times more infrared absorption capacity, or global warming potential, than CO$_2$ on a molecular basis (Rath et al., 1999). Rice paddy wetlands are a key source of CH$_4$ emissions, which are estimated to be 10 to 20% of total global CH$_4$ emissions (Sass et al., 1994). Therefore, it is necessary to understand which agricultural farming system has the best potential to mitigate the paddy CH$_4$ emissions contributing to global warming. In order to develop an improved conservation technology for mitigating CH$_4$ emissions that offers multiple benefits to the economy, environment, food security, and agricultural sustainability, rice–duck farming systems have been reexamined in recent years, especially in South China.

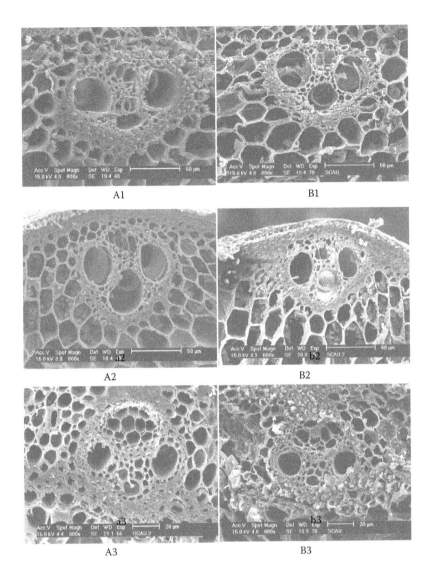

Figure 5.14 Microstructures in the outer ring of the vascular bundles for N4 (A1), N3 (A2), and N1 (A3) of the rice culm for the RD treatment. Microstructures in the outer ring of the vascular bundles for N4 (B1), N3 (B2), and N1 (B3) of the rice culm for the RND treatment. Amplifying factor of the microscope was 600 for A1, A2, B1, and B2 and 800 for A3 and B3.

Several studies have investigated the emissions of greenhouse gases in rice–duck farming systems in recent years. Researchers measured the CH_4 emissions from a wetland rice–duck ecosystem in China and found that rice–duck co-culture can mitigate CH_4 emissions to some extent (Y. Huang et al., 2005; Zhan et al., 2008). Cumulative emissions of CH_4 (the amount of total CH_4 emissions over a given experimental period) for the three cultivation treatments in our experiment are given in Figure 5.16. The overall emissions of CH_4 from the paddy field were 62,810, 38,583, and 41,375 mg m^{-2} for the organic fertilizer and duck treatment (OF+D), the chemical fertilizer and duck treatment (CF+D), and the control treatments, respectively. A 6.7% reduction of CH_4 emissions was observed in the CF+D treatment as compared to the control treatment. Results imply that, although the impacts of ducklings on CH_4 emissions depended on the rice growth stages, the

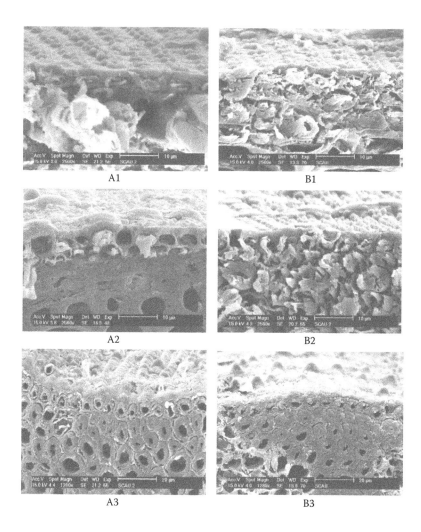

Figure 5.15 Epidermis layers for internodes N4 (A1), N3 (A2), and N1 (A3) of the rice culm for the RD treatment. Epidermis layers for internodes N4 (B1), N3 (B2), and N1 (B3) of the rice culm for the RND treatment. Amplifying factor of the microscope was 2560 for A1, A2, B1, and B2 and 1280 for A3 and B3.

introduction of ducks into the rice farming system generally mitigated some of the emissions of CH_4 into the atmosphere (J.E. Zhang et al., 2011a). Global warming potential (GWP) is a measure of how much a given mass of greenhouse gas is estimated to contribute to global warming. Results showed that the GWP based on the 20- or 100-year time horizon was the highest in the OF+D treatment, lowest in the CF+D treatment, and in the middle for the control treatment. A 6.7% decrease in GWP based on the 20- or 100-year time horizon was observed for the CF+D treatment compared to that of the control treatment without ducks. Statistical analysis with an F test demonstrates that such a decrease is within a 1% level of significance (i.e., $p = 0.01$). Results also show that the introduction of ducks into a rice farming system reduces the emission of CH_4 into the atmosphere as compared to conventional rice farming (i.e., control treatment) in South China. Table 5.9 further reveals that the GWP with the use of organic fertilizer was 65% higher than with the use of chemical fertilizer in the rice–duck farming system. It is apparent that the use of organic fertilizer would enhance the GWP through the increase of CH_4 emission.

Table 5.8   Contents of Leaf Abscisic Acid (ABA) under Different Treatments

Treatment	Measured on September 25, 2004 (ng g⁻¹)	Measured on October 5, 2004 (ng g⁻¹)	Measured on October 15, 2004 (ng g⁻¹)
Control (no touch)	4440.02 ± 466.64[b]	4041.68 ± 296.75[a]	4484.22 ± 371.09[b]
Touch for 5 seconds per day	4852.13 ± 569.83[b]	5030.80 ± 1024.27[a]	6009.33 ± 1447.33[a]
Touch for 10 seconds per day	6576.36 ± 323.13[a]	5191.14 ± 1526.31[a]	6083.97 ± 1067.12[a]
Touch for 15 seconds per day	4963.42 ± 119.68[b]	5382.30 ± 947.60[a]	7486.23 ± 744.11[a]

Note: Experiment began on September 5, 2014, and ended on October 15, 2004. Same letters after the values in the same column in the table indicate no statistical significance.

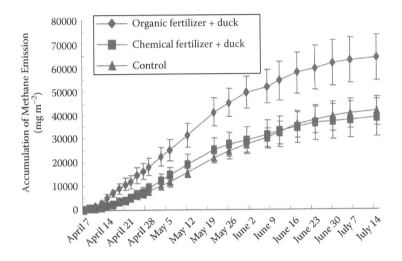

Figure 5.16   Accumulation of CH_4 emissions during the experiment.

Table 5.9   Impact of CH_4 Emissions from the Rice–Duck Farming Ecosystem on Global Warming Potential (GWP)

Treatment[a]	Average Hourly CH_4 Emission (mg m⁻² hr⁻¹)	Cumulative CH_4 Emission (mg m⁻²)	GWP Based on 20 Years	GWP Based on 100 Years
Organic fertilizer + duck (OF+D)	26.32	63,810.48	3,956,249.76	146,7641.04
Chemical fertilizer + duck (CF+D)	15.91	38,583.98	2,392,206.76	88,7431.54
Chemical fertilizer (control)	17.06	41,375.88	2,565,304.56	951,645.24

Note: The GWP values based on 20- and 100-year time horizons were calculated by multiplying the cumulative CH_4 emissions by 62 and 23, respectively.

[a] The values among different treatments are statistically different at $p = 0.01$ by F test.

5.8.2   Mechanism of Rice–Duck Co-Culture on CH_4 Emissions

Usually, paddy fields are wetland systems with a reducing environment rather than oxidizing one, which promotes more CH_4 formation; however, when ducks are introduced into the paddy field, their muddying and stirring activities can change the aquatic and soil environments, especially their redox states. Kumaraswamy et al. (2000) demonstrated that the level of CH_4 emissions declined in

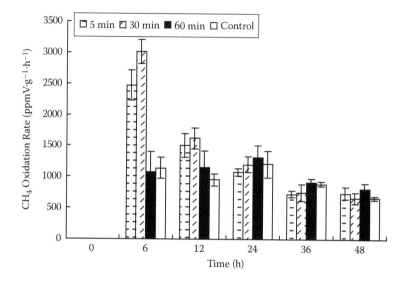

Figure 5.17 CH_4 oxidation rates of rice roots in the control and in the 5-, 30-, and 60-min/day treatments at different time intervals after mechanical stimulus.

the rice–duck farming system due to the increase in dissolved oxygen (DO) concentration, which is a result of the frequent movement of ducks. Our study also revealed another mechanism by which duck activities can stimulate rice root growth and thereby enhance O_2 secretions and CH_4 oxidation processes in the rice rhizosphere. The experiment was a laboratory simulation, and the results showed that continuous mechanical stimuli led to diverse responses in rice at the physiological and morphological levels. Rice height was decreased significantly in 30-min/day stimulation treatments. Biomass, radial oxygen loss (ROL), and methane oxidation rates (6 hours after stimulation) of the rice roots were all significantly improved for 5-min/day and 30-min/day stimulation treatments (Figure 5.17). The length, surface area, and volume of rice roots in the 30-min/day stimulation treatment all improved significantly. Stomata numbers were significantly higher for the 30-min/day and 60-min/day stimulation treatments than for the control. This result confirms the importance of duck stimulation on rice in rice–duck farming systems and also supplies another favorable explanation for the reduction in methane emissions in rice–duck farming systems. Furthermore, this study emphasizes the importance of animal activities, such as raising ducks and fish in paddy fields, on the methane release process (Zhao et al., 2013).

5.9 ECOSYSTEM SERVICE ASSESSMENT OF RICE–DUCK CO-CULTURE SYSTEMS

5.9.1 Roles and Functions of Ducks in Paddy Fields

As described above, ducks play a key role and do many kinds of work in the paddy field. Many ecosystem functions would be lost without ducks in the rice–duck co-culture system. Generally, duck activities include walking, swimming, eating, grooming, paddling, and rubbing. They can provide various combined effects, such as full-time weeding, plowing and muddying, nutrient supplying, pest controlling, mechanical stimulating, microclimate improvement, and rice disease prevention (Figure 5.18). Based on the ecosystem functions of ducks mentioned above, rice–duck co-culture can provide not only healthy products and good economic income but also various ecological services and social benefits.

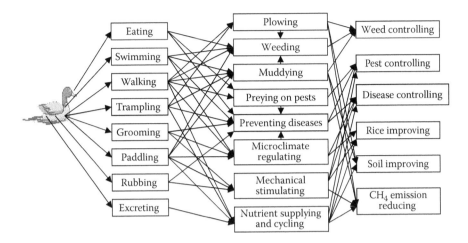

Figure 5.18   Full-time work and activities of ducks in the paddy fields

5.9.2 Ecosystem Service Assessment of Rice–Duck Co-Culture

Ecosystem service assessment has been a hot topic in ecology and ecological economics since the publication of Robert Costanza's famous article in *Nature* in 1997. Ecosystem services can be defined as the conditions and processes through which natural ecosystems, and the species that comprise them, sustain and fulfill human life, or the goods and service provided by ecosystems which contribute to human welfare, both directly or indirectly (Costanza et al., 1997; Daily, 1997).

So, in order to further explore the comprehensive benefits relevant to humans in the integrated rice–duck farming system, its ecological services were identified and analyzed on the basis of principles of ecological economics and relevant theories of ecosystem services. The ecological services analyzed include agricultural production, air regulation, water conservation, soil organic matter accumulation, and primary nutrient elements maintenance (Qin et al., 2010).

Results showed that the total ecological services value of the rice–duck farming system was roughly 18,104.82 yuan per hectare, about 15.64% higher than that of the conventional rice monocropping system. For the rice–duck farming system, the agricultural production and gas regulation function value were 6720 and 5928.72 yuan per hectare, accounting for 37.12% and 32.75% of the total ecological service value, respectively. The water conservation function value in the rice–duck integrated system was found to be the lowest, merely accounting for 14.16% of the total ecological service value. The ecological value of soil organic matter accumulation and main nutrient elements maintenance in the rice–duck farming system amounted to 2892.40 yuan per hectare, about 3.2 times that in conventional rice farming systems. Comprised mainly of air regulation, soil organic matter accumulation, and main nutrient elements maintenance, the environmental regulation function service values in the rice–duck co-culture system and conventional rice monoculture system were estimated to be 11,384.82 and 9885.62 yuan per hectare, respectively, or about 1.69 and 1.84 times greater than the value of the agriculture product (Table 5.10). Consequently, it can be concluded that the paddy rice–duck co-culture provides not only steady yields of grain production function but also significant environmental regulation services function. Effective in reducing greenhouse gas emissions as well as conserving soil fertility and quality, ducks raised in rice cropping fields could be conducive to improving the ecological and economic benefits of the paddy rice system. Even though a certain disagreement exists regarding the environmental and economic value of agroecosystem services because of the lack of acknowledged evaluation methods and variations in the cropping system, these assessment results can inform decision making around the improvement and extension of sustainable rice farming technologies, strategic planning, and policies for regional agricultural development.

Table 5.10   Ecological Service Value Items and Their Proportions for Rice–Duck Co-Culture System and Conventional Rice Cropping System

Service Functions	Rice–Duck Co-Culture		Conventional Rice Cropping	
	Service Value (yuan hm^{-2})	Percentage (%)	Service Value (yuan hm^{-2})	Percentage (%)
Agriculture production	6720.00	37.12	5387.50	37.60
Air regulation	5928.72	32.75	6459.65	45.09
O_2 emission and CO_2 absorption	6987.99	38.60	7764.44	54.20
Environmental costs for CH_4 and N_2O emissions	(1059.27)	(5.85)	(1304.79)	(9.11)
Soil organic matter accumulation and main nutrient element maintenance	2892.40	15.97	903.99	6.31
Soil organic matter accumulation	2793.76	15.43	823.14	5.75
Nutrient element maintenance	98.64	0.54	80.85	0.56
Water conservation	2563.70	14.16	2521.98	17.60
Total value of ecosystem service	18,104.82	—	15,273.12	—

5.10   PERSPECTIVES ON RICE–DUCK CO-CULTURE DEVELOPMENT IN CHINA

5.10.1 Current Problems

The rice–duck co-culture system can be very friendly and beneficial to humans and the environment, but the techniques involved are not ideal. Some innovations are still required for improving its production technology and application.

5.10.1.1 Technological Innovations Required

The first necessary innovation is related to the integrated controlling techniques of rice diseases and pests. Although duck activities can prevent or control major rice pests and diseases to a certain extent, there are still some diseases and pests—such as dwarf wilt, rice blast, and rice leafroller— against which ducks seem to be powerless, especially when disease and pest outbreaks occur during unusual weather and result in rice plant damage and grain yield loss. Applying chemical pesticides would be very harmful to safe grain production, duck growth, and environmental quality. This is a dilemma. Therefore, it is very necessary and urgent to develop and incorporate a series of new highly effective and environmentally friendly technologies for controlling rice diseases and pests in the rice–duck co-culture system, so as to maintain stability and sustainability in grain production.

The second set of technological innovations involves good nursing and management for ducks in paddy fields. Although ducks have a strong survival rate in the wild, it is still difficult for them to live in the relatively challenging paddy field environment long term, where occasional storms, typhoons, strong midday sunlight, and cold night temperatures can make them prone to disease, such as influenza. For this reason, there is a need to improve epidemic prevention and field management techniques, which would include the breeding and selection of duck species with high resistance to stress in a wild environment, the optimal design and construction of resting sheds and shelters for ducks, disease prevention, and nursing countermeasures.

Third, it is necessary to develop new types of low-cost healthy feedstuffs for ducks. It is known that ducks have good appetites. Usually, weeds, pests, and certain aquatic organisms in paddy fields cannot grow fast enough to meet duck consumption needs during the co-culture period of nearly 2 months. Supplemental external feedstuffs should be provided for ducks in rice–duck co-culture systems. Currently, during the rice–duck co-culture period, commercial corn and grains and even some chemical additives are often used as food for ducks; however, adding these feed inputs every day would increase production costs and may cause secondary pollution from the additive agents. Hence, this is still a pending problem to be resolved (J.E. Zhang et al., 2005).

5.10.1.2 Problems Faced by Extension

The rice–duck co-culture has proven to be an ecological and environmentally friendly farming model, and it can be found in many countries or regions of the world; however, its actual extension and applied area are very limited worldwide, because certain obstacles and challenges make it difficult to popularize such a farming system (J.E. Zhang et al., 2005, 2006). First, comparatively speaking, agriculture is a relatively long-cycle, high-risk, and low-income industry. Over the past several decades in China farmers have chosen to leave the farm for urban jobs that provide more money. Even if new farming innovations are becoming available, farmers are usually not interested in them, as farming cannot provide benefits comparable to what they can get working in the city. Despite the good income per unit area that farmers can earn from ecological farming, they still might not make a large profit because of the limited farmland their family manages.

Second, there is a significant market risk for agricultural products. In China, families, as the dominant basic farming units, usually do not have the money, time, and capacity to develop original brands for their agricultural products and market them, which makes it difficult to avoid market fluctuation risks. In contrast, agricultural companies often enjoy such advantages as adequate funding and marketing management resources; however, they do not have the farmland necessary to generate more profit. Companies could try to organize farmers' lands through the "company plus farmers" model, but they would have to obtain land users' rights transfer agreements from tens, hundreds, or even thousands of farmers, which would require significant negotiations, transactions, and management costs. In addition to the difficulties associated with acquiring more land, companies also face market risks. Thus, market risks act as barriers or bottlenecks to the development of agricultural and industrialization processes.

Third, farmers usually prefer to practice farming that is easy and simple and saves labor and time. Ecological agriculture is a somewhat comprehensive and time-consuming type of farming; for example, rice–duck co-culture often requires additional labor to build fence enclosures and to feed and nurse ducks in the field from time to time. This farming system, however, may save the time required to apply chemical pesticides and fertilizers in the field compared to conventional rice monoculture. There is clearly a need to further improve and simplify the production techniques to ensure the extension and further application of rice–duck co-culture.

5.10.2 Perspectives on Future Development

Overall, rice–duck co-culture, as an ecologically friendly technique, has a great potential for long-term application, extension, and market prospects because it meets the demands of food security, environmental protection, biodiversity maintenance, and sustainable development.

5.10.2.1 Technological Innovation and Research Perspective

As discussed above, there is still room for innovation to overcome technical weaknesses and to develop more effective technologies for use in rice–duck co-culture, including controlling rice pests and diseases, field nursing and management, providing low-cost healthy feedstuff for ducks, and developing relevant labor-saving and time-saving techniques that can be incorporated into the rice–duck co-culture system. Fortunately, in recent years in China, researchers have made many technical innovations and some good progress in rice–duck co-culture systems, such as rice–duck farming with direct seeding, rice–duck farming with zero-tillage, rice and two-batch ducks co-culture, rice intercropped with other water plants and duck polyculture, etc. (J.E. Zhang et al., 2013). China has expansive land for rice planting, from south to north and east to west. Because of the various seasonal climates and water conditions, as well as different cropping systems used and socioeconomic development situations in the various provinces, farming management procedures for rice–duck co-culture can vary greatly. It is necessary to formulate a series of technical regulations and standards for rice–duck co-culture systems that meet the demands of different regions. This would allow farmers from all across China to learn and better use rice–duck farming systems.

5.10.2.2 Management and Marketing Perspective

Compared with conventional rice monoculture, rice–duck co-culture per unit area yields a good economic profit, but even with a good farming system the benefits of economies of scale cannot be obtained without large-scale production. For this reason, future development of rice–duck co-cultures should consider larger scale production techniques. Over the past several years, various kinds of cooperative organizations involving agricultural companies and farmers, farmers and farmers, or even agricultural producers and consumers have come to the fore in China. These patterns include company plus farmer, farmer cooperatives, contract farming, and agricultural production plus relevant distribution companies, among others. These agricultural organizations have the potential to provide win–win benefits for the different stakeholders, but some of the bottleneck problems discussed above must be resolved among farmers and other stakeholders. Today, safe and healthy agricultural products are favored by consumers; however, because of the asymmetry of market information, as well as consumer behavior, high-quality products cannot always bring good prices and market shares, particularly organic or green foods. It is important to build product branding, explore market shares, distribute product information, gain credibility among consumers, and establish a quality tracing system for products. A stronger market orientation and growth in market share will ensure wider practice of ecological farming systems such as rice–duck co-cultures.

5.10.2.3 Comprehensive Development Strategy

Modern agriculture is a multifunctional industry; it not only directly provides products for human life but also contributes various ecosystem services, including climate regulation, pollution buffering and purification, biodiversity maintenance, and water and soil conservation. Beyond promoting only the economic benefits of agricultural products, we can also make full use of agroecosystem services to develop agricultural ecotourism and training programs, such as family farm restaurants and inns, agricultural fairs, farming experience activities, do-it-yourself programs, leisure and vacation services, agricultural culture festivals, and so on. Such a comprehensive approach would promote agricultural economic benefits in many ways. Currently, agroecotourism is booming in China. Ecological farming such as the rice–duck co-culture serves as a good representative of ecotourism and offers significant market potential.

REFERENCES

Costanza, R., d'Arge, R., de Groot, R., Farber, S., Grasso, M. et al. 1997. The value of the world's ecosystem services and natural capital. *Nature*, 387: 253–260.

Daily, G.C. 1997. *Nature's Services: Societal Dependence on Natural Ecosystems*. Island Press: Washington, DC.

Huang, H., Wang, H., Hu, Z. et al. 2003. Theoretic analysis and practice of the integrated rice–duck farming engineering. *Crop Res.*, 17(4): 189–191.

Huang, Y., Wang, H., Huang, H., Feng, Z.W., Yang, Z.H., and Luo, Y.C. 2005. Characteristics of methane emission from wetland rice–duck complex ecosystem. *Agric. Ecosyst. Environ.*, 105(1-2): 181–193.

Huang, Z.X., Zhang, J., Liang, K.M., Quan, G.M., and Zhao, B.L.. 2012. Mechanical stimulation of duck on rice phyto-morphology in rice–duck farming system. *Chin. J. Eco-Agric.*, 20(6): 717–722.

Kumaraswamy, S., Rath, A.K., Ramakrishnan, B., and Sethunathan, N. 2000. Wetland rice soils as sources and sinks of methane: a review and prospects for research. *Biol. Fertil. Soils*, 31(6): 449–461.

Liang, K.M., Zhang, J.E., Lin, T.N., Quan, G.M., and Zhao, B.L. 2012. Control effects of two-batch-duck raising with rice farming on rice diseases, insect pests and weeds in paddy field. *Adv. J. Food Sci. Technol.*, 4(5): 309–315.

Lin, Z.R. and Jin, Z.Z. 2002. Preliminary study on the control of rice pests by releasing ducks in rice field. *Chin. J. Biol. Contr.*, 18(2): 94–95.

Liu, X.Y., Yang, Z.P., Huang, H., Hu, L.D., Chen, Y.F., and Wen, Z.Y. 2004. Developing roles of field weeds in wetland rice–duck compounded system. *J. Hunan Agric. Univ.*, 30(3): 292–294.

Qin, Z., Zhang, J., Luo, S.M., Xu, H.Q., and Zhang, J. 2010. Estimation of ecological services value for the rice–duck farming system. *Resour. Sci.*, 32(5): 864–872.

Qin, Z., Zhao, B.L., Zhang, J.E., and Luo, S.M.. 2011. Study on diversity of arthropod community in a rice–duck integrated farming system in South China. *J. Resour. Ecol.*, 2(2): 151–157.

Quan, G.M., Zhang, J.E., Huang, Z.X., and Xu, R.B. 2005a. Review on the ecological effects of integrated rice–duck farming system. *Chin. Agric. Sci. Bull.*, 21(5): 360–364.

Quan, G.M., Zhang, J.E., Xu, R.B., Liu, J.L., and Huang, Z.X. 2005b. Effects of biological control of pests by raising ducks in the paddy fields. *Ecol. Sci.*, 24(4): 356–358.

Quan, G.M., Zhang, J.E., Chen, R., and Xu, R.B. 2008. Effects of rice–duck farming on paddy field water environment. *Chin. J. Appl. Ecol.*, 19(9): 2023–2028.

Rath, A.K., Mohanty, S.R., Mishra, S., Kumaraswamy, S., Ramakrishnan, B., and Sethunathan, N. 1999. Methane production in unamended and rice-straw-amended soil at different moisture levels. *Biol. Fertil. Soils*, 28: 145–149.

Sass, R.L., Fisher, F.M., Lewis, S.T., Jund, M.F., and Turner, F.T. 1994. Methane emission from rice yields: effects of soil properties. *Global Biogeochem. Cycles*, 8(2): 135–140.

Shen, X.K. 2003. *New Production Technology of Pollution-Free and Organic Rice in Rice–Duck Farming*. China Agricultural Science & Technology Press: Beijing.

Takao, F. 2001. *The Power of Duck: Integrated Rice and Duck Farming*. Tagari Publications: Tasmania, Australia.

Tong, Z.X. 2002. Preliminary study on the relationships between raising ducks and biological populations in the paddy fields. *China Rice*, (1): 33–34.

Wang, J.P., Cao, C.G., Li, C.F., Jin, H., Yuan, W.L., and Zhan, M. 2009. Algal change in paddy water under rice–duck complex ecosystem. *Acta Ecol. Sin.*, 29(8): 4353–4360.

Wang, Q.S., Huang, P.S., Zhen, R.H., Jin, L.M., Tang, H.B., and Zhang, C.Y. 2004. Effect of rice–duck mutualism on nutrition ecology of paddy field and rice quality. *Chin. J. Appl. Ecol.*, 15: 639–645.

Wang, Q.S., Zhen, R.H., Ding, Y.F., and Wang, S.H. 2008. Strong stem effect and physiological characteristics of rice plant under rice–duck farming. *Chin. J. Appl. Ecol.*, 19(12): 2661–2665.

Wei, S.H., Qiang, S., Ma, B., Wei, J.G., Chen, J.W., Wu, J.Q., Xie, T.Z., and Shen, X.K. 2006. Influence of long-term rice–duck farming systems on the composition and diversity of weed communities in paddy fields. *Acta Phytoecol. Sin.*, 30: 9–16.

Xiong, G.Y. and Zhu, X.B. 2003. Preliminary discussion on the technique of the integrated rice–duck farming. *Fodder Ind.*, 24(4): 28–30.

Yang, Z.H., Huang, H., and Wang, H. 2004. Paddy soil quality of a wetland rice–duck complex ecosystem. *Chin. J. Soil Sci.*, 35(2): 117–121.

Zhan, M., Cao, C.G., Wang, J.P., Yuan, W.L., Jiang, Y., and Gao, D.W. 2008. Effects of rice–duck farming on paddy field's methane emission. *Chin. J. Appl. Ecol.*, 19(12): 2666–2672.

Zhang, J., Zhang, J.E., Qin, Z., Fu, L., and Liang, K.M. 2012. Effect of integrated rice–duck farming on rice canopy structure index. *Chin. J. Eco-Agric.*, 20(1): 1–6.

Zhang, J.E. 2013. Progresses and perspective on research and practice of rice–duck farming in China. *Chin. J. Eco-Agric.*, 21(1): 70–79.

Zhang, J.E., Lu, J.X., Zhang, G.G., and Luo, S.M. 2002. Study on the function and benefit of rice–duck agro-ecosystem. *Ecol. Sci.*, 21(1): 6–10.

Zhang, J.E., Lu, J.X., Huang, Z.X., Xu, R.B., and Zhao, B.L. 2005. Discussion on practical and theoretic issues of integrated rice–duck farming system. *Ecol. Sci.*, 24(1): 49–51.

Zhang, J.E., Lu, J.X., Zhang, G.H., and Huang, Z.X. 2006. Discussion on rice–duck integrated farming eco-system and the related technology innovation. *Syst. Sci. Compr. Stud. Agric.*, 22(2): 94–97.

Zhang, J.E., Xu, R.B., Quan, G.M., and Chen, R. 2007. Effects of rice–duck farming system on physiological characters of rice. *Chin. J. Appl. Ecol.*, 18(9): 1959–1964.

Zhang, J.E., Xu, R.B., Chen, X., and Quan, G.M. 2009a. Effects of duck activities on a weed community under a transplanted rice–duck farming system in southern China. *Weed Biol. Manage.*, 9: 250–257.

Zhang, J.E., Xu, R.B., Quan, G.M., Xu, H.Q., Qin, Z. 2009b. Effects of integrated rice–duck farming on soil microbial quantity and functional diversity. *Resour. Sci.*, 31(1): 56–62.

Zhang, J.E., Zhao, B.L., Chen, X., and Luo, S.M. 2009c. Insect damage reduction while maintaining rice yield in duck–rice farming compared with mono rice farming. *J. Sustain. Agric.*, 33: 801–809.

Zhang, J.E., Ouyang, Y., Huang, Z.X., and Quan, G.M. 2011a. Dynamic emission of CH_4 from a rice–duck farming ecosystem. *J. Environ. Prot.*, 2: 537–544.

Zhang, J.E., Xu, R.B., Quan, G.M., and Zhao, B.L. 2011b. Influence of rice–duck integrated farming on rice growth and yield characteristics. *Resour. Sci.*, 33(6): 1053–1059.

Zhang, J.E., Quan, G.M., Huang, Z.X., and Quan, G. 2013. Evidence of duck activity induced anatomical structure change and lodging resistance of rice plant. *Agroecol. Sustain. Food Syst.*, 37: 975–984.

Zhao, B.L., Zhang, J.E., Lv, X.H., Peng, L., and Padilla, H. 2013. Methane oxidation enhancement of rice roots with stimulus to its shoots. *Plant Soil Environ.*, 59(4): 143–149.

Zheng, J.C., Tan, S.H., Liu, H.H., Feng, J.X., and Zhang, W.J. 2005. Present status and development strategy of theory and technology of integrated rice–duck farming system in China. *Jiangsu Agric. Sci.*, (5): 1–5.

CHAPTER 6

Study of Non-Point-Source Pollution Control in Integrated Rice–Frog Agroecosystems

Cao Linkui and Zhang Hanlin

CONTENTS

6.1 OVERVIEW OF INTEGRATED RICE–FROG AGROECOSYSTEMS

At present, maintaining the security of the food supply of China has become an important goal and responsibility for both central and local government agencies. It is necessary to produce grain in suburban areas to achieve a certain level of self-sufficiency, even for a large city such as Shanghai. Paddy fields are generally considered artificial wetland ecosystems, and their environmental purification and flood buffering functions are important; therefore, paddy wetland systems are also indispensable in urban ecosystems, and they are crucial to the development of urban economies and societies. However, with urbanization and the rapid development of urban agriculture, the ecological functions performed by paddy wetland systems have been somewhat weakened, especially with the use of large amounts of chemical fertilizers, pesticides, and other agricultural chemicals (W.L. Zhang et al., 2004). The result is agricultural non-point-source pollution, which has a serious impact on the sustainable development of urban people's lives and economy (G. Zhang et al., 2008).

Integrated rice–frog agroecosystems were first tested and demonstrated in 2007, in the Qingpu Modern Agricultural Park in Shanghai, where the stocking density of tiger frogs, reasonable planting structures, and pest control technologies were explored. In 2009, with the technical support of the School of Agriculture and Biology at Shanghai Jiao Tong University, further research was conducted on integrated rice–frog agroecosystems. Recently, "frog rice" marketed by Zizaiyuan Agricultural Development Co., Ltd., was awarded green and organic certification, which indicates that integrated rice–frog agroecosystems have demonstrated good pilot and demonstration results and offer significant ecological and economic benefits.

Integrated rice–frog agroecosystems (Figure 6.1) can address the problem of excessive applications of paddy fertilizers and chemical pesticides, as well as the consequences of serious nutrient loss and pollution. This technology could also be used to promote sustainable urban agricultural development. Integrated rice–frog agroecosystems have modified traditional rice production methods by reducing chemical fertilizers and pesticides to control non-point-source pollutants. These modifications include using ecological ditches and paddy wetland techniques to intercept and absorb runoff pollutants from river water and to block runoff pollution from paddy systems (Jiang et al., 2004). The rational allocation of agricultural production factors, combined with the use of integrated rice–frog agroecosystems, helps to achieve the goal of a synergistic reduction of nitrogen and phosphorus pollution. By combining source control, process blocking, and ecological planting and breeding, integrated rice–frog agroecosystems are ideal for urban agricultural non-point-source pollution control.

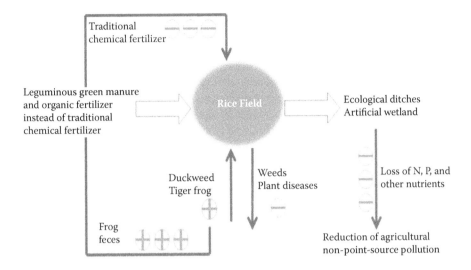

Figure 6.1 Integrated rice–frog agroecosystem.

6.1.1 Research Objectives

6.1.1.1 Construction of Integrated Rice–Frog Agroecosystems to Promote a Virtuous Cycle in Paddy Ecosystems

In integrated rice–frog agroecosystems, a paddy field is considered to be an agricultural ecosystem that can be regulated and the rice is an ecological factor in that agricultural ecosystem. To maintain system stability, this research examined species selection, chemical input management, improved cultivation techniques, and the use of animals for their symbiotic effects. The aim was to develop a technical package to create a rice paddy environment suitable for rice products—but not for grass weeds and pests—in order to achieve ecological control of pests in paddy fields and produce safe, high-quality rice.

6.1.1.2 Construction of Interception Systems (Ecological Ditches) to Reduce Non-Point-Source Pollution Generated by Paddy Fields

In the ecological drainage ditches and wetlands downstream of various integrated rice–frog agroecosystems, aquatic plants were arranged to block, absorb, deposit, or transform nitrogen, phosphorus, and other chemical substances in paddy runoff. The nutrient loss from paddy fields was effectively intercepted, and the goals of controlling paddy nutrient loss, recycling nutrients, and reducing water pollution substances were achieved. Integrated rice–frog agroecosystems can reduce the risk of non-point-source pollution from paddy fields and enhance biodiversity in paddy field systems, ultimately maintaining the stability of paddy ecosystems.

6.1.2 System Features

Integrated rice–frog agroecosystems in paddy fields can be designed to combine chemical fertilizer and pesticide source control technologies, internal process blocking, and output interception technology for pollutants using various planting and breeding strategies. Such systems can improve paddy ecosystem environments, maintain biodiversity, and produce safe, high-quality rice.

6.1.2.1 Controlling Chemical Fertilizers and Pesticides, Constructing Interception Systems, and Reducing Paddy Field Non-Point-Source Pollution

Based on the principles of ecological agriculture and by replacing conventional chemical fertilizers and chemical pesticides with bio-organic fertilizers and biopesticides, integrated rice–frog agroecosystems can provide relatively good environments for rice growth and reduce the non-point-source pollution from paddy fields. Various types of aquatic plants are planted in drainage ditches and wetlands to take up nitrogen, phosphorus, and other elements as a natural source of nutrition. Such an ecosystem ensures the natural growth of plants, and at the same time not only does it enhance plant diversity such that the paddy system can be ornamental but it also reduces the risk of agricultural non-point-source pollution generated by the rice fields.

6.1.2.2 Using Ecological Planting and Breeding to Reduce Pollution and Pests in Paddy Fields

In traditional rice cultivation, excessive fertilizers are often used to increase rice yields, but the fertilizers have a negative impact on the surrounding soil, water, and air. Pest control in paddy fields has mainly relied on chemical pesticides in traditional systems, but the chemical pesticide residues can run off or leach into water around the paddy field. This runoff not only harms the rural ecological environment but also reduces the quality of the rice. In contrast, ecological planting and breeding techniques can be used to control pests and reduce environmental risks, as well as achieve high-quality products at competitive prices, important goals of rice production in urban areas.

6.1.2.3 Implementation of Integrated Rice–Frog Agroecosystems to Reduce Agricultural Ecological Risk and Improve Rice Quality

In order to guarantee high yields of rice, conventional agriculture must use pesticides to reduce the incidence of pests and diseases. Introducing tiger frogs into paddy fields can greatly reduce the dependence on pesticides during rice growth. Tiger frogs, as natural enemies of rice pests (e.g., planthoppers), protect the rice as it grows and contribute to a rice–frog–pest ecological balance, reducing damage to the environment caused by spraying pesticides. Leguminous crops such as milk vetch can also be planted in integrated rice–frog agroecosystems. Tiger frog manure also serves as a supplement to organic fertilizer. Through such management techniques, ecological efficiency can be achieved and rice quality improved.

6.2 SOURCE CONTROL TECHNOLOGY FOR FERTILIZER AND PESTICIDE POLLUTION IN PADDY FIELDS

The application of alternative technologies, such as bio-organic fertilizer, leguminous green manure crops, biological controls, and integrated rice–frog agroecosystems, reduces the inputs of fertilizers and chemical pesticides, and these technologies have been very effective in achieving source control of non-point-source pollution in paddy fields.

6.2.1 Fertilizer Reduction Technologies in Paddy Fields

An experiment was conducted using a randomized block design with six different treatments and three replicates. The tested rice variety was Xiushui-128, and the rice planting density was 15 × 20 cm. The bacterial agents tested came from the Institute of Microbiology, Hebei Academy of Sciences, and were primarily combined with silicate bacterium and nitrogen-fixing bacteria that were

Table 6.1 Design of Fertilizer Reduction Experiment

Treatment		N (kg ha^{-2})	P$_2$O$_5$ (kg ha^{-2})	K$_2$O (kg ha^{-2})	Agent (kg ha^{-2})
CK	No fertilizer	0	0.0	0.0	0
T1	Chemical fertilizer	120	96.0	84.0	0
T2	Fertilizer reduced 20% + agent	96	76.8	67.2	15
T3	Fertilizer reduced 20% + agent reduced 40%	—	—	—	9
T4	Fertilizer reduced 40% + agent	72	57.6	50.4	15
T5	Fertilizer reduced 40% + agent reduced 40%	—	—	—	9

isolated and purified in the laboratory. These bacterial agents functioned to dissolve phosphorus and potassium and to fix nitrogen. The experiment was conducted from June to November 2011. Before and after the trial, soil properties and enzyme activity were measured to test the effect of bacterial agents on paddy soil. The design of this fertilizer reduction experiment is shown in Table 6.1.

6.2.1.1 Impact of Fertilizer Reduction on Paddy Soil Nutrients

The soil test had six treatments (Table 6.1): CK (no fertilizer), T1 (100% chemical fertilizer), T2 (80% chemical fertilizer + 100% microbial agents), T3 (80% chemical fertilizer + 60% microbial agents), T4 (60% chemical fertilizer + 100% microbial agents), and T5 (60% chemical fertilizer + 60% microbial agents). The results after rice harvest showed that chemical fertilizer alone (T1) could not provide a sufficient supply of soil-available phosphorus and potassium, but the nutrient supply in the treatments with microbial agents lasted longer. T2 had the highest soil phosphorus content (33.24 mg kg^{-1}). T4 improved soil potassium and organic matter content, but T1 decreased the soil organic matter content. Experimental results of the effect of microbial agents on paddy soil enzyme activities showed that the application of microbial agents could improve soil enzyme conversion activity and accelerate the decomposition and transformation of soil organic matter; however, enzyme conversion activity in T1 was reduced. In the rice-growing period, enzyme conversion activity increased in all the treatments. Before the heading stage, T2 was the highest, and after that T4 was the highest. T1 was always the lowest.

6.2.1.2 Effect of Fertilizer Reduction on Growth and Yield of Rice

The experiment showed that microbial agents play a significant role in promoting rice growth (Figure 6.2). Rice leaf length, number of tillers, and plant height all exhibited significant differences among the treatments, being the highest in T2 (Table 6.2). Grain yield results are shown in Table 6.3. The grain yield in T3 was the highest, at 9447.00 kg ha^{-2}. The no-fertilizer treatment (CK) had the lowest economic coefficient. The economic coefficient in T3 was 0.482, the highest among all treatments. Adding microbial agents can reduce the amount of chemical fertilizer by 20 to 40% without either a reduction or an increase in rice production.

6.2.2 Technology to Reduce Chemical Pesticides in Paddy Fields

Another experiment was carried out from June to October 2012. The rice variety was late Japonica "Baonong 34," which was being promoted in the Shanghai area. Four treatments were set up: (1) ecological frog paddy fields, (2) organic frog paddy fields, (3) green frog paddy fields, and (4) chemical control paddy fields. In the treatment of ecological frog paddy fields, no fertilizer or pesticide was applied; only 1500 kg ha^{-2} microbial fertilizer was applied, and the parasitoid wasp *Trichogramma* was released twice. In the treatment of organic frog paddy fields, no chemical fertilizer or chemical

Figure 6.2   Rice plant morphology under different fertilization treatments.

Table 6.2   Rice Growth Traits (August 26, 2011)

Treatments	T1	T2	T3	T4	T5	CK
Flag leaf length (cm)	35.3[b]	40.8[a]	38.7[ab]	35.8[b]	33.4[a]	29.4[c]
Tiller number	14.5[ab]	15.1[a]	13.2[b]	13.2[b]	12.4[b]	10.0[c]
Plant height (cm)	67.7[ab]	69.7[a]	67.2[ab]	68.8[ab]	67.0[ab]	62.3[b]

Note: Superscript letters indicate significant difference ($p < 0.05$).

Table 6.3   Analysis of Grain Yield under Different Treatments

Treatment	Effective Panicle (10,000 ha^{-2})	1000 Grain Weight (g)	Straw Yield (kg ha^{-2})	Grain Yield (kg ha^{-2})	Economic Coefficient
CK	578.29[a]	28.78[a]	8404.57[a]	5939.58[a]	0.414[a]
T1	722.36[b]	25.34[b]	11052.91[b]	9293.39[b]	0.457[b]
T2	733.70[b]	25.32[b]	10,341.13[c]	9395.80[b]	0.476[c]
T3	678.34[b]	26.64[b]	10,149.47[c]	9447.00[b]	0.482[c]
T4	722.36[b]	26.06[b]	10,546.33[bc]	9283.15[b]	0.468[bc]
T5	689.01[c]	28.48[a]	10,311.21[c]	8960.57[c]	0.465[bc]

Note: Superscript letters indicate significant difference ($p < 0.05$).

pesticide was applied; instead, biopesticides were applied that included two applications of matrine and two applications of azadirachtin, as well as 3750-kg ha^{-2} oil seed cake. In the treatment of green frog paddy fields, in keeping with national green food standards, only limited chemical fertilizer was applied (chemical pesticides twice and biopesticides three times). In the chemical control paddy fields, chemical fertilizers and chemical pesticides were applied nine times. The rice was sown in mid-May and transplanted in early June. Machine land preparation was used, and 750 frogs per ha^2 were released after transplantation in the treatments with frogs. By the end of June to early July, tiger frogs had been placed in the paddy fields and frog weight was an average of 20 g each. The tiger frogs were captured 20 to 25 days before rice harvest. The previous crop in the paddy field was leguminous green manure (milk vetch, *Astragalus sinicus* L.), which was intercropped in paddy fields by sowing before the rice ripened in mid-September. After rice harvest, practices were used to make the milk vetch grow better during the winter. The milk vetch was plowed and composted by the end of April the following year, allowing it to be used as a source of organic fertilizer in the paddy fields.

Table 6.4 shows the effects of the treatments on pesticide use reduction. Application of the chemical pesticides in the chemical control treatment was done according to the standard Baonong 34 high-yield production technical package (2971.5 g pesticides ha^{-2}). In the treatments with frogs,

Table 6.4 Effects of Paddy Pesticide Reduction

Treatment	No. of Times Pesticide Used	Pesticide Reduction (%)	Amount of Pesticide Reduction (g ha^{-2})	Grain Yield (kg ha^{-2})	Income (10,000 yuan ha^{-2})	Cost (10,000 yuan ha^{-2})	Cost-to-Income Ratio
Ecological frog paddy field	0	100.0	2971.5	6817.5	23.87	1.65	1:14
Organic frog paddy fields[a]	4	55.5	1648.5	5625.0	19.69	2.23	1:9
Green frog paddy fields	5	44.4	1318.5	6750.0	14.18	2.25	1:6
Chemical control	9	0	0	7725.0	6.50	1.23	1:5

[a] The organic treatment of frog rice had four applications of biological pesticides.

7500 tiger frogs ha^{-2} were put in the paddy fields. The total cost of frogs was 12,000 yuan ha^{-2}. The market price of ecological and organic frog rice was 50 yuan kg^{-1}, the market price of green frog rice was 30 yuan kg^{-1}, and the chemical control rice cost 12 yuan kg^{-1}. As compared with the chemical control, the three frog paddy field treatments averaged a 66% pesticide reduction, thus decreasing the average chemical dosage to 1980 g ha^{-2}. The average input–output ratio also increased 94%. The increase in ecological and economic benefits in the treatments with reduced pesticide was significant.

6.3 BLOCKING AND INTERCEPTION TECHNOLOGY FOR POLLUTANT MIGRATION AND PROLIFERATION IN PADDY RUNOFF

The blocking, interception, and proliferation of paddy runoff pollutants, including nitrogen and phosphorus, were achieved through screening with aquatic plants, constructing ditches, and using artificial wetlands to adsorb and intercept pollutants from drainage water (Figure 6.3).

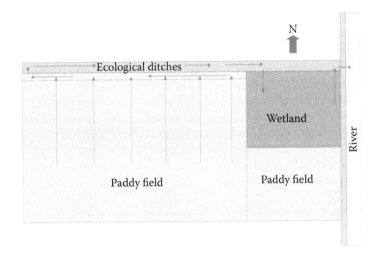

Figure 6.3 Paddy runoff pollutant blocking and interception system.

Figure 6.4 Schematic of aquatic screening pool.

6.3.1 Screening and Breeding Technology for Aquatic Plant Species

Experiments were carried out in aquatic pools (Figure 6.4) to evaluate the capacity of aquatic plant species to absorb nitrogen and phosphorus. A screening aquatic pool was built in 2010 in the Qingpu Modern Agricultural Park in Shanghai. It was divided into 12 areas, and each cell was 8.5 m long, 10.75 m wide, and 1.0 m high. The soil depth on the bottom of the pool was 50 cm. After the aquatic plants were transplanted, the water level was controlled at 50 cm. The experiments had four treatments: sweet flag (*Acorus calamus*), cattail (*Typha orientalis* Presl.), iris (*Iris tectorum* Maxim.), and no plants (as a control). There were three replicates of each treatment in a randomized block design. By using river water from the Qingpu Modern Agricultural Park and adding different dosages of nitrogen and phosphorus fertilizer, high, medium, and low concentration experiments were set up. Each test had a period of 1 month. At days 1, 3, 5, 7, 9, 12, 15, 20, and 25, the aquatic absorption effects were tested by measuring the ammonium, nitrate, total nitrogen (TN), and total phosphorus (TP) levels in the water. The biomass, nitrogen, and phosphorus content in plants before and after the experiment were also determined.

The results showed that different aquatic plants had different capacities for nitrogen and phosphorus removal in water. Iris possessed the highest potential to remove total nitrogen at low and medium concentrations, and cattails had the highest potential to reduce total nitrogen at high concentrations. In the high, medium, and low phosphorus concentrations, iris treatments had the best total phosphorus removal rates. The removal rates of cattail treatments in the low and medium phosphorus concentrations were the lowest (Figures 6.5 and 6.6). In sum, iris had the best ability to remove nitrogen and phosphorus from water, followed by *Acorus calamus*.

The levels of nitrogen and phosphorus absorbed by different aquatic plants were not the same. At low concentrations of nitrogen, iris and *Acorus calamus* could absorb more nitrogen, whereas for medium and high nitrogen concentrations the nitrogen absorption by iris was the highest. At all three concentrations, iris was the best at absorbing phosphorus. Thus, iris and *A. calamus* could absorb more nitrogen and phosphorus from the water. In conclusion, iris and *A. calamus* are recommended for use in aquatic ecology ditches.

6.3.2 Construction Technology for Ecological Ditches

Ecological ditches have proved to be the most important element of blocking and interception systems for paddy runoff pollutants. As a transition zone between farmland and river, farmland drainage ditches can reduce nitrogen and phosphorus runoff from farmland into rivers. The processes

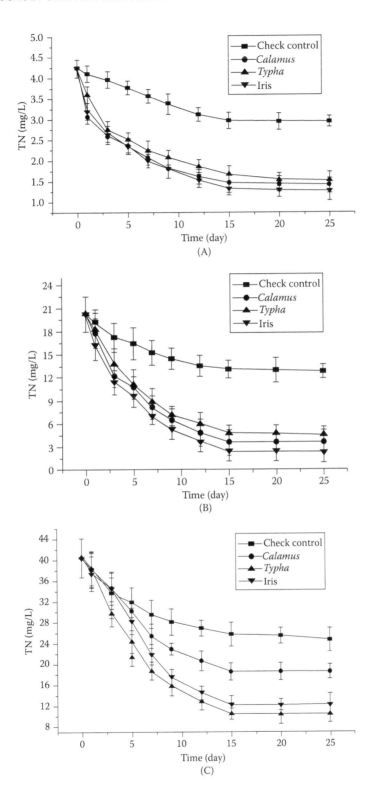

Figure 6.5 Changes of water TN for different treatments: (A) low concentration, (B) mid-concentration, (C) high concentration.

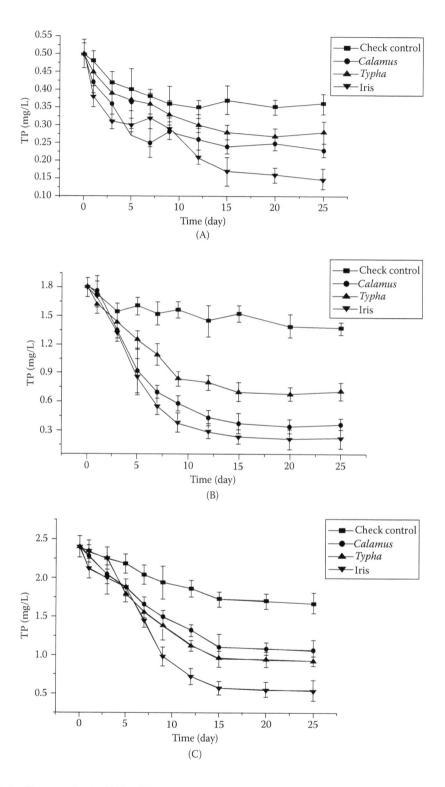

Figure 6.6 Changes of water TP for different treatments (A) low concentration, (B) mid-concentration, (C) high concentration.

Figure 6.7 An ecological ditch.

involved include soil adsorption, plant uptake, and biodegradation. Farmland drainage ditches are semi-natural ecosystems impacted by human activities. They play a role in irrigation and drainage in addition to serving as a wetland. The soil of the test area was purple clay, which is prone to erosion. The use of conventional drainage ditches in such areas could result in soil erosion and trench wall collapse, which would have a negative impact on irrigation, drainage, and the environment. Thus, cement porous plates were used in the construction of trench walls and ditch bottoms in the experimental area. Cement porous plates can prevent trench collapse and prolong the life of trench walls. The diameter of the porous plate was about 0.1 m. Aquatic plants were planted in the bottom of the ditch, so the ditches could be used for irrigation and drainage as well as environmental protection.

Based on the aquatic screening test results described above, one 125-m-long ecological ditch planted with 20 m of *Acorus calamus* and 105 m of iris was built in the test area (Figure 6.7). Two monitoring points were set up, one in the entrance point and one at the exit point of the ditch. After the paddy fields began to generate runoff, the water initially present in the ecological ditch was discharged, and the paddy runoff water then ran into the ecological ditch. Water samples were collected every day for the first 7 days, and the concentrations of ammonium, nitrate, total nitrogen, and total phosphorus in the water were determined.

In the treatment with iris and *Acorus calamus* (both offering good nitrogen and phosphorus absorption capacity) planted as a mixture in the ecological ditches, the ditches not only effectively reduced the concentration of nitrogen and phosphorus in paddy runoff but also had good effects on the ecological landscape. The results showed that iris and *A. calamus* planted in the ecological ditches had grown very well throughout the rice growing season, and the operation of ecological ditches was stable.

In the rice growing season of 2011, runoff occurred four times. The water from each runoff was kept in the ecological ditches for 7 days. The samples were collected at 2 p.m., and the TN and TP concentrations of water samples were tested every day. The results showed that ecological ditches had a very good rate of blocking and removal (Figures 6.8 and 6.9). The removal rates of ammonium, nitrate, total nitrogen, and total phosphorus were 50.82%, 44.71%, 45.08%, and 30.18%, respectively. In short, ecological ditches could remove 30% or more of the nitrogen and phosphorus present in paddy runoff.

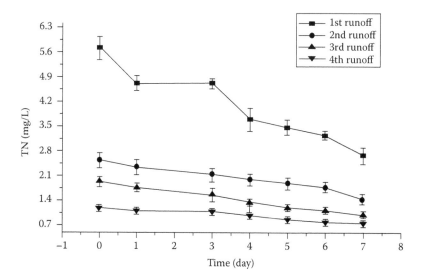

Figure 6.8   Changes of total nitrogen concentration in ecological ditches.

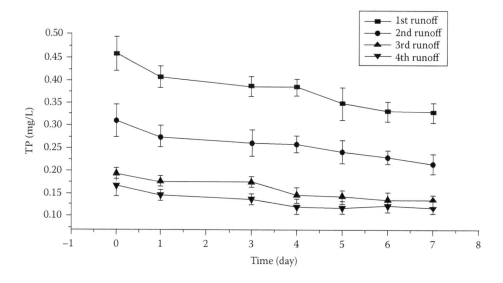

Figure 6.9   Changes of total phosphorus concentration in ecological ditches.

6.3.3 Wetland System Construction Technology

Wetlands are known as the "kidneys of the Earth." Soils, water, aquatic plants, microorganisms, and the atmosphere of wetlands make a good recycling system. Wetlands play an important role in improving local microclimates. An artificial wetland system 1776 m² in area was designed as the last processing unit of runoff water from the paddy fields. The design combined the functions of purification, ecological restoration, and cultural communication and made full use of the rich local aquatic resources. Wetland construction was integrated into the rice field landscape planning, which helped to improve the negative attitude often demonstrated by conventional agriculture toward such innovations. Currently, wetlands are primarily used in the construction of sewage treatment sites

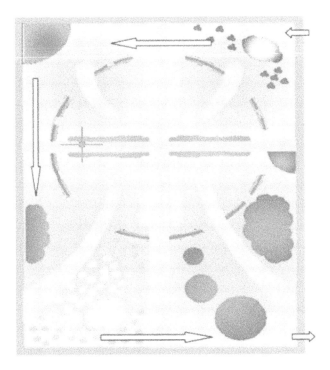

Figure 6.10   Water flow in artificial wetland systems (arrows represent direction of flow).

and for landscape restoration research, and not so much in the field of agricultural production. Upon completion of these particular wetlands, the system integrated ecological restoration, water purification, farming culture, and corporate brand-building, making it possible to promote low-cost ecological pollution control within rural agricultural development (Figure 6.10).

The principles of artificial wetland design include the following:

1. *Ecological priorities*—Try to use biological methods rather than engineering methods whenever possible in order to achieve ecological pollution control, increased biodiversity, and maintenance of ecological balance.
2. *Unimpeded flow*—The function of wetlands as flood storage should ensure that drainage from farm- land is unobstructed even during heavy rains.
3. *Balanced amount of cut and fill during construction of the landscape*—Try to make the excavated soil equal to the volume of the mound of soil, maintaining the balance of cut and fill.
4. *Minimize the economic costs*—An effort to use existing local aquatic plants should be made first in order to control costs in operation and management.

These artificial wetland systems can contibute to ecological pollution control, water soil con- servation, microclimate regulation, and increased biodiversity without an additional power supply. Specifically, drainage ditches funneling drainage water to the wetland system can function as nutri- ent capture and collection sites and play a role in purifying the water before it reaches the wetlands. A wetland is divided into four parts. Water flows in from the southwest to the southeast, then to the northeast, and ultimately discharges at the northeast corner through overflow. In order to achieve good rates of gravity flow and oxygenation, four levels rotating counterclockwise with the water flow are formed from the inlet in the northwest. The elevation of the wetland (ridge side) is set as 0; the elevation of the northwest part is –50 cm, and the elevations of the southwest, southeast, and northeast parts are –55 cm, –60 cm, and –70 cm, respectively. Hydraulic drop is 5 to 10 cm

Figure 6.11 Aquatic plant configurations in artificial wetland.

between successive parts. The aquatic plants, including emergent, floating, and submerged plants, are planted in an artificial wetland with a gravel bed. The plants can remove nutrients efficiently and reduce the content of heavy metals and phenols. The water system design in the ring ditch can be described as follows: Water lettuce, algae, and other aquatic plants in the trench play the role of scenic landscape on normal days. When the weather becomes extreme, the trenches can withstand heavy rain and quickly divert water from farmland drainage. Cash crops, such as seed lotus, arrowhead, and taro, can be added to the plant configuration to generate agricultural output (Figure 6.11).

6.4 SYNERGISM AND POLLUTION CONTROL TECHNOLOGY OF INTEGRATED RICE–FROG AGROECOSYSTEMS THROUGH COMBINATIONS OF PLANTING AND BREEDING

The rational allocation of agricultural production factors and the use of integrated rice–frog agroecosystems that combine planting and breeding techniques have made the goal of achieving nitrogen and phosphorus pollution control in paddy fields possible (Wu et al., 2008).

6.4.1 Frog Cultivation Techniques in Paddy Fields

6.4.1.1 Breeding Tiger Frogs

The tiger frog (*Rana tigrinarugulosa*) is an amphibious cold-blooded animal so climate and temperature have significant impacts on its habitat, feeding, growth, and reproduction (Xiang et al., 2010). The suitable temperature for growth and reproduction is between 20 and 30°C. To overcome the unfavorable factors of low temperatures in winter and spring in Shanghai, adjusting the nutrition provided and the sheltering facilities could help to break the frogs' dormancy and keep the frogs safe during winter to ensure enough brood hatching out the next year. The frog-breeding period occurs from the end of April to early July each year. Recent experiments have shown that temperature and water are the two key hatchery factors for ensuring that the requirements for frog tadpoles in paddy fields are met (Figure 6.12).

Figure 6.12 Breeding and wintering area for tiger frog.

6.4.1.2 Basic Settings for Frog Paddy Fields

The area of each field is approximately 1334 m², and frog ditches are built 0.5 m wide and 0.3 m deep. Three to four feeding platforms (Figure 6.13) are set on each side of the paddy field as places for the tiger frogs to rest and feed. Feeding platforms are 4 to 5 cm higher than the ditch. Ten to 15 days after the rice seedlings are transplanted, frog stocking is carried out on sunny mornings or evenings. The frogs need to be strong and clean, with no residual disease, and must weigh 15 to 20 g each. The appropriate number of tiger frogs is 7500 to 15,000 per hectare. Tiger frogs are stocked in cages in the paddy for 5 to 7 days before being released, which helps them to adapt to the paddy field environment and improve their survival rate.

Figure 6.13 Feeding station surrounding paddy fields.

6.4.1.3 Feed Rationing

To improve the survival rate of tiger frogs in paddy fields, they need to be properly fed. Stopping feedings when the tiger frogs weigh about 50 g encourages them to prey on rice pests and other insects. Feed must be clean, fresh, and provided regularly. Feeding frogs decreasing amounts each time and more frequently each day can improve feed utilization rates.

6.4.2 Ecological Weed Management Techniques

Weeds are important components of plant diversity in farmland ecosystems and have a significant impact on crop production. Weeds absorb light and water, take up crop fertilizer, and act as shelters for pest breeding. Current physical and chemical methods to control weeds are less than ideal, so it is imperative to find a more effective way to control weeds in paddy fields. Duckweed (*Lemna minor*) has been used in paddy fields to control weeds through the shade it provides and other inhibition effects. Due to its rapid growth and reproduction rates, duckweed is often used in sewage treatment. Its leaves are rich in nitrogen, phosphorus, potassium, calcium, and other nutrients that can be used as green manure or animal feed. Numerous studies have examined these aspects of duckweed, but the ability of duckweed to quickly cover water surfaces and control grass weeds in paddy fields had not yet been investigated. Therefore, a study was conducted of the effects of duckweed in paddy fields and its control of grass weeds in an effort to reduce the use of fertilizer, purify agricultural drainage water, and control agricultural non-point-source pollution.

To reduce the seed bank of weeds in the paddy, soil preparation and field irrigation were done as usual before rice planting. Then, excess water in the field was drained in order to induce weed seed germination. After weed emergence, weed seedlings were removed by plowing. Then, duckweed and nitrogen-fixing cyanobacteria were used to block and control weeds after the rice was planted (Figure 6.14). This approach effectively inhibited the germination and growth of weeds in paddy fields; it not only decreased the number of weed individuals but also resulted in a reduction in the total weight of each individual weed. Experimental results showed that the total weights of weeds in paddy fields were reduced effectively—between 74% and 100%—as compared to controls. Moreover, this ecotechnology could have a significant effect on carbon sequestration and nitrogen fixation. It also improves paddy soil permeability and enhances microbial activities. The biological debris of duckweed that settled in paddy soil could improve the water storage capacity of paddy soil and increase soil fertility. The monthly soil carbon sequestration was found to be 43.32 kg ha^{-2}.

Figure 6.14 Test of stocking duckweed to control weeds in paddy field: (left) dry algae plus duckweed, (middle) microbial agents plus duckweed, (right) fresh algae plus duckweed.

6.5 AGRICULTURAL NON-POINT-SOURCE POLLUTION CONTROL IN INTEGRATED RICE–FROG AGROECOSYSTEMS

Based on field experiments, the patterns of nitrogen and phosphorus loss from paddy fields were further investigated to explore the effects of integrated rice–frog agroecosystems. The three treatments were conventional paddy field, green frog paddy field, and organic frog paddy field. The experiment was conducted in 2013 from May to November at the Qingpu Modern Agricultural Park in a water source protection area around the Taihu River region in the upper part of the Huangpu River. The rainfall of the region within a year and between years varies greatly, with an average annual rainfall of 1044.7 mm. The average annual temperature is 15.5°C, with a highest monthly mean temperature of 27.5°C in July and lowest of 3.2°C in January. The soil is purple paddy soil derived from a swamp, which consists of silt loam soil.

In this experiment, the green frog paddy field and organic frog paddy field were the two treatments that stocked tiger frogs for pest control, and the conventional paddy field without frogs acted as a control. A rice–clover crop rotation was used in organic and green frog paddy fields, and the clover was directly plowed in as basal manure, combined with organic fertilizer or chemical fertilizer. In the conventional paddy field, a rice–wheat crop rotation was applied and chemical fertilizer used. The nitrogen levels of the three treatments were all 300 kg N ha^{-2}, but the application time and type of fertilizer used in each of the treatments differed. Each treatment used four plots to monitor nitrogen and phosphorus loss. The arrangement of the plots was random, and the size of each plot was 1600 m^2. Zero to 20-cm topsoil samples were collected before rice planting in order to measure soil properties. Surface water and leaching were sampled on days 1, 3, 5, 7, 11, 15, 20, and 30 after fertilization and once every two weeks after the first 30 days. Samples of runoff were collected whenever runoff occurred. Data and charts were analyzed using Microsoft® Excel® (2010) and SPSS 17.0 statistical software. One-way ANOVA analysis was used for significant difference tests in the various treatments, using a level of significance of $p < 0.05$.

The results showed that the order of decreasing average concentration of total nitrogen (TN) in surface water was conventional paddy field > green frog paddy field > organic frog paddy field. This implies that the conventional paddy field treatment had a higher risk of nitrogen loss in runoff. Compared with the conventional paddy field, the amount of TN loss in the green frog paddy field and organic frog paddy field decreased 15.27% and 25.76%, respectively. Ammonium (NH$_4^+$-N) was the main source of nitrogen loss in runoff, and nitrate (NO$_3$-N) was the main source of nitrogen loss through leaching. The order of total phosphorus (TP) loss was organic frog paddy field > green frog paddy field > conventional paddy field, but the ratio of the total TP loss to TP in fertilization was green frog paddy field > conventional paddy field > organic frog paddy field. The soluble phosphorus (DP) was the main source of TP. In sum, the organic frog paddy field and green frog paddy field could more effectively control nitrogen loss at the same input level as the conventional paddy field.

6.5.1 Nitrogen Loss in Paddy Runoff under Different Treatments

The amount of nitrogen loss in runoff under the three treatments is shown in Table 6.5. TN losses in paddy runoff in the conventional paddy field, green frog paddy field, and organic frog paddy field were 13.56 kg ha^{-2}, 13.85 kg ha^{-2}, and 11.61 kg ha^{-2}, respectively. Although the green frog paddy field had the highest TN loss, there was no significant difference between the green frog paddy field and the conventional paddy field, and the loss in the organic frog paddy field was significantly lower than in the other treatments. Because fertilization schedules differed among the treatments, rain had a variable impact on nitrogen loss in these treatments. The first runoff in the green frog paddy field occurred the second day after fertilization, which may be why it had higher TN loss

Table 6.5   Amount of Nitrogen Loss in Runoff under Three Treatments

Treatment	NO_3^--N (kg ha^{-2})	NH_4^+-N (kg ha^{-2})	Total Nitrogen (TN) (kg ha^{-2})	TN Loss/TN in Fertilization (%)
Conventional paddy field	2.27[a]	3.01[a]	13.56[a]	4.52[a]
Green frog paddy field	4.17[b]	5.64[b]	13.85[a]	4.72[a]
Organic frog paddy field	1.80[a]	3.54[a]	11.61[b]	3.68[b]

Note: Superscript letters indicate significant difference ($p < 0.05$).

Table 6.6   Amount of Nitrogen Loss in Leaching under Three Treatments

Treatment	NO_3^--N (kg ha^{-2})	NH_4^+-N (kg ha^{-2})	Total Nitrogen (TN) (kg ha^{-2})	TN Loss/TN in Fertilization (%)
Conventional paddy field	2.40[a]	0.27[a]	7.13[a]	2.38[a]
Green frog paddy field	2.12[a]	0.13[b]	3.68[b]	1.25[b]
Organic frog paddy field	2.77[a]	0.48[c]	5.75[a]	1.82[b]

Note: Superscript letters indicate significant difference ($p < 0.05$).

Table 6.7   Amount of Phosphorus Loss in Runoff under Three Treatments

Treatment	Soluble Phosphorus (DP) (kg ha^{-2})	Total Phosphorus (TP) (kg ha^{-2})	TP Loss/TP in Fertilization (%)
Conventional paddy field	0.35[a]	0.59[a]	2.24[a]
Green frog paddy field	0.91[b]	1.38[b]	3.02[a]
Organic frog paddy field	1.09[b]	1.56[b]	0.67[b]

Note: Superscript letters indicate significant difference ($p < 0.05$).

in runoff. The nitrogen concentrations of surface water in the organic frog paddy field were lower than in the other two treatments. Thus, the organic frog paddy field had a lower risk of nitrogen loss in runoff. Compared with the green frog paddy field and conventional paddy field, the organic frog paddy field could reduce TN loss in runoff by 1.44% and 16.17%, respectively. In addition, the ratios of TN loss to TN in fertilizer in the conventional paddy field, green frog paddy field, and organic frog paddy field were 4.52% , 4.72%, and 3.68%, respectively.

6.5.2 Nitrogen Loss through Paddy Leaching in Different Treatments

The amount of nitrogen loss through leaching in the three treatments is shown in Table 6.6. TN losses through paddy leaching in the conventional paddy field, green frog paddy field, and organic frog paddy field were 7.13 kg ha^{-2}, 3.68 kg ha^{-2}, and 5.75 kg ha^{-2}, respectively. TN leaching loss in the green frog paddy field was significantly lower than in the other treatments. Compared to the conventional paddy field, TN leaching losses in the organic frog paddy field and green frog paddy field were reduced 19.35% and 48.67%, respectively. In addition, the ratios of TN loss to TN in fertilizer varied from 1.25% to 2.38%, values that were all less than 3% and much lower than the ratio found in the runoff water.

6.5.3 Phosphorus Loss in Paddy Runoff in Different Treatments

The amount of phosphorus loss in runoff in the three treatments is shown in Table 6.7. TP losses in paddy runoff in the conventional paddy field, green frog paddy field, and organic frog paddy field were 0.59 kg ha^{-2}, 1.38 kg ha^{-2}, and 1.56 kg ha^{-2}, respectively. TP loss in runoff was highest in the organic frog paddy field, which was mainly due to using only organic fertilizer. In order to obtain

Table 6.8 Amount of Phosphorus Loss in Leaching under Three Treatments

Treatment	Soluble Phosphorus (DP) (kg ha^{-2})	Total Phosphorus (TP) (kg ha^{-2})	TP Loss/TP in Fertilization (%)
Conventional paddy field	0.21[a]	0.25[a]	1.12[a]
Green frog paddy field	0.31[a]	0.39[b]	1.00[a]
Organic frog paddy field	0.42[a]	0.53[c]	0.26[b]

Note: Superscript letters indicate significant difference ($p < 0.05$).

equal nitrogen input levels in all treatments, the organic frog paddy field had the highest phosphorus input with organic fertilizer. Although the phosphorus input in the green frog paddy field was lower than in the organic frog paddy field, there was no significant difference between TP losses in runoff in both treatments. This may be due to the use of different types of fertilizer. TP losses in runoff in the organic frog paddy field were the highest, but the ratio of phosphorus loss to phosphorus input was the lowest, which means that in addition to the TP absorbed by plants and loss in runoff water the majority of TP remained in the soil in the organic frog paddy field. Compared to the organic frog paddy field, the green frog paddy field and conventional paddy field could reduce TP loss in runoff by 11.54% and 62.18%, respectively, with the ratios of total TP loss to TP in fertilizer being 0.67% and 3.02%, respectively. DP was the main part of TP loss, accounting for 59.32% to 69.88%. Therefore, the main source of phosphorus loss in surface water was due to DP.

6.5.4 Phosphorus Loss through Paddy Leaching in Different Treatments

The amount of phosphorus loss through leaching water in each of the three treatments is shown in Table 6.8. TP losses through paddy leaching in the conventional paddy field, green frog paddy field, and organic frog paddy field were 0.25 kg ha^{-2}, 0.39 kg ha^{-2}, and 0.53 kg ha^{-2}, respectively. The TP leaching loss in the organic frog paddy field was the highest, and there were significant differences between the three treatments. Compared to the organic frog paddy field, TP leaching losses in the green frog paddy field and conventional paddy field were reduced 26.42% and 52.83%, respectively. The ratios of TP loss to TP in fertilizer in the conventional paddy field, green frog paddy field, and organic frog paddy field were 1.12%, 1.00%, and 0.26%, respectively. All of the ratios were below TP losses through runoff, which showed that phosphorus loss through leaching was less than runoff loss.

6.6 DEMONSTRATION AND GENERALIZATION OF INTEGRATED RICE–FROG AGROECOSYSTEMS

Integrated rice–frog agroecosystems are highly integrated technological and agricultural models for pollution prevention and control. Through the establishment of norms and standards, efficient and controllable production can be achieved. In order to ensure better monitoring and control of the planting and breeding processes in urban integrated rice–frog agroecosystems and of the quality of rice–frog products, production technology specifications and standards have been developed, including *Operation Rules for Organic Rice–Frog Products, Quality Management Handbook of Organic Rice–Frog Products, Production Technology Rules for Green Rice–Frog*, and *Quality Management Handbook of Green Rice–Frog Products*, which are being developed by the Zizaiyuan Agricultural Development Company. Such guides will help sustain further development of the model and ensure the quality and safety of products from rice–frog agroecosystems.

Table 6.9 Analysis of Yield and Economic Benefits of Rice–Frog System

Type of Rice	Grain Yield (kg ha^{-2})	Rice Yield (kg ha^{-2})	Rice Price (yuan kg^{-1})	Income (10,000 yuan ha^{-2})	Net Profits (10,000 yuan ha^{-2})	Area (ha^2)
Green frog rice	5769	3750	43	11.29	5.4	100
Organic frog rice	4615	3000	100	21.00	10.2	36

Note: Rice yield equals 65% of the grain yield; outputs were calculated according to wholesale price, which was 30% lower than retail.

6.6.1 Economic Benefits

The demonstration and promotion of integrated rice–frog agroecosystems have allowed us to achieve our goal for clean rice production, resulting in high-quality rice and high production efficiency. Integrated rice–frog agroecosystems have improved upon the current situation of high input, high pollution, and low output systems. An analysis of the yield and economic benefits of "frog rice" is shown in Table 6.9. In 2011, the output levels of green and organic "frog rice" reported by Zizaiyuan Agricultural Development reached 112,900 yuan ha^{-2} and 210,000 yuan ha^{-2}, respectively, much higher than the output of conventional paddy fields.

6.6.2 Ecological and Social Benefits

By breeding tiger frogs and building new water-saving, pollution-control ditches in paddy fields, integrated rice–frog agroecosystems can offer stable production, reduced pesticide and chemical fertilizer use, decreased soil nitrogen and phosphorus loss, and enhanced environmental conservation functions of urban agriculture systems. Promoting integrated rice–frog agroecosystems could be conducive to encouraging the use of fertilizer- and water-saving techniques and bio-organic fertilizers and biopesticides, which would reduce agricultural non-point-source pollution (Tables 6.10 and 6.11). The emissions of greenhouse gases were measured and analyzed in integrated rice–frog agroecosystems in 2012. The results showed that organic frog paddy fields and organic paddy fields had similar trends in three types of greenhouse gas emissions, and the overall amount of emissions was greater than that of conventional paddy fields. The CH_4 and CO_2 emissions of organic frog paddy fields were higher than

Table 6.10 Nitrogen Load in Paddy Fields Under Different Treatments

Type of Paddy Field	Leakage		Runoff		Total Load	
	Load (kg N ha^{-1})	Ratio of N Loss to Fertilization	Load (kg N ha^{-1})	Ratio of N Loss to Fertilization	Load (kg N ha^{-1})	Ratio of N Loss to Fertilization
Conventional	7.13	2.38%	13.56	4.52%	20.69	6.90%
Green frog	3.68	1.25%	13.85	4.72%	17.53	5.97%
Organic frog	5.75	1.82%	11.61	3.68%	17.36	5.50%

Table 6.11 Phosphorus Load in Paddy Fields Under Different Treatments

Type of Paddy Field	Leakage		Runoff		Total Load	
	Load (kg N ha^{-1})	Ratio of P Loss to Fertilization	Load (kg N ha^{-1})	Ratio of P Loss to Fertilization	Load (kg N ha^{-1})	Ratio of P Loss to Fertilization
Conventional	0.25	1.12%	0.59	2.24%	0.84	3.36%
Green frog	0.39	1.00%	1.38	3.02%	1.77	4.02%
Organic frog	0.53	0.26%	1.56	0.67%	2.09	0.93%

Table 6.12 Global Warming Potential (GWP) of Greenhouse Gas Emissions Under Different Treatments

Type of Paddy Field	CH$_4$		CO$_2$		N$_2$O		Total
	Amount (kg ha^{-2})	GWP (kg ha^{-2})	Amount (kg ha^{-2})	GWP (kg ha^{-2})	Amount (kg ha^{-2})	GWP (kg ha^{-2})	GWP (kg ha^{-2})
Organic frog	75.71	1589.91	2583.09	2583.05	0.87	269.7	4442.70
Organic paddy	70.79	1486.59	2334.25	2334.25	0.99	306.9	4127.73
Conventional	27.27	572.77	2611.74	2611.74	1.34	415.52	3600.03

Note: The GWP factors of CO$_2$, CH$_4$, and N$_2$O were 1:21:310.

those of organic paddy fields, but the N$_2$O emissions were the opposite. The N$_2$O and CO$_2$ emissions of conventional paddy fields were higher than those of organic frog paddy fields and organic paddy fields, but the CH$_4$ emissions were significantly lower than in the other two treatments (Table 6.12).

Due to the application of a large volume of organic fertilizers, organic frog paddy fields and organic paddy fields provided a large source of produced CH$_4$, resulting in greater CH$_4$ emissions, but these two treatments effectively reduced N$_2$O emissions. From the 1-year experimental results, the global warming potential (GWP) of organic frog paddy fields and organic paddy fields was significantly higher than for conventional paddy fields. This finding remains to be fully examined and requires more years of testing; however, it is safe to say that organic frog paddy fields and organic paddy fields can reduce the loss of nitrogen and phosphorus in water and improve rice quality. For this reason, these two methods are still worthy of further demonstration and promotion.

6.6.3 Promotion Prospects

6.6.3.1 Demonstrations

A 266.67-ha^2 demonstration rice–frog farm was built in Qingpu District, Shanghai, with the cooperation of the School of Agriculture and Biology at Shanghai Jiao Tong University, Qingpu Modern Agricultural Park, and Zizaiyuan Agricultural Development under the national science and technology support program. The organic rice–frog system received national organic certification in 2010 and good agricultural practices (GAP) certification in 2011. During the 12th Five-Year Plan period, the integrated rice–frog agroecosystem demonstration areas will grow to 666.67 ha^2. Zizaiyuan Agricultural Development also operates a rice–frog farm featuring a 4800-m^2 ecotourism showroom as the core facility. The farm combines harvesting, tourism, exhibitions, catering, accommodations, entertainment, and stage performances within one complex. A visit to the farm can lead to a fresh and tranquil state of mind while visitors enjoy the clean eco-agricultural environment. The farm highlights breeding frogs in paddy fields and using paddy ecosystem food chains and food webs for pest control. Such rice–frog farms utilize eco-agricultural technologies to reduce environmental pollution, ensure the safety of rice products, and achieve added value. Control of paddy non-point-source pollution is achieved through the use of ecological planting and a breeding technology package. Agroecosystems with ecological ditches and wetlands are decorated with a variety of aquatic plants that have a high capacity for nitrogen and phosphorus absorption. These ecological ditches and wetlands intercept chemical contaminants, thus achieving the goal of controlling agricultural non-point-source pollution.

6.6.3.2 Development and Market Analysis

Integrated rice–frog agroecosystems are low-input, high-yield systems offering lower energy consumption and less pollution, thus meeting the goals of developing efficient ecological agriculture and high-quality, safe food production. The possibility of large-scale applications and the promising market prospects of these agroecosystems should continue to attract considerable interest (Xie et al., 2011).

6.6.3.3 *Suitable Agroecosystem Areas*

Integrated rice–frog agroecosystems can be initiated in virtually all single-rice cropping areas in southern China; other single-rice cropping areas may also be suitable for adopting integrated rice–frog agroecosystems.

REFERENCES

Jiang, C.L., Cui, G.B., Fan, X.Q., and Zhang, Y.B. 2004. Purification capacity of ditch wetland to agricultural non-point pollutants. *Environ. Sci.*, 25(2): 125–128.

Wu, J.P. 2008. Study on eco-agricultural mode of "*Orgza sativa–Rana tigrina.*" *Fujian Sci. Technol. Rice Wheat*, 26(1): 1–5.

Xiang, J.J., Wang, H.J., and Lu, J.H. 2010. Study on matching technology of integrated rice–frog ecosystems. *Shanghai Agric. Sci. Technol.*, 4: 65–66.

Xie, J., Hu, L.L., Tang, J.J., Wu, X., Li, N., Yuan, Y.G., Yang, H.S., Zhang, J.E., Luo, S.M., and Chen, X. 2011. Ecological mechanisms underlying the sustainability of the agricultural heritage rice–fish co-culture system. *Proc. Natl. Acad. Sci. U.S.A.*, 108(50): 1381–1387.

Zhang, G., Wang, D.J., and Chen, X.M. 2008. Effects of reduced fertilizer application on environmental quality of paddy field. *Chin. J. Eco-Agric.*, 16(2): 327–330.

Zhang, W.L., Xu, A.G., Ji, H.J., and Kolbe, H. 2004. Estimation of agricultural non-point source pollution in China and the alleviating strategies. III. A review of policies and practices for agricultural non-point source pollution control in China. *Sci. Agric. Sin.*, 37(7): 1026–1033.

CHAPTER 7

Research on and Application of Rice Allelopathy and Crop Allelopathic Autotoxicity in China

Lin Wenxiong, Fang Changxun, Wu Linkun, and Lin Sheng

CONTENTS

Since allelopathy was defined by Elroy Rice (1984), important research progress has been made on plant allelopathy. At the beginning, allelopathy was widely understood only for the harmful effects of one plant on another plant due to allelochemicals the first plant releases into the environment. However, the definition of allelopathy has been expanded to include inhibitory, autotoxic, and stimulatory effects. In general, there are four types of plant allelopathy: amensalism, autotoxicity, autostimulation, and facilitation. Rice allelopathy, defined by amensalism, has

First 30 days

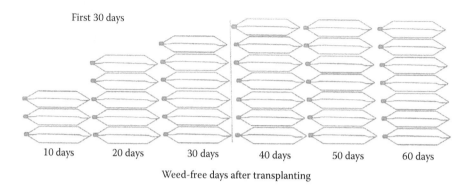

10 days   20 days   30 days   40 days   50 days   60 days

Weed-free days after transplanting

Figure 7.1   Important stages for weed control during rice production.

been on the research agenda for decades. The amensalism of rice allelopathy has been applied to weed control, and great achievements have been made in the process. Allelopathic autotoxicity, which can come about as a consequence of consecutive monoculture, has been shown to be a serious soil disease in plants, especially in consecutively cropped Chinese medicinal herbs, resulting in a decline in biomass and plant quality. With the in-depth study of rhizosphere ecology, research on plant allelopathy gradually came to focus on rhizospheric biological processes. The "cross-talk" between plants and microbes is one of the key factors of plant allelopathy. Increasingly, studies have found that various positive and negative plant–plant interactions within or among plant populations—such as amensalism, autotoxicity, autostimulation, and interspecific facilitation—have all resulted from the integrative effects of plant–microbe interactions mediated by root exudates. This deep understanding of plant–soil–microbe interactions mediated by root exudates has important implications for elucidating the mechanisms of different types of plant allelopathy and the functions of rhizosphere microecology, in addition to providing practical guidelines in agriculture production.

7.1 RICE ALLELOPATHY AND ITS APPLICATIONS FOR WEED CONTROL

As a vital Chinese industry, agriculture employs over 650 million farmers. About 75% of China's cultivated area is used for food crops. Rice is one of the most important food crops in China, accounting for about 30% of the cultivated area. Weeds, including *Echinochloa crus-galli*, *Cyperus difformis*, and *Leptochloa chinensis*, are the causes of serious yield reduction in rice production in China (Figure 7.1). More specifically, about 20 million tons of rice are lost annually due to weed interference. That amount of rice could feed at least 50 million people for a year. To ensure a consistently high production yield, millions of tons of herbicides and pesticides are used every year. In the 21st century, the global use of herbicides for weed control has increased from 650,000 tons to 2,000,000 tons, an increase of 207.69%, while grain yield increased by only 12.28% (from 1.889 billion tons to 2.121 billion tons). In China, 1,799,823.17 tons of herbicides were produced in 2013, reaching a new high of 30% of the total pesticides produced in China (Figure 7.2); however, the negative impacts of commercial herbicide use on the environment make it undesirable in diverse weed management options. Using crop allelopathy to control weeds is now considered one of the most environmentally friendly approaches in sustainable agriculture, which has been one of the focuses of agroecology (Figure 7.3).

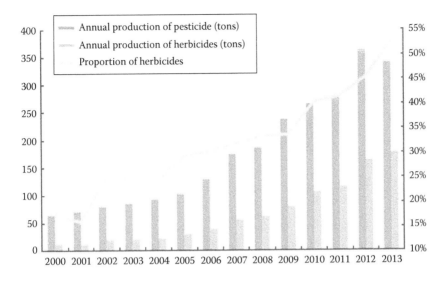

Figure 7.2 Annual production of pesticides and herbicides from 2000 to 2013 in China (http://www.askci.com search results).

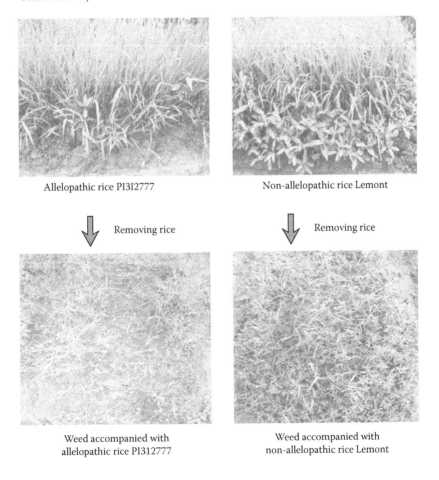

Figure 7.3 Weed occurrence with different allelopathic rice accessions in the field.

Figure 7.4   The phenomenon of rice allelopathy in the field.

7.1.1 Concept of Rice Allelopathy

Rice varieties with allelopathic inhibition on accompanied weeds are regarded as allelopathic rice accessions (Figure 7.4). The relevant studies on rice allelopathy began about 20 years ago in the United States. Dilday first observed allelopathy in rice in field examinations in Arkansas, where about 191 of 5000 rice accessions inhibited the growth of *Heteranthera limosa* (Dilday et al., 1989). A large field-screening program was conducted after this finding. More than 16,000 rice accessions from 99 countries in the U.S. Department of Agriculture's Agricultural Research Service (USDA-ARS) germplasm collections have been screened. Of these, 412 accessions inhibited the growth of *H. limosa* and 145 accessions inhibited the growth of *Ammannia coccinea* (Dilday et al., 1994, 1998). In Egypt, 1000 rice varieties were screened for a suppressive ability against *Echinochloa crus-galli* and *Cyperus difformis* under field conditions, and more than 40 of them were found to have a weed-suppressive ability (Hassan et al., 1998). Similar investigations have been conducted in other countries, and many rice varieties were found to inhibit the growth of different weeds (K.U. Kim and Shin, 1998; Olofsdotter et al., 1999; Pheng et al., 1999). A large number of rice varieties were found to inhibit the growth of several plant species when these rice varieties were cultured together with these plants in the field and/or in laboratory conditions (Dilday et al., 1994, 1998; K.U. Kim et al., 1999; Azmi et al., 2000; Gealy et al., 2003; Seal et al., 2004; S.Y. Kim et al., 2005).

Researchers successfully used relay seeding in agar (RSA) to evaluate the variability in allelopathic potential of 57 rice accessions on barnyardgrass. Five rice cultivars—Iguape Cateto, PI312777, Azucena, Taichung Native1, and IAC25—showed over 50% inhibitory rates (IRs) on barnyardgrass root growth. The IRs of 12 cultivars ranged from 40 to 50%, 21 cultivars had 30 to 40% IRs, and 13 cultivars exhibited 20 to 30% IRs on the target weed. The IRs of 6 cultivars were less than 20% (Figure 7.5) (Shen et al., 2004). The National Rice Research Institute of China reported that the rice varieties I-Kung-Pao, Parahainakoru, and HB-1 have valuable rice germplasm with high allelopathic potential. The inhibitory rates of these rice accessions on the length of barnyardgrass root were 57%, 64%, and 55%, respectively. The allelopathic indices (AIs) were 0.61, 0.56, and 0.59, respectively, similar to the allelopathic rice P1312777 developed in the United States and considered as a control in the allelopathy experiment (Zhou et al., 2005). Screening of allelopathic rice varieties has also been conducted in various Asian countries, including Japan, Bangladesh, and Korea.

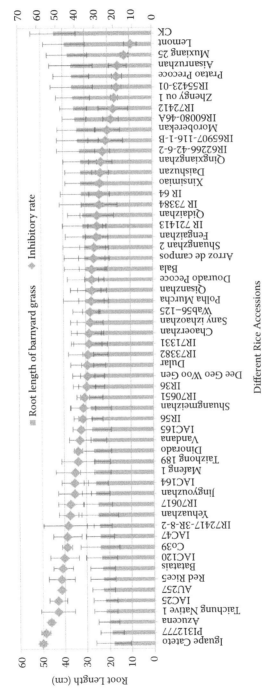

Figure 7.5 Inhibitory rates of 57 rice accessions on the roots of barnyardgrass by relay seeding in agar (RSA).

These rice accessions are important germplasm resources for use in further study on the mechanism of allelopathic weed inhibition. Using the allelopathy and competition separation (ACS)-based approach to successfully separate allelopathic effects from competition, we found that allelopathic rice accession PI312777 exhibited higher allelopathic potential to suppress the growth of accompanying weeds, especially when they were exposed to stressful conditions. These results show that the total biointerference (TB) of allelopathic rice PI312777 on barnyardgrass (BYG) was about twice that of the non-allelopathic rice Lemont (LE), and the allelopathic effect (AE) of PI312777 on BYG was about four times that of LE in rice–BYG mixed-cultures. It was also found that the AE of PI312777 on BYG was approximately 3 to 4 times that of LE under normal or higher nitrogen supply treatments (10 to 20 mg/L), but it was 15 times that of LE under low N treatment (5 mg/L) (Song et al., 2008; He et al., 2012b). These findings suggested that rice allelopathy could be regulated by factors in the external environment, especially by stressful conditions; however, the problem of determining whether the trait of rice allelopathy could be improved by genetic engineering has attracted the attention of many scientists.

7.1.2 Mechanism of Heredity and Action Mode in Rice Allelopathy

7.1.2.1 Mapping of Quantitative Trait Loci for Rice Allelopathy

Evidence in earlier studies has indicated that rice allelopathy is a quantitative trait that is mediated by both genetic effects and environmental conditions (Putnam and Tang, 1986; Dilday et al., 1998, 2000; W.X. Lin et al., 2003; Xiong et al., 2007). The genetic control of allelopathy in rice has been assessed. Several studies have mapped quantitative trait loci (QTL) for allelopathy in rice. Jensen et al. (2001) identified four QTL on three chromosomes, which collectively explained 35% of the total phenotypic variation of allelopathic activity in the population. Ebana et al. (2001) also identified several QTL, among which one QTL on chromosome 6 had the largest effect, explaining 16.1% of the phenotypic variation. Other studies were also reported by Zeng et al. (2003), Z.H. Xu et al. (2003), and S.B. Lee et al. (2005). More recently, we investigated the digenic epistatic effects and their interactions with the environment for allelopathy in rice (*Oryza sativa* L.) (Xiong et al., 2007). Two additive-effect QTL were mapped on chromosomes 5 and 10, which explained 6.95% and 4.35% of phenotypic variation, respectively, and one of them interacted with the environment significantly. Three pairs of QTL with significant additive–additive epistatic effect interactions were detected on chromosomes 1, 3, 4, 5, and 10, and one of them exhibited significant interaction with the environment. The findings suggested that a "favorable" gene with positive additive effects could become "unfavorable" (i.e., deleterious) in a new genetic background due to negative epistatic effects. Therefore, for allelopathy in rice, QTL mapping and selection experiments should place more emphasis on identifying the best multi-locus allelic combinations, instead of pyramiding individual favorable QTL alleles.

7.1.2.2 Rice Allelochemicals with Weed-Suppressive Effects

As mentioned above, rice allelopathy results from allelochemicals released from the donor plants of rice varieties into paddy fields, which inhibit the neighboring target plants known as receiver species, such as barnyardgrass. Previous studies showed that rice allelochemicals could be divided into two chemical species: the compounds of phenolic acids and terpenoids. However, which of them plays the main role in this process of weed suppression is still disputed.

7.1.2.2.1 Allelopathic Activity of Terpenoids in Rice Accessions

The allelopathic potential of rice is dependent mainly on the target species and the content of allelochemicals. Although there is disagreement among researchers regarding which compounds should be classified as rice allelochemicals, a number of putative candidates, such as momilactone, terpenoids, steroids, and phenolic acids, have been reported. Kato and Takahashi first isolated momilactone A and B from rice husks as growth inhibitors (Kato et al., 1973; Takahashi et al., 1976), and the two metabolites were later found in rice leaves and straw as phytoalexins (Cartwright et al., 1977, 1981; Kodama et al., 1988; C.W. Lee et al., 1999). A comparison of the growth inhibitory activity of momilactone A and B showed that momilactone B was much more effective than momilactone A, and it was found that some rice accessions exhibited stronger allelopathic effects than others due to their higher concentrations of the two momilactone compounds (Takahashi et al., 1976; Kato et al., 1977). However, investigation of the allelopathic inhibition mechanism of momilactones is still limited.

Momilactone A and B were also reported to inhibit the growth of *Amaranthus lividus*, *Digitaria sanginalis*, *Poa annua*, *Echinochloa crus-galli*, and *E. colonum*, which are the most troublesome weeds in rice fields (Chung et al., 2005). The effectiveness of momilatone B on growth inhibition is much greater than that of momilactone A at the same concentration. The growth inhibitory activities of momilactome B are also greater than those of momilactone A for other bioassay systems (Takahashi et al., 1976; Kato et al., 1977; Fukuta et al., 2007; Toyomasu et al., 2008). The inhibitory activities of momilactone A and B on the root and shoot growth of rice seedlings were 1 to 2% and 0.6 to 2%, respectively, of those on the root and shoot growth of *E. crus-galli* and *E. colonum*. Thus, the effectiveness of momilactone A and B on the growth of rice seedlings was much less than on the growth of *E. crus-galli* and *E. colonum*.

These results suggest that the toxicities of momilactone A and B to rice seedlings are much less than those to the two weed species (Kato-Noguchi et al., 2008). Momilactone A and B were secreted from rice plants into the rhizosphere throughout all life-cycle stages of rice (Kato-Noguchi et al., 2003, 2008, 2010). The momilactones are labdane-related diterpenoids, whose biosynthesis is initiated by dual cyclization reactions sequentially catalyzed by class II diterpene cyclases and class I diterpene synthases (Peters, 2006). In particular, initial cyclization of the general diterpenoid precursor (*E,E,E*)-geranylgeranyl diphosphate (GGPP) to *syn*-copalyl diphosphate (*syn*-CPP) is catalyzed by the rice CPP synthase OsCPS4 (Otomo et al., 2004b; M.M. Xu et al., 2004). This is followed by further cyclization of *syn*-CPP to *syn*-pimaradiene by the rice kaurene synthase-like OsKSL4 (Otomo et al., 2004a; Wilderman et al., 2004). Notably, the genes encoding OsCPS4 and OsKSL4 are found close together in the rice genome (Wilderman et al., 2004). Furthermore, this region also contains a dehydrogenase that catalyzes the final step in the production of momilactone A (OsMAS) and two closely related cytochromes P450 (CYP99A2 and CYP99A3), one or both of which are required for momilactone biosynthesis (Shimura et al., 2007). More specifically, CYP99A3 has recently been shown to act as a *syn*-pimaradiene oxidase that produces *syn*-pimaradien-19-oic acid, presumably as an intermediate enroute to the 19,6-clactone ring of the momilactones (Q. Wang et al., 2011).

Weidenhamer et al. (1993) also determined the aqueous solubility of 31 biologically active monoterpenes. They found that whereas hydrocarbons had a low solubility (<35 ppm) the oxygenic monoterpenes exhibited solubilities one or two orders of magnitude higher, ranging from 155 to ~6990 ppm for ketones and 183 to ~1360 ppm for alcohols. Many monoterpenes are phytotoxic in concentrations below 100 ppm. In our previous studies, some oxygenic terpenoids, such as limonene dioxide, menthol, carvone oxide, and cedrol, were detected in the diethyl ether extracts of root exudates from the allelopathic rice variety PI312777 cultured in paddy soil and sand media (He et al., 2005a,b). The phytotoxicity of five analogs—(+)-carvone, (+)-cedrol, (–)-carveol, (–)-carvyl acetate, and (–)-menthone—were further evaluated in bioassay. The results showed that four of the five analogs had inhibitory effects on the root growth of *Echinochloa crus-galli*. An inhibitory rate (IR) of 20 to 41% at a dosage of 15 ppm was observed, except for (+)-carvone, which was a stimulant at 0.1 to 15 ppm (He et al., 2006). Another compound, an oxygenic sesquiterpenoid called (+)-cedrol, was detected in rice root exudates (He et al., 2005a,b).

From the gyrating regression design experiment in the investigation of allelopathic effects of oxygenic terpenoids on barnyardgrass, the results showed that the main effects of (–)-carveol, (+)-carvone, (–)-menthone, and (–)-carvyl acetate had a downward parabolic curve, whereas (+)-cedrol had the opposite. Boundary effects of single factors indicated that, when single-factor concentration fell below –2 (0.001 mmol/L), the inhibition trend of the oxygenic terpenoids on barnyardgrass root length was as follows: (–)-menthone > (–)-carveol > (+)-carvone > (–)-carvyl acetate > (+)-cedrol. When the single-factor concentration was above +2 (0.230 mmol/L), the order of inhibition trend of the oxygenic terpenoids followed the sequence (+)-cedrol > (+)-carvone > (–)-carvyl acetate > (–)-carveol > (–)-menthone. Furthermore, when the concentrations of (–)-carveol, (+)-carvone, (–)-menthone, (–)-carvyl acetate, and (+)-cedrol were 0.033, 0.030, 0.080, 0.020, and 0.001 mmol/L, respectively, the inhibition rate could theoretically reach up to 93.69%. The optimum combination of the five terpenoids was then determined by orthogonal rotatable central composite design for five variables and five concentration levels in the greenhouse. The result obtained from hydroponic experiments showed that the optimum combination of the five terpenoids significantly inhibited root length, plant height, and plant dry weight of *Echinochloa crus-galli*, causing a considerable decline in the superoxide dismutase (SOD), peroxidase (POD), and catalase (CAT) activities in the roots and leaves of *E. crus-galli*. In paddy soil experiments, the joint action of the added compounds was much higher than the effect of any single compounds used in suppressing *E. crus-galli* development in monoculture and grown with rice. At the same time, no inhibitory effects were observed on rice accessions in hydroponics of rice mixed with barnyardgrass (He et al., 2009). It is therefore suggested that rice allelopathy might result from the joint action of terpenoids.

When we further investigated why the effects of rice allelopathy could be significantly increased by stressful conditions, as mentioned above, it became clear that the increased impact of rice allelopathy was not due to the enhancement of terpenoid synthesis. Rather, it can be attributed to the upregulated expressions of genes coding for the main enzymes related to phenolic synthesis in allelopathic rice under lower nitrogen supply in the hydroponics of rice–barnyardgrass mixtures, as shown in Figure 7.6. The results showed that among the 12 key genes coding for the enzymes involved in terpenoid synthesis, six genes coding for 3-hydroxy-3-methylglutaryl-CoA synthase (HMGS), 3-hydroxy-3-methylglutaryl-CoA reductase (HMGR), mevalonate kinase (MK), geranylgeranyl diphosphate synthetase (GGDS), prenyltransferase (PT), and phytoene synthase (PS) were upregulated. The others, including geranyl diphosphate synthetase (GDS), farnesyl diphosphate synthetase (FDS), monoterpenes cyclase (MC), sesquiterpene cyclase (SC), squalene synthase (SS), and diterpene cyclase (DC), were downregulated in allelopathic rice accession.

A similar tendency was also observed in the non-allelopathic rice variety Lemont in the same nitrogen-stressed condition, suggesting that the two rice accessions have the same molecular behavior in terpenoid metabolism in response to lower nitrogen supply. No significant difference in the concentrations of terpenoids was observed in the comparison of the two rice accessions under the same stress conditions (Figures 7.6 and 7.7). Further analysis showed that low nitrogen supply led to significantly declined expressions of the key genes related to monoterpene, sesquiterpene, diterpene, and triterpene. However, the opposite was true in the case of phenolic metabolism in the two rice accessions under the N-stressed condition, thus indicating that all nine genes coding for the key enzymes participating in a phenolic synthesis pathway were significantly upregulated in allelopathic rice, but the reverse was true in the case of non-allelopathic rice. The exception is that the phenylalanin ammonia lyase (PAL) gene showed a slight upregulation in response to N stress (Figure 7.6),

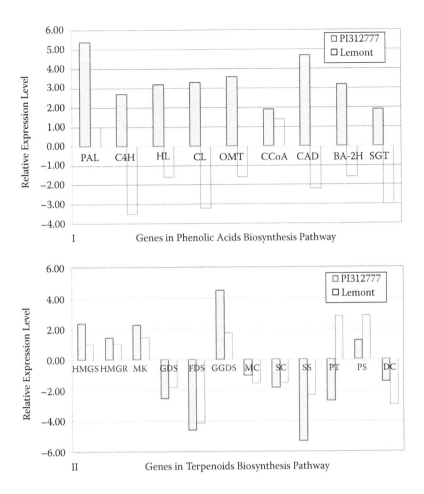

Figure 7.6 The changes in gene expression of the 9 enzymes in phenolic metabolism (I) and 12 enzymes in terpenoid metabolism (II) in the roots of the two rice accessions under low nitrogen treatment (low nitrogen treatment [1/4 N] vs. normal nitrogen treatment [1 N]).

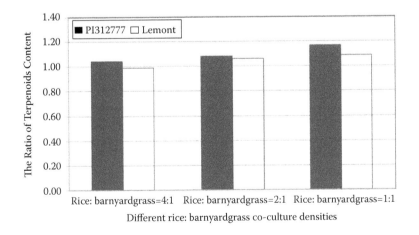

Figure 7.7 The changes in relative contents of terpenoid compounds in rice root exudates under different rice–barnyardgrass co-culture densities (co-culture rice vs. monoculture rice).

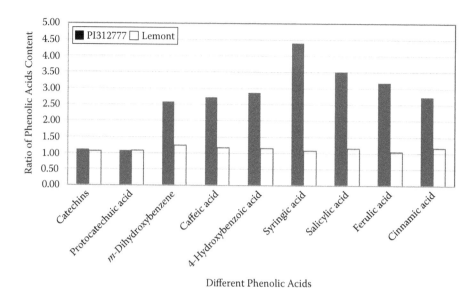

Figure 7.8 The changes in relative contents of phenolic acids in root exudates of the two rice accessions co-cultured with barnyardgrass under different nitrogen treatments (low nitrogen treatment [1/4 N] vs. normal nitrogen treatment [1 N]).

thereby resulting in significantly lower concentrations of phenolic acids in non-allelopathic rice than in its counterpart (Figure 7.8). Accordingly, phenolic acids were considered to be allelochemicals, especially under lower N stress and higher densities of target weeds (e.g., barnyardgrass) (Song et al., 2008; Fang et al., 2010).

7.1.2.2.2 Allelopathic Activity of Phenolic Acids in Rice Accessions

Plant phenolic compounds are derived from cinnamic acid, which is formed from phenylalanine by phenylalanin ammonia lyase (PAL). This key enzyme in the biosynthesis of phenolic compounds catalyzes the transition from primary (shikimate pathway) to secondary (phenylpropanoid pathway) metabolism (Dixon and Paiva, 1995; Silverman et al., 1995; Buchanan et al., 2000; Sawada et al., 2006). Cinnamic acid is a branch point in the phenylpropanoid pathway. One branch is to benzoic acid derivatives (molecules with a C6–C1 skeleton), such as benzoic acid and salicylic acid. Another branch leads to cinnamic acid derivatives (molecules with a C6–C3 skeleton)—for example, to *p*-coumaric acid by cinnamate 4-hydroxylase (C4H), to caffeic acid by ferulic acid 5-hydroxylase (F5H), to ferulic acid by caffeic acid *O*-methyltransferases (COMT), and to 5-hydroxyferulic acid by F5H. As mentioned above, the allelopathic potential of rice could be enhanced under abiotic stress, accompanied by an increase of phenolic compounds and the upregulation of related biosynthesis genes (Shin et al., 2000; H.B. Wang et al., 2008, 2010; Xiong et al., 2010).

Activation of the genes encoding key enzymes involved in the phenylpropanoid pathway leads to the increased release of allelochemicals, which in turn suppresses the growth of accompanying weeds. Elucidation of the functional genes of allelopathic rice in regulating the adaptation of plants to stressful conditions will help develop a possible method of enhancing the allelopathic potential of rice using biotechnology. Song et al. (2008) found that the allelopathic rice varietal PI312777 exhibited greater allelopathic potential than the non-allelopathic rice Lemont in suppressing the growth of accompanying weeds, especially under low nitrogen levels (5 mg/L of nitrogen content) in an experimental culture solution. The result could be attributed to higher gene transcript levels of key enzymes, such as PAL and cytochrome P450, in the phenylalanine metabolism pathway. However, in the rice–weed co-culture mixture, there were two stress factors: low nitrogen content and target weeds (barnyardgrass). These stress factors may evoke different response mechanisms that enhance the allelopathic potential or plant defenses, and the response mechanisms of PI312777 to these two stress factors remain unknown. Bi et al. (2007) found that transcription of the *PAL* and *C4H* genes in allelopathic rice leaves was significantly increased after their exposure to methyl jasmonate and methyl salicylate.

Similar results presented in the studies of Song et al. (2008) and Fang et al. (2009) showed that the level of *PAL* gene expression was significantly increased in the allelopathic rice line PI312777 compared with its counterpart, the non-allelopathic rice Lemont, when the nitrogen supply was restricted or when the leaves were sprayed with exogenous salicylic acid, resulting in an enhanced inhibitory effect on target weeds. Other genes that are involved in the biosynthesis of phenolic compounds, including *C4H*, *F5H*, and *COMT*, were also found to be more strongly upregulated in the allelopathic rice varietal PI312777 than in the non-allelopathic rice varietal Lemont under different rice–weed ratio conditions. This was attributed to the higher levels of phenolic acids in PI312777 (He et al., 2012a), but a study by Olofsdotter et al. (2002) pointed out that phenolic acids released from living allelopathic rice roots were unlikely to reach phytotoxic levels. The study argued that, because the phytotoxicity of these putative phenolic allelochemicals was evaluated in the laboratory, it was not representative of actual environmental conditions. Therefore, the identity of the allelochemicals of allelopathic rice is still being disputed.

Recently, many scientists have documented that plant allelochemicals released from roots could be transformed by surrounding microorganisms. The degraded products of this process might play a key role in weed growth inhibition, which implies that plant allelopathy might result from plant–microorganism interactions in the rhizosphere (Bertin et al., 2003; Inderjit, 2005). Phenolic acids, flavonoids, and momilactones released from plants have all been found to be degraded and transformed in the soil of paddy fields within 2 to 4 weeks (Blum, 2011; R.Y. Lin et al., 2011; Kato-Noguchi and Peters, 2013; Weston and Mathesius, 2013). So, it is very crucial to investigate the fate of allelochemicals and their ecological processes, including how allelochemicals influence the physical, chemical, and biological characteristics in rhizosphere soil. Such research will further our understanding of the mechanisms of rice allelopathy in the field.

In a word, it is still too early to reach a conclusion regarding the main allelochemicals in rice. Among the putative allelochemicals, terpenoids—especially momilactones—are recognized as rice allelochemicals, due to their lower bioactive dosage with a high inhibitory impact on weeds in laboratory conditions; however, there are few reports about the bioactivity of these terpenoids on weeds under actual field soil conditions. There is also still a lack of field trial evidence to support a simple conclusion regarding the vital role they play in rice allelopathic processes. For phenolic acids, higher concentrations are required to affect the same inhibitory rate on weeds, compared to concentrations of terpenoids detected in laboratory conditions. However, trial evidence obtained from laboratory tests and field experiments support the fact that any increased inhibitory rate is due to enhanced concentrations of phenolic acids, which results from the upregulation of the genes encoding the key enzymes associated with phenolic synthesis in allelopathic rice, as a result of exposure to stressful conditions such as lower nitrogen supply or higher accompanying weed (barnyardgrass) density. It is worth mentioning that both phenolic acids with higher bioactive dosage and terpenoids with lower bioactive dosage could be degraded by soil microorganisms, and this process might mediate the changes in microbial diversity in the rhizosphere. Further study on the rhizospheric process of the interactions between putative allelochemicals and soil microorganisms would help to uncover the underlying mechanisms of rice allelopathy.

7.1.2.3 *Role of Rhizospheric Microflora in Rice Allelopathy*

It has been observed that most of the rice accessions exhibit inhibitory effects on weeds in the laboratory; however, these effects become insignificant in field trials. It is yet unknown whether this phenomenon is caused by the antagonistic effect from rhizospheric microbes in the field or by the loss of allelopathic compounds in the soil. We conducted a study to investigate the dynamics of microbial populations and their functional diversities in the seedling rhizosphere soils of rice accessions with different allelopathic activities. The results showed that rice cultivars significantly affected the microbial carbon content in rhizospheric soil (R.Y. Lin et al., 2007). Recently, many technologies have been used to explain the diversity of soil microorganisms on the molecular level, such as terminal restriction fragment length polymorphism (T-RFLP), denaturing gradient gel electrophoresis (DGGE), and single-strand-conformation polymorphism (SSCP). In our previous study, the T-RFLP technique was employed to investigate microbial community structure and composition in the rhizosphere of weak and strong allelopathic rice accessions. The rice accessions had been directly sown on dry fields at the five-leaf stage and the seven-leaf stage. These soils along with the control soil (without rice) were tested to further understand the relationship between the rice allelopathic inhibitory effect and microbial flora in rhizosphere soil. The results showed that the rank-abundance of terminal restriction fragments (TRFs) from different samples showed extremely significant differences, indicating that the relative abundance of rhizosphere bacteria was markedly different.

The diversity indices, including the species abundance (S), Shannon–Wiener diversity index (H), Pilieu diversity index (D), and evenness index (E), were used to analyze the diversity of TRFs in six samples. The Shannon–Wiener diversity in the sample of allelopathic rice accession PI3127777 was the highest out of all six samples. Cluster analysis and principle component analysis (PCA) suggested that the similarity among the three samples of the five-leaf stage was greater than that for the seven-leaf stage samples. In particular, the three samples of seven-leaf stage were separated by not only principal component I but also principal component II, which implies that the composition and structure of the soil microorganism community could be regulated by both the growth stage of rice and the rice variety. It was found that 34 bacteria were unique to PI3127777, 7 genera of which belong to myxobacteria. Some sliding bacteria and N-fixing bacteria were also found as unique genera associated with PI3127777. It has been reported that secondary metabolites of myxobacteria may improve the allelopathic potential of allelopathic rice by inhibiting germination of weed seeds in the soil seed bank and decreasing the density of weeds in rice fields (Xiong et al., 2012).

We also used T-RFLP to analyze and compare the differences in microorganism diversity in the rhizospheres of allelopathic rice PI312777 and its counterpart Lemont under low nitrogen conditions. The results showed that some microorganisms functioning in the nitrogen cycle, nitrification, nitrogen fixation, and exotoxin production were significantly increased in the rhizosphere of PI312777 compared with those in the rhizosphere of its counterpart. These microorganisms contribute to enhancing tolerance in nutrient-limited conditions via the regulatory machinery that enhances nitrogen utilization efficiency and the allelopathic effect in suppression of the target weed. So, it is critical to actual use to further investigate how the root exudates from the allelopathic rice mediate the chemotaxis aggregation of the specific microorganisms in rhizosphere soil. This in turn may reveal the process and mechanism of rice allelopathy in the suppression of weeds in paddy fields, especially under stressful conditions.

7.1.3 Application Prospects of Rice Allelopathy in Sustainable Agriculture

At present, sustainable agriculture has received increased attention from agricultural scientists, ecologists, and social economists. Sustainable agriculture requires making efficient use of a farm's internal resources and relying on a minimum of essential external inputs (Tesio and Ferrero, 2010). Putnam and Duke (1974) first evaluated the possibility of using allelopathic crops to manage weeds

Allelopathic rice PI312777   Non-allelopathic rice Lemont

Harvesting rice   Harvesting rice

Weed residue   Weed residue

Figure 7.9 Study on the accompanying weed biomass among different allelopathic rice accessions under field conditions.

in agricultural sites in order to minimize serious problems in current agricultural production, including environmental pollution, human health concerns, and depletion of biodiversity. Allelopathy may represent a new frontier for the implementation of applicable practices based on integrated weed management strategies that use suppressive cover crops, crop rotation, and the selection of varieties with strong allelopathic potential to biologically reduce weed intensity. We used five different allelopathic rice accessions (allelopathic rice of PI-1, Taichung Native 1, Azucena, IAC47, and non-allelopathic rice Lemont) to analyze accompanying weed biomass and rice grain yield under field conditions from 2008 to 2010 (Figure 7.9).

The results indicated that weed biomass was the highest under the non-allelopathic rice Lemont and lowest under the allelopathic rice PI-1. Namely, allelopathic rice PI-1 had the highest inhibitory effect on weeds, and non-allelopathic Lemont had the lowest. Thus, allelopathic rice PI-1 had the

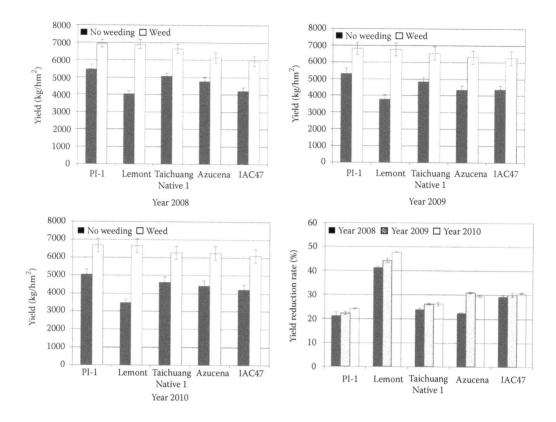

Figure 7.10   Rice yield indices of different allelopathic rice accessions under weeding and non-weeding treatments.

highest grain yield; the reverse was true for Lemont under the non-weeding treatment. Allelopathic rice PI-1 also had the highest grain yield, and IAC47 had the lowest yield under the weeding treatment. The correlation analysis on weed biomass, rice allelopathic potential, and rice yield indicated that grain yields of different allelopathic rice accessions showed a significant negative correlation with weed biomass while showing a significant positive correlation with allelopathic potential. Based on the result, allelopathic rice PI-1 had the highest inhibitory effect on paddy field weeds; the reverse was true in the case of non-allelopathic Lemont in a comparison of five different allelopathic rice accessions over 3 years. Without the weeding treatment, the PI-1 grain yield also exceeded the yields of the other rice accessions (Figure 7.10). Thus, rice accession PI-1 was the strongest allelopathic accession with the highest grain yield among the five rice accessions; the reverse was true in the case of Lemont (H.B. Wang et al., 2012).

Z.H. Xu et al. (2013a,b) also carried out a study to investigate the allelopathic abilities of tested rice materials on the three main paddy field weeds. The results showed that the superb allelopathic rice variety Xiayitiao had the strongest ability to decrease the aboveground dry weight of barnyardgrass. Early indica hybrid rice Zhong 9A/602 and early rice Zaoxianzhe 101 also exhibited inhibitory effects on the aboveground dry weight of barnyardgrass. Rice allelopathic variety TN1, which has an allelopathic effect on several weeds such as *Echinochloa crus-galli*, *Trianthema portulacastrum*, *Heteranthera limosa*, and *Ammannia coccinea*, could also effectively control the aboveground dry weight of barnyardgrass. These findings indicated that rice materials could effectively inhibit the aboveground dry weight of barnyardgrass mainly by suppressing its tillering ability.

Another important approach is the application of breeding and transgenic techniques to strengthen allelopathic traits into useful crops. Making progress in understanding the mechanisms of allelochemicals, physiological modes of action, and genetic regulation of biosynthesis should be the basis for manipulation of germplasm resources. In our study, gene expression of *PAL* in allelopathic rice PI312777 was inhibited by RNA interference (RNAi); the transgenic rice showed lower levels of *PAL* gene expression and PAL activity than wild-type (WT) rice. Concentrations of phenolic compounds were lower in the root tissues and root exudates of transgenic rice than in those of wild-type plants. When barnyardgrass (BYG) was used as the receiver plant, the allelopathic potential of transgenic rice decreased (Figure 7.11).

The sizes of bacterial and fungal populations in rice rhizospheric soil at the three-, five-, and seven-leaf stages were estimated using real-time reverse transcription–polymerase chain reaction (qRT-PCR), which showed a decrease in both populations at all stages of leaf development; however, PI312777 had a larger microbial population than transgenic rice. In addition, in T-RFLP studies, 14 different groups of bacteria were detected in WT and only 6 were detected in transgenic rice. This indicates that there was less rhizospheric bacterial diversity associated with transgenic rice than with WT. These findings collectively suggest that *PAL* functions as a positive regulator of rice allelopathic potential. The evidence presented here provides a relevant molecular target for breeding and/or metabolic engineering efforts in rice. In particular, increased allelochemical production levels can serve as a target for selective molecular breeding and/or genetic and metabolic engineering.

7.2 ALLELOPATHIC AUTOTOXICITY OF CHINESE MEDICINAL HERBS AND ITS CONTROL STRATEGY

7.2.1 Concept of Allelopathic Autotoxicity

Allelopathic autotoxicity, a special allelopathic phenomenon (also known as a consecutive monoculture problem, replanting disease, or soil sickness) refers to a chemoecological phenomenon of plant growth dysplasia, pest and disease problems, and decline in yield and quality caused by consecutively planting the same plant for many years in the same land. Many food crops (e.g., wheat, potato), cash crops (e.g., tobacco, cotton), oil crops (e.g., soybean, peanut), vegetables (e.g., cucumber, tomato), horticultural crops (e.g., watermelon, strawberry), medicinal plants (e.g., *Panax ginseng*, *Rehmannia glutinosa*, *Panax notoginseng*), and plantations (e.g., poplar, fir) experience various degrees of consecutive monoculture problems. However, the crops whose production is most limited by consecutive monoculture problems are Chinese medicinal herbs. About 70% of medicinal plant species with tuber roots have various degrees of consecutive monoculture problems (Z.Y. Zhang and Lin, 2009), including *Rehmannia glutinosa* (Figure 7.12), *Panax notoginseng*, *Angelica sinensis*, *Panax ginseng*, and other medicinal plants.

As is well known, high-quality herbal medicine production is conducted in so-called authentic production areas. Under specific natural conditions and ecological environments, the same medicinal plants in different regions will synthesize different chemical ingredients, thus forming a genuine medicine with obvious regional characteristics. Therefore, sustainable herbal agricultural production should be in what can be termed benign "climate–soil–medicinal plants–agronomic measures" agroecosystems. By regulating the controlled ecological factors and taking appropriate cultivation and management measures, farmers can cultivate medicinal plants in a suboptimal healthy growth environment and achieve a safe, effective, stable, and controllable quality of product (Z.Y. Zhang and Lin, 2009). But, with the increasing market demand for traditional Chinese medicine in recent years, Chinese medicinal agricultural production has gradually begun to exhibit simplification, larger scales, and intensive cultivation practices. This shift is resulting in habitat fragmentation,

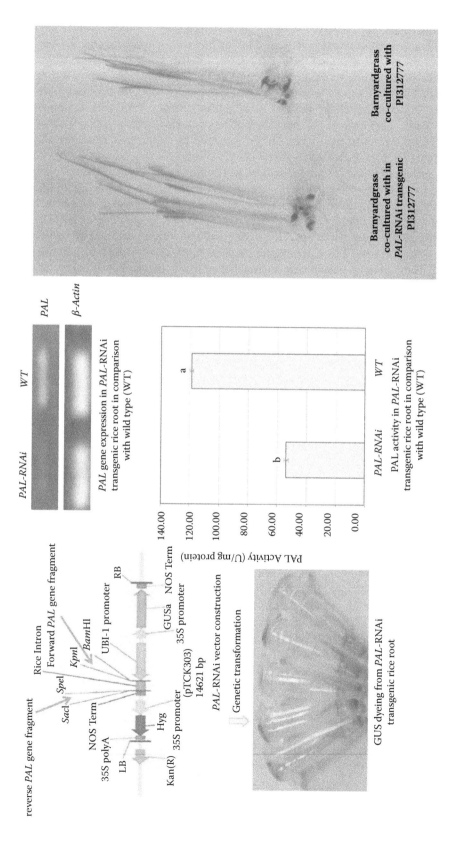

Figure 7.11   Allelopathic potential of PAL-RNAi transgenic PI312777 and its wild type on barnyardgrass.

Newly planted *R. glutinosa*

Consecutively monocultured *R. glutinosa*

Figure 7.12 The above- and below-ground growth status of newly planted (top) and consecutively monocultured (bottom) *Rehmannia glutinosa*.

lower levels of biodiversity, more serious pests and diseases, and a significant decline in yield and quality. In order to supply the market demand, the planting region is constantly being extended, but these regions are not authentic production areas and have unsuitable environmental conditions that seriously compromise the production and quality assurance of Chinese medicines (Z.Y. Zhang et al., 2003; C.J. Guo et al., 2004). Moreover, lacking knowledge of the mechanism of consecutive monoculture problems, most farmers try to maintain production by increasing fertilizer and pesticide inputs, which not only increases input expenses but also leads to environmental pollution, excessive pesticide residues, and degraded functioning of the farmland ecosystem, thus turning the production of traditional Chinese medicine into a vicious cycle. Further, development of Chinese medicinal resources and the related industrial system is facing the double limitation of prime farmland protection policies and food security policies (Huang et al., 2010; Q.G. Zhao et al., 2011). Therefore, research on consecutive monoculture problems plays an important role in current medicinal plant ecology, which has become an international focus of study.

7.2.2 Research Advances in Allelopathic Autotoxicity

Based on the latest research results around the world, the main causes of consecutive monoculture problems include the following:

1. Abnormal gene expression and regulation of consecutively monocultured plants, resulting in adverse physiological responses (W.X. Lin et al., 2011; Yang et al., 2011)
2. Deteriorating soil physiochemical properties, resulting in nutrient deprivation or disorders (Jian et al., 2011)

3. Allelopathic autotoxicity of root exudates (Du et al., 2009)
4. Structural imbalance of soil microecology, decreasing levels of microbial diversity, fragile relationships of biological interaction, reduced automatic regulation ability, and increase of pathogenic microorganisms and pests (Qi et al., 2009; Latz et al., 2012)

7.2.2.1 *Gene Expression Disorder and Consecutive Monoculture Problems*

Under a consecutive monoculture regime, plants often show growth dysplasia, such as decreases in the growth rate, shortening of the growth period, and declines in yield and quality related to particular physiological responses and to gene expression and regulation (G.Y. Guo et al., 2012). *Rehmannia glutinosa* Libosch, an important Chinese medicinal plant belonging to the family of Scrophulariaceae, suffers significant decline under a consecutive monoculture regime. So far, more than 70 compounds including iridoids, amino acids, saccharides, inorganic ions, and other trace elements have been identified in this herb. Pharmacological analyses showed that it can be useful for ailments of the endocrine, cardiovascular and nervous systems, as well as the immune system (R.X. Zhang et al., 2008). Research has shown that the consecutive monoculture problems of *R. glutinosa* become apparent at the early stages of growth and carry over through the entire growth period. A deficiency of source capacity, one of several reasons for this issue, is an important factor in restraining the root development of consecutively monocultured *R. glutinosa*. Furthermore, the accumulation of free radicals in plant cells, increase of malondialdehyde (MDA) content, and damage to the membrane structure caused by the stressful conditions of consecutive cropping resulted in decreased chlorophyll content and stomata closure (W.J. Yin et al., 2009), which in turn led to a decrease in photosynthetic capacity and a shortage of source supply (Z.Y. Zhang et al., 2010a).

Additionally, studies have found that the decrease of root activity and ATPase activity, as well as the small capacity and low activity of the sink, caused the poor growth of consecutively monocultured *Rehmannia glutinosa* (Yin et al., 2009; Z.Y. Zhang et al., 2010a). The contents of endogenous hormones (e.g., IAA, ABA) of consecutively monocultured *R. glutinosa* were found to change significantly (Niu et al., 2011). At the seedling stage, the abscisic acid (ABA) content was significantly higher in the consecutively monocultured *R. glutinosa* than in the newly planted *R. glutinosa*. The high content of ABA, a stress hormone, in leaves at the early stage gives rise to stoma closure and a decrease in plant photosynthesis; however, at the root elongation stage, the indole-3yl-acetic acid (IAA) content of consecutively monocultured *R. glutinosa* was significantly lower than that of the newly planted *R. glutinosa*. The higher level of IAA in leaves helps to delay leaf senescence, resulting in more photosynthetic products being transported to the roots. Thus, consecutive monoculture destroys the balance of endogenous hormones in *R. glutinosa*, leading to physiological metabolic disorders that affect the normal growth and development of plants.

The changes in specific physiological metabolisms and endogenous hormones are responses to environmental stress that regulate the expression of related genes and proteins. Fan et al. (2012) used the suppression subtractive hybridization (SSH) technique to construct the forward and reverse subtractive cDNA libraries of *Rehmannia glutinosa*. They found that key genes that regulate essential metabolic pathways (e.g., cyclin D, RNA polymerases, RNA replication enzyme, RNA-binding proteins) were restrained or shut down, disrupting their normal expressions in consecutively monocultured *R. glutinosa*. Conversely, calcium signaling systems (e.g., calcium-dependent protein kinase, calcium channel protein, calmodulin) and ethylene biosynthesis-related genes (e.g., ACC oxidase, *S*-adenosylmethionine synthetase) have specific expression that results in disruptions of normal metabolic processes. M.J. Li et al. (2013) applied high-throughput Solexa/Illumina sequencing to generate a transcript library of *R. glutinosa* transcriptome and degradome in order to identify the key miRNAs and their target genes implicated in the replanting disease. The results showed that the miRNAs involved in the regulation of signal transduction, ion transport, and cell division (e.g., miR2931, miR1861, miR7811) were upregulated in consecutively monocultured *R. glutinosa*, which suppressed the normal expressions of these target genes and protein functions. In contrast, miRNAs associated with regulating the formation of fibrous roots and early flowering (such as miR165, miR408, miR156/157) were downregulated, leading to the upregulation of target genes, promoting the formation of fibrous roots, early flowering, and a shortened growth period, resulting in typical symptoms of consecutive monoculture problems. It is obvious that *R. glutinosa* possesses a unique set of perception, conduction, and response systems against environmental stress, including the calcium signaling system in particular.

Another study found that two calcium channels away from the cytoplasm (plasma membrane calcium-transporting ATPase 13 and sarcoplasmic reticulum calcium-transporting ATPase 3) had downregulated expression in consecutively monocultured *Rehmannia glutinosa*, while the calcium channels toward the cytoplasm (calcium-dependent protein kinase, calcium-binding protein [CBP], calcineurin B-like protein [CBL], calcium-ion-binding protein [CIBP], and phospholipase C [PLC]) had upregulated expression (G.Y. Guo et al., 2013). Therefore, *R. glutinosa* consecutive monocultures could lead to a great increase in cytoplasmic calcium concentration. When consecutively monocultured *R. glutinosa* was treated with different levels of calcium signal blockers, the results indicated that calcium signal blockers could inhibit the gene expression of the calcium signaling pathway (CBP, CBL, CIBP, and PLC) and thus relieve consecutive monoculture problems to some extent.

7.2.2.2 Soil Physicochemical Properties and Consecutive Monoculture Problems

The physical and chemical properties of soil, including soil texture and structure, water, air, temperature, pH, organic matter, and inorganic elements, have a close relationship with plant growth and development. J.J. Wang et al. (2013) found that the consecutive monoculture of cotton had a significant effect on the composition of soil aggregates and the distribution of organic matter. To be more specific, long-term consecutive monoculture reduced the mechanical stability of soil aggregates; however, reasonable crop rotations (such as soybean and corn) could effectively promote the formation of aggregates, improve soil texture and structure, and enhance soil fertility, thus promoting cotton growth.

Plants have certain selectivities and preferences for absorbing soil nutrients, so different plants take up different mineral nutrients, particularly with regard to certain trace elements. Studies suggested that under long-term monoculture, because of the selective absorption of mineral elements by crops—in addition to improper fertilization, water management, and other agronomic practices—there was a decline in some nutrients and an accumulation of others, causing an imbalance of soil nutrients that triggered stunted plant growth (Peng et al., 2009). M. Yu et al. (2004) found that long-term monoculture of lilies caused serious soil deficiency of available potassium, decreased organic matter, and soil acidification, which became the major limiting factors of high yields. As for the decline of soil fertility, some research failed to alleviate the problems through increased fertilizer application. Consecutively monocultured wheat, for example, shows decreased yields even with adequate fertilizer levels (X.L. Zhang et al., 2007). Z.L. Zhao et al. (2006) also found that using potassium fertilizer had no significant effect on alleviating the consecutive monoculture problem of *Capsicum* peppers.

Many researchers, however, have found that consecutive monoculture does not lead to a decrease in major mineral levels, instead increasing many nutrients in the soil. With increasing years of cucumber monoculture, soil fertility was found to have increased (Liang et al., 2004). Similarly, with extended monoculture, the contents of total phosphorus (P) and available manganese (Mn) and zinc (Zn) in the soil significantly increased; the contents of available boron (B), calcium (Ca), copper (Cu), and magnesium (Mg) were enhanced to some extent; and no significant changes were observed in the contents of total potassium (K) and nitrogen (N) or on available iron (Fe). The results suggested that the main reason for consecutive monoculture problems is not the lack of nutrients in the soil; rather, the plants have difficulty absorbing soil nutrients under replanting stress (Pei,

2010). L. Liu et al. (2013) found that the levels of soil nutrients (e.g., total N and P, available P and K) and mineral ions (HCO_3^-, Cl^-, NO_3^-, SO_4^{2-}, Ca^{2+}, Mg^{2+}) in the soil of consecutively monocultured hot peppers were significantly higher than those in control soil, and there was an accumulation of most of the soil nutrients and mineral ions. Yet, soil pH decreased with the increasing years of monoculture. L.K. Wu et al. (2013) also indicated that many soil nutrients such as organic matter, total N, available N, and available K were significantly higher in the consecutively monocultured soil of *Rehmannia glutinosa* than in newly planted *R. glutinosa* soil. Therefore, it is still highly debatable whether the decline in soil nutrients is the main factor leading to consecutive monoculture problems.

7.2.2.3 *Allelopathic Autotoxins and Consecutive Monoculture Problems*

In recent years, research on consecutive monoculture problems has been more focused on the separation of phytotoxic substances, component identification, content determination, and biological activity evaluation. Researchers have isolated and identified several autotoxic allelochemicals such as *p*-hydroxybenzoic acid, coumaric acid, benzoic acid, vanillic acid, vanillin, ferulic acid, and cinnamic acid from the tissues, organs, and root exudates of *Rehmannia glutinosa*, cucumbers, watermelons, peanuts, and tomatoes. These allelochemicals could affect water utilization, nutrient uptake, photosynthesis, and the process of gene and protein expression to inhibit normal plant growth and development (Friebe et al., 1997; Yu et al., 1997; Blum et al., 1999; Z.Y. Zhang et al., 2011). The root exudates of cucumber contained 11 phenolic acids, including 2,5-dihydroxybenzoic acid, *p*-hydroxybenzoic acid, benzoic acid, and cinnamic acid, all of which (except 2-hydroxybenzothiazole) were toxic to the growth of receiver plants (Yu and Matsui, 1994). H.Y. Liu et al. (2006) found that the root exudates and leachates from leaves and tubers of *R. glutinosa* could inhibit the tuberous root growth of the plant itself. The most significantly inhibitory effect on the tuberous root expansion was from root exudates. Meanwhile, the phytotoxic substances were separated from *R. glutinosa* fibrous roots using a different polar organic solvent. The ethyl acetate extracts with the highest inhibition on receiver plants were subjected to gas chromatography/mass spectrometry (GC/MS) analysis. Several phenolic compounds were identified, such as ferulic acid, vanillin, *p*-hydroxybenzoic acid, benzoic acid, protocatechuic acid, and gallic acid (Z.F. Li et al., 2012a). The levels of coumaric acid, *p*-hydroxybenzoic acid, syringic acid, and ferulic acid in soil had a notably negative correlation with tuberous root growth of *R. glutinosa* under consecutive monoculture (Du et al., 2009).

Some researchers believe, however, that when toxic substances are secreted by plant roots into the soil, they are bound to undergo a series of changes in physical, chemical, and biological processes, including soil adsorption, microbial decomposition, and transformation (Inderjit et al., 2010). In other words, the direct phytotoxic effects of root exudates on plants might not happen in a consecutive monoculture regime, but the root exudates themselves may be inducing factors that indirectly influence plant performance by changing soil microbial communities. More and more scholars believe that the indirect ecological effects of root exudates and soil microecological imbalance are the major factors in the formation of consecutive monoculture problems (Z.W. Wu et al., 2009; W.X. Lin et al., 2012).

7.2.2.4 *Rhizosphere Microflora and Consecutive Monoculture Problems*

Root exudates have specialized roles in nutrient cycling and signal transduction between root systems and soil, as well as in plant responses to environmental stress. They are the key regulators in rhizosphere communication and can modify the biological and physical interactions between roots and soil organisms. Organic carbon fixed by plants through photosynthesis can be released into the soil by root secretion, providing material and energy for the surrounding microorganisms. Moreover, soil microorganisms gather in the rhizosphere and rhizoplane where there is abundant rhizodeposition by chemotaxis response. In recent years, scholars have determined group specific-ity in microbial utilization of root exudate compounds and whole rhizodeposition and have found that the different components and proportions of root exudates of different plants give soil microbial community structure certain specificity and representations (Paterson et al., 2007). Gschwendtner et al. (2011) even found that plant genotype determined the quality and quantity of root exudates, in turn influencing microbial community structure in the rhizosphere. Chaparro et al. (2013) also found a significant change in components and content of root exudates of *Arabidopsis* at differ-ent stages of development. Further metatranscriptome analysis revealed significant correlations between microbial functional genes involved in the metabolism of root exudates; corresponding compounds were released by the roots at particular stages of plant development.

Conversely, the structural and functional diversity of soil microbial communities also affect plant growth and development, including root secretion, nutrient absorption and utilization, and stress/defense responses (Eisenhauer et al., 2012). As assessed by GC/MS analysis, soil microbial communities could affect the biosynthesis of leaf metabolites of host plants, which in turn impacts feeding behaviors of insects (Badri et al., 2013). Lakshmanan et al. (2012) demonstrated that foliar infection by pathogens induced malic acid transporter expression, leading to increased malic acid titers in the *Arabidopsis* rhizosphere. Malic acid secretion in the rhizosphere increased beneficial rhizobacteria titers, causing an induced systemic resistance response in plants against pathogens. Zolla et al. (2013) found that under drought conditions a sympatric microbiome (i.e., having a his-tory of exposure to *Arabidopsis* at a natural site) significantly increased *Arabidopsis* biomass, while the non-sympatric soils did not affect plant biomass. This was related to plant-growth-promoting rhizobacteria (PGPR) in the soil (e.g., *Bacillus*, *Burkholderia*, *Acinetobacter*), which could modify the plant's ability to sense abiotic stress and increase plant biomass.

The rhizosphere, as the most active region of microbial activity, is a platform of the frequent material exchange and signal transmission between plants and soil microorganisms, which are closely related to plant growth both underground and aboveground. The collective genomes of rhi-zosphere microbial communities can be seen as a second genome of each plant (Berendsen et al., 2012). Similarly, East (2013) observed that Daniele Daffonchio and colleagues believe that the plant should not be seen merely as a single organism but as a meta-organism, referring to the plant and the surrounding soil microbiome as a whole. Mendes et al. (2011) used PhyloChip-based metagenomics to analyze the microbial community structure in disease-conducive soil and disease-suppressive soil. β-Proteobacteria, γ-Proteobacteria, and Actinobacteria are consistently associated with disease suppression. In particular, the number of *Pseudomonas* spp. was significantly higher in disease-suppressive soil than in disease-conducive soil. H.F. Lin (2010) analyzed the diversity of the rhizo-sphere microbial community in allelopathic rice and non-allelopathic rice using terminal restriction fragment length polymorphism (T-RFLP). The results showed that the allelopathic rice rhizosphere contained 30 kinds of specific microorganisms, including seven kinds of myxobacteria that could significantly suppress the growth of weeds (such as barnyardgrass) as mentioned earlier. Moreover, these myxobacteria have a quorum-sensing system, with obvious chemotactic responses to the phe-nolic acids in root exudates of rice (Y.Z. Li, 2013).

Qu and Wang (2008) used denaturing gradient gel electrophoresis (DGGE) to detect the effects of different phenolic acids of soybean root exudates on soil microbial populations. DGGE analysis revealed that the two phenolic acid (vanillic acid and 2,4-di-*tert*-butyl phenol) applications—vanil-lic acid, in particular—could have considerable impacts on microbial communities, especially by causing specific microorganisms (e.g., *Hymenagaricus* sp., *Cyathus striatus*) to become the domi-nant population. Similarly, X.G. Zhou and Wu (2012) indicated that *p*-coumaric acid, an autotoxin in root exudates, played a role in the allelopathic autotoxicity of cucumber by influencing soil micro-bial communities. *p*-Coumaric acid could change the structure and composition of bacterial and fungal communities in the rhizosphere, increasing the relative abundances of *Betaproteobacteria* and *Firmicutes* and decreasing the relative abundances of *Deltaproteobacteria*, *Bacteroidetes*, and

Planctomycetes. In addition, *p*-coumaric acid increased *Fusarium oxysporum* population densities in the soil. Increasingly, studies have found that various positive and negative plant–plant interactions within or among plant populations, such as amensalism, autotoxicity, stimulation, and interspecific facilitation, were all the results of the integrative effect of plant–microbe interactions mediated by root exudates.

With in-depth study of rhizosphere ecology, research on the consecutive monoculture problem has gradually focused on rhizospheric biological processes. "Cross-talk" between plants and microbes is the key factor for the allelopathic autotoxicity of medicinal plants. L.K. Wu et al. (2013) found that *Rehmannia glutinosa* consecutive monoculture led to an alteration of rhizospheric microbial community composition and activity, with distinct separations between the control, newly planted plots, and second- and third-year consecutive monocultured plot soils. Under consecutive monoculture, the number of bacteria in the rhizosphere of *R. glutinosa* decreased, but fungi and actinomycetes populations increased (Z.F. Li et al., 2012b). Furthermore, the genetic diversity analysis of soil microbial populations demonstrated that *R. glutinosa* consecutive monoculture led to a decline in the Shannon diversity index and the Margalef index and simplified the bacterial community structure. The dominant group in the newly planted soil was the class Bacilli, while the dominant group in the 2-year monocultured soil was comprised of ε-Proteobacteria (Z.Y. Zhang et al., 2010b). Lin Maozi et al. (2012a,b) found that *Pseudostellaria heterophylla* consecutive monoculture led to a significant decline in the amounts of free-living bacteria and aerobic nitrogen-fixing bacteria in the rhizosphere, but significantly increased the amounts of anaerobic cellulose-decomposing bacteria, actinomycetes, fungi, and *Fusarium oxysporum*. Other research on medicinal plants also showed that consecutive monoculture resulted in an increase in soilborne pathogens such as *Fusarium solani*, *Fusarium tricinctum*, *Aspergillus calidoustus*, *Phytophthora cactorum*, and *Pythium irregulare*, among others (Nicol et al., 2003; Duan, 2013). The findings encouraged us to establish a scientific systems approach to effectively overcome the consecutive monoculture problem.

7.2.3 Control Strategy of Allelopathic Autotoxicity and Its Application

7.2.3.1 Chemical Control

Chemical remediation agents, which are low cost and high performance, not only can improve soil quality by killing or inhibiting soilborne pathogens and pests but can also promote plant root growth and nutrient absorption. At present, several popular fumigants are available, including 1,3-dichloropropene, chloropicrin, methyl isothiocyanate, metam sodium, metam potassium, methyl iodide, sulfuryl fluoride, dazomet, propylene oxide, sodium azide, furfural, and dimethyl disulfide (Rosskopf et al., 2005; Cao et al., 2006; Sydorovych et al., 2006; UNEP, 2015). Studies have shown that some fumigation programs using 1,3-dichloropropene + chloropicrin, chloropicrin + metam sodium, or fosthiazate + chloropicrin can successfully protect strawberries from soilborne diseases and nematodes in Spain (Gilreath et al., 2008). Gao et al. (2006) also reported that chloropicrin could significantly improve the survival rate of American ginseng and reduce the occurrence of root diseases but, due to a combination of environmental and ecological factors, these fumigants are being restricted for use. Other approaches using agroecological or eco-agriculture practices must be found. Fulvic acid, a short-carbon-chain molecule, is extracted from natural humic acid. It has high loading capacity and biological activity. Fulvic acid can chelate metal cations and micronutrient matter, making them more available for plant utilization; improve soil structure and fertilizer; protect plants from diseases and pests; and enhance plant germination and growth. Xie et al. (2011) reported that potassium fulvic acid had a distinct inhibitory effect on root rot in consecutively monocultured cowpeas. J. Li et al. (2011) found that lime nitrogen could alleviate the consecutive monoculture problems of *Rehmannia glutinosa* to some extent and promote its yield.

7.2.3.2 *Agricultural Measures*

7.2.3.2.1 *Farming Mode Improvement*

Row intercropping, mixed cropping, relay strip intercropping, and rotations (all key practices in Chinese traditional agriculture) have a long history in China and can effectively improve the field crop community structure, nutrient absorption, and soil microflora (L. Zhang et al., 2008). A previous study by Y.F. Wu et al. (2008) showed that crop rotation could significantly reduce soil salt accumulation, increase the number of soil microbes, inhibit the proliferation of *Fusarium*, and increase the yield and income of consecutively monocultured cucumbers. Y.F. Wu et al. (2008) also indicated that intercropping cucumbers with wheat and *Vicia villosa* could increase the soil microbial community diversity of cucumbers, inhibit the growth of fungal pathogens (i.e., *Fusarium oxysporum*), and improve cucumber yields. Similarly, *Fusarium oxysporum* caused serious watermelon *Fusarium* wilt and drastically reduced yields under a consecutive monoculture regime, but a watermelon–aerobic rice intercropping system significantly reduced the content of *F. oxysporum* in the rhizosphere of watermelons while increasing the amounts of bacteria and actinomycetes, which effectively prevented the occurrence of *Fusarium* wilt (Su et al., 2008). J. Li et al. (2011) also found that *Rehmannia glutinosa* intercropped with *Achyranthes bidentata*, a plant suitable for consecutive monoculture, in combination with *A. bidentata* green manure, could alleviate the consecutive monoculture problems of *R. glutinosa* to some extent and improve its yield.

7.2.3.2.2 *High-Temperature Sterilization*

Currently, serious replanting diseases exist in greenhouse cultures because of large-scale monocultures of single species, as well as poor air circulation and lack of natural rain. Solar heat sterilization is a convenient and pollution-free way to effectively control the growth of soilborne pathogens. High-temperature agriculture in a closed environment is simple: plow deeply, irrigate, then seal the greenhouse. A high-temperature and high-radiation environment can greatly increase the temperature in the greenhouse, thus achieving high-temperature sterilization. A study has shown that high temperatures in a closed environment could significantly decrease the proportion of non-capillary voids and the value of the soil redox potential, as well as lower soil pH and the disease indices of root rot, damping off, bacterial wilt, banded sclerotial blight, and powdery mildew (Q.S. Guo, 2004). Similarly, Sugimura et al. (2001) found that soil solarization using mulching and tunnel-covering combined with solar-heated water irrigation could effectively control *Fusarium* wilt in strawberries.

7.2.3.2.3 *Organic Amendments*

Organic amendments, such as animal manure, green manure, and compost, are commonly used in agricultural production to address various management issues, including soil nutrition, water retention, disease control, and soil health. Brassicaceae plant residues have been studied widely for their potential to control a wide range of plant pests, such as pathogens, insects, and weeds (Brown and Morra, 1997). The mechanism of Brassicaceae residue-based amendments has been attributed to the release of biologically active products such as isothiocyanates (ITCs), organic cyanides, and oxazolidinethiones (OZTs) (Brown and Morra, 1997). Studies have shown that Brassicaceae seed meal amendments could effectively control a number of soilborne plant pathogens, including *Aphanomyces euteiches* f. sp. *pisi* on peas and *Rhizoctonia* spp. and *Pratylenchus penetrans* on apples (Smolinska et al., 1997; Mazzola et al., 2001). Mazzola et al. (2012) reported that different Brassicaceae seed meals have different effects on the *Pythium* densities in soil; a reasonable selection of Brassicaceae seed meal types was regarded as an effective method to improve soil fertility and control soilborne diseases in plant production systems. The application of organic soil amendments is also a traditional control method for plant-parasitic nematodes. Possible mechanisms involved in nematode suppression include the generation and release of nematicidal compounds, the introduction and enhancement of antagonistic microorganisms, and changes in soil properties that are unsuitable to nematode behavior (Oka, 2010).

7.2.3.2.4 Plant Origin Biopesticides

Plants are natural sources of bioactive compounds, with tens of thousands of secondary metabolites including flavonoids, terpenes, hesperidin, alkaloids, and annonaceous acetogenins with strong antibacterial and insecticidal activity (S.T. Wang et al., 2006a). These plant extracts play an important role in inhibiting the presence of continuous cropping soil pests. S.T. Wang et al. (2006b) determined the inhibition of 126 kinds of Chinese herb extracts (CHEs) against two plant pathogenic fungi and found that 16 species of CHEs had significant inhibitory effects against the mycelium growth of *Fusarium graminearum*, and 23 species of CHEs had inhibitory effects against the mycelium growth of *Rhizoctonia solani*. The inhibition rates of methanol extract from *Justicia procumbens* against *Colletotrichum gloeosprioides*, *Phomopsis asparagi*, and *Botrytis cinerea* were over 50%, which was positively correlated with the extract concentration (M.C. Guo et al., 2013). Wei et al. (2014) found that *Cercidiphyllum japonicum* extracts had certain antimicrobial activities against plant pathogenic fungi (e.g., *Caralluma fimbriata*, *Magnaporthe grisea*). In particular, the inhibitory rate of 50-mg/mL petroleum ether extract against *C. fimbriata* reached up to 100%.

7.2.3.3 Biocontrol

Bioremediation is a biological method for improving the soil environment by isolating and restoring soil beneficial strains with antagonistic activity against specific pathogens (Table 7.1). Boby and Bagyaraj (2003) found that inoculation with *Trichoderma viride* + *Glomus mosseae* yielded the best

Table 7.1   Soilborne Pathogens, Their Host Medicinal Plants, and Their Corresponding Antagonistic Microorganisms

Pathogen	Reported Host Plants	Reported Antagonists	Refs.
Fusarium chlamydosporum	*Coleus forskohlii*	*Pseudomonas fluorescens*, *Glomus mosseae*, *Trichoderma viride*	Boby and Bagyaraj (2003)
Pythium irregulare	*Panax quinquefolius*	*Bacillus subtilis*, *Bacillus polymyxa*, *Trichoderma harzianum*	Hwang et al. (1996), Ivanov et al. (2012), Naar (2006)
Rhizoctonia solani	*Panax ginseng*	*Paenibacillus polymyxa*, *Bacillus* sp., *Pseudomonas poae*	Cho et al. (2007), Reeleder et al. (1994)
Fusarium solani	*Hibiscus sabdariffa*, *Rehmannia glutinosa*	*Pseudomonas stutzeri*, *Bacillus subtilis*, *Trichoderma viride*, *Trichoderma harzianum*	Z.F. Li et al. (2012b), Lim et al. (1991), Morsy et al. (2009), Rojo et al. (2007)
Cylindrocarpon destructans	*Panax quinquefolius*	*Bacillus* sp., *Trichoderma* sp.	Rahman et al. (2005), L. Wu et al. (1998)
Fusarium oxysporum	*Hibiscus sabdariffa*, *Rehmannia glutinosa*, *Pseudostellaria heterophylla*	*Paenibacillus polymyxa*, *Pseudomonas aeruginosa*, *Bacillus subtilis*	Dijksterhuis et al. (1999), Gupta et al. (1999), X. Yu et al. (2011), Z.F. Li et al. (2012b), M.Z. Lin et al. (2012a)
Macrophomina phaseolina	*Angelica dahurica*	*Pseudomonas* sp. EM85, *Bacillus subtilis* BN1	Pal et al. (2001), Rani et al. (2014), Singh et al. (2008)
Sclerotium rolfsii Sacc.	*Atractylodes lanceal*	*Trichoderma* sp.	B. Liu et al. (2007), Shaigan et al. (2008)

results in disease control and led to maximum growth, yield, and root forskolin concentration of *Coleus forskohlii*. Effective microorganisms (EMs) are biological agents developed by Teruo Higa (University of the Ryukyus, Japan) that utilize 80 species of anaerobic or aerobic microorganisms belonging to 10 genera. Studies have shown that EMs can effectively overcome consecutive mono-culture problems and increase the total amounts of available nutrients and soil microbial activity (Sun et al., 2001). The microbial fertilizer Yalian 1 (AsiaLink Enterprise Group Corp., Dallas, TX) and bio-organic fertilizer developed by the College of Agriculture and Biotechnology, Zhejiang University, could play an important role in improving soil pH and nutrients and preventing disease and pest damage, as well as affecting the soil remediation of consecutively monocultured cowpeas (J. Li et al., 2011). Microbial remediation agents (Shanghai Chuangbo Biotechnology Co., Ltd.) contain beneficial microorganisms including *Bacillus*, *Lactobacillus*, yeast, and their active metabolites, which can improve the soil environment and increase the height, leaf area, and yield of cucumbers (Lu et al., 2010). Gu et al. (2013) found that bio-organic fertilizer could effectively improve the soil organic matter, available nutrients, and microbial biomass, resulting in an increase in the yield and quality of ginseng. Our research has demonstrated that microbial fertilizer treatments by BioStem Technologies (BSEM) could significantly improve the growth of *Physalis heterophylla* (Figure 7.13). Among them, the yield of the BSEM treatment was almost equivalent to the newly planted treatment and then the ESK treatment (see Table 7.2). However, the positive effect was not significant when using a single beneficial antagonistic microorganism (i.e., *Bacillus amyloliquefaciens*). Other studies also demonstrated that microbial fertilizer and biological control agents were experiencing a resur-gence as efficient ways to restore the soil microenvironment, improve soil quality, control diseases and pests, and alleviate the consecutive monoculture problems widely existing in medicinal plants such as *Rehmannia glutinosa* and *Dioscorea opposite* (Z.G. Li et al., 2008; Y.G. Liu and Li, 2010).

7.3 CHALLENGES AND PERSPECTIVES IN ALLELOPATHY RESEARCH

7.3.1 Challenges and Perspectives in Rice Allelopathy

The interactions between allelochemicals and rhizospheric microbes are the dominant processes in plant allelopathy. Further research should be conducted in this field to investigate microbial func-tions and to separate related microbes in order to assist in developing natural microbial herbicides for weed control in sustainable agriculture. The next substantial challenge is annotating microbial diversity with functional details in the microbial ecosystems mediated by allelochemicals. For this purpose, we have attempted to use transcriptomics to decipher gene expression in the soil microbial systems mediated by allelopathic rice accessions, particularly during exposure to stressful conditions. However, transcriptomics is still limited in its applications for understanding protein abundances—and thus cellular and community functions—because mRNA abundance is not always related to protein abundance, and post-translational modifications cannot be currently predicted by mRNA. Proteins play the primary role in carrying out cellular functions encoded by the genome, so the large-scale study of protein abundances in soil should provide a significant leap in the understanding of soil ecological processes and the environmental factors that govern microbial activity and survival (Taylor and Williams, 2010). Elucidating the changing properties in the expression abundance of soil proteins is an effective way to understand the function of soil microorganisms in response to environ-mental factors, such as the allelochemicals released from allelopathic rice accessions.

More recently, we have developed an effective sequential extraction and separation method to extract proteins from different agricultural soil samples and in turn to separate the proteins by sodium dodecyl sulfate polyacrylamide gel electrophoresis (SDS-PAGE) or two-dimensional gel electro-phoresis (2-DE). This method could be used successfully for the large-scale study of soil metapro-teomics using a mass-spectrometric technique combined with bioinformatics. Matrix-assisted laser

Figure 7.13   The growth status of *Pseudostellaria heterophylla* under consecutive monoculture and microbial fertilizer treatments.

Table 7.2   Effects of Various Microbial Fertilizers on Yield and Components of *Pseudostellaria heterophylla*

Treatments	Root Length (cm)	Number of Roots per m²	Yield (kg per 666.7 m²)
NP	16.24[a]	1269.33[b]	285.30[a]
CM	8.51[d]	967.00[d]	176.77[c]
BSEM	13.59[b]	1419.67[a]	284.19[a]
ESK	12.56[b]	1112.33[c]	263.81[b]
BA	10.39[c]	604.33[e]	180.37[c]

Note: NP, newly planted; CM, 2-year consecutively monocultured; BSEM, 2-year consecutively monocultured treated by microbial fertilizer no. 1; ESK, 2-year consecutively monocultured treated by microbial fertilizer no. 2; BA, 2-year consecutively monocultured treated by *Bacillus amyloliquefaciens*. Different letters within a column indicate different levels of significance ($p \leq 0.05$, Tukey's test, $n = 3$).

desorption/ionization (MALDI) time of flight (TOF)/TOF analysis by our group indicated that the proteins from allelopathic rice soils function in xenobiotic detoxification, conversion of allelochemicals, chemotaxis, and signal transduction. These results suggested that the function of soil proteins mediated by the allelochemicals was complex. Further work is needed to deeply examine the metaproteome change and their function in the rhizospheric soil of allelopathic rice. From our previous studies, it is clear that the myxobacterial population in allelopathic rice rhizospheres is significantly higher than that of non-allelopathic rice rhizospheres; however, the mechanism of quick aggregation in allelopathic rice rhizospheres is still not understood. The chemotaxis from allelopathic rice root exudates to myxobacteria would be available for the aggregation; whether or not the quorum sensing of the specific myxobacteria could exist in allelopathic rice rhizospheres, the search for the signal molecules in this mechanism will help reveal the process. On the other hand, other studies have shown that low concentrations of terpenoids and momilactones had great allelopathic activity in laboratory conditions; however, little research has been conducted in field soil conditions, in which biodegradation of these metabolites would be much quicker. The study of terpenoid residuals, including momilactones, and their relationship with microbial flora in soil conditions would clarify the real role of such metabolites in rice allelopathic processes, further clearing up confusion around which kinds of allelochemicals act as dominant factors in rice allelopathy.

7.3.2 Challenges and Perspectives in Allelopathic Autotoxicity

Allelopathic autotoxicity, a complex chemical–biological phenomenon, exists widely in the cultivation of medicinal plants, horticultural crops, and so on. Due to increases in population and food demand, intensive cropping has been widely adapted in agricultural production; however, it results in lost farmland biodiversity and serious replanting problems. Earlier studies have mainly focused on the separation, identification, and autotoxic potential detection of phytotoxic substances, but the concentration of autotoxins used in laboratory studies is based only on the dosage of extracts in root exudates or plant tissues, which is usually higher than the actual soil concentration. In general, the field used for *Rehmannia glutinosa* cultivation can be replanted once every 8 to 10 years. Considering the phytotoxic effect of root exudates, it is unreasonable to expect these autotoxins to last 8 to 10 years because they are easily degraded and transformed by soil microbes. R.Y. Lin et al. (2011) found that the concentration of exogenous phenolic acid added to the soil was decreased by 45 to 75% on the third day and further decreased by 55 to 98% on the ninth day. More scholars now consider allelopathic autotoxicity to be a result of the comprehensive effect of the rhizospheric "talk" between plants and microbes; however, soil, figuratively known as a black box, is a complex matrix.

Although research on the allelopathic autotoxicity of different medicinal plants shows that all plants displayed an increase in soilborne pathogens but a decline in beneficial microbes under consecutive monoculture, many of the studies focused on only one kind of microbe rather than the microbial community *in situ*. As for root exudates, many scholars have also focused only on the phenolic acids and their ecological effect on soil microflora, but more attention should be paid to the other compounds involved, such as amino acids and fatty acids. A deeper understanding of plant–microbe interactions mediated by root exudates would provide a theoretical basis and technological support for alleviating allelopathic autotoxicity. Biocontrol, harnessing disease-suppressive microorganisms to improve plant health, is an efficient way to improve the soil microenvironment and control soilborne pathogens. Types of antagonistic bacteria (e.g., *Pseudomonas* sp., *Burkholderia* sp., *Bacillus subtilis*) and fungi (e.g., *Trichoderma*, *Gliocladium*) have been isolated from soil and then used to inhibit phytopathogens. The antagonistic effect of beneficial microbes on pathogens is significant in the lab, but many of them cannot successfully colonize in rhizospheric soil, suppress soilborne diseases, or improve plant growth using a single beneficial microorganism. Therefore, the combination of beneficial microorganisms with organic amendments as well as the application of various probiotics should be considered in order to effectively employ bio-organic fertilizer as fertility management or as a soilborne disease control strategy in plant production systems.

REFERENCES

Azmi, M., Abdullah, M.Z., and Fujii, Y. 2000. Exploratory study on allelopathic effect of selected Malaysian rice varieties and rice field weed species. *J. Trop. Agric. Food Sci.*, 28: 39–54.

Badri, D.V., Zolla, G., Bakker, M.G., Manter, D.K., and Vivanco, J.M. 2013. Potential impact of soil microbiomes on the leaf metabolome and on herbivore feeding behavior. *New Phytol.*, 198: 264–273.

Berendsen, R.L., Pieterse, C.M., and Bakker, P.A. 2012. The rhizosphere microbiome and plant health. *Trends Plant Sci.*, 17: 478–486.

Bertin, C., Yang, X.H., and Weston, L.A. 2003. The role of root exudates and allelochemicals in the rhizosphere. *Plant Soil*, 256: 67–83.

Bi, H.H., Zeng, R.S., Su, L.M., An, M., and Luo, S.M. 2007. Rice allelopathy induced by methyl jasmonate and methyl salicylate. *J. Chem. Ecol.*, 33: 1089–1103.

Blum, U. 2011. *Plant–Plant Allelopathic Interactions*. Springer: Heidelberg, pp. 85–149.

Blum, U., Shafer, R., and Lehmen, M.E. 1999. Evidence for inhibitory allelopathic interactions involving phenolic acids in field soils: concepts vs. an experimental model. *Crit. Rev. Plant Sci.*, 18: 673–693.

Boby, V.U. and Bagyaraj, D.J. 2003. Biological control of root-rot of *Coleus forskohlii* Briq. using microbial inoculants. *World J. Microbiol. Biotechnol.*, 19: 175–180.

Brown, P.D. and Morra, M.J. 1997. Control of soil-borne plant pests using glucosinolate containing plants. *Adv. Agron.*, 61: 167–231.

Buchanan, B.B., Gruissem, W., and Jones, R.L. 2000. *Biochemistry & Molecular Biology of Plants*. American Society of Plant Physiologists Press: Rockville, MD.

Cao, A.C., Guo, M.X., Duan, X.Y., and Zhang, W.J. 2006. Alternatives to methyl bromide and reducing its emission technology on strawberry. *J. Plant Prot.*, 33(3): 291–297.

Cartwright, D.W., Langcake, P., Pryce, R.J., Leworthy, D.P., and Ride, J.P. 1977. Chemical activation of host defence mechanisms as a basis for crop protection. *Nature*, 267: 511–513.

Cartwright, D.W., Langcake, P., Pryce, R.J., Leworthy, D.P., and Ride, J.P. 1981. Isolation and characterization of two phytoalexins from rice as momilactones A and B. *Phytochemistry*, 20: 535–537.

Chaparro, J.M., Badri, D.V., Bakker, M.G., Sugiyama, A., and Vivanco, J.M. 2013. Root exudation of phytochemicals in *Arabidopsis* follows specific patterns that are developmentally programmed and correlate with soil microbial functions. *PLoS ONE*, 8: e55731.

Cho, K.M., Hong, S.Y., Lee, S.M., Kim, Y.H., Kahng, G.G., Lim, Y.P., Kim, H., and Yun, H.D. 2007. Endophytic bacterial communities in ginseng and their antifungal activity against pathogens. *Microbial Ecol.*, 54: 341–351.

Chung, I.M., Hahn, S.J., and Ahmad, A. 2005. Confirmation of potential herbicidal agents in hulls of rice, *Oryza sativa*. *J. Chem. Ecol.*, 31: 1339–1352.

Dijksterhuis, J., Sanders, M., Gorris, L.G.M., and Smid, E.J. 1999. Antibiosis plays a role in the context of direct interaction during antagonism of *Paenibacillus polymyxa* towards *Fusarium oxysporum*. *J. Appl. Microbiol.*, 86: 13–21.

Dilday, R.H., Nastasi, P., and Smith, Jr., R.J. 1989. Allelopathic observations in rice (*Oryza sativa* L.) to duck-salad (*Heteranthera limosa*). *Arkansas Acad. Sci.*, 43: 21–22.

Dilday, R.H., Lin, J., and Yan, W. 1994. Identification of allelopathy in the USDA–ARS rice germplasm collection. *Aust. J. Exp. Agric.*, 34: 907–910.

Dilday, R.H., Yan, W.G., Moldenhauer, K.A.K., and Gravois, K.A. 1998. Allelopathic activity in rice for controlling major aquatic weeds. In: Olofsdotter, M., Ed., *Allelopathy in Rice*. International Rice Research Institute: Manila, pp. 7–26.

Dilday, R.H., Mattice, J.D., and Moldenhauer, K.A. 2000. An overview of rice allelopathy in the USA. In: Kim, H.U. and Shin, D.H., Eds., *Proceedings of International Workshop in Rice Allelopathy*. Kyungpook National University: Taegu, pp. 15–26.

Dixon, R.A. and Paiva, N.L. 1995. Stress-induced phenylpropanoid metabolism. *Plant Cell*, 7: 1085–1097.

Du, J.F., Yin, W.J., Li, J., and Zhang, Z.Y. 2009. Dynamic change of phenolic acids in soils around rhizosphere of replanted *Rehmannia glutinosa*. *China J. Chin. Mater. Med.*, 34: 948–952.

Duan, J.L. 2013. Studies on Microecological Mechanism of *Salvia miltiorrhiza* Bge. Root Diseases and Growth–Promoting Effect of Antimicrobial Actinomycetes, PhD Thesis, Northwest Agriculture and Forestry University, Yangling.

East, R. 2013. Microbiome: soil science comes to life. *Nature*, 501: S18–S19.

Ebana, K., Yan, W.G., Dilday, R.H., Namai, H., and Okuno, K. 2001. Analysis of QTL associated with the allelopathic effect of rice using water-soluble extracts. *Breeding Sci.*, 51: 47–51.

Eisenhauer, N., Scheu, S., and Jousset, A. 2012. A bacterial diversity stabilizes community productivity. *PLoS ONE*, 7: e34517.

Fan, H.M., Li, M.J., Zheng, H.Y., Yang, Y.H., Gu, L., Wang, F.Q., Chen, X.J., and Zhang, Z.Y. 2012. Spatiotemporal expression and analysis of responding consecutive monoculture genes in *Rehmannia glutinosa*. *China J. Chin. Mater. Med.*, 37: 3029–3035.

Fang, C.X., He, H.B., Wang, Q.S., Qiu, L., Wang, H.B., Zhuang, Y.E., Xiong, J., and Lin, W.X. 2010. Genomic analysis of allelopathic response to low nitrogen and barnyardgrass competition in rice (*Oryza sativa* L.). *J. Plant Growth Regul.*, 61: 277–286.

Fang, C.X., Xiong, J., Qiu, L., Wang, H.B., Song, B.Q., He, H.B., Lin, R.Y., and Lin, W.X. 2009. Analysis of gene expressions associated with increased allelopathy in rice (*Oryza sativa* L.) induced by exogenous salicylic acid. *J. Plant Growth Regul.*, 57: 163–172.

Friebe, A., Roth, U., Kuek, P., Schnabl, H., and Schulz, M. 1997. Effects of 2,4-dihydroxy-1,4-benzoxazin-3-ones on the activity of plasma membrane H$^+$-ATPase. *Phytochemistry*, 44: 979–983.

Fukuta, M., Xuan, T.D., Deba, F., Tawata, S., Khanh, T.D., and Chung, I.M. 2007. Comparative efficacies *in vitro* of antibacterial, fungicidal, antioxidant, and herbicidal activities of momilatones A and B. *J. Plant Interact.*, 2: 245–251.

Gao, W.W., Chen, Z., Zhang, L.P., Ma, X.J., and Zhao, Y.J. 2006. Study on the effect of disinfection in rhizospheric microorganism and root disease from the West ginseng. *China J. Chin. Mater. Med.*, 31: 684.

Gealy, D.R., Wailes, E.J., Estorninos, L.E., and Chavez, R.S.C. 2003. Rice cultivar differences in suppression of barnyardgrass (*Echinochloa crus-galli*) and economics of reduced propanil rates. *Weed Sci.*, 51: 601–609.

Gilreath, J.P., Santos, B.M., and Motis, T.N. 2008. Performance of methyl bromide alternatives in strawberry. *HortTechnology*, 18: 80–83.

Gschwendtner, S., Esperschütz, J., Buegger, F., Reichmann, M., Müller, M., Munch, J.C., and Schloter, M. 2011. Effects of genetically modified starch metabolism in potato plants on photosynthate fluxes into the rhizosphere and on microbial degraders of root exudates. *FEMS Microbiol. Ecol.*, 76: 564–575.

Gu, H.P., Yuan, X.X., Chen, X., Cui, X.Y., Chen, H.T., and Zhu, L.L. 2013. Effect of high temperature soaking soil on continuous cropping soil remediation and disease control. *Jiangsu Agric. Sci.*, 41: 348–351.

Guo, C.J., Guan, Z.H., Li, Y.W., Yang, X.S., and Wang, F.Y. 2004. Effects of bio-organic fertilizer on soil microenvironment of ginseng under consecutive monoculture. *Biotechnology*, 14: 55–56.

Guo, G.Y., Wang, F.Q., Fan, H.M., Li, M.J., Zheng, H.Y., Li, J., Chen, X.J., and Zhang, Z.Y. 2012. Advances in allelopathic autotoxicity and monoculture cropping problem of *Rehmannia glutinosa* Libosch. *Mod. Chin. Med.*, 14: 35–39.

Guo, G.Y., Li, M.J., Wang, P.F., Wang, F.Q., He, H.Q., Li, J., Zheng, H.Y., Chen, X.J., and Zhang, Z.Y. 2013. Abnormal change of calcium signal system on consecutive monoculture problem of *Rehmannia gluti-nosa. China J. Chin. Mater. Med.*, 38: 1471–1478.

Guo, M.C., Li, B.T., Tang, L.M., Ping, X.L., Li, X.H., and Liang, T.J. 2013. Fungicide and insecticide activity of *Justicia procumbens* extracts. *Chin. J. Eco-Agric.*, 02: 212–216.

Guo, Q.S. 2004. *Cultivation of Medicinal Plants.* Higher Education Press: Beijing.

Gupta, C.P., Sharma, A., Dubey, R.C., and Maheshwari, D.K. 1999. *Pseudomonas aeruginosa* (GRC~1) as a strong antagonist of *Macrophomina phaseolina* and *Fusarium oxysporum. Cytobios.*, 99: 183–189.

Hassan, S.M., Aidy, I.R., Bastawisi, A.O., and Draz, A.E. 1998. Weed management using allelopathic rice vari-eties in Egypt. In: Olofsdotter, M., Ed., *Allelopathy in Rice: Proceedings of Workshop on Allelopathy in Rice.* International Rice Research Institute (IRRI): Manila, pp. 27–37.

He, H.B., Chen, X.X., Lin, R.Y., Lin, W.X., He, H.Q., Jia. X.L., Xiong, J., Shen, L.H., and Liang, Y.Y. 2005a. Chemical components of root exudates from allelopathic rice accession PI312777 seedlings. *Chin. J. Appl. Ecol.*, 16: 2383–2388.

He, H.B., Lin, W.X., Chen, X.X., He, H.Q., Xiong, J., Jia, X.L., and Liang, Y.Y. 2005b. The differential analy-sis on allelochemicals extracted from root exudates in different allelopathic rice accessions. In: Haper, J.D.I., An, M., Wu, H., and Kent, J.H., Eds., *Proceedings of the Fourth World Congress on Allelopathy.* International Allelopathy Society: Hisar, India.

He, H.B., Lin, W.X., Wang, H.B., Fang, C.X., Gan, Q.F., Wu, W.X., Chen, X.X., and Liang, Y.Y. 2006. Analysis of metabolites in root exudates from allelopathic and nonallelopathic rice seedlings. *Allelopathy J.*, 18: 247–254.

He, H.B., Wang, H.B., Fang, C.X., Lin, Y.Y., Zeng, C.M., Wu, L.Z., Guo, W.C., and Lin, W.X. 2009. Herbicidal effect of a combination of oxygenic terpenoids on *Echinochloa crus-galli. Weed Res.*, 49: 183–192.

He, H.B., Wang, H.B., Fang, C.X., Wu, H.W., Guo, X.K., Liu, C.H., Lin, Z.H., and Lin, W.X. 2012a. Barnyardgrass stress up regulates the biosynthesis of phenolic compounds in allelopathic rice. *J. Plant Physiol.*, 169: 1747–1753.

He, H.B., Wang, H.B., Fang, C.X., Lin, Z.H., Yu, Z.M., and Lin, W.X. 2012b. Separation of allelopathy from resource competition using rice/barnyardgrass mixed–cultures. *PLoS ONE*, 7: e37201.

Huang, L.Q., Li, J.D., Li, Z., and Lu, J.S. 2010. The development trends and countermeasures of Chinese mod-ern medicine industry chain. *Chin. Technol. Invest.*, 5: 67–69.

Hwang, S.F., Chang, K.F., Howard, R.J., Deneka, B.A., and Turnbull, G.D. 1996. Decrease in incidence of *Pythium* damping-off of field pea by seed treatment with *Bacillus* spp. and metalaxyl. *Z. Pflanzenkr. Pflanzenschutz*, 103: 31–41.

Inderjit. 2005. Soil microorganisms: an important determinant of allelopathic activity. *Plant Soil*, 274: 227–236.

Inderjit, Bajpai, D., and Rajeswari, M.S. 2010. Interaction of 8-hydroxyquinoline with soil environment medi-ates its ecological function. *PLoS ONE*, 5: e12852.

Ivanov, D.A. and Bernards, M.A. 2012. Ginsenosidases and the pathogenicity of *Pythium irregulare. Phytochemistry*, 78: 44–53.

Jensen, L.B., Courtois, B., Shin, L.S., Li, Z.K., Olofsdotter, M., and Mauleon, R.P. 2001. Locating genes con-trolling allelopathic effects against barnyardgrass in upland rice. *Agron. J.*, 93: 21–26.

Jian, Z.Y., Wang, W.Q., Meng, L., Wang, D., You, P.J., and Zhang, Z.L. 2011. Analysis of element contents in soil for continuous cropping ginseng. *Chin. J. Soil Sci.*, 42: 369–371.

Kato, T., Kabuto, C., Sasaki, N., Tsunagawa, M., Aizawa, H., Fujita, K., Kato, Y., and Kitahara, Y. 1973. Momilactones, growth inhibitors from rice, *Oryza sativa* L. *Tetrahedron Lett.*, 39: 3861–3864.

Kato, T., Tsunakawa, M., Sasaki, N., Aizawa, H., Fujita, K., Kitahara, Y., and Takahashi, N. 1977. Growth and germination inhibitors in rice husks. *Phytochemistry*, 16: 45–48.

Kato-Noguchi, H. and Peters, R.J. 2013. The role of momilactones in rice allelopathy. *J. Chem. Ecol.*, 39: 175–185.

Kato-Noguchi, H., Ino, T., and Ichii. M. 2003. Changes in release level of momilactone B into the environment from rice throughout its life cycle. *Funct. Plant Biol.*, 30: 995–997.

Kato-Noguchi, H., Ota, K., and Ino, T. 2008. Release of momilactone A and B from rice plants into the rhizo-sphere and its bioactivities. *Allelopathy J.*, 22: 321–328.

Kato-Noguchi, H., Hasegawa, M., Ino, T., Ota, K., and Kujime, H. 2010. Contribution of momilactone A and B to rice allelopathy. *J. Plant Physiol.*, 167: 787–791.

Kim, K.U. and Shin, D.H. 1998. *Rice Allelopathy Research in Korea.* Manila: International Rice Research Institute Press.

Kim, K.U., Shin, D.H., Kim, H.Y., Lee, Z.L., and Olofsdotter, M. 1999. Evaluation of allelopathic potential in rice germplasm. *Korean J. Weed Sci.*, 19: 1–9.

Kim, S.Y., Madrid, A.V., Park, S.T., Yang, S.J., and Olofsdotter, M. 2005. Evaluation of rice allelopathy in hydroponics. *Weed Res.*, 45: 74–79.

Kodama, O., Suzuki, T., Miyakawa, J., and Akatsuka, T. 1988. Ultraviolet-induced accumulation of phytoalex-ins in rice leaves. *Agric. Biol. Chem.*, 52: 2469–2473.

Lakshmanan, V., Kitto, S.L., Caplan, J.L., Hsueh, Y.H., Kearns, D.B., Wu, Y.S., and Bais, H.P. 2012. Microbe-associated molecular patterns-triggered root responses mediate beneficial rhizobacterial recruitment in *Arabidopsis. Plant Physiol.*, 160: 1642–1661.

Latz, E., Eisenhauer, N., Rall, B.C., Allan, E., Roscher, C., Scheu, S., and Jousset, A. 2012. Plant diversity improves protection against soil-borne pathogens by fostering antagonistic bacterial communities. *J. Ecol.*, 100: 597–604.

Lee, C.W., Yoneyama, K., Takeuchi, Y., Konnai, M., Tamogami, S., and Kodama, O. 1999. Momilactones A and B in rice straw harvested at different growth stages. *Biosci. Biotechnol. Biochem.*, 63: 1318–1320.

Lee, S.B., Seo, K.I., Koo, J.H., Hur, H.S., and Shin, J.C. 2005. QTLs and molecular markers associated with rice allelopathy. In: Haper, J.D.I., An, M., Wu, H., and Kent, J.H., Eds., *Proceedings of the Fourth World Congress on Allelopathy.* International Allelopathy Society: Hisar, India.

Li, J., Huang, J., Zhang, Z.Y., Niu, M.M., Fan, H.M., and He, H.Y. 2011. Research on abatement measures of allelopathic autoxicity of *Rehmannia glutinosa. China J. Chin. Mater. Med.*, 36: 405–408.

Li, M.J., Yang, Y.H., Chen, X.J., Wang, F.Q., Lin, W.X., Yi, Y.J., Zeng, L., Yang, S.Y., and Zhang, Z.Y. 2013. Transcriptome/degradome-wide identification of *R. glutinosa* miRNAs and their targets: the role of miRNA activity in the replanting disease. *PLoS ONE*, 8: e68531.

Li, Y.Z. 2013. The Dynamic Changes of Rice Allelochemical Phenolic Acids and Their Effects on Specific Rhizosphere Microorganisms, master's thesis, Fujian Agriculture and Forestry University, Fuzhou.

Li, Z.G., Wang, X.M., Liu, T.Y., Zhang, X.G., Jie, X.L., and Zhao, Y.J. 2008. Restoration of continuous crop-ping obstacles of *Rehmannia glutinosa* Libosch by applying compound bacterial manure. *J. Henan Agric. Sci.*, 5: 62–65.

Li, Z.F., Yang, Y.Q., Xie, D.F., Zhu, L.F., Zhang, Z.G., and Lin, W.X. 2012a. Identification of autotoxic com-pounds in fibrous roots of Rehmannia (*Rehmannia glutinosa* Libosch.). *PLoS ONE*, 7: e28806.

Li, Z.F., Yang, Y.Q., Xie, D.F., Zhu, L.F., Zhang, Z.G., Huang, M.J., Liu, Z.Q., Zhang, Z.Y., and Lin, W.X. 2012b. Effects of continuous cropping on the quality of *Rehmannia glutinosa* L. and soil micro-ecology. *Chin. J. Eco–Agric.*, 20: 217–224.

Liang, Y.L., Chen, Z.J., Xu, F.L., Zhang, C.E., Du, S.N., and Yan, Y.G. 2004. Soil re-cropping obstacles in facility agriculture on Loess Plateau. *J. Soil Water Conserv.*, 18: 134–136.

Lim, H.S., Kim, Y.S., and Kim, S.D. 1991. *Pseudomonas stutzeri* YPL-1 genetic transformation and anti-fungal mechanism against *Fusarium solani*, an agent of plant root rot. *Appl. Environ. Microbiol.*, 57: 510–516.

Lin, H.F. 2010. Analysis of the Function of Microorganisms in the Rhizosphere of Allelopathic Rice Under Different Water Conditions, master's thesis, Fujian Agriculture and Forestry University, Fuzhou.

Lin, M.Z., Hua, S.H., Chen, Q.Q., and Cai, Y.F. 2012a. Studies on continuous cropping obstacle of *Pseudostelariae heterophylla* and the change of *Fusarium oxysporum* numbers in its rhizosphere. *J. Yunnan Agric. Univ.*, 27: 716–721.

Lin, M.Z., Wang, H.B., and Lin, H.F. 2012b. Effects of *Pseudostellariae heterophylla* continuous cropping on rhizosphere soil microorganisms. *Chin. J. Ecol.*, 31: 106–111.

Lin, R.Y., Rong, H., Zhou, J.J., Yu, C.P., Ye, C.Y., Chen, L.S., and Lin, W.X. 2007. Impact of rice seedling allelop-athy on rhizospheric microbial populations and their functional diversity. *Acta Ecol. Sin.*, 27: 3644–3654.

Lin, R.Y., Wang, H.B., Guo, X.K., Ye, C.Y., He, H.B., and Lin, W.X. 2011. Impact of applied phenolic acids on the microbes, enzymes and available nutrients in paddy soils. *Allelopathy J.*, 28: 225–236.

Lin, W.X., He, H.Q., and Kim, K.U. 2003. The performance of allelopathic heterosis in rice (*Oryza sativa* L.). *Allelopathy J.*, 2: 179–188.

Lin, W.X., Fang, C.X., Wu, L.K., Li, G.L., and Zhang, Z.Y. 2011. Proteomic approach for molecular physiological mechanism on consecutive monoculture problems of *Rehmannia glutinosa*. *J. Integr. Omics*, 1: 287–296.

Lin, W.X., Chen, T., and Zhou, M.M. 2012. New dimensions in agroecology. *Chin. J. Eco-Agric.*, 20: 253–264.

Liu, B., Gumpertz, M.L., Hu, S., and Ristaino, J.B. 2007. Long-term effects of organic and synthetic soil fertility amendments on soil microbial communities and the development of southern blight. *Soil Biol. Biochem.*, 39: 2302–2316.

Liu, H.Y., Wang, F., Wang, Y.P., and Lu, C.T. 2006. The causes and control of continuous cropping barrier in Dihuang (*Rehmannia glutinosa* Libosch.). *Acta Agric. Boreali Sin.*, 21: 131–132.

Liu, L., Sun, J., Guo, S.R., Huang, B.J., Guo, H.W., and Li, L.Q. 2013. Relationship between changes of nutrients, ions and soil acidification in different continuous cropping years of hot pepper greenhouse soils. *Chin. Agric. Sci. Bull.*, 29: 100–105.

Liu, Y.G. and Li, Z.G. 2010. Study on the repair mechanism of complex micro-biological additives to continuous cropping obstacle of *Dioscorea opposita* Thunb. *J. Henan Agric. Sci.*, 11: 90–93.

Lu, P., Wu, J., Chen, X.F., and Rao, Y.M. 2010. Study on the application of microbial soil improvement agent in cucumber. *Shanghai Agric. Sci. Technol.*, 4: 109–110.

Mazzola, M., Granatstein, D.M., Elfving, D.C., and Mullinix, K. 2001. Suppression of specific apple root pathogens by *Brassica napus* seed meal amendment regardless of glucosinolate content. *Phytopathology*, 91: 673–679.

Mazzola, M., Reardon, C.L., and Brown, J. 2012. Initial *Pythium* species composition and Brassicaceae seed meal type influence extent of *Pythium*-induced plant growth suppression in soil. *Soil Biol. Biochem.*, 48: 20–27.

Mendes, R., Kruijt, M., de Bruijn, I., Dekkers, E., van der Voort, M. et al. 2011. Deciphering the rhizosphere microbiome for disease-suppressive bacteria. *Science*, 332: 1097–1100.

Morsy, E.M., Abdel-Kawi, K.A., and Khalil, M.N.A. 2009. Efficiency of *Trichoderma viride* and *Bacillus subtilis* as biocontrol agents against *Fusarium solani* on tomato plants. *Egypt. J. Phytopathol.*, 37: 47–57.

Nicol, R.W., Yousef, L., Traquair, J.A., and Bernards, M.A. 2003. Ginsenosides stimulate the growth of soilborne pathogens of American ginseng. *Phytochemistry*, 64: 257–264.

Naar, Z. 2006. Effect of cadmium, nickel and zinc on the antagonistic activity of *Trichoderma* spp. against *Pythium irregulare* Buisman. *Acta Phytopathol. Entomol. Hung.*, 41: 193–202.

Niu, M.M., Li, J., Du, J.F., Yin, W.J., Yang, Y.H., Chen, X.J., and Zhang, Z.Y. 2011. Changes in source–sink relationship of photosynthate in *Rehmannia glutinosa* L. and their relations with continuous cropping obstacle. *Chin. J. Ecol.*, 30: 248–254.

Oka, Y. 2010. Mechanisms of nematode suppression by organic soil amendments: a review. *Appl. Soil Ecol.*, 44: 101–115.

Olofsdotter, M., Navarez, D., Rebulanan, M., and Streibig, J.C. 1999. Weed suppressing rice cultivars: does allelopathy play a role? *Weed Res.*, 39: 441–454.

Olofsdotter, M., Rebulanan, M., Madrid, A., Wang, D.L., Navarez, D., and Olk, D.C. 2002. Why phenolic acids are unlikely primary allelochemicals in rice. *J. Chem. Ecol.*, 28: 229–242.

Otomo, K., Kanno, Y., Motegi, A., Kenmoku, H., Yamane, H. et al. 2004a. Diterpene cyclases responsible for the biosynthesis of phytoalexins, momilactones A, B, and oryzalexins A–F in rice. *Biosci. Biotechnol. Biochem.*, 68: 2001–2006.

Otomo, K., Kenmoku, H., Oikawa, H., Konig, W.A., Toshima, H., Mitsuhashi, W., Yamane, H., Sassa, T., and Toyomasu, T. 2004b. Biological functions of *ent*- and *syn*-copalyl diphosphate synthases in rice: key enzymes for the branch point of gibberellin and phytoalexin biosynthesis. *Plant J.*, 39: 886–893.

Pal, K.K., Tilak, K.V.B.R., Saxena, A.K., Dey, R., and Singh, C.S. 2001. Suppression of maize root diseases caused by *Macrophomina phaseolina*, *Fusarium moniliforme* and *Fusarium graminearum* by plant growth promoting rhizobacteria. *Microbiol. Res.*, 156: 209–223.

Paterson, E., Gebbing, T., Abel, C., Sim, A., and Telfer, G. 2007. Rhizodeposition shapes rhizosphere microbial community structure in organic soil. *New Phytol.*, 173: 600–610.

Pei, G.P. 2010. Physiological and Rhizosphere Soil Nutrients Changes of Continuous Cropping in Potato, master's thesis, Gansu Agricultural University, Lanzhou.

Peng, Y.C., Liu, T., Zhao, J.J., Sun, S.G., Gao, J., Wu, F.R., Liu, G.S., and Ye, X.F. 2009. Research advances in effect of continuous cropping on soil characteristics. *Acta Agric. Jiangxi*, 21: 100–103.

Peters, R.J. 2006. Uncovering the complex metabolic network underlying diterpenoid phytoalexin biosynthesis in rice and other cereal crop plants. *Phytochemistry*, 67: 2307–2317.

Pheng, S., Adkins, S., Olofsdotter, M., and Jahn, G. 1999. Allelopathic effects of rice (*Oryza sativa* L.) on the growth of awnless barnyardgrass (*Echinochloa colona* (L.) Link): a new form for weed management. *Cambodian J. Agric.*, 2: 42–49.

Putnam, A.R. and Duke, S.O. 1974. Biological suppression of weeds: evidence for allelopathy in accessions of cucumber. *Science*, 185: 370–372.

Putnam, A.R. and Tang, C.S. 1986. Allelopathy: state of the science. In: Putnam, A.R. and Tang, C.S., Eds., *The Science of Allelopathy*. John Wiley & Sons: New York.

Qi, J.J., Yao, H.Y., Ma, X.J., Zhou, L.L., and Li, X.N. 2009. Soil microbial community composition and diversity in the rhizosphere of a Chinese medicinal plant. *Commun. Soil Sci. Plant Anal.*, 40: 1462–1482.

Qu, X.H. and Wang, J.G. 2008. Effect of amendments with different phenolic acids on soil microbial biomass, activity, and community diversity. *Appl. Soil Ecol.*, 39: 172–179.

Rahman, M. and Punja, Z.K. 2005. Factors influencing development of root rot on ginseng caused by *Cylindrocarpon destructans*. *Phytopathology*, 95: 1381–1390.

Rani, M.U., Arundhath, A., and Reddy, G. 2014. Screening of rhizobacteria containing plant growth promoting (PGPR) traits in rhizosphere soils and their role in enhancing growth of pigeon pea. *Afr. J. Biotechnol.*, 11: 8085–8091.

Rice, E.L. 1984. *Allelopathy*, 2nd ed. Academic Press: New York.

Rani, M.U., Arundhath, A., and Reddy, G. 2014. Screening of rhizobacteria containing plant growth promoting (PGPR) traits in rhizosphere soils and their role in enhancing growth of pigeon pea. Afr. J. Biotechnol., 11: 8085–8091.

Reeleder, R.D. and Brammall, R.A. 1994. Pathogenicity of *Pythium* species, *Cylindrocarpon destructans*, and *Rhizoctonia solani* to ginseng seedlings in Ontario. *Can. J. Plant Pathol.*, 16: 311–316.

Rojo, F.G., Reynoso, M.M., Ferez, M., Chulze, S.N., and Torres, A.M. 2007. Biological control by *Trichoderma* species of *Fusarium solani* causing peanut brown root rot under field conditions. *Crop Protect.*, 26: 549–555.

Rosskopf, E.N., Chellemi, D.O., Kokalis-Burelle, N., and Church, G.T. 2005. Alternatives to methyl bromide: a Florida perspective. *APSnet Feature* (http://www.apsnet.org/publications/apsnetfeatures/Pages/MethylAlternatives.aspx).

Sawada, H., Shim, I.S., and Usui, K. 2006. Induction of benzoic acid 2-hydroxylase and salicylic acid biosynthesis modulation by salt stress in rice seedlings. *Plant Sci.*, 171: 263–270.

Seal, A.N., Pratley, J.E., Haig, T., and An, M. 2004. Identification and quantitation of compounds in a series of allelopathic and non-allelopathic rice root exudates. *J. Chem. Ecol.*, 30: 1647–1662.

Shaigan, S., Seraji, A., and Moghaddam, S.A.M. 2008. Identification and investigation on antagonistic effect of *Trichoderma* spp. on tea seedlings white foot and root rot (*Sclerotium* rolfsii Sacc.) *in vitro* condition. *Pak. J. Bio. Sci.*, 19: 2346–2350.

Shen, L.H., Liang, Y.Y., He, H.Q., He, J., Liang, K.J., and Lin, W.X. 2004. Comparative analysis on the evaluation efficiency of allelopathic potential in rice (*Oryza sativa* L.) by using different bioassay methods and its application. *Chin. J. Appl. Ecol.*, 9: 1575–1579.

Shimura, K., Okada, A., Okada, K., Jikumaru, Y., Ko, K.W., Toyomasu, T., Sassa, T., Hasegawa, M., Kodama, O., Shibuya, N., Koga, J., Nojiri, H., and Yamane, H. 2007. Identification of a biosynthetic gene cluster in rice for momilactones. *J. Biol. Chem.*, 282: 34013–34018.

Shin, D.H., Kim, K.U., Sohn, D.S., Kang, S.G., Kim, H.Y., Lee, I.J., and Kim, M.Y. 2000. Regulation of gene expression related to allelopathy. In: Kim, K.U. and Shin, D.H., Eds., *Proceedings of the International Workshop in Rice Allelopathy*. Institute of Agricultural Science and Technology, Kyungpook National University: Taegu, pp. 109–124.

Silverman, P., Seskar, M., Kanter, D., Schweizer, P., Metraux, J.P., and Raskin, I. 1995. Salicylic acid in rice: biosynthesis, conjugation, and possible role. *Plant Physiol.*, 108: 633–639.

Singh, N., Pandey, P., Dubey, R.C., and Maheshwari, D.K. 2008. Biological control of root rot fungus *Macrophomina phaseolina* and growth enhancement of *Pinus roxburghii* (Sarg.) by rhizosphere competent *Bacillus subtilis* BN1. *World J. Microbiol. Biotechnol.*, 24: 1669–1679.

Smolinska, U., Morra, M.J., Knudsen, P.D., and Brown, G.R. 1997. Toxicity of glucosinolate degradation products from *Brassica napus* seed meal toward *Aphanomyces euteiches* f. sp. *pisi*. *Phytopathology*, 87: 77–82.

Song, B.Q., Xiong, J., Fang, C.X., Qiu, L., Lin, R.Y., Liang, Y.Y., and Lin, W.X. 2008. Allelopathic enhancement and differential gene expression in rice under low nitrogen treatment. *J. Chem. Ecol.*, 34: 688–695.

Su, S.M., Ren, L.X., Huo, Z.H., Yang, X.M., Huang, Q.W., Xu, Y.C., Zhou, J., and Shen, Q.R. 2008. Effects of intercropping watermelon with rain fed rice on *Fusarium* wilt and the microflora in the rhizosphere soil. *Sci. Agric. Sin.*, 41: 704–712.

Sugimura, T., Masahiro, N., and Keiichi, H. 2001. Control of *Fusarium* wilt of strawberry by soil solarization using mulching and tunnel-covering combining with irrigation of solar-heated water. *Bull. Nara Prefect. Agric. Exp. Stn.*, 32: 1–7.

Sun, H.X., Wu, Q., Zheng, G.X., and Wang, Z.Z. 2001. Effect of EM on continuous cropping and soil biological activity of eggplant and cucumber. *Soil*, 5: 264–267.

Sydorovych, O., Safley, C.D., Ferguson, L.M., Poling, B.E., Fernandez, G.E., Brannen, P.M., Monks, D.M., and Louws, F.J. 2006. Economic evaluation of methyl bromide alternatives for the production of strawberries in the southeastern United States. *HortTechnology*, 16: 1–11.

Takahashi, N., Kato, T., Tsunagawa, M., Sasaki, N., and Kitahara, Y. 1976. Mechanisms of dormancy in rice seeds. II. New growth inhibitors, momilactone-A and -B isolated from the hulls of rice seeds. *Jpn. J. Breed.*, 26: 91–98.

Taylor, E.B. and Williams, M.A. 2010. Microbial protein in soil: influence of extraction method and C amendment on extraction and recovery. *Microb. Ecol.*, 59: 390–399.

Tesio, F. and Ferrero, A. 2010. Allelopathy, a chance for sustainable weed management. *Int. J. Sustain. Dev. World Ecol.*, 17: 377–389.

Toyomasu, T., Kagahara, T., Okada, K., Koga, J., Hasegawa, M., Mitsuhashi, W., Sassa, T., and Yamane, H. 2008. Diterpene phytoalexins are biosynthesized in and exuded from the roots of rice seedlings. *Biosci. Biotechnol. Biochem.*, 72: 562–567.

UNEP. 2015. *Handbook for the Montreal Protocol on Substances that Deplete the Ozone Layer.* United Nations Environmental Program: Geneva (http://ozone.unep.org/en/treaties-and-decisions/montreal-protocol-substances-deplete-ozone-layer).

Wang, H.B., He, H.B., Xiong, J., Qiu, L., Fang, C.X., Zeng, C.M., Yan, L., and Lin, W.X. 2008. Effects of potassium stress on allelopathic potential of rice (*Oryza sativa* L.). *Acta Ecol. Sin.*, 28: 6219–6127.

Wang, H.B., He, H.B., Ye, C.Y., Lu, J.C., Chen, R.S., Liu, C.H., Guo, X.K., and Lin, W.X. 2010. Molecular physiological mechanism of increased weed suppression ability of allelopathic rice mediated by low phosphorus stress. *Allelopathy J.*, 25: 239–248.

Wang, H.B., Yu, Y.M., He, H.B., Guo, X.K., Huang, J.W., Zhou, Y., Xu, Z.B., and Lin, W.X. 2012. Relationship between allelopathic potential and grain yield of different allelopathic rice accessions. *Chin. J. Eco-Agric.*, 20: 75–79.

Wang, J.J., Zhang, F.H., Kang, M.D., Zhang, A.X., and Gong, X. 2013. Long-term continuous cropping soil aggregate composition and distribution of organic carbon under different planting patterns. *Agric. Eng. Technol. (New Energy Ind.)*, 4: 15–18.

Wang, Q., Hillwig, M.L., and Peters, R.J. 2011. CYP99A3: functional identification of a diterpene oxidase from the momilactone biosynthetic gene cluster in rice. *Plant J.*, 65: 87–95.

Wang, S.T., Cao, K.Q., Hu, T.L., and Zhang, F.Q. 2006a. Inhibition of *Anemarrhena asphodeloides* Bunge extract (AAE) against *Phytophthora infestans* and control efficacy on potato late blight. *Acta Phytopathol. Sin.*, 36: 267–272.

Wang, S.T., Zhang, F.Q., Gao, R.P., Zhen, W.C., Bao, S.X., Jiang, J.H., and Cao, K.Q. 2006b. Inhibition of 126 kinds of Chinese herb extracts against two plant pathogenic fungi. *J. Henan Agric. Sci.*, 10: 62–65.

Wei, Z.X., Yang, C.B., He, H., and Tang, J.Y. 2014. Inhibition activity of extracts of *Cirsium japonicum* DC on plant pathogenic fungi. *J. Yunnan Agric. Univ. (Nat. Sci.)*, 1: 140–143.

Weidenhamer, J.D., Macias, F.A., Fischer, N.H., and Williamson, G.B. 1993. Just how insoluble are nonoterpenes? *J. Chem. Ecol.*, 19: 1799–1807.

Weston, L.A. and Mathesius, U. 2013. Flavonoids: their structure, biosynthesis and role in the rhizosphere, including allelopathy. *J. Chem. Ecol.*, 39: 1–15.

Wilderman, P.R., Xu, M., Jin, Y., Coates, R.M., and Peters, R.J. 2004. Identification of synpimara-7,15-diene synthase reveals functional clustering of terpene synthases involved in rice phytoalexin/allelochemical biosynthesis. *Plant Physiol.*, 135: 2098–2105.

Wu, L., Guo, J., Liu, J., Yang, Y., and Wu, X. 1998. Studies of the control of *Cylindrocarpon destructans* of ginseng by soil antagonistic microbes. *Chin. J. Biol. Control*, 15: 166–168.

Wu, L.K., Li, Z.F., Li, J., Khan, M.A., Huang, W.M., Zhang, Z.Y., and Lin, W.X. 2013. Assessment of shifts in microbial community structure and catabolic diversity in response to *Rehmannia glutinosa* monoculture. *Appl. Soil Ecol.*, 67: 1–9.

Wu, Y.F., Zhang, X.Y., Li, Y., Wei, W.J., and Gao, L.H. 2008. Influence of rotation on continuous cropping soil environment and cucumber yield. *Acta Hort. Sin.*, 35: 357–362.

Wu, Z.W., Wang, M.D., Liu, X.Y., Chen, H.G., and Jia, X.C. 2009. Phenolic compounds accumulation in continuously cropped *Rehmannia glutinosa* soil and their effects on *R. glutinosa* growth. *Chin. J. Ecol.*, 28: 660–664.

Xie, Z.M., Tan, X.L., and Qu, Y.M. 2011. Effects of different soil restoration agents on root rot of consecutively monocultured cowpea. *Chin. Hort. Abstr.*, 12: 37–38.

Xiong, J., Jia, X.L., Deng, J.Y., Jiang, B.Y., He, H.B., and Lin, W.X. 2007. Analysis of epistatic effect and QTL interactions with environment for allelopathy in rice (*Oryza sativa* L.). *Allelopathy J.*, 20: 259–268.

Xiong, J., Wang, H.B., Qiu, L., Wu, H.W., Chen, R.S., He, H.B., Lin, R.Y., and Lin, W.X. 2010. qRT-PCR analysis of key enzymatic genes related to phenolic acid metabolism in rice accessions (*Oryza sativa* L.) exposed to low nitrogen treatment. *Allelopathy J.*, 25: 345–356.

Xiong, J., Lin, H.F., Li, Z.F., Fang, C.X., Han, Q.D., and Lin, W.X. 2012. Analysis of rhizosphere microbial community structure of weak and strong allelopathic rice varieties under dry paddy field. *Acta Ecol. Sin.*, 32: 6100–6109.

Xu, M.M., Hillwig, M.L., Prisic, S., Coates, R.M., and Peters, R.J. 2004. Functional identification of rice syn-copalyl diphosphate synthase and its role in initiating biosynthesis of diterpenoid phytoalexin/allelopathic natural products. *Plant J.*, 39: 309–318.

Xu, Z.H., Yu, L.Q., and Zhao, M. 2003. Rice allelopathy to barnyardgrass. *Chin. J. Appl. Ecol.*, 14: 737–740.

Xu, Z.H., Xie, G.X., Zhou, Y.J., and Gao, S. 2013a. Interference of rice with different morphological types and allelopathy on barnyardgrass under three planting patterns. *Acta Agron. Sin.*, 39: 537–548.

Xu, Z.H., Xie, G.X., Zhou, Y.J., and Gao, S. 2013b. Interference of different morphological types and allelopathic rice materials with three principal paddy weeds. *Acta Agron. Sin.*, 39: 1293–1302.

Yang, Y.H., Chen, X.J., Chen, J.Y., Xu, H.X., Li, J., and Zhang, Z.Y. 2011. Differential miRNA expression in *Rehmannia glutinosa* plants subjected to continuous cropping. *BMC Plant Biol.*, 11: 53.

Yin, W.J., Du, J.F., Li, J., and Zhang, Z.Y. 2009. Effects of continuous cropping obstacle on growth of *Rehmannia glutinosa*. *China J. Chin. Mater. Med.*, 34: 18–21.

Yu, J.Q. and Matsui, Y. 1994. Phytotoxic substances in root exudates of cucumber (*Cucumis sativus* L.). *J. Chem. Ecol.*, 20: 21–31.

Yu, J.Q. and Matsui, Y. 1997. Effect of root exudates of cucumber (*Cucumis sativus*) and allelochemicals on ion uptake by cucumber seedlings. *J. Chem. Ecol.*, 23: 817–827.

Yu, M., Yu, J.W., Cao, P.G., Liang, H.D., Xiao, H.D., Wang, Y.B., and Cui, Z.X. 2004. Agrochemical characteristics of soil for continuous cropping lily. *Chin. J. Soil Sci.*, 35: 377–379.

Yu, X., Ai, C., Xin, L., and Zhou, G. 2011. The siderophore-producing bacterium, *Bacillus subtilis* CAS15, has a biocontrol effect on *Fusarium* wilt and promotes the growth of pepper. *Eur. J. Soil Biol.*, 47: 138–145.

Zeng, D., Qian, Q., Teng, S., Dong, G.J., Fujimoto, H., Yasufumi, K., and Zhu, L.H. 2003. Genetic analyses on rice allelopathy. *Chin. Sci. Bull.*, 48: 70–73.

Zhang, L., Spiertz, J.H.J., Zhang, S., Li, B., and van der Werf, W. 2008. Nitrogen economy in relay intercropping systems of wheat and cotton. *Plant Soil*, 303: 55–68.

Zhang, R.X., Li, M.X., and Jia, Z.P. 2008. *Rehmannia glutinosa*: review of botany, chemistry and pharmacology. *J. Ethnopharmacol.*, 117: 199–214.

Zhang, X.L., Pan, Z.G., Zhou, X.F., and Ni, W.H. 2007. Autotoxicity and consecutive monoculture problems. *Chin. J. Soil Sci.*, 38: 781–784.

Zhang, Z.Y. and Lin, W.X. 2009. Continuous cropping obstacle and allelopathic autotoxicity of medicinal plants. *Chin. J. Eco-Agric.*, 17: 189–196.

Zhang, Z.Y., Li, P., Qi, H., and Li, J. 2003. Analysis on the geologic background and physicochemical properties of cultivated soil of Flos Lonicerae in the geo-authentic and non-authentic producing areas. *China J. Chin. Mater. Med.*, 28: 114–117.

Zhang, Z.Y., Yin, W.J., Li, J., Du, J.F., Yang, Y.H., Chen, X.J., and Lin, W.X. 2010a. Physio-ecological properties of continuous cropping *Rehmannia glutinosa. Chin. Jour. Plant Ecol.*, 34: 547–554.

Zhang, Z.Y., Chen, H., Yang, Y.H., Chen, T., Lin, R.Y., Chen, X.J., and Lin, W.X. 2010b. Effects of continuous cropping on bacterial community diversity in rhizosphere soil of *Rehmannia glutinosa. Chin. J. Appl. Ecol.*, 21: 2843–2848.

Zhang, Z.Y., Fan, H.M., Yang, Y.H., Li, M.J., Li, J., Xu, H.X., Chen, J.Y., and Chen, X.J. 2011. Construction and analysis of suppression subtractive cDNA libraries of continuous monoculture *Rehmannia glutinosa. China J. Chin. Mater. Med.*, 36: 276–280.

Zhao, Q.G., Yang, J.S., and Zhou, H. 2011. "Ten words" strategic policy for ensuring red line of farmland and food security in China. *Soils*, 5: 1–7.

Zhao, Z.L., Shi, L.L., Yan, Y.R., Gong, Z.H., Wu, Q.Q., and Guo, J.W. 2006. Effects of fertilization on continuous cropping obstacle in pepper. *Agric. Res. Arid Areas*, 24: 77–80.

Zhou, X.G. and Wu, F.Z. 2012. *p*-Coumaric acid influenced cucumber rhizosphere soil microbial communities and the growth of *Fusarium oxysporum* f. sp. *cucumerinum* Owen. *PLoS ONE*, 7: e48288.

Zhou, Y.J., Li, D., Lu, Y.L., Yu, L.Q., and Chen, M.X. 2005. Evaluation of allelopathic potential in rice germplasm. *Acta Ecol. Sin.*, 25: 1599–1603.

Zolla, G., Badri, D.V., Bakker, M.G., Manter, D.K., and Vivanco, J.M. 2013. Soil microbiomes vary in their ability to confer drought tolerance to Arabidopsis. *Appl. Soil Ecol.*, 68: 1–9.

CHAPTER 8

Structure and Function of Grass-Covered Orchards in Fujian Province, China

Weng Boqi and Wang Yixiang

CONTENTS

The application of living mulch to orchard floor management was widely accepted until the 1950s in many developed countries (Faust, 1979). It is estimated that, around the beginning of this century, living mulch covered about 55 to 70% of the total area of orchards in the world (Gregoriou and Rajkumar, 1984; Lanini et al., 1988). Living mulch has been proven to be one of the most efficient means of controlling water erosion (Glover et al., 2000; Mathews et al., 2002; Wang et al., 2010), increasing levels of soil organic matter, and enhancing soil biological activity (Hoagland et al., 2008; Peck et al., 2011; Xu et al., 2012). In China, experimental utilization of living mulch in orchard floor management began in the early 1980s, and in 1998 the method was adopted by the Green Food Development Center of China and promoted as part of "green fruit" production systems in the provinces of Fujian, Guangdong, and Shandong (Zhao et al., 2000).

Fujian Province is located in the southeast of China. It has 10.1 million km² of mountainous area, accounting for 84.1% of the total land area. There are 541,000 hm² of orchards, of which about 90% are distributed in mountainous areas, where red and yellow soils dominate (Guo et al., 2010). Most of these orchards were initially converted from original natural woodland and have historically been threatened by serious soil erosion and degradation (Q.X. Chen et al., 1996; N.W. Chen et al., 2009). In addition, red soil itself is not healthy for fruit trees due to its high acidity, poor fertility, and low water-holding capacity (Weng et al., 2006a). Conventional clean tillage has made the situation worse by removing all soil cover and thus reducing the orchard floor roughness, vegetation diversity, and soil quality (You et al., 2004; F. Zhang et al., 2004). This has led to accelerated soil erosion and nutrient loss (Liu et al., 2012). According to a remote sensing survey carried out in 2000, about 174,000 hectares of orchard plantings in Fujian Province (31.46% of the total orchard area in the province) were exposed to soil erosion (M.H. Chen, 2004a,b). In 2003, soil and water loss occurred on 45.8% of the orchard lands in the Fujian Province, and the soil organic matter content was less than 1% in 70% of the low-yield orchards according to statistics. However, no living mulch technique had ever been tried for hilly red soil orchards in Fujian until the 1980s. It was very important to develop a more sustainable, advanced agricultural management method to improve this situation.

This research has focused on the investigation of the effectiveness of living mulch, especially the species *Arachis pintoi*, on soil fertility and fruit productivity. Field comparisons were made between terrace orchards with living mulch and terrace/slope orchards with clean tillage. Values of ecosystem services were estimated to help develop a comprehensive understanding of the effects. Furthermore, it was also hoped that this research would demonstrate that crop cultivation with living mulch might contribute to countermeasures against global warming.

8.1 SELECTION OF MULCHING PLANTS FOR ORCHARDS ON HILLY RED SOIL

Forage species resources are quite rich in Fujian, but most of them do not satisfy the requirements for orchard mulching on hilly red soils. They are low yielding, of poor quality, or have a long dormant period. Temperate species do not grow well or survive the hot summers, while tropical species usually do not survive the winter. This has led to the need to introduce and select new forage species that are adapted to the local climate and suitable for hilly orchards with red soils. In addition, the forage selected should possess the following characteristics: (1) short stems, large biomass, and high coverage above ground; (2) mainly fibrous roots below ground: (3) no pests in common with fruit trees; (4) extended green groundcovering period and short vigorous growing period; and (5) resistance to shading and trampling.

8.1.1 Temperate Forage Species

Between 1990 and 1998, 25 such species, involving 88 strains or varieties, were introduced by the Fujian Academy of Agricultural Sciences (FAAS) as living mulches for use on local hilly, red-soil orchards. We found that *Eragrostis curvula* (Consol), *Triticale* sp., *Trifolium repens* (Haifa), and *Dactylis glomerata* (Porto) were particularly adaptive to the environment of Fujian Province's hilly, red-soil orchards, especially with regard to speed of establishing groundcover, stress resistance, production, and nutrition value. These species were rarely seen in Northern Fujian red soils before this study but came to be regarded as dominant mulching plant species afterwards. Table 8.1

Table 8.1   Main Characteristics of Dominant Species

Representatives	Growth Characters
Eragrostis curvula (Family: Gramineae)	Establish slowly, strong vitality; grow vigorously in summer; production lowered in autumn and winter; high tillering; used for soil and water conservation
Dactylis glomerata (Family: Gramineae)	Establish rapidly with rich roots; high nutrition and fine palatability as animal feed
Triticale sp. (Family: Gramineae)	Establish more rapidly than *D. glomerata*, with strong resistance to acidity, aluminum, and disease; crude protein content 8.44 to 10.44%; high nutrition and fine palatability as animal feed
Trifolium repens (Family: Leguminosae)	Establish more rapidly than *D. glomerata*; stems prostrate; roots with nitrogen-fixing nodules after artificial inoculation; crude protein content 28.7%; excellent quality.

provides a summary of their growth characteristics. Unfortunately, most of the introduced temperate forage species failed to survive during the drought period in autumn or the cold period in winter; therefore, further introduction and selection of tropical species were carried out.

8.1.2 Tropical Forage Species

Forty tropical species were introduced to Fujian Province for the present study. Except for the annual species, most of the introduced tropical accessions were able to survive through the cold winter in Fujian, with *Digitaria smutsii* (Figure 8.1A) and *Arachis pintoi* (Figure 8.1B) exhibiting better resistance to cold than the others. *Chamaecrista rotundifolia* (Figure 8.1C) had poor overwintering ability, with only 10% of the plants surviving the winter, but they regenerated in the second year by shedding seeds.

(A)   (B)

(C)

Figure 8.1   (A) *Digitaria smutsii*, (B) *Arachis pintoi*, and (C) *Chamaecrista rotundifolia*.

Arachis pintoi has strong resistance to high aluminum levels, shade, and cold (Luo et al., 1994). Its ability to fix atmospheric nitrogen, produce dense cover with little maintenance, and protect soil from erosion makes it an appealing groundcover. The stems of *A. pintoi* creep over the ground with many branches, which extend new roots from the stem nodes. The plants grow rapidly and have an average height of 15 cm, thus covering the ground in a short time with coverage rates as high as 95%. It is useful for reducing soil erosion and water loss.

Chamaecrista rotundifolia, in general, also offers good features for mulching, such as suitability for interplanting in newly cultivated orchards, soil and water conservation, and soil fertility improvement. Moreover, it is particularly resistant to barren and dry environments. Unfortunately, this introduced species did not resist the frosts that frequently occur in northern Fujian.

In addition, *Digitaria smutsii* has been appreciated for being highly adaptive to different soil conditions (e.g., barren, acidic), to hot and wet climates, and to drought and cold (Luo et al., 1994). In the present experiment, it survived well during short-term frosty periods. Individuals of this species have strong regeneration ability and are usually harvested three or four times each year. Bearing rich root systems, *D. smutsii* is certainly a potential candidate for soil conservation; however, *D. smutsii* had an average height of 100 cm and easily forms a hedge. This attribute has limited its utilization in orchard floor management.

8.2 LIVING MULCH EFFECT ON FRUIT GROWTH, YIELD, AND QUALITY

The successful use of *Arachis pintoi* mulch in orchard management has been reported worldwide (Hogue and Neilsen, 1987; Firth and Wilson, 1995). *A. pintoi* has proven to be successful as an orchard mulch for the following reasons: rapid growth, protection of the ground during torrential rain, improvement of soil fertility, tolerance to intense shading, reduction of weed populations, and the ability to not compete with the main orchard plants for light, space, and nutrients (Firth et al., 2002; Doanh and Tuan, 2004; Candog-Bangi et al., 2007). But, few literature citations discuss the effect of *A. pintoi* mulch on fruit growth and output in hilly, red-soil orchards. A split-plot design was applied to investigate the effect of *A. pintoi* mulch on fruit tree growth, fruit yield, and quality. In this study, there were three treatments: T1 (sloping orchard with clean tillage), T2 (terraced orchard with clean tillage), and T3 (terraced orchard with *A. pintoi* mulch) (Figure 8.2). In T1 and T2, unexpected weeds were removed by hand-pulling three or four times annually. T2 and T3 involved eight terraces along the hill slope. Peach trees were planted along the middle line of each terrace. In T3, *A. pintoi* was also planted on the terraces.

As shown in Figure 8.3, by the end of the experiment, the average trunk diameter was larger in T3 than in T2, with both doing better than T1. The average size of the tree crowns did not differ significantly between T2 and T3. Compared with clean tillage management, the peach orchard with *Arachis pintoi* mulch had, on average, 9.4% higher fruit production. Kartika et al. (2009) noticed that the plants of *A. pintoi* were able to fix atmospheric nitrogen and thus improved soil fertility. In the program report given by the College of Tropical Agriculture and Human Resources (CTAHR, 2009), fruit tree growers who employed *A. pintoi* for orchard floor covering reported better growth of the trees than when clean tillage management was used.

The fruit sugar/acid ratios and vitamin C content have been considered important indices for peach quality assessment (Peck et al., 2006). The values of both were significantly higher in fruits from the orchard with living mulch, based on chemical analysis data obtained in this research. This could be the result of a chain reaction beginning with the application of *Arachis pintoi* mulch on the orchard floor and involving the reduction of soil bulk density, the enhancement of soil porosity and water retention, the moderation of diurnal fluctuations in soil temperature, and the increase of soil organic matter content and plant nutrient availability (Pinamonti et al., 1995; Buban et al., 1996; Glover et al., 2000).

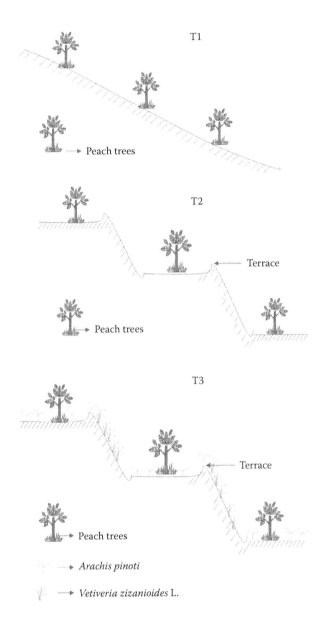

Figure 8.2 An illustration of the landscape treatments to research the effects of living mulch in peach orchards.

8.3 LIVING MULCH EFFECT ON SOIL PHYSICAL AND CHEMICAL PROPERTIES

Arachis pintoi mulch in orchards is reported to be effective in preventing fertilizer loss as well as water and soil erosion (Glover et al., 2000; Mathews et al., 2002; Wang et al., 2010). In addition, it increases soil organic matter (SOM) content and carbon (C) and nitrogen (N) levels; regulates pH values; activates the availability of N, phosphorus (P), and potassium (K); increases microbial biomass and respiration; and enhances soil biological activity (Hoagland et al., 2008; Peck et al., 2011; Xu et al., 2012). *A. pintoi* mulch effects on red soils in peach orchards in the mountainous area of northern Fujian Province were evaluated for soil physical and chemical properties, soil organic carbon in aggregates, and water and soil conservation.

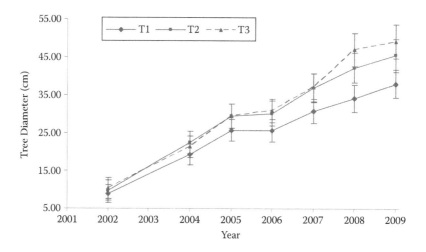

Figure 8.3   Variations in peach tree trunk diameters with different types of orchard floor management.

8.3.1 Soil Bulk Density and Porosity

Usually, a lower soil bulk density is associated with a higher material porosity and water-holding capacity. Soil bulk density, as dry weight, includes air space and organic material in the soil volume. The bulk density (BD) of cultivated loam is about 1.1 to 1.4 g cm^{-3} (Miller, 1977). A high BD indicates either compaction of the soil or high sand content. In this work, the peach orchard soils had a relatively low BD, varying around 1.0 g cm^{-1}. With *Arachis pintoi* mulch, the soil BD became even lower (Table 8.2). Meanwhile, the soil water content did not increase in the same group in spite of the porosity enhancement. It is therefore difficult to comment on the effect of *A. pintoi* mulch on the soil physical conditions at this stage.

Pore space is that part of the bulk volume that is not occupied by either mineral or organic matter but is open space occupied by either air or water. Ideally, the total pore space should be 50% of the soil volume. Apparently, the values of soil porosity for the present peach orchard were low. Having large pore spaces that allow rapid air and water movement is superior to smaller pore space but with a greater percentage of pore space (Miller, 1977). Tillage has the short-term benefit of temporarily increasing the number of pores of largest size, but in the end those will be degraded by the destruction of soil aggregates. It will still take many years for a living mulch to function in the improvement of soil porosity.

Table 8.2   Soil Bulk Density and Porosity Variations for Different Tillage Patterns (Sample Size $n = 3$)

Parameters	Soil Depth (cm)	Experiment Groups		
		T1	T2	T3
Bulk density (g cm^{-3})	0–10	1.08 ± 0.04[a]	1.05 ± 0.05[a]	0.96 ± 0.04[b]
	10–20	1.15 ± 0.06[a]	1.08 ± 0.05[a]	0.97 ± 0.04[b]
Porosity (%)	0–10	17.5 ± 0.77[c]	20.1 ± 0.89[b]	26.6 ± 1.12[a]
	10–20	12.2 ± 0.56[c]	17.1 ± 0.77[b]	26.1 ± 1.04[a]

Note: T1, sloping orchard with clean tillage; T2, terrace orchard with clean tillage; T3, terrace orchard with *Arachis pintoi* mulch. Within the same row, means without a common superscript letters differ significantly ($p < 0.05$).

Table 8.3 Content of >0.25-mm Water-Stable Aggregates, Mean Weight Diameter, Geometric Mean Diameter, and Fractal Dimension under Different Tillage Systems

Treatment	Soil Depth (cm)	Water-Stable Aggregates ($R_{0.25}$)	Mean Weight Diameter (MWD)	Geometric Mean Diameter (GMD)	Fractal Dimension (D)
T1	0–20	83.5 ± 1.26[b]	1.08 ± 0.199[ab]	0.791 ± 0.144[a]	2.18 ± 0.091[a]
	20–40	83.4 ± 2.03[b]	0.89 ± 0.179[bc]	0.644 ± 0.112[b]	2.21 ± 0.053[a]
T2	0–20	85.2 ± 1.47[a]	1.19 ± 0.101[a]	0.834 ± 0.080[a]	2.15 ± 0.029[a]
	20–40	76.0 ± 2.94[c]	0.72 ± 0.019[c]	0.481 ± 0.002[c]	2.41 ± 0.013[a]
T3	0–20	88.4 ± 0.25[a]	1.26 ± 0.194[a]	0.897 ± 0.168[a]	2.12 ± 0.078[a]
	20–40	86.5 ± 2.34[a]	1.08 ± 0.231[ab]	0.773 ± 0.176[a]	2.18 ± 0.092[a]

Note: T1, sloping orchard with clean tillage; T2, terrace orchard with clean tillage; T3, terrace orchard with *Arachis pintoi* mulch. Within the same row, means without a common superscript letters differ significantly ($p < 0.05$).

8.3.2 Soil Aggregates

Soil aggregates are groups of soil particles that bind to each other more strongly than to adjacent particles. Aggregate stability refers to the ability of soil aggregates to resist disintegration when disruptive forces associated with tillage and water or wind erosion are applied. In our study, the results indicated that, compared with the T1 treatments, the content of >0.25-mm water-stable aggregates ($R_{0.25}$), mean weight diameter (MWD), and geometric mean diameter (GMD) of soil aggregates in orchards under the living mulch (T3) treatment increased by 3.78 to 5.90%, 16.82 to 20.94%, and 5.86 to 50.31%, respectively. Compared to the T2 treatment, these values increased by 3.81 to 13.82%, 13.33 to 19.95%, and 7.50 to 60.63%, respectively (Table 8.3).

The quantity, distribution, and stability of aggregates with living mulch were better than for clean tillage. There are several reasons for these results. First, non-tillage and *Arachis pintoi* mulch reduced soil disturbance. Second, organic matter produced by *A. pintoi* cover continuously added to the soil enhanced soil fertility and pore structure. Finally, enrichment of organic matter promoted the formation of large-sized aggregates, thereby enhancing the stability of aggregates. The study of Tisdall et al. (1997) indicated that large-sized aggregates (>250 μm) formed primarily due to the cementation of soil roots with mycelium, whereas microaggregates (<250 μm) were formed by multivalent cation bridges and polysaccharide. When forages are interplanted in the orchard, root density in the soil improves. Large roots cross and meet with each other, multiplying the combinations of mineral particles with organic matter, and large sized aggregates are formed. Labile organic matter incorporated into the soil from the decomposition of dead roots of the mulch cover promoted the accumulation of more water into soil capillaries. With continuous infiltration of capillary water, soil density with mulch coverage was lower than with clean tillage and formed a better soil structure (Zheng et al., 2003).

Aggregates also store, stabilize, and protect organic carbon (Tan et al., 2006) and are tightly associated with the decomposition and transformation of soil organic carbon (Whalen et al., 2003). Buyanovsky et al. (1994) studied the average dwell time of newly formed organic carbon using the ^{14}C-labeled method, and their results showed that the dwell duration was 1 year for 1- to 2-mm soil aggregates, but was 6 years for the 100- to 250-μm class. In other words, organic carbon in large microaggregates possesses higher bioavailability; in smaller microaggregates it has a longer replacement cycle and better stability (Buyanovsky et al., 1994). Our results indicated that organic carbon stocks and the distribution of aggregates with different sizes were affected by tillage. The distribution of large aggregates for the *Arachis pintoi* mulch treatment (T3) increased by 7.95% and 0.65% compared to the sloping peach orchard without conservation measures (T1) and the

Table 8.4   Reserves and Distribution Proportion of Organic Carbon for Various Sizes of Water-Stable Soil Aggregates

Soil Depth (cm)	Particle Size (mm)	Organic Carbon Storage (t ha^{-2})			Ratio (%)		
		T1	T2	T3	T1	T2	T3
0–20	>2	9.67	15.0	19.20	25.90	34.50	43.20
	1–2	3.45	3.14	3.40	9.27	7.21	7.63
	0.5–1	6.65	6.70	4.81	17.80	15.40	10.80
	0.25–0.5	4.34	5.33	3.64	11.60	12.30	8.16
	0.106–0.25	1.25	3.51	1.68	3.36	8.06	3.76
	<0.106	11.90	9.84	11.80	32.00	22.60	26.50
TOC		37.30	43.5	44.50	100.00	100.00	100.00
20–40	>2	5.93	3.22	10.10	17.70	8.72	26.60
	1–2	3.84	3.07	3.74	11.50	8.29	9.91
	0.5–1	7.36	6.86	5.73	22.0	18.60	15.20
	0.25–0.5	6.56	8.03	5.05	19.60	21.70	13.40
	0.106–0.25	1.53	2.49	1.17	4.56	6.730	3.11
	<0.106	8.24	13.30	12.00	24.60	36.00	31.80
TOC		33.50	37.00	37.76	100.00	100.00	100.00

Note: TOC, total organic carbon; T1, sloping orchard with clean tillage; T2, terrace orchard with clean tillage; T3, terrace orchard with *Arachis pintoi* mulch.

terraced peach orchard without conservation measures (T2), respectively. The distribution of micro-aggregates decreased by 14.5% and 1.48%, respectively, compared to T1 and T2 (Table 8.4). There are several reasons for this result. First, there was much more external organic carbon input from interplanted *A. pintoi* in T3. This helped form more large-sized aggregates because the formation of large aggregates was faster than that of the microaggregates. Second, microaggregates were depleted by forming large aggregates through the cementation effect of organic matter.

8.3.3 Soil Chemical Properties

In this study, soil samples from a peach orchard with *Arachis pintoi* mulch (T3) had better fertility conditions than samples from the other two treatments (T1 and T2) in terms of SOM; total N, P, and K; and available N, P, and K. These results point out the effectiveness of the living mulch for soil fertility improvement. In addition, the soil pH was slightly higher in T3. The value of SOM was 26.1% higher in T3 than in T1, and 18.4% higher than in T2. The humic and fulvic acid contents in T3 were also higher (Table 8.5). According to Kartika et al. (2009), *A. pintoi* provides a dense soil cover that can reduce erosion and leaching of some soil chemical components, and by fixing nitrogen from the atmosphere it helps to improve nitrogen availability. Additionally, the soil microbes decompose the plant and animal residues in the soil which in turn influences the soil physical, chemical, and biological properties (Chhotaray et al., 2011). As is well known, orchard soils kept bare through the use of herbicides become physically and chemically degraded over time (Haynes, 1981; Hipps and Samuelson, 1991) and are prone to erosion, especially in tropical environments subject to intense rainfall. The use of living mulches is an innovative management strategy that can improve soil quality, reduce grower reliance on synthetic inputs, and improve the overall sustainability of tropical fruit orchards (Mullahey et al., 1994; Shen and Chu, 2004). *Arachis pintoi* has been considered the most promising legume groundcover for such orchards.

Table 8.5   Chemical Analysis Data Obtained in September 2006 for
Soil Samples

Data	T1	T2	T3
SOM (g kg^{-1})	18.4 ± 0.230^c	19.6 ± 0.210^b	23.2 ± 0.200^a
TP (g kg^{-1})	0.130 ± 0.002^c	0.150 ± 0.004^b	0.180 ± 0.006^a
TK (g kg^{-1})	55.5 ± 0.502^b	55.8 ± 0.212^b	56.7 ± 0.472^a
TN (g kg^{-1})	1.01 ± 0.025^c	1.09 ± 0.022^b	1.478 ± 0.018^a
AN (mg kg^{-1})	69.9 ± 0.210^c	71.8 ± 0.220^b	105 ± 0.282^a
AP (mg kg^{-1})	7.80 ± 0.201^b	7.90 ± 0.230^b	10.4 ± 0.262^a
AK (mg kg^{-1})	35.8 ± 0.410^c	57.8 ± 0.215^b	131 ± 0.445^a
pH	4.65 ± 0.050^b	4.90 ± 0.040^a	4.95 ± 0.030^a
Humic acid (g kg^{-1})	0.445 ± 0.006^c	0.454 ± 0.007^b	1.01 ± 0.006^a
Fulvic acid (g kg^{-1})	5.37 ± 0.046^c	5.54 ± 0.024^b	6.70 ± 0.016^a

Note: SOM, soil organic matter; TP, total phosphorus; TK; total potassium; TN, total nitrogen; AN, available nitrogen; AP, available phosphorus, AK, available potassium; T1, sloping orchard with clean tillage; T2, terrace orchard with clean tillage; T3, terrace orchard with *Arachis pintoi* mulch. Within the same row, means without a common superscript letter differ significantly ($p < 0.05$).

8.4 LIVING MULCH EFFECT ON WATER AND SOIL CONSERVATION

The hilly, red-soil region covers a wide area of southern China where the hydrothermal conditions are favorable for developing agriculture. Many subtropical and tropical fruits can grow well here, but orchard reclamation causes environmental problems due to the relatively fragile ecology of the mountain environment. According to a report by Liang et al. (2008), 1243 km^2 of red soil in the hilly area of southern China are facing extremely serious erosion and fertility depletion. Soil erosion not only reduces soil fertility but also causes pollution of water bodies. Therefore, urgent action is needed by farmers to improve their tillage practices and conservation practices in order to improve the environment of these hilly, red-soil regions.

The results from our long-term experiments of water and soil conservation showed that the order of runoff coefficients was T2 > T1 > T3. *Arachis pintoi* mulch not only retarded the velocity of surface runoff but also prevented the sand and sediments from being transported down slopes. As a result, the sediment yields of T3 were significantly lower than those of T1 and T2 ($p < 0.05$). No erosion was observed in all T3 treatments, whereas the annual average erosion means of T1 and T2 reached 23.9 t ha^{-1} and 28.8 t ha^{-1}, respectively. The effectiveness in reducing soil erosion by living mulch can be explained by increases in (1) percent soil surface covered by vegetation, (2) surface roughness and ridges, (3) soil incorporated residue, and (4) soil detachability. The interaction of these variables accounts for either a reduction in sediment concentration or volume of runoff water, or both. Living mulch can protect the soil from raindrop impact and splash, tends to slow down the movement of surface runoff, and allows excess surface water to infiltrate.

In addition, living mulch improved the soil structure and increased the soil organic matter and nutrient content, which in turn increased soil corrosion resistance and reduced soil erosion. The soil erodibility index (K) is a parameter for evaluating the capability of a soil to withstand erosion. This index is also sensitive to the ability of a system to dissipate, separate, or transfer external forces of erosion. Declining penetration ability and anti-corrosion of soil eventually lead to an increasing soil erodibility index (K). As can be seen in Figure 8.4, the soil erodibility K value for the T3 treatment was the lowest (<0.20) and was 17.6% and 17.4% less than for the T1 and T2 treatments, respectively.

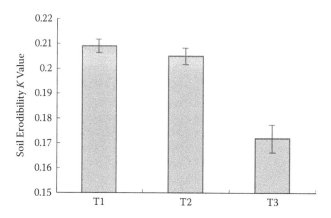

Figure 8.4   Erodibility *K* value of soil under different tillage.

8.5 LIVING MULCH EFFECT ON ORCHARD BIODIVERSITY

In orchard ecosystems, biodiversity is an important provider of ecosystem services, in addition to the food, fiber, fuel, and income provided. Examples include recycling of nutrients, control of the local microclimate, regulation of local hydrological processes, regulation of the abundance of undesirable organisms, and detoxification of noxious chemicals. These renewal processes and ecosystem services are largely biological; therefore, their persistence depends on the maintenance of biological diversity (Altieri, 1994). When these natural services are lost due to biological simplification, the economic and environmental costs can be quite significant (Altieri, 1999). Thus, orchard design contributes to plant diversity within agricultural areas and therefore to an increase in resources for animal communities such as arthropods and birds (Boller et al., 2004), among which are pest antagonists, provided that the cultural practices (namely, pesticide use) are not disruptive.

Compared to simplified orchard ecosystems, cover crops in an orchard affect arthropod populations in both positive and negative ways. Piercing and sucking insects have been found to be attracted by leguminous plants on the orchard floor and may cause crop damage. *Lygus* bugs and brown marmorated stink bugs cause cat-facing damage in apples and pears (Alston and Reding, 2003) by feeding on developing fruit, causing a scar that inhibits fruit growth relative to the surrounding tissue. Two-spotted spider mites prefer broadleaf plants. Mowing operations in an orchard during the hot summer months could drive the mites into the trees and hence increase potential damage (Hale and Williams, 2003; Alston and Reding, 2006). Kong et al. (2001) studied the regulation and control of forage and pests in an apple orchard. Their results showed that the forage could provide good habitat conditions for natural enemies of pests and thus promote the early population development of natural enemies. Apple and alfalfa intercropping could increase the richness of arthropods, diversity index, and evenness index. In their work, the number of natural enemies increased 200%. Although no absolute conclusions have been drawn as to whether enhanced biodiversity can improve the stability of ecosystems, a lot of research has been conducted that demonstrates the decrease in pest populations in a polyculture orchard (Hooks and Johnson, 2003). In addition, cover crops in orchards take up resources and light, thereby shading the soil and reducing the opportunity for weeds to establish themselves. The soil-loosening effect of deep-rooting covercrops also reduces weed populations that thrive in compacted soils.

8.5.1 Arthropod Diversity

To control the abundance of pests, it is important to know their ecology. In order to take full advantage of the abilities of natural enemies for the control of potential pests and to develop insect-resistant agroecosystems, it is important to study insect community composition, community structure, and intraspecific and interspecific relations (Pimentel et al., 1989; Whitten, 1993). Recovery and reconstruction of vegetation in orchards can increase biodiversity effectively by improving insect community evenness and stability through the protection of natural enemies of insect pests. The experiments described in this section were carried out to investigate the variations in arthropod community structure in loquat orchards with the presence of different living mulches, including the plant species *Vigna sinensis* (VS), *Chamaecrista nictitans* (CN), *Chamaecrista rotundifolia* (CR), and *Arachis pintoi* (AP). It was anticipated that such an investigation would improve our understanding of the feasibility of applying living mulch techniques for orchard environment protection and restoration. The extent to which living mulch would support sustainable orchard pest control through biodiversity improvement was also evaluated.

8.5.1.1 Arthropod Population Structure

Plant cover not only provides shelter for a more diversified arthropod community but also provides alternative prey or hosts for arthropods such as aphids and mites (Meagher and Meyer, 1990a), phytophagous mirids (Fye, 1980), leafhoppers (Meagher and Meyer, 1990b), tortricids (Brown, 2001), and *Coleoptera* (Wyss et al., 1995). For arthropod fauna, the presence of a grassy groundcover within the orchard may increase the diversity of beneficial species (Altieri and Schmidt, 1985; Wyss et al., 1995). In our work, the analysis of arthropod community composition showed that the proportion of pest groups under VS living mulch and the control area (CK) were higher than other treatments, accounting for more than 80% of the entire community, but the proportion of natural enemies in the community accounted for only 3.02% and 3.28%, respectively. In the other three plots, the proportion of natural enemies increased, accounting for more than 12.0%, and the proportions of pest groups were lower than in the control area. This indicated that CN, CR, and AP living mulches could help to optimize the arthropod community composition in loquat orchards. Especially in the case of CN living mulch, the proportions of what can be called neutral groups and pest groups were 56.1% and 31.6%, respectively, which showed that CN living mulch had the greatest effect on changing the arthropod community composition. The effects of CR and AP living mulches were ranked next.

8.5.1.2 Arthropod Population Diversity

The effect of plant diversity on the arthropod populations of pests and natural enemies relies on several complex mechanisms (Russell, 1989) such as plant–insect relationships, prey–predator and host–parasitoid interactions, population dynamics, and the structure and organization of arthropod communities (Liss et al., 1986). Consequently, the results might vary according to the host fruit species, the pest, and the plant species composition (Simon et al., 2010). In this study, a comparison among different living mulches showed that four groundcovers differed in the diversity indices for the arthropod community in the orchards (Figures 8.5 and 8.6). In treatments with *Chamaecrista nictitans* (CN), *Chamaecrista rotundifolia* (CR), and *Arachis pintoi* (AP) cover, the diversity index (H') and the evenness index (J) of the arthropod communities maintained higher levels than the control area (CK) throughout the experimental period. However, in *Vigna sinensis* (VS) coverage of the loquat orchard, although the species richness (S) and number of natural enemies had increased, the species richness and number of phytophagous pests also significantly

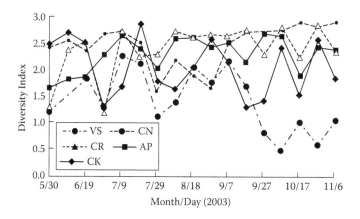

Figure 8.5 Diversity index (*H*⊠) of arthropod communities in a loquat orchard under different legume living mulches. (From Zhan, Z.X. et al., *J. Fujian Agric. Forestry Univ.*, 34(2), 162–167, 2005.)

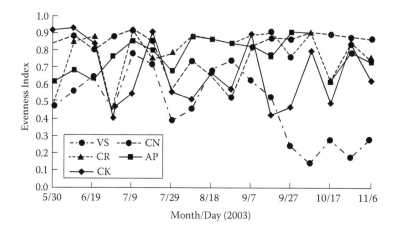

Figure 8.6 Evenness index (*J*) of arthropod communities in a loquat orchard under different legume living mulches. (From Zhan, Z.X. et al., *J. Fujian Agric. Forestry Univ.*, 34(2), 162–167, 2005.)

increased (Table 8.6). For example, populations of herbivorous insects, especially *Enterococcus flavescens* and *Prodenia litura*, broke out in autumn in the VS treatment. The *H'* and *J* of VS in Figures 8.5 and 8.6 were unstable, and the community structure was highly variable. Several aspects including pest–antagonist coupling, microclimate condition, and tested plant assemblage could explain such variability in the results. Plant diversity may directly affect the diversity of arthropod predators and parasitoids by providing needed structural complexity, cover, or nectar

Table 8.6 Numbers of Arthropod Groups in a Loquat Orchard for Different Living Mulches

Arthropod Category	*Vigna sinensis* Qty.	%	*Chamaecrista nictitans* Qty.	%	*Chamaecrista rotundifolia* Qty.	%	*Arachis pintoi* Qty.	%	Control (CK) Qty.	%
Phytophagous	6330	86.0	675	31.6	955	56.7	738	58.2	2197	81.8
Natural enemy	222	3.0	262	12.3	217	12.9	161	12.7	88	3.3
Neutral group	808	10.9	1199	56.1	511	30.4	370	29.1	400	14.9
Total	7360	100.0	2136	100.0	1583	100.0	1269	100.0	2685	100.0

and pollen for omnivorous species (Sirrine and Letourneau, 2009). In addition to the direct effects of plant diversity, microsite temperature could mediate the effect of food availability on arthropod communities by controlling access to those resources (Cerdá et al., 1998; Lessard et al., 2009). Thus, the different microenvironments associated with the canopies of *V. sinensis*, *C. nictitans*, *C. rotundifolia*, and *A. pintoi* are an important factor influencing arthropod diversity. Therefore, before choosing varieties of mulches, we should make thorough studies of soil fertility, orchard economics, community ecology, and other disciplines in order to avoid the adverse ecological consequences of environmental manipulations.

In conclusion, effective legume living mulch in an orchard agroecosystem can be conducive to protecting against natural enemies, enhancing biodiversity, and effectively protecting the orchard's ecosystem balance. Interplanting *Chamaecrista nictitans*, *Chamaecrista rotundifolia*, and *Arachis pintoi* in loquat orchards is beneficial to the thriving of natural enemies and neutral insects. They can significantly improve the arthropod community richness, diversity, and evenness indices and can play an important role in optimizing community structure. Although *Vigna sinensis* living mulch can improve the species richness of the arthropod community and increase the quantity of natural enemies, it was found that the number of phytophagous pests increased sharply and the indices of diversity evenness decreased, indicating a loss of community stability.

8.5.2 Soil Microbial Diversity

Microbial communities perform essential ecosystem services, including nutrient cycling, pathogen suppression, and stabilization of soil aggregates and degradation of xenobiotics. Changes in microbial community structure or function can result in detectable changes in soil chemical and physical properties, thus providing an early sign of soil improvement or an early warning of soil degradation (Pankhurst et al., 1995). Soil microorganisms themselves are important components of an orchard ecosystem. Their diversity may represent the stability of the microbial community, and they have an important role in the development of plants and community development.

8.5.2.1 Soil Microbial Community Structure

Soil microbial community composition and populations are affected by a number of edaphic factors, plant species, and soil management practices (Hooper et al., 1995; Marschner et al., 2001). Root exudates and organic matter decomposition in soil are important sources of energy and nutrients for soil microorganism (Saetre and Baath, 2000). Various studies on both managed and natural ecosystems have reported diverse relationships between soil microbial communities and soil physical and chemical properties (Zhong and Cai, 2007; Slabbert, 2008). Zuo and Wang (1995) reported that both living mulch and mowed grasscover in orchards could promote a change in the dominant soil microorganism populations. Interplanted living mulch can significantly increase the number of soil microbial groups in an orchard, especially free-living bacteria and nitrogen-fixing bacteria (Zheng et al., 1998). According to the results revealed by phospholipid fatty acid (PLFA) analysis, the microbial community in the rhizosphere soils of a hilly, red-soil peach orchard with living mulch trials was comprised of bacteria, fungi, actinomycetes, and protozoa. The biomass of bacteria and fungi in soil covered with *Arachis pintoi* (T3) was higher than that of clean tillage (T1, T2) (Figure 8.7). This study also showed that long-term living mulch enhanced soil fertility, soil biological activity, organic matter, and soil available nutrients. These may also result in an increased ratio of bacteria to fungi in the rhizospheric soil. The biomass of methane-oxidizing bacteria in soils with *A. pintoi* mulch was significantly higher than that of clean tillage. This might explain why root exudates and residual roots from the legume helped to form improved soil aggregate structure, decrease soil bulk density, increase soil porosity, and improve soil aeration. They were also helpful in inhibiting methane emissions.

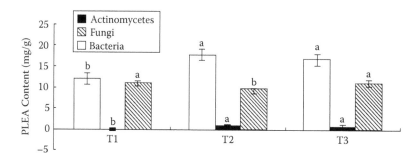

Figure 8.7 Microbial community structure in rhizosphere soils.

Table 8.7 Species Number and Diversity Index of Rhizospheric Soils for Different Mulch Treatments

Treatment	Number of Fatty Acids	Simpson Index (D)	Shannon Index (H′)	Pielou's Index (J′)	Brillouin Index (H₂)	McIntosh Index (H₃)
T1	19	0.914	3.80	0.893	3.41	0.769
T2	23	0.915	3.86	0.854	3.46	0.770
T3	22	0.919	3.90	0.875	3.47	0.778

Note: T1, sloping orchard with clean tillage; T2, terrace orchard with clean tillage; T3, terrace orchard with *Arachis pintoi* mulch.

8.5.2.2 *Microbial Diversity Index*

Keeping long-term living mulch in orchards can improve soil physical structure and soil available nutrients, provide nutrition and space for microbial activity, and thereby increase soil microbial diversity and populations. The Simpson index (D), Shannon index (H'), Brillouin index (H_2), and McIntosh index (H_3) of soils covered with *Arachis pintoi* (T3) were all higher than those of clean tillage (T1, T2) (Table 8.7). This was consistent with previous studies by Bucher and Lanyon (2005) and Tu et al. (2006), who found that organic orchard management enhanced soil microbial diversity and functional diversity; however, the differences among the various treatments did not reach statistically significant levels. More research is needed to monitor the direction of shifts in the microbial community structure and composition with a long-term mulch system, as well as a progressive evaluation of the soil properties.

8.6 LIVING MULCH EFFECT ON CARBON SEQUESTRATION OF ORCHARD ECOSYSTEMS

Orchard ecosystems can make an important contribution to the role that terrestrial ecosystems play in carbon sequestration by absorbing carbon dioxide from the atmosphere and the subsequent fixing of carbon dioxide in vegetation or soil. Carbon sequestration is related not only to land use types but also to soil management practices of orchards (e.g., no tillage, straw, living mulch). Traditionally, many orchards have adopted conventional tillage measures with a lack of soil cover in order to reduce competition between trees and weeds for water and nutrient uptake, but this tillage can induce surface roughness and may have a serious impact on vegetation and soil carbon storage, thereby increasing the emission of soil carbon to the atmosphere (Freibauer et al., 2004; Francia Martínez et al., 2006). However, reducing tillage intensity (such as using no tillage) can have positive effects on soil organic carbon (SOC) sequestration (Zibilske et al., 2002; Ramos et al., 2011), and conservative tillage systems, such as clover and straw mulching, are more effective in retaining SOC content, thereby maintaining a relatively higher level of SOC than conventional tillage systems (G.S. Zhang et al., 2005; Ramos et al., 2011; D.M. Wu et al., 2011).

Carbon sequestration by orchards is affected by soil type, topography, climatic conditions, and especially management practices (e.g., pruning, fertilizing, harvesting) (Vesterdal et al., 2002; Homann et al., 2004; Shukla and Lal, 2005; Zinn et al., 2005), thus increasing the complexity and uncertainty of the ecosystem carbon flow. Despite the efforts made in recent years to determine the process of carbon sequestration, there is still a lot of uncertainty about the effects of different management practices on carbon sequestration of hilly, red-soil peach orchards. This study estimated the carbon stock and distribution in vegetation, litter, and soil pools in peach orchards under different land management practices. The aim was to assess the effects of different management practices on orchard carbon sequestration in the hilly, red-soil region of China.

8.6.1 Carbon Storage in Soil

Soil carbon storage capacity can be quickly reduced by the processes of land use changes that combine agricultural activities, deforestation, and degradation of forest ecosystems (Evrendilek et al., 2004; Lal, 2004a,b; L.D. Chen et al., 2007; Ordóñez et al., 2008). However, the management strategies that increase SOC should be directed toward increasing residue inputs or decreasing decomposition rates (Batjes and Sombroek, 1997; Lal and Kimble, 2000). For example, conservation tillage practices, such as no tillage and mulch coverage, increase SOC content because they reduce the disturbance caused by tillage and the risks of erosion, enhance organic residues added to soil, and improve soil carbon sequestration (Sainju et al., 2002; Pulleman et al., 2005; Castro et al., 2008; Weng et al., 2006b). On the other hand, plant communities covering the soil surface contribute to soil carbon sequestration through the deposition of litter, dead root material, and rhizo-deposition (Saner et al., 2007; Peri et al., 2012). Our results (Figures 8.8 and 8.9) showed that SOC content and soil organic carbon density (SOC_D) in soil layers above 60 cm were higher than those in soil layers below 60 cm. The highest SOC content and SOC_D appeared in the T3 treatment with living *Arachis pintoi* mulch and were significantly higher than those of clean tillage (T1 and T2).

The $\delta^{13}C$ values of soil organic carbon reflect the relative contribution of plant species to the community net primary productivity (Boutton et al., 1998). As a result, pedological, biogeochemical, and ecological information can be obtained from stable isotope analyses of soils (Boutton and Yamasaki, 1996). When *Arachis pintoi* with a low $\delta^{13}C$ value was planted in peach orchards, the soil organic carbon $\delta^{13}C$ value for the 0- to 20-cm soil layer was significantly reduced compared to that of clean tillage. No significant difference was found between the two clean tillage practices (T1 and T2). The results indicated that organic matter from dead root, litter, and root exudates of *A. pintoi* were the major reason for reduced $\delta^{13}C$ values and increased SOC content and SOC density. The higher total organic carbon than that of bare ground implied that living *A. pintoi* mulch played an important role in the stability of the soil organic carbon pool during the conversion process of bare ground in the orchard. (See Table 8.8.)

8.6.2 Carbon Sequestration of Fruit Tree Vegetation

In our work, we found that the carbon storage in peach tree vegetation ranged from 13.0 to 14.7 t ha^{-2}. Regarding the distribution of carbon storage in vegetation, some studies showed that stems and roots are the major storage sites for young orchards, while the leaves, pruning branches, and fruits were more important for old orchards. In this study, 62.8 to 67.0% of vegetative organic carbon was stored in the branches (Figure 8.10), which was different than the result

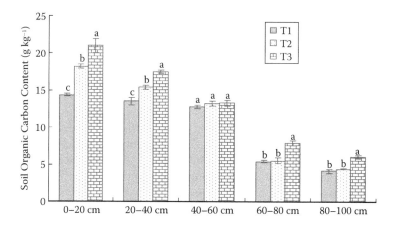

Figure 8.8 Variation of soil organic carbon (SOC) content for different soil layers and mulch treatments.

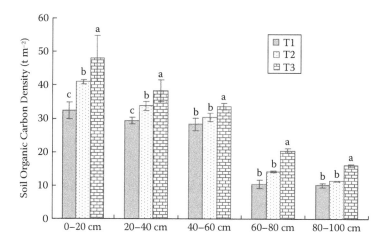

Figure 8.9 Soil organic carbon density (SOC$_D$) for different soil layers and mulch treatments.

of Z.D. Wu et al. (2008). The results also showed that living mulch did not significantly affect carbon sequestration of the fruit tree itself, but increased by 5120 kg ha^{-2} the carbon sequestered in the mulch and soil layer, which indicated that constructing a fruit–mulch system could effectively increase the carbon sink of an orchard agroecosystem. Litterfall is a metabolic by-product

Table 8.8 δ^{13}C Values in Soil and Soil Organic Carbon (SOC) for Different Types of Orchard Management

Treatment	δ^{13}C (‰)	Total SOC (g kg^{-1})	SOC from Wild Grassland (g kg^{-1})	SOC from Peach Orchard (g kg^{-1})	SOC from Groundcover (g kg^{-1})
T1	−23.7 ± 0.07	14.8 ± 0.15	8.95 ± 0.04	5.87 ± 0.04	—
T2	−24.1 ± 0.04	17.9 ± 0.21	9.88 ± 0.07	7.98 ± 0.02	—
T3	−25.6 ± 0.06	21.1 ± 0.32	7.80 ± 0.08	7.98 ± 0.02	5.32 ± 0.04
Unreclaimed weed land	−20.5 ± 0.03	19.5 ± 0.24	19.49 ± 0.02	—	—

Note: T1, peach orchard with no cover on sloped land; T2, peach orchard on terraces with no cover; T3, peach orchard on terraces with *Arachis pintoi* living mulch.

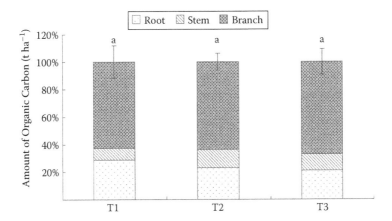

Figure 8.10   Distribution of carbon in orchard vegetation for different landscapes and mulch covers.

during plant growth and development and an important component of any ecosystem. It is also an important link between the soil carbon pool and the plant carbon pool and has an important influence on the carbon cycle of terrestrial ecosystems. Previous studies have shown that the increase of litter accumulation was consistent with the improvement of soil organic carbon (Saner et al., 2007). Different management measures not only significantly affected litter accumulation but also significantly affected the soil organic carbon accumulation rate (Evrendilek et al., 2004). In our study, the annual litter biomass for the three treatments ranged from 1.32 to 1.45 kg per fruit tree without significant difference; however, a significant difference in soil organic carbon content between these treatments existed. It was the living mulch that caused the difference.

8.6.3 Carbon Emission of Orchard Soil

Human activities significantly change the characteristics of soil respiration in terrestrial ecosystems. It was generally believed that minimum tillage and zero tillage would reduce greenhouse gas emissions (Lal, 2004a,b). Some research has shown that soil carbon storage increased when conventional tillage was changed to no tillage, but the effect on soil respiration rates by no tillage and conventional tillage were not consistent. Reicosky et al. (1999) studied CO_2 emissions in soil covered with crimson clover with conventional tillage and no tillage in the winter. Their short-term trials showed that the cumulative CO_2 emissions with conventional tillage were three times those of the no-tillage soil. Quantification of soil respiration without roots, which is known as soil heterotrophic respiration, is an important component to estimating an ecosystem carbon balance. Based on short-term measurement, our study showed the soil heterotrophic respiration in the peach orchard under living mulch management was significantly higher than that with clean tillage management, and changes in the microenvironment (moisture and temperature) caused by the living mulch were an important reason for soil respiration changes. These results were in agreement with previous reports (Siele et al., 2008). In addition, after interplanting legumes in the orchard, the litter from the mulch that returned to the soil had a priming effect, affecting the turnover of soil organic carbon and soil respiration rates, but the increase of soil respiration rates caused by the living mulch did not result in an increase of net emissions of CO_2 from the orchard ecosystem. Living mulch increased carbon sequestered in the mulch biomass by 5120 kg ha^{-2}. In addition to the carbon sequestration from aboveground mulch biomass, belowground root biomass from the legume was also an important factor enhancing SOC content and carbon sequestration.

8.7 CONCLUSIONS ON THE USE OF LIVING MULCH IN ORCHARDS

Red–yellow soil dominates in Fujian Province, which is located in the subtropics of China. In tropical and subtropical regions, the topsoil can be easily eroded because of the enhanced soil weathering that causes nutrient leaching and destruction of aggregate structure in the soil. These soil characteristics are found widely in the subtropics throughout the world. It is commonly recognized that good soil management is extremely important for agriculture in these regions, but scientific research focused on sustainable cultivation management is lacking.

On the slopes of mountainous areas in northern Fujian Province, the orchard floor management system described in our studies can have a positive effect on the growth of trees, as well as the yield and quality of fruit. For orchard cultivation along mountain slopes, conventional clean tillage used to be the most common weed control practice. Consequently, tillage erosion delivers topsoil, together with soil nutrient elements, down the slope. As the barren exposed subsoil does not support tree growth, the soil becomes less resistant to water loss and less productive and develops an even poorer structure. In contrast, more sustainable soil management not only helps prevent soil erosion but also enhances soil fertility. Especially in red–yellow soils, soil aggregation is less developed because of low microbial activity and low organic matter accumulation, resulting in the decline of nutrient and water-holding capacities and permeability in such soils.

In order to suppress soil erosion and increase soil fertility, we successfully established a more sustainable red–yellow soil land management system by introducing living mulches. Such techniques are extremely useful for managing the cultivation of orchards and for more sustainable soil management in red–yellow soil in general. Moreover, this research also focused on carbon sequestration, which could contribute in a small way to combating global warming.

Living mulches suppress weeds by competition or their allelopathic properties. They also benefit the trees by protecting soil from water and wind erosion and can sometimes increase the populations of natural enemies of crop pests. Legumes used as living mulches can be expected to provide nitrogen fixation, thus reducing the need for external fertilizer. The addition of organic matter derived from living mulch into the soil increases carbon accumulation, resulting in the enhancement of soil fertility.

In orchards with *Arachis pintoi* mulch, the largest proportion of tree roots occurred in the 20- to 80-cm soil layer. This decreased the competition for water and fertilizer between the trees and the legumes. Benefiting from healthier root systems, the soils showed an improved porosity that enabled more efficient nutrient cycling and, as a result, enhanced soil fertility and productivity of the orchard. The presence of *A. pintoi* mulch also inhibited soil erosion; for example, higher soil porosity was shown to be associated with a lower soil erodibility index K ($K < 0.20$, belonging to a class of lower erodibility).

Compared with clean tillage, living mulch showed a distinct advantage with regard to raising orchard biodiversity. It was demonstrated that the values of arthropod community richness, diversity, and evenness with *Arachis pintoi* mulch were 30.73%, 40.10%, and 35.55% higher, respectively, than those displayed in orchards without cover. The Simpson, Shannon, Brillouin, and McIntosh indices of the rhizosphere microbial community in orchards with living mulch also increased by 0.50% to 0.55%, 0.93% to 2.77%, 0.35% to 1.91%, and 1.07% to 1.17%, respectively, compared with clean tillage.

With *Arachis pintoi* mulch, the soil organic carbon (SOC) content and soil organic carbon density (SOC_D) increased from 7.4% to 46.48% and from 0.42% to 26.99%, respectively. The soil organic carbon turnover rate with living mulch showed a mean of 63.05% of replacement in the top 20-cm soil layer after 14 years, which was relatively faster than the rate on slope orchards (39.63% of replacement) and terrace orchards (44.69% of replacement) with clean tillage. In living mulch treatments, soil organic carbon from legume residues accounted for 25.22% of the amount of total soil organic carbon, implying an increase of exogenous carbon input to the soil and an improvement of stability in the soil organic carbon pool.

In conclusion, the vegetation of *Arachis pintoi* mulch on red soil in the current terrace peach orchard management system exerted significantly positive effects on orchard productivity by improving soil physical and chemical properties, biodiversity, and organic carbon sequestration. Results of this research provide a pioneering and successful example of the application of living mulch for orchard floor management in northern Fujian mountainous areas. Its wider application among local agricultural practices is highly expected. Furthermore, this research showed that crop cultivation with more sustainable management using living mulch can contribute to mitigation measures against global warming.

REFERENCES

Alston, D. and Reding, M.E. 2003. *Cat-Facing Insects*. Utah State University: Logan.

Alston, D. and Reding, M.E. 2006. *Web Spinning Spider Mites*. Utah State University: Logan.

Altieri, M.A. 1994. *Biodiversity and Pest Management in Agroecosystems*. Haworth Press: New York, p. 185.

Altieri, M.A. 1999. The ecological role of biodiversity in agroecosystems. *Agric. Ecosyst. Environ.*, 74: 19–31.

Altieri, M.A. and Schmidt, L.L. 1985. Cover crop manipulation in Northern California orchards and vineyards: effects on arthropod communities. *Biol. Agric. Hort.*, 3: 1–24.

Batjes, N.H. and Sombroek, W.G. 1997. Possibilities for carbon sequestration in tropical and subtropical soils. *Global Change Biol.*, 3: 161–173.

Boller, E.F., Häni, F., and Poehling, H.M., Eds. 2004. *Ecological Infrastructures: Ideabook on Functional Biodiversity at the Farm Level*. Switzerland: Lindau, Switzerland: Landwirtschaftliche Beratungszentrale Lindau.

Boutton, T.W. and Yamasaki, S.I. 1996. *Mass Spectrometry of Soils*. Marcel Dekker: New York, p. 492.

Boutton, T.W., Archer, S.R., Midwood, A.J., Zitzer, S.F., and Bol, R. 1998. $\delta^{13}C$ values of soil organic carbon and their use in documenting vegetation change in a subtropical savanna ecosystem. *Geoderma*, 82: 5–41.

Brown, M.W. 2001. Flowering ground cover plants for pest management in peach and apple orchards. *Bull. IOBC/Wprs*, 24: 379–382.

Buban, R., Helmeczi, B., Papp, J., Dorgo, E., Jakab, I., Kajati, I., and Merwin, I. 1996. IPF-compatible ground-cover management systems in a new-planted apple orchard. *Acta Hort.*, 422: 263–267.

Bucher, A.E. and Lanyon, L.E. 2005. Evaluating soil management with microbial community-level physiological profiles. *Appl. Soil Ecol.*, 29: 59–71.

Buyanovsky, G.A., Aslam, M., and Wagner, G.H. 1994. Carbon turnover in soil physical fractions. *Soil Sci. Soc. Am. J.*, 58: 1167–1173.

Candog-Bangi, J. and Cosico, W.C. 2007. Corn yield and soil properties in Cotabato as influenced by the living mulch *Arachis pintoi*. *Philippine J. Crop Sci.*, 32(3): 59–68.

Castro, J., Fernández-Ondoño, E., Rodríguez, C., Lallena, A.M., Sierra, M., and Aguilar, J. 2008. Effects of different olive-grove management systems on the organic carbon and nitrogen content of the soil in Jaen (Spain). *Soil Tillage Res.*, 98: 56–67.

Cerdá, X., Retana, J., and Manzaneda, A. 1998. The role of competition by dominants and temperature in the foraging of subordinate species in Mediterranean ant communities. *Oecologia*, 117: 404–412.

Chen, L.D., Gong, J., Fu, B.J., Huang, Z.L., Huang, Y.L., and Gui, L.D. 2007. Effect of land use conversion on soil organic carbon sequestration in the Loess hilly area, Loess Plateau of China. *Ecol. Res.*, 22: 641–648.

Chen, M.H. 2004a. Soil erosion status and prevention measures in mountain orchards of Fujian province [in Chinese with English abstract]. *Fujian Soil Water Conserv.*, 16(2): 21–24.

Chen, M.H. 2004b. Status and prevention measures of soil erosion in mountain orchard, Fujian [in Chinese with English abstract]. *Fujian Soil Water Conserv.*, 16(2): 25–34.

Chen, N.W., Li, H.C., and Wang, L.H. 2009. A GIS-based approach for mapping direct use value of ecosystem services at a county scale: management implications. *Ecol. Econ.*, 68: 2768–2776.

Chen, Q.X., Liao, J.S., Zheng, G.H., and Liu, S. 1996. The effects of grass living mulch on soil fertility and fruit growth in young Longyan orchard [in Chinese with English abstract]. *J. Fujian Agric. Univ.*, 25(4): 429–432.

Chhotaray, D., Mohapatra, P.K., and Mishra, C.S.K. 2011. Soil macronutrient availability and microbial population dynamics of organic and conventional agroecosystems. *Eur. J. Biol. Sci.*, 3(2): 44–51.

CTAHR. 2009. *Perennial Peanut: Sustainable Agriculture Cover Crop Database*. College of Tropical Agriculture and Human Resources, University of Hawaii: Manoa.

Doanh, L.Q. and Tuan, H.D. 2004. Improving indigenous technologies for sustainable land use in northern mountainous areas of Vietnam. *J. Mountain Sci.*, 1(3): 270–275.

Evrendilek, F., Celik, I., and Kilic, S. 2004. Changes in soil organic carbon and other physical soil properties along adjacent Mediterranean forest, grassland and cropland ecosystems in Turkey. *J. Arid Environ.*, 59: 743–752.

Faust, M. 1979. Evolution of fruit nutrition during the 20th century. *HortScience*, 14: 321–325.

Firth, D.J. and Wilson, G.P.M. 1995. Preliminary evaluation of species for use as permanent ground cover in orchards on the north coast of New South Wales. *Trop. Grasslands*, 29: 18–27.

Firth, D.J., Jones, R.M., McFayden, L.M., Cook, B.G., and Whalley, R.D.B. 2002. Selection of pasture species for groundcover suited to shade in mature macadamia orchards in subtropical Australia. *Trop. Grasslands*, 36: 1–12.

Francia Martínez, J.R., Durán Zuazo, V.H., and Martínez Raya, A. 2006. Environmental impact from mountainous olive orchards under different soil-management systems (SE Spain). *Sci. Total Environ.*, 358: 46–60.

Freibauer, A., Rounsevell, M.D.A., Smith, P., and Verhagen, J. 2004. Carbon sequestration in the agricultural soils of Europe. *Geoderma*, 122(1): 1–23.

Fye, R.E. 1980. Weed sources of *Lygus* bugs in the Yakima valley and Columbia basin in Washington. *J. Econ. Entomol.*, 73: 469–473.

Glover, J.D., Reganold, J.P., and Andrews, P.K. 2000. Systematic methods for rating soil quality of conventional, organic, and integrated apple orchards in Washington State. *Agric. Ecosyst. Environ.*, 80: 29–45.

Gregoriou, C. and Rajkumar, D. 1984. Effects of irrigation and mulching on shoot and root growth of avocado (*Persea americana* Mill.) and mango (*Mangifera indica* L.). *J. Hort. Sci.*, 59(1): 109–117.

Guo, L.R., Lin, Q.H., and Weng, B.Q. 2010. Research on ecological cultivation of fruit trees in red soil hills of Fujian [in Chinese with English abstract]. *Acta Agric. Jiangxi*, 22(7): 52–55.

Hale, F.A. and Williams, H. 2003. *Two Spotted Spider Mites*, Bull. SP290-D. University of Tennessee, Cooperative Extension: Knoxville.

Haynes, R.J. 1981. Soil pH decrease in the herbicide strip of grassed-down orchards. *Soil Sci.*, 132: 274–278.

Hipps, N.A. and Samuelson, T.J. 1991. Effects of long-term herbicide use, irrigation and nitrogen fertilizer on soil fertility in an apple orchard. *J. Sci. Food Agric.*, 55: 377–387.

Hoagland, L., Carpenter-Boggs, L., Granatstein, D., Mazzola, M., Smith, J., Peryea, F., and Reganold, J.P. 2008. Orchard floor management effects on nitrogen fertility and soil biological activity in a newly established organic apple orchard. *Biol. Fertil. Soils*, 45(1): 11–18.

Hogue, W.A. and Neilsen, G.H. 1987. Orchard floor vegetation management. *Hort. Rev.*, 9: 377–430.

Homann, P.S., Remillard, S.M., Harmon, M.E., Bormann, B.T. 2004. Carbon storage in coarse and fine fractions of Pacific Northwest old-growth forest soils. *Soil Sci. Soc. Am. J.*, 68: 2023–2030.

Hooks, C.R.R. and Johnson, M.W. 2003. Impact of agricultural diversification on the insect community of cruciferous crops. *Crop Protect.*, 22: 223–228.

Hooper, D.U., Bignell, D.E., Brown, V.K., Brussaard, L., Dangerfield, J.M., Wall, D.H., Wardle, D.A., Coleman, D.C., Giller, K.E., Lavelle, P., Van der Putten, W.H., De Ruiter, P., and Houghton, R.A. 1995. Changes in the storage of terrestrial carbon since 1950. In: Lal, R., Ed., *Soils and Global Change*. CRC Press: Boca Raton, FL, pp. 45–65.

Kartika, J.G., Reyes, M.R., and Susila, A.D. 2009. Review of Literature on Perennial Peanut (*Arachis pintoi*) as Potential Cover Crop in the Tropics. North Carolina A&T State University: Greensboro (http://sanrem.cals.vt.edu/1123/IkaRevision.doc).

Kong, J., Wang, H.Y., Zhao, B.G., Ren, Y.D., Liu, Y.X., Chen, H.J., Shan, L.N., and Wang, A.C. 2001. Study on ecological regulation system of pest control in apple orchards [in Chinese with English abstract]. *Acta Ecol. Sinica*, 21(5): 789–794.

Lal, R. 2004a. Carbon emission from farm operations. *Environ. Int.*, 30: 981–990.

Lal, R. 2004b. Soil carbon sequestration to mitigate climate change. *Geoderma*, 123(1-2): 1–22.

Lal, R. and Kimble, J.M. 2000. Tropical ecosystems and the global carbon cycle. In: Lal, R., Kimble, J.M., and Stewart, B.A., Eds., *Global Climate Change and Tropical Ecosystems*. CRC Press: Boca Raton, FL, pp. 3–32.

Lanini, W.T., Shribbs, J.M., and Elmore, C.E. 1988. Orchard floor mulching trials in the USA. *Components*, 56(3): 228–249.

Lessard, J.P.R., Dunn, R., and Sanders, N.J. 2009. Temperature-mediated coexistence in forest ant communities. *Insectes Soc.*, 56: 149–156.

Liang, Y., Zhang, B., Pan, X.Z., and Shi, D.M. 2008. Current status and comprehensive control strategies of soil erosion for hilly region in the Southern China [in Chinese with English abstract]. *Sci. Soil Water Conserv.*, 6(1): 22–27.

Liss, W.J., Gut, L.J., Westigard, P.H., and Warren, C.E. 1986. Perspectives on arthropod community structure, organization, and development in agricultural crops. *Ann. Rev. Entomol.*, 31: 455–478.

Liu, Y., Tao, Y., Wan, K.Y., Zhang, G.S., Liu, D.B., Xiong, G.Y., and Chen, F. 2012. Runoff and nutrient losses in citrus orchards on sloping land subjected to different surface mulching practices in the Danjiangkou Reservoir area of China. *Agric. Water Manage.*, 10: 34–40.

Luo, T., Liu, C.H., Xie, F.X., Xie, S.H., and Jiang, Q.J. 1994. Introduction to several forages for soil and water conservation. *Fujian Soil Water Conserv.*, 1: 32–33.

Marschner, P., Yang, C.H., Lieberei, R., and Crowley, D.E. 2001. Soil and plant specific effects on bacterial community composition in the rhizosphere. *Soil Biol. Biochem.*, 33: 1437–1445.

Mathews, C.R., Bottrell, D.G., and Brown, M.W. 2002. A comparison of conventional and alternative understory management practices for apple production, multi-trophic effects. *Appl. Soil Ecol.*, 21(3): 221–231.

Meagher, R.L. and Meyer, J.R. 1990a. Influence of ground cover and herbicide treatments on *Tetranychus urticae* populations in peach orchards. *Exp. Appl. Acarol.*, 9: 149–158.

Meagher, R.L. and Meyer, J.R. 1990b. Effect of ground cover management on certain abiotic and biotic interactions in peach orchard ecosystems. *Crop Protect.*, 9: 65–72.

Miller, R.W. 1977. *Soils: An Introduction to Soils and Plant Growth*. Prentice Hall: Englewood Cliffs, NJ, pp. 59–71.

Mullahey, J.J., Rouse, R.E., and French, E.C. 1994. Perennial peanut in citrus groves—an environmentally sustainable agricultural system. In: *Environmentally Sound Agriculture: Proceedings of the Second Conference*, Orlando, FL, April 20–22, pp. 479–483.

Ordóñez, J.A.B., De Jong, B.H.J., García-Oliva, F., Aviña, F.L., Pérez, J.V., Guerrero, G., Martínez, R., and Maser, O. 2008. Carbon content in vegetation, litter, and soil under 10 different land-use and land-cover classes in the Central Highlands of Michoacan, Mexico. *Forest Ecol. Manage.*, 255: 2074–2084.

Pankhurst, C.E., Hawke, B.G., McDonald, H.J., Kirkby, C.A., Buckerfield, J.C., Michelsen, P., O'Brien, K.A., Gupta, W.S.R., and Doube, B.M. 1995. Evaluation of soil biological properties as potential bioindicators of soil health. *Aust. J. Exp. Agric.*, 35: 1015–1028.

Peck, G.M., Andrews, P.K., and Reganold, J.P. 2006. Apple orchard productivity and fruit quality under organic, conventional, and integrated management. *HortScience*, 41(1): 99–107.

Peck, G.M., Merwin, I.A., Thies, J.E., Schindelbeck, R.R., and Brown, M.G. 2011. Soil properties change during the transition to integrated and organic apple production in a New York orchard. *Appl. Soil Ecol.*, 48: 18–30.

Peri, P.L., Ladd, B., Pepper, D.A., Bonser, S.P., Laffan, S.W., and Amelung, W. 2012. Carbon (δ^{13}C) and nitrogen (δ^{15}N) stable isotope composition in plant and soil in Southern Patagonia's native forests. *Global Change Biol.*, 18: 311–321.

Pimentel, D., Culliney, T.W., Buttler, I.W., Reinemann, D.J., and Beckman, K.B. 1989. Low-input sustainable agriculture using ecological management practices. *Agric. Ecosyst. Environ.*, 27(1–4): 3–24.

Pinamonti, F., Zorzi, G., Gasperi, F., Silvestri, S., and Stringari, G. 1995. Growth and nutritional status of apple trees and grapevines in municipal solid-waste-amended soil. *Acta Hort.*, 383: 313–321.

Pulleman, M.M., Six, J., Van Breemen, N., and Jongmans, A.G. 2005. Soil organic matter distribution and microaggregate characteristic as affected by agricultural management and earthworm activity. *Eur. J. Soil Sci.*, 56: 453–467.

Ramos, M.E., Robles, A.B., Sánchez-Navarro, A., and González-Rebollar, J.L. 2011. Soil responses to different management practices in rainfed orchards in semiarid environments. *Soil Tillage Res.*, 112(1): 85–91.

Reicosky, D.C., Reeves, D.W., Prior, H.H., Runionc, G.B., Rogersb, H.H., and Raperb, R.L. 1999. Effects of residue management and controlled traffic on carbon dioxide and water loss. *Soil Tillage Res.*, 52(3-4): 153–165.

Russell, E.P. 1989. Enemies hypothesis, a review of the effect of vegetational diversity on predatory insects and parasitoids. *Environ. Entomol.*, 18: 590–599.

Saetre, P. and Baath, E. 2000. Spatial variation and patterns of soil microbial community structure in a mixed spruce–birch stand. *Soil Biol. Biochem.*, 32: 909–917.

Sainju, U.M., Sing, B.P., and Whitehead, W.F. 2002. Long-term effects of tillage, cover crops, and nitrogen fertilization on organic carbon and nitrogen concentrations in sandy loam soils in Georgia. *Soil Tillage Res.*, 63: 167–179.

Saner, T.J., Cambardella, C.A., and Brandle, J.R. 2007. Soil carbon and tree litter dynamics in a red cedar-scotch pine shelterbelt. *Agroforestry Syst.*, 71(3): 163–174.

Shen, Q.R. and Chu, G.X. 2004. Bi-directional nitrogen transfer in an intercropping system of perennial peanut with rice cultivated in aerobic soil. *Biol. Fertil. Soils*, 40(2): 81–87.

Shukla, M.K. and Lal, R. 2005. Erosional effects on soil organic carbon stock in an on-farm study on Alfisols in west central Ohio. *Soil Tillage Res.*, 81: 173–181.

Siele, M.P., Mubyana-John, T., and Bonyongo, M.C. 2008. The effects of soil cover on soil respiration and microbial population in the mopane (*Colophospermum mopane*) woodland of northwestern Botswana. *Dyn. Soil Dyn. Plant*, 2(2): 61–68.

Simon, S., Bouvier, J.C., Debras, J.F., and Sauphanor, B. 2010. Biodiversity and pest management in orchard systems: a review. *Agron. Sustain. Dev.*, 30: 139–152.

Sirrine, J.R. and Letourneau, D.K. 2009. The effect of groundcover management on arthropod diversity and abundance in an orchard agroecosystem. In: *Proceedings of Great Lakes Fruit Workers Meeting*, Fishkill, NY, October 25–28, pp. 19–20.

Slabbert, E. 2008. Microbial Diversity of Soils of the Sand Fynbos, master's thesis, Stellenbosch University, South Africa.

Tan, W.F., Zhu, Z.F., Liu, F., Hu, R.G., and Shan, S.J. 2006. Organic carbon distribution and storage of soil aggregates under land use change in Jianghan plain, Hubei Province [in Chinese with English abstract]. *Nat. Resour. J.*, 21(6): 973–980.

Tisdall, J.M., Smith, S.E., and Rengasamy, P. 1997. Aggregation of soil by fungal hyphae. *Aust. J. Soil Res.*, 35: 55–60.

Tu, C., Louws, F.J., Creamer, N.G., Mueller, J.P., Brownie, C., Fager, K., Bell, M., and Hu, S.J. 2006. Responses of soil microbial biomass and N availability to transition strategies from conventional to organic farming systems. *Agric. Ecosyst. Environ.*, 113: 206–215.

Vesterdal, L., Ritter, E., and Gundersen, P. 2002. Change in soil organic carbon following afforestation of former arable land. *Forest Ecol. Manage.*, 162: 137–147.

Wang, L., Tang, L.L., Wang, X., and Chen, F. 2010. Effects of alley crop planting on soil and nutrient losses in the citrus orchards of the Three Gorges Region. *Soil Tillage Res.*, 110(2): 243–250.

Weng, B.Q., Huang, Y.B., Ying, Z.Y., Luo, T., and Wang, Y.X. 2006a. The technology and efficient analysis of ecological orchard in red soil hills [in Chinese with English abstract]. *Chin. Agric. Sci. Bull.*, 22(12): 465–470.

Weng, B.Q., Ying, Z.Y., Huang, Y.B., Wang, Y.X., and Fang, J.M. 2006b. Study on construction and application of comprehensive utilization mode and ecological rehabilitation of empoldered red soil mountains in northern Fujian [in Chinese with English abstract]. *J. Soil Water Conserv.*, 20(1): 147–150.

Whalen, J.K., Hu, Q.C., and Liu, A.G. 2003. Compost applications increase water stable aggregates in conventional and no-tillage systems. *Soil Sci. Soc. Am. J.*, 67(6): 1842–1847.

Whitten, W.J. 1993. Pest management in 2000: what we might learn from the twentieth century. In: Aziz, A., Kadir, S.A., and Barlow, H.S., Eds., *Pest Management and the Environment in 2000*. Oxford University Press: Oxford, pp. 9–44.

Wu, D.M., Yu, Y.C., Xia, L.Z., Yin, S.X., and Yang, L.Z. 2011. Soil fertility indices of citrus orchard land along topographic gradients in the Three Gorges area of China. *Pedosphere*, 21(6): 782–792.

Wu, Z.D., Wang, Y.X., Weng, B.Q., Cai, Z.J., and Wen, S.X. 2008. Organic carbon and nitrogen storage in 7 year old citrus orchard ecosystem in Fuzhou, China [in Chinese with English abstract]. *J. Fujian Agric. Forestry Univ.*, 37(3): 316–319.

Wyss, E., Niggli, U., and Nentwig, W. 1995. The impact of spiders on aphid populations in a strip-managed apple orchard. *J. Appl. Entomol.*, 119: 473–478.

Xu, Q.X., Wang, T.W., Cai, C.F., Li, Z.X., and Shi, Z.H. 2012. Effects of soil conservation on soil properties of citrus orchards in the Three Gorges area, China. *Land Degrad. Dev.*, 23(1): 34–42.

You, M.S., Liu, Y.F., and Hou, Y.M. 2004. Biodiversity and integrated pest management in agroecosystems [in Chinese with English abstract]. *Acta Ecol. Sin.*, 24(1): 117–122.

Zhan, Z.X., Qiu, L.M., Fu, J.W., Wei, H., Ying, Z.Y., and Weng, B.Q. 2005. Structure and dynamics of arthropod community in the loquat orchards covered with various pastures [in Chinese with English abstract]. *J. Fujian Agric. Forestry Univ.*, 34(2): 162–167.

Zhang, F., Yang, Z.G., Wen, S.L., and Su, Z.Y. 2004. Ecological effect of planting forage *Lotononis* in orchards [in Chinese with English abstract]. *J. Northwest Agric. Forestry Univ.*, 32(7): 104–106.

Zhang, G.S., Huang, G.B., and Chan, Y. 2005. Soil organic carbon sequestration potential in cropland [in Chinese with English abstract]. *Acta Ecol. Sin.*, 25(2): 352–357.

Zhao, Z.F., Chen, A.C., Shi, F.P., and Zhao, C.L. 2000. Extension of ground cover with sod grass in fruit and nut orchards [in Chinese with English abstract]. *Econ. Forest Res.*, 18(2): 37.

Zheng, Z.D., Huang, X.S., Cai, Z.F., Luo, T., Zhang, M.H., and Fang, J.M. 1998. Study on the effect of herbaceous vetiver grass on soil water conservation in hilly red earth land [in Chinese with English abstract]. *Fujian J. Agric. Sci.*, 18(Suppl.): 69–75.

Zheng, Z.D., Huang, Y.B., Wenig, B.Q., Luo, T., and Yu, J.W. 2003. Studies on the comprehensive development of hilly land in Fujian. I. Effects of different planting methods on the orchard ecosystem [in Chinese with English abstract]. *Chin. J. Eco-Agric.*, 11(3): 149–151.

Zhong, W.H. and Cai, Z.C. 2007. Long-term effects of inorganic fertilizers on microbial biomass and community functional diversity in a paddy soil derived from quaternary red clay. *Appl. Soil Ecol.*, 36: 84–91.

Zibilske, L.M., Bradford, J.M., and Smart, J.R. 2002. Conservation tillage induced changes in organic carbon, total nitrogen and available phosphorus in a semi-arid alkaline subtropical soil. *Soil Tillage Res.*, 66: 153–163.

Zinn, Y.L., Lala, R., and Resck, D.V.S. 2005. Changes in soil organic carbon stocks under agriculture in Brazil. *Soil Tillage Res.*, 84: 28–40.

Zuo, H.Q. and Wang, Z.S. 1995. Population dynamics of citrus rhizosphere microorganisms and rhizosphere effect [in Chinese with English abstract]. *Eco-Agric. Res.*, 3(1): 39–47.

CHAPTER 9

Ecological Effects of No-Tillage Rice in Middle and Lower Reaches of the Yangtze River

Li Chengfang and Cao Cougui

CONTENTS

9.1 INTRODUCTION

China surpassed the United States as the largest emitter of anthropogenic greenhouse gases (GHGs) globally in 2012 (Leggett, 2012). Atmospheric GHGs are generated largely by the agricultural sector. Specifically, agricultural activities in China contribute 39% of the methane (CH_4) and 1% of the carbon dioxide (CO_2) present in global emissions (Reiner and Milkha, 2000). Because rice is the most important food crop in China, the country's rice production exceeds that of any

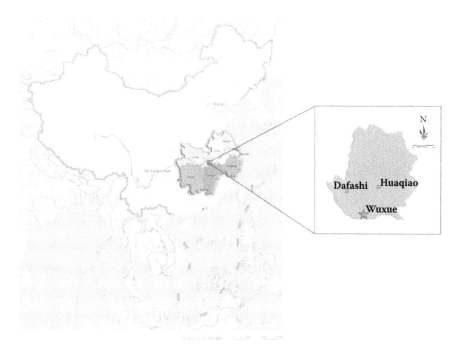

Figure 9.1   Experimental sites and middle and lower reaches of the Yangtze River.

other country and accounts for 30% of the world's total rice production (IRRI, 2004). In 2011, the planting area of China was approximately 30 million hectares, which accounted for approximately 20% of the world's total planting area (Ministry of Agriculture, 2012a). Given such characteristics, China has a large number of rice paddies that are important sources of anthropogenic GHGs. The annual amount of CH_4 emissions from Chinese rice fields is estimated to be 7.4 Tg, which is approximately 29% of the global CH_4 emissions from rice cultivation (X.Y. Yan et al., 2009). Meanwhile, the annual amount of N_2O emissions from Chinese rice fields is estimated to be 91 Gg N, of which 50 Gg N are emitted during rice growing seasons (Zheng et al., 2004). Therefore, mitigating GHG emissions from paddy fields in China is important in addressing the issue of climate change and the development of a more sustainable agriculture.

The middle and lower reaches of the Yangtze River include the provinces of Hunan, Hubei, Jiangxi, Anhui, Zhejiang, and Jiangsu, as well as the city of Shanghai (Figure 9.1). These areas are important sites of rice production in China. In 2011, the total planting area in these sites was estimated to be 18 million hectares, which accounts for approximately 59% of the total rice cultivation area in China (Ministry of Agriculture, 2012a); however, several problems occur during rice production in these areas. First, large amounts of GHGs are emitted from these paddy fields during rice growing seasons. Wang and Li (2002) estimated that, in 1994, CH_4 emissions from the rice fields in this Chinese region ranged from 3.547 to 5.327 Tg yr^{-1}. Field baking, the process of keeping the field drained for almost a week during the vegetative stage of rice, is a common agricultural management technique in the region. This method can reduce CH_4 emissions from paddy fields, but it inevitably increases nitrous oxide (N_2O) emissions. Zou et al. (2007) estimated that N_2O emissions from Chinese rice fields in 2007 accounted for 7 to 11% of the total N_2O emissions from agricultural fields in China. Second, rice production in China consumes approximately 37% of the total nitrogen (N) fertilizer used for rice production worldwide (Peng et al., 2002b). The middle and lower reaches of the Yangtze River are economically advanced areas of China. The N fertilizer application rate for rice production (300 kg N ha^{-1}) in these areas is higher than the average N application rate in China (215 kg N ha^{-1}) (F.S. Zhang et al., 2008). Moreover, the efficiency of N fertilizer use in these areas is lower than that in other major rice growing regions in China (Shi et al., 2010). Hence, large quantities of N applied in rice fields are lost, causing severe environmental pollution.

China's rapid economic development has driven the movement of rural labor forces to its developed coastal areas. This development has caused rice planting activities to decrease over the past three decades (L.Y. Chen et al., 2007; M. Huang et al., 2011). This change is partly due to limited availability of the labor force due to an increasing number of young farmers choosing to work in cities and thus leaving older farmers behind (Derpsch and Friedrich, 2009). In traditional rice production, conventional tillage (CT) and transplanting require a large workforce (Bhushan et al., 2007; S. Chen et al., 2007), but such a workforce has become increasingly scarce and expensive because of economic development and urbanization in China. For this reason, the development of simple, low-carbon cultivation technologies is urgently needed to reduce GHG emissions, fertilizer N loss, and labor requirements. Low-carbon agriculture is a new type of agriculture against the background of a low-carbon economy. It is an innovative pattern of modern agricultural development (Cao and Li, 2014). Compared with conventional high-energy consumption and high-input agriculture, low-carbon agriculture is aimed at saving energy, reducing GHG emissions from agriculture and rural production and living, increasing agricultural soil organic carbon (SOC), and maintaining the ecosystem balance between the agricultural carbon cycle and biodiversity (L.D. Yan et al., 2010).

Less intensive rice cultivation technologies are commonly achieved by simplifying land preparation and crop establishment; several less intensive cultivation technologies, such as reduced or no-tillage (NT), seedling throwing, and direct seeding, have been developed in China (M. Huang et al., 2011). Seedling throwing is a simplified cultivation technology where rice seedlings with soil on their roots, grown either on dry beds or in trays, are thrown aslant by hand or a throwing machine into a paddy field with shallow water (M. Huang et al., 2011). NT is a simple cultivation technology that has attracted considerable attention since the establishment of a government policy favoring the adoption of NT farming (Derpsch et al., 2010). In China, research on and the application of NT have developed quickly since the 1970s. By the end of 2008, NT had been applied to approximately 1.33 million hectares of land (Tang and Zhang, 1996; Derpsch and Friedrich, 2009). The implementation of NT in the middle and lower reaches of the Yangtze River has recently become popular; for example, NT rice production was implemented in approximately 231,590 hectares of land in Hubei Province in 2011 (Ministry of Agriculture, 2012b). Studies on NT rice production in this region have made considerable progress, as evidenced by their focus on three simple cultivation technologies: direct seeding, seedling throwing, and crop residue mulching (Yao et al., 2011). Based on our previous reports, this chapter reviews the effects of these NT cultivation technologies on GHG emissions, organic carbon sequestration, organic carbon fractions, N cycling and microbial communities, and rice grain yields from rice fields in this region. It also discusses the practicality and effectiveness of different NT cultivation technologies in this region.

9.2 EFFECTS OF NO-TILLAGE CULTIVATION TECHNOLOGIES ON GREENHOUSE GAS EMISSIONS FROM PADDY FIELDS

Global surface air temperatures have increased by 0.88°C since the late 19th century (IPCC, 2007). This increase has become a public concern. Climate change is induced by GHG emissions mainly through anthropogenic activities. CH_4 and CO_2, the most important GHGs, contribute 15% and 60%, respectively, to the anthropogenic GHG effect (Reiner and Milkha, 2000). Rice fields are an important source of atmospheric CH_4; in fact, CH_4 emissions from paddy fields account for 10 to 20% of total CH_4 emissions (Reiner and Milkha, 2000). Agricultural management practices affect GHG emissions from paddy fields by influencing soil biological, chemical, and physical properties (Oorts et al., 2007a,b) and through returning crop residue to the fields (C.F. Li et al., 2012). Hence, studies on GHG emissions from paddy fields under different agricultural management practices in China are important when considering the practicality and effectiveness of low-carbon rice farming systems.

Agricultural soil, with a complex carbon pool, is significantly affected by anthropogenic activities. GHG emissions from agricultural soil are the results of complex interactions between climate and the biological, chemical, and physical properties of the soil (Davidson and Janssens, 2006; Oorts et al., 2007b). Moreover, these emissions are closely related to microbial turnover and the effects of enzymes on soil organic matter (Paustian et al., 2000). Agricultural management practices, such as fertilization and tillage, affect the soil microenvironment, alter soil microbial activities, and change enzymatic substrates, thereby influencing soil organic matter decomposition and GHG emissions (Cao and Li, 2014). No consensus on the differences in GHG emissions between NT and CT has been reached. For example, Aslam et al. (2000) and Elder and Lal (2008) observed similar CO_2 emissions from NT and CT. By contrast, M. Liu et al. (2006) and Oorts et al. (2007b) found larger soil CO_2 emissions in NT than in CT. Ball et al. (1999) also recorded larger soil CO_2 emissions for some periods and smaller emissions for other periods in NT than in CT. Hence, further research is needed to clarify the effects of NT on GHG emissions from agricultural soil, particularly from paddy soils. In the next section, we review the effects of NT on GHG emissions from the paddy fields of Wuxue City (Figure 9.1) based on our previous studies.

9.2.1 Effects of No-Tillage on Greenhouse Gas Emissions from Direct-Seeding Paddy Fields

We conducted a field experiment from 2008 to 2010 to investigate the effects of combining NT with direct seeding on GHG emissions from paddy fields in Zhonggui County (29°55′ N latitude, 115°30′ E longitude), Dafashi Town, Wuxue City, Hubei Province, China (Ahmad et al., 2009; C.F. Li et al., 2012). The experimental soil is a clay loam soil classified as an Anthrosol (FAO, 2014). The experiment included NT and CT with or without NPK fertilizers (NT0, NTC, CT0, and CTC). Immediately after rape was harvested, the CT plots were moldboard plowed to a depth of 30 cm using a SNH554 tractor. No soil disturbance occurred in the NT plots. The chemical fertilizers were applied on the soil surface of the paddy fields. The CH_4 and N_2O fluxes were measured using the static chamber technique, and the CO_2 fluxes were determined using a portable photosynthesis analyzer. The GHGs were sampled according to fertilization and climatic conditions during rice growing seasons.

During the 2008 rice growing season, CO_2 fluxes ranged from 113.3 to 520.8 mg m^{-2} hr^{-1} in NT treatments and from 114.2 to 395.4 mg m^{-2} hr^{-1} in CT treatments. Moreover, CH_4 and N_2O fluxes ranged from –2.41 to 40.42 mg m^{-2} hr^{-1} and from –40.92 to 731.2 μg m^{-2} hr^{-1} in NT treatments, respectively, and from –2.50 to 52.01 mg m^{-2} hr^{-1} and from –52.31 to 663.3 μg m^{-2} hr^{-1} in CT treatments, respectively (Figure 9.2). In addition, cumulative CO_2 emissions did not differ significantly between tillage treatments. Although tillage practices did not affect CH_4 emissions, NTC significantly increased N_2O emissions by 32% from CTC levels. NT significantly decreased CH_4 emissions by 21 to 22% compared with CT.

We observed similar CO_2 emissions between NT and CT treatments during the 2008 rice growing season (Figure 9.2). Specifically, significantly greater CO_2 emissions were recorded under NT than under CT during the 2009 and 2010 rice growing seasons; this result might be attributed to the differences in the soil carbon mineralization between the NT and CT treatments at the experiment site (C.F. Li et al., 2012). Higher soil carbon mineralization under NT than under CT might result in greater CO_2 emissions from the former than from the latter. Our results are in contrast to the reports of Ball et al. (1999), Lal (2004), and Liang et al. (2007). They thought that CT ensures soil inversion, breaks down soil aggregates, and exposes protected organic matter to microbial decomposition, thereby increasing soil CO_2 emissions (Reicosky et al., 1997, 1999; Ball et al., 1999). Meanwhile, more surface crop residues in NT than in CT probably serve as a barrier for CO_2 emissions from the

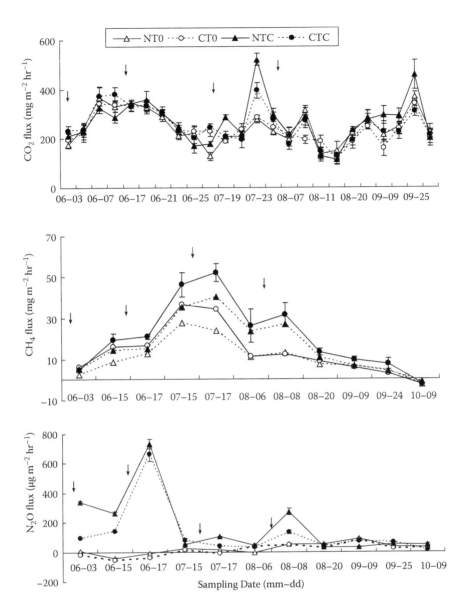

Figure 9.2 Changes in CO_2, CH_4, and N_2O fluxes from paddy fields during the 2008 rice growing season under different tillage treatments. The arrows indicate N fertilization in fertilized treatments during the rice growing season. NT0, NT without NPK fertilizers; CT0, CT without NPK fertilizers; NTC, NT with NPK fertilizers; CTC, CT with NPK fertilizers. (From Ahmad, S. et al., *Soil Tillage Res.*, 106, 54–61, 2009.)

soil to the atmosphere, and the crop residue decomposition rate decreases as the contact between soil and crop residues is reduced (Lal, 2004). The lack of consensus on the tillage effects on soil CO_2 emissions suggests that tilling soil is not the only factor that affects CO_2 emissions.

In our study, NT inhibits CH_4 emissions from paddy fields during rice growing seasons (Figure 9.2) (Ahmad et al., 2009; C.F. Li et al., 2012, 2013). This finding is in accordance with that of Harada et al. (2007) and Liang et al. (2007). Other researchers have found similar results during rice growing seasons in this region (F.L. Wu et al., 2007, 2008; Xiao et al., 2007; Bai et al., 2010; D.M. Li et al., 2011). CH_4 is produced by bacterial activities under extremely anaerobic conditions; therefore, paddy fields are an important source of atmospheric CH_4. CH_4 flux is a result of the production, oxidation, and transport of CH_4 (Wang and Li, 2002). Tilling soil results in a considerable amount of CH_4 fixed in soil being emitted into the atmosphere. By contrast, NT can improve and maintain the continuity of soil macroporosity (Ball et al., 1999). This improvement in soil macroporosity might increase CH_4 uptake and thus decrease CH_4 emissions. Moreover, soil surface compaction after NT (Ahmad et al., 2009) prolongs the travel time of CH_4 toward the atmosphere (Ball et al., 1999). This process also prolongs the response time of microbial consumers to CH_4 (Pandey et al., 2012), thereby increasing CH_4 uptake and decreasing CH_4 emissions. Ali et al. (2009) pointed out that iron oxide strongly affects CH_4 production from paddy soils because of iron acting as an acceptor of electrons generated from organic matter decomposition under aerobic conditions. In another study, we observed greater concentrations of total iron oxide under NT (9.29 g kg^{-1}) than under CT (7.01 g kg^{-1}). A large total amount of iron oxide under NT may suppress methanogen activity and, consequently, CH_4 emissions.

Our study in this region shows that cumulative N_2O emissions under NT are 1.32 times those under CT with NPK fertilizer (Figure 9.2). In our study, NT was characterized by large aggregates and high denitrification rates of paddy soils (H. Zhang et al., 2011), possibly leading to the production of a large amount of N_2O from fertilizer N through denitrification. X.J. Liu et al. (2006) observed a similar result. Large aggregates form hot spots of denitrification because of the anaerobic conditions inside the aggregates (Six et al., 2002b). Such formation leads to large N_2O emissions. Ball et al. (1999) indicated that low diffusivity near the NT soil surface does not block rapid N_2O emission because N_2O production sites are close to the soil surface. Possibly produced by rape residues on the soil surface, the high SOC at a depth of 0 to 5 cm under NT in our study (J.S. Zhang et al., 2011) could provide the reactive substances of denitrification; the use of fertilizer N on the soil surface in our study increased total N in the NT soils as well as the amount of N_2O under NT.

A long-lasting GHG, N_2O is produced as a result of microbial nitrification and denitrification (Beheydt et al., 2008; Mkhabela et al., 2008; Ussiri et al., 2009). Soil N_2O emissions are affected by several important factors, including soil bulk density, large aggregates, N source, and available carbon (Aulakh et al., 1984; Six et al., 2002a,b, 2004; Oorts et al., 2007a,b). Tillage may affect these soil properties and therefore influence N_2O emissions. However, the ultimate effects of NT on N_2O emissions from rice fields remain unclear (Chatskikh and Olesen, 2007; Xue et al., 2013). Bai et al. (2010) reported higher N_2O emissions from rice fields under NT than under CT. Some studies reported lower N_2O emissions from NT rice fields than from CT rice fields (Xiao et al., 2007; Pandey et al., 2012; J.K. Zhang et al., 2012). H.L. Zhang et al. (2013) stated that differences in N_2O emissions from rice fields between NT and CT vary with variations in weather.

As a simplified index, the global warming potential (GWP) indicates the ability of GHGs to trap heat relative to a unit mass of CO_2 (Shine et al., 1995); it is introduced to assess the potential impacts of GHG emissions on the climate system (Lashof and Ahuja, 1990). To estimate GWP, CO_2 is typically regarded as the reference gas, and an increase or decrease in CH_4 and N_2O emissions is converted into "CO_2 equivalents" through their GWPs. A positive GWP represents a net source of CO_2 equivalents, whereas a negative value indicates a net sink of atmospheric GHGs (Shang et al., 2011). Therefore, comparing GWPs calculated by GHG fluxes under different treatments can help assess the possibility of mitigating GHG emissions through NT adoption (Six et al., 2002a).

Table 9.1 shows the changes in GWPs under different tillage treatments in the 2008 and 2009 rice growing seasons. Despite the absence of significant differences in the GWPs between NT and CT without NPK fertilizer, the GWPs under NT were significantly lower than those under CT when NPK fertilizer was applied. Grace et al. (2003) estimated the GWPs of an irrigated rice–wheat system in the Indo-Gangetic Plains to range from 12,806 to 16,143 kg CO_2 ha^{-1}. These estimates are lower than those found in our study; however, the authors estimated only CH_4 and N_2O emissions and not CO_2 emissions. Our study found that relatively large amounts of CO_2 contribute to GWPs (Ahmad et al., 2009; Cao and Li, 2014). Although our study indicated that NT promotes N_2O and

TABLE 9.1 Global Warming Potential of Different Tillage Treatments in the 2008 and 2009 Rice Growing Seasons

Treatment	Global Warming Potential (GWP) (kg CO_2 ha^{-1})	
	2008	2009
NT0	16,409[c]	19,411[c]
CT0	18,837[c]	20,907[c]
NTC	23,361[b]	26,366[b]
CTC	26,012[a]	27,617[a]

Note: Different letters in a column mean significant difference at the 5% level. NT0, NT without NPK fertilizers; CT0, CT without NPK fertilizers; NTC, NT with NPK fertilizers; CTC, CT with NPK fertilizers.

CO_2 emissions, the inhibitory effects of NT on CH_4 emissions can offset the reducing effects of NT on N_2O and CO_2 emissions because, among the GHGs, CH_4 makes the greatest contribution to GWPs (Ahmad et al., 2009; Cao and Li, 2014). Thus, NT is an effective strategy to reduce GHG emissions from rice fields in central China and thus can help alleviate global warming.

The differences in GHG emissions among different studies are likely the result of differences in paddy conditions, weather, and crop rotation (Xue et al., 2013). Hence, extensive monitoring of N_2O emissions from paddy fields under different tillage treatments comprising different crop rotations, regions, and soil qualities is needed to adopt an effective region-specific management method (Pandey et al., 2012). Moreover, many of the recent studies on the effects of NT on GHG emissions from rice fields did not consider the effects of farm operations (e.g., seedbed preparation, tillage, fertilizer, irrigation, harvest). Farm operations can affect CO_2 emissions through fossil fuel usage and agricultural inputs, thereby impacting non-CO_2 GHG emissions from rice fields. The changes in GHG emissions from farm operations can partly or even fully offset the mitigation benefits of SOC sequestration as a result of NT implementation (West and Marland, 2002) when considering the net GWPs from NT rice fields. Hence, carbon emissions associated with farm operations should be incorporated comprehensively into net GWPs from NT rice fields.

Rice production begins with the seedling stage in flooded nurseries for about a month and proceeds to the growing stage in paddy fields for about 3 to 4 months. Both stages feature waterlogged conditions that are conducive to CH_4 formation (Smith et al., 2007). In the last several decades, considerable research has been conducted to estimate GHG emissions from transplanting to harvesting (Z.C. Cai et al., 1997; Zou Jianwen et al., 2005; D.M. Li et al., 2011; Shang et al., 2011). Meanwhile, only a few studies have estimated GHG emissions from rice nurseries. Recently, S.W. Liu et al. (2012) and Y.C. Ma et al. (2012) revealed that GHG emissions from rice nurseries are often ignored because the GWPs of CH_4 and N_2O from the rice seedling stage are comparable to those from the post-transplanting period. These studies suggested that incorporating GHG emissions from rice nurseries into an overall estimate of GHGs emissions from rice cropping systems is important in the comprehensive assessment of NT effects on GHG emissions and GWP.

9.2.2 Effects of Tillage and Seeding Methods on Carbon Emissions from Double Rice Systems

Conventionally, transplanting rice seedlings in paddy soils requires a large workforce, which is becoming scarce and expensive (Bhushan et al., 2007). Hence, simplified rice seeding technologies, such as direct seeding and seedling throwing (Figure 9.3), have become increasingly attractive in China (M. Huang et al., 2011). Seedling throwing is a simplified cultivation technology in which

Figure 9.3 Seedling throwing. The top photograph shows the seedlings being thrown, and the bottom photograph shows the seedlings on the paddy surface after throwing.

rice seedlings with soil on roots, grown either on dry beds or in trays, are thrown aslant by hand or throwing machine into a paddy field with shallow water (Figure 9.3) (M. Huang et al., 2011). Direct seeded rice refers to rice production through direct seeding in the production field rather than transplanting rice seedlings grown in a nursery. It involves sowing pre-germinated rice seeds into a paddy soil surface, with or without standing water (M. Huang et al., 2011). Although some studies have investigated the effects of different seeding technologies on carbon emissions from paddy fields, most research has focused on the differences in the carbon emissions between transplanted rice and direct-seeded rice (Ko and Kang, 2000; Singh et al., 2009; E.D. Ma et al., 2010; Pathak et al., 2013). Singh et al. (2009) found that transplanting seedlings results in greater CH_4 and CO_2 emissions compared with direct seeding; however, few studies have investigated the differences in CH_4 and CO_2 emissions between transplanting and seedling throwing (E.D. Ma et al., 2010). Therefore, the present work investigates the effects of tillage practices (NT and CT) and seeding methods (seedling transplant and seedling throwing) on CH_4 and CO_2 emissions from double rice cropping systems in central China.

We conducted the experiment at an experimental farm in Lanjie Village, Huaqiao Town, Wuxue City, Hubei Province, China (29°51′ N, 115°33′ E). The experimental soil is sandy loam, containing 78% sand, 10% silt, and 12% clay. The experiment included three treatments: (1) early rice (seedling throwing) with NT–late rice (seedling throwing) with NT (NTST); (2) early rice (seedling throwing) with CT–late rice (seedling throwing) with CT (CTST); and (3) early rice (transplanting seedling) with CT–late rice (transplanting seedling) with CT (CTTPS). The CH_4 and N_2O fluxes were measured using the static chamber technique, and the CO_2 fluxes were determined using a portable photosynthesis analyzer. The GHGs were sampled at intervals of approximately 10 days during the early and late rice growing seasons.

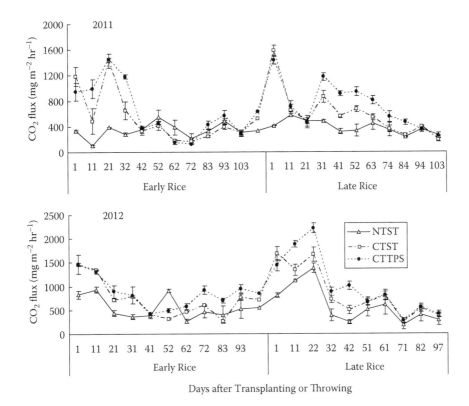

Figure 9.4 Changes in CO_2 fluxes under different treatments during the 2011 and 2012 early and late rice growing seasons. NTST, early rice (seedling throwing) with NT–late rice (seedling throwing) with NT; CTST, early rice (seedling throwing) with CT–late rice (seedling throwing); CTTPS, early rice (transplanting seedlings) with CT–late rice (transplanting seedlings) with CT. (From Li, C.F. et al., *Atmos. Environ.*, 80, 438–444, 2013.)

For early rice in this study, the CO_2 fluxes ranged from 99.3 to 1445.2 mg m^{-2} hr^{-1} in 2011 and from 259.8 to 1463.6 mg m^{-2} hr^{-1} in 2012. For late rice, the CO_2 fluxes varied from 180.6 to 1585.6 mg m^{-2} hr^{-1} in 2011 and from 188.0 to 2219.0 mg m^{-2} hr^{-1} in 2012 (Figure 9.4). In addition, higher CO_2 emissions were found from CT than from NT. This result is inconsistent with our previous finding (C.F. Li et al., 2012) that CO_2 emissions are higher in NT than in CT in another experimental site in Wuxue City. This discrepancy may be attributed to the different soil types used. The present experimental soil is sandy loam, whereas the previous soil is silt clay loam. Feiziene et al. (2012) showed similar results and stated that NT can mitigate CO_2 emissions from sandy loam soils but NT mitigation is negligible on loam soils.

Figure 9.5 shows that the CH_4 fluxes for early rice ranged from –2.52 to 125.0 mg m^{-2} hr^{-1} in 2011 and from –1.95 to 102.2 mg m^{-2} hr^{-1} in 2012. For late rice, the CH_4 fluxes varied from –7.22 to 242.3 mg m^{-2} hr^{-1} in 2011 and from –5.82 to 223.7 mg m^{-2} hr^{-1} in 2012. Moreover, we observed that the seasonal total CH_4 emissions from CTST were 1.75 to 2.10 times the total CH_4 emissions from NTST for early rice and 1.64 to 1.79 times for late rice. NT significantly decreased the CH_4 emissions from the paddy fields relative to CT, as demonstrated in our previous studies (Ahmad et al., 2009; C.F. Li et al., 2012). Moreover, NTST significantly lowered the seasonal total CO_2 emissions by 19% down to 33% for early rice and by 27% to 31% for late rice compared with CTST (C.F. Li et al., 2013). The significantly higher CO_2 emissions from CT than from NT are consistent with the result of Liang et al. (2007). This result may be attributed to the good mixing

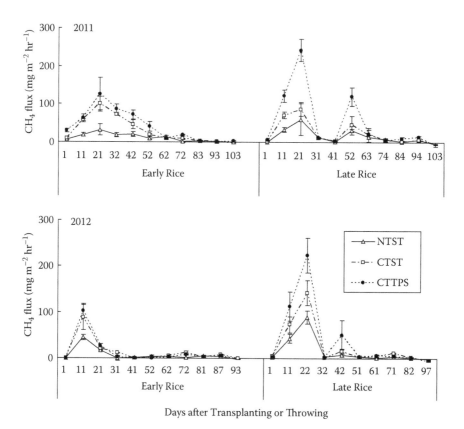

Figure 9.5 Changes in CH_4 fluxes under different treatments during the 2011 and 2012 early and late rice growing seasons. NTST, early rice (seedling throwing) with NT–late rice (seedling throwing) with NT; CTST, early rice (seedling throwing) with CT–late rice (seedling throwing); CTTPS, early rice (transplanting seedlings) with CT–late rice (transplanting seedlings) with CT. (From Li, C.F. et al., *Atmos. Environ.*, 80, 438–444, 2013.)

of residues as a result of the frequent tilling of soil. This method promotes soil CO_2 emissions (Lal, 2004). Tilling soil also increases the aeration of the topsoil and exposes organic matter to oxidizing conditions.

We also found that CTST significantly decreased ($p < 0.05$) the seasonal total CH_4 emissions and CO_2 emissions by 15% down to 40% and 18% to 21%, respectively, for early rice and by 38% down to 47% and 19% to 22%, respectively, for late rice, in comparison with CTTPS (Figures 9.4 and 9.5). Unlike seedling transplanting, seedling throwing considerably reduced the CH_4 and CO_2 emissions from the paddy fields; the reduced root weight from seedling throwing might be responsible for the reduced CH_4 emissions (C.F. Li et al., 2013). Reduced root weight under seedling throwing indicates a decrease in plant-mediated CH_4 transport because of the reduction in the surface area and the depth of root penetration into the soil (Pandey et al., 2012). It also indicates reduced conductivity of CH_4 from the root rhizosphere to the atmosphere (Ali et al., 2009). In addition, a reduction in root weight is indicative of the limited availability of substrates from root exudation, thereby restricting the supply of substrate for methanogens.

We observed that NTST can effectively reduce CH_4 and CO_2 emissions. Compared with CTST and CTTPS, NTST significantly reduced the seasonal total CH_4 and CO_2 emissions by 43% down to 71% and 19% to 47%, respectively, for early rice and by 39% down to 70% and 27% to 46%, respectively, for late rice (Figures 9.4 and 9.5). This result may be attributed to the suppression of

CH_4 and CO_2 emissions by NT as described above and to the reduced rice aboveground biomass and root growth under seedling throwing. The dual suppression of carbon emissions under NTST resulted in the lowest carbon emissions. Similar grain yields were observed among the treatment groups (C.F. Li et al., 2013). Unlike CT with seedling transplanting, NT with seedling throwing considerably reduces the labor requirement and production costs while increasing the economic benefit. Hence, our results suggest that the agricultural economic viability and mitigation of CH_4 and CO_2 from paddy fields in central China can be achieved simultaneously by NT combined with seedling throwing.

9.3 EFFECTS OF NO-TILLAGE ON SOIL ORGANIC CARBON FRACTIONS AND CARBON SEQUESTRATION FROM PADDY FIELDS

The global soil carbon pool (approximately 2500 Gt) is composed of approximately 1550 Gt of SOC (Lal, 2004). In terrestrial ecosystems, the soil carbon pool is an important component of the global carbon cycle. K.R. Li et al. (2003) estimated that China has a soil carbon storage of 82.65 Gt, which makes up approximately 3 to 4% of the global soil carbon storage. Paddy soil is the largest cultivated soil type in China, with an average topsoil carbon density of 46.91 to 25.73 t ha^{-1} (Song et al., 2005). G.X. Pan et al. (2004) claimed that the carbon sequestration potential of paddy soil under present conditions can reach 0.7 Pg and even 3.0 Pg. Y. Huang and Sun (2006) analyzed the literature and found significant increases in the topsoil SOC storage of Chinese farmlands over the past 20 years. Compared with that of other soil types, the SOC content of paddy field soil has significantly increased. The authors inferred that the application of NT technology will significantly increase organic carbon.

As a sensitive indicator of changes in soil quality and health, SOC and its fractions significantly contribute to the chemical, physical, and biological properties of soil (Haynes, 2005); thus, an improvement in SOC is necessary for developing sustainable agriculture. However, changes in SOC are difficult to detect in the short and medium term because of the high temporal and spatial variability of recalcitrant carbon (Blair et al., 1995). By contrast, labile soil organic carbon fractions (i.e., microbial biomass carbon, dissolved organic carbon, particulate organic carbon, and easily oxidizable carbon) can turn over quickly, thereby responding more rapidly to soil management practices than SOC (Blair et al., 1995; Ghani et al., 2003; Haynes, 2005). Thus, several researchers have suggested that these labile fractions be considered as early sensitive indicators of changes in soil quality because they are affected by soil management practices (Blair et al., 1995; Rudrappa et al., 2006; C.M. Yang et al., 2005). We have discussed the effects of NT on soil labile organic fractions and SOC sequestration from the same experimental paddy fields in Zhonggui County, Dafashi Town, Wuxue City (C.F. Li et al., 2012).

Our study found that short-term NT has no significant effect on SOC, suggesting that SOC is not sensitive to changes in soil quality caused by short-term tillage or crop management possibly because of the complexity of soil characteristics (Cao and Li, 2014). This result agrees with the results reported by T.Y. Cai et al. (2011) and Zhan et al. (2009). Long-term NT can improve SOC, and its enhancement pattern follows an S-shaped curve, in which SOC remains unchanged 2 to 5 years after NT implementation, changes greatly 5 to 10 years after, and thereafter continues to grow more slowly the next 25 to 50 years (Freibauer et al., 2004).

The components of SOC have different stabilities. These components include labile, chronic (slow), and inert organic carbon (Parton et al., 1993). The labile organic carbon that turns over quickly is most easily utilized by living organisms. Labile organic carbon fractions, such as dissolved organic carbon and particulate organic carbon, can be sensitive enough to reflect changes in soil quality induced by soil management practices (H.Q. Chen et al., 2009).

Table 9.2   Changes in Labile Soil Organic Carbon (SOC) Fractions in 0- to 20-cm Soil Layer and SOC Sequestration in Different Soil Layers Under Different Tillage Treatments

Treatment	Soil Organic Carbon (g kg⁻¹)	Dissolved Organic Carbon (g kg⁻¹)	Easily Oxidizable Carbon (g kg⁻¹)	Microbial Biomass Carbon (g kg⁻¹)	Particulate Organic Carbon (g kg⁻¹)	Organic Carbon Sequestration (kg C ha⁻¹ season⁻¹)	
						0- to 5-cm Layer	0- to 20-cm Layer
No-tillage (NT)	19.50[a]	0.72[a]	13.1[a]	1.42[a]	7.74[a]	1537[a]	2878[a]
Conventional tillage (CT)	19.07[a]	0.54[b]	10.7[b]	1.12[b]	7.44[a]	697[b]	2666[a]

Note: Different letters in a column mean significant difference at the 5% level.

In the present study, NT significantly affected soil labile organic carbon (Table 9.2). Unlike CT, NT significantly increased dissolved organic carbon by 33%, easily oxidizable carbon by 22%, and microbial organic carbon by 27%; however, NT did not affect particulate organic carbon. Similar results were observed by Jacobs et al. (2009), Roper et al. (2010), and J.B. Zhang et al. (2005). These results may be attributed to the fact that NT avoids soil disturbance and thus hinders the release of substances protected by soil microorganisms (H.Q. Chen et al., 2009). In NT systems, crop residues covering the soil surface increase organic matter and thus promote microbial activity; therefore, the by-products of the microbial decomposition of organic matter serve as the main sources of soil labile organic carbon (M. Liu et al., 2006).

Dissolved organic carbon is derived from humified organic matter, plant litter, root exudates, and microbial biomass (Zhao et al., 2003). In the present study, the dissolved organic carbon in the paddy soils ranged from 0.54 to 0.72 g kg⁻¹, which is well within the values reported by Zhan et al. (2010) for paddy soils in this region (0.44 to 0.58 g kg⁻¹) but exceeds the values reported by J.B. Zhang et al. (2005) for forest fields (0.22 to 0.24 g kg⁻¹) and those by X.Y. Wu et al. (2013) for corn fields (0.07 to 0.08 g kg⁻¹). This difference is attributable to the differences in land use. Flooding can decrease soil organic carbon decomposition under paddy soils relative to upland soils (Shang et al., 2011) and thus improve dissolved organic carbon. In the present study, tilling soil lowered dissolved organic carbon. Haynes (2005) reported a positive relationship between dissolved organic carbon and soil tillage. Delprat et al. (1997) and Y.F. Wang and Chen (1998) indicated that long-term cultivation decreases dissolved organic carbon in soil.

Easily oxidizable carbon oxidized by neutral $KMnO_4$, or permanganate-oxidizable carbon, is a relatively new method that can characterize the organic matter of various soils and evaluate changes in organic matter affected by soil management (Tirol-Padre and Ladha, 2004; Culman et al., 2012). In the present study, NT significantly increased easily oxidizable carbon compared with CT (Table 9.2). Similar results were found by many researchers. For example, Quincke et al. (2007) found that stocks of easily oxidizable carbon are lower under CT than under NT in the Rogers Memorial Farm in Eastern Nebraska. Melero et al. (2009) evaluated the short- and long-term effects of NT on soil organic carbon fractions in semi-arid southwest Spain and found that, compared with CT, NT produced higher easily oxidizable carbon in the long-term and short-term trials in the 0- to 5-cm soil layer. After 20 years of consecutive observation in Chongqing, X.Y. Wu et al. (2013) found that tilling soil can significantly increase easily oxidizable carbon in the 0- to 40-cm layer of paddy soil compared with NT soil.

Balesdent (1996) indicated that, although soil microbial biomass carbon only accounts for 2% to 5% of total organic carbon, it responds more sensitively to changes in soil quality than SOC. In our study, NT significantly increased soil microbial biomass carbon compared with CT (Table 9.2). Many studies have discussed the relationships between soil microbial biomass carbon and tillage practices in different regions (Dou et al., 2008; H.Q. Chen et al., 2009; Shao et al., 2009; Gajda,

2010; Kang et al., 2010). Our result is inconsistent with the result reported by Kang et al. (2010), possibly because of differences in the soil types, cropping systems, and duration of the experiments between the two studies.

Soil particulate organic carbon can be used as a sensitive indicator of the long-term variations of SOC (H.Q. Chen et al., 2009). In our study, short-term NT had no significant effect on particulate organic carbon (Table 9.2). This result is inconsistent with the results reported by Dou et al. (2008), Franzluebbers and Arshad (1997), and C.R. Yan et al. (2010). Franzluebbers and Arshad (1997) reported that the average particulate organic carbon under CT is 0.63 kg m^{-2}, which is lower than that under NT (0.76 kg m^{-2}). Moreover, the ratio of specific particulate organic carbon mineralization to specific whole-SOC mineralization under NT is 23% greater than that under CT, suggesting that particulate organic carbon is more active (i.e., more mineralizable) under NT relative to SOC. As mentioned above, the diverse results suggest that tillage effects on soil labile organic carbon fractions depend on regional climate, soil type, residue management practice, crop rotation, and length of study under consideration (Puget and Lal, 2005). Moreover, the effects of tillage practices on soil labile organic carbon fractions include not only soil disturbance and the distribution of crop residues but also the changes in the physical, chemical, and biological properties of the soil that cause long-term effects.

Our study showed that the application of N fertilizer significantly increased the concentration of SOC by 4% up to 9% and SOC sequestration by 21% to 94% (C.F. Li et al., 2012). N fertilization increased the biomass of rice, thus causing a large amount of rice stubble to return to the soil surface (Lu et al., 2009). Lu et al. (2009), Nayak et al. (2009), Shang et al. (2011), and C.J. Wang et al. (2010) reported similar results, but some studies have claimed that N fertilization does not increase and even reduces SOC concentration and SOC sequestration (Halvorson et al., 2002; Khan et al., 2007; J.T. Li and Zhang, 2007; López-Bellido et al., 2010). These adverse results may be related to the different climate conditions, crop residue management, rotation systems, and test duration (Lou et al., 2011).

Conventional tillage frequently disturbs soil and destroys the structure of soil aggregates, thereby reducing macroaggregates (Wright and Hons, 2005a; Fabrizzi et al., 2009). Moreover, CT can improve soil aeration and increase the contact area between the soil organic matter and the atmosphere, thereby accelerating the degradation of soil organic matter. Therefore, CT is not conducive to soil carbon sequestration (Pendell et al., 2007; Fabrizzi et al., 2009; Mishra et al., 2009), whereas NT can increase SOC sequestration (Six et al., 2000a,b, 2002a,b), particularly on the soil surface. Similarly, conclusions about soil carbon sequestration under NT paddy soils are relatively consistent, with most research indicating that NT can sequester more carbon in paddy soils compared with CT (Y.J. Gao et al., 2000; Tang et al., 2007; Bai et al., 2009; Lu et al., 2009). Tang et al. (2007) indicated that NT can sequester 112.3 kg C ha^{-1} yr^{-1} in the 0- to 20-cm soil layer 13 years after the conversion from CT to NT of purple paddy soil in the Beipei District of Chongqing City, China. Gao et al. (2000) conducted a long-term experiment (growing 25 crops) in Zhangjiagang City and found that NT can sequester 26.68 kg C ha^{-1} yr^{-1} in gray fluvoaguic paddy soils in the 0- to 30-cm soil layer; however, Six et al. (2002b) indicated that the effects of NT on SOC sequestration depend on soil types. Moreover, the sampling protocol may have biased the results (Baker et al., 2007; He et al., 2010). In the study of He et al. (2010), NT did not increase the SOC sequestration in the 10- to 20-cm soil layer after a 5-year experiment in a sandy loam soil in Ningxiang Country, Hunan Province. Baker et al. (2007) analyzed the effect of sampling depth on the potential of SOC sequestration under conservation tillage and found that deep sampling generally shows no carbon sequestration advantage for conservation tillage. In addition, the duration time of experiments leads to different results. Tian et al. (2013) analyzed nearly 30 years of results related to China's farmland conservation tillage and found that the rate of increase of SOC in short-term field experiments (<5 years) is approximately 1.75 times as high as that in long-term ones (>5 years). Therefore, they suggested the possible overestimation of the SOC sequestrating potential of conservation tillage in short-term field experiments.

Our study showed that, unlike CT, NT did not affect SOC concentration in the 0- to 20-cm soil layer but significantly increased SOC concentration by 12 to 15% in the 0- to 5-cm soil layer (Table 9.2). Similar observations have been reported in some studies (Causarano et al., 2006; Dolan et al., 2006). In NT paddy fields, crop residues accumulate on the soil surface, leading to high SOC concentrations; however, CT incorporates residues into a greater soil volume (Six et al., 1999; Wright et al., 2008), resulting in a relatively high SOC concentration at deeper depths than that in NT (C.F. Li et al., 2010). In this manner, low SOC concentrations in the subsurface soil under NT offset the impact of tillage on the surface soil.

Our study also showed that the sampling protocol affected SOC sequestration (Table 9.2). Specifically, NT significantly increased SOC sequestration by 102 to 270% in the 0- to 5-cm soil layer compared with CT, but it did not affect SOC sequestration in the entire 0- to 20-cm soil layer. This result is attributable to the accumulation of high crop residues on the soil surface under NT. Wright et al. (2008) and Wright and Hons (2005b) reported similar results. Deen and Kataki (2003) suggested that, although NT considerably increases SOC sequestration in the 0- to 5-cm soil layer compared with CT, low SOC sequestration under NT can be observed in the 0- to 40-cm soil layer. Christopher et al. (2009) found that in the 0- to 60-cm soil profile no significant difference in SOC sequestration can be observed between NT and CT. We thus speculated that when considering a deep soil profile (>20 cm) the potential of SOC sequestration under NT paddy fields may be overestimated. Hence, further studies are needed to assess the effects of sampling protocols on SOC concentration and SOC sequestration in NT paddy soils.

9.4 EFFECTS OF NO-TILLAGE ON THE SOIL MICROBIAL COMMUNITY IN PADDY FIELDS

Soil microorganisms are known to significantly affect agroecosystem health because of their roles in residue decomposition and nutrient cycling and their associations with other organisms (Ge et al., 2008; F.P. Li et al., 2008). J. Li et al. (2008) suggested that soil microbial properties, such as microbial biomass and population activity, have strong correlations with soil health and quality. The abundance, structure, and diversity of soil microbial communities have been widely used to indicate soil quality changes because of their sensitivity to environmental changes (Kennedy and Smith, 1995; Marschner et al., 2003; Chu et al., 2007). Filip et al. (2002) pointed out that the productivity and health of agricultural systems are partly dependent on the functional processes of soil microbial communities. Tilling soil, which is a common agricultural management practice, has complex effects on the soil's physical, chemical, and biological environment (Tilman et al., 2002; M.Q. Liu et al., 2009). Lehmann et al. (2011) stated that tilling soil causes shifts in the physicochemical properties of soil pH and bulk density, in the availability of C and nutrients, or in the accessibility to water and nutrients and could potentially result in changes in microbial community structure for resource competition. By contrast, NT greatly reduces soil physical disturbance, increases soil aggregation, and results in the enrichment of organic matter in the soil surface, therefore providing a beneficial environmental condition for soil biota organisms (Jacobs et al., 2009; Helgason et al., 2010; Sundermeier et al., 2011).

Recently, the effects of different tillage practices on microbial communities have been widely studied, with a particular focus on the importance of soil microbes in geochemical nutrient turnover and soil health. Numerous studies have found significant shifts in the structure and function of microbial communities along with changes in tillage practices (Cheneby et al., 2009; Vargas et al., 2009; Attard et al., 2010). However, little attention has been paid to the effects of tillage practices on the microbial communities in paddy soils, especially those in China. This research gap thus presents the need to carry out further investigations to understand the changes in soil microbial properties under NT in Chinese rice-based cropping systems and to clarify the mechanisms of the NT effect on soil microbial properties (Chaparro et al., 2012).

9.4.1 Effects of Short-Term No-Tillage on the Soil Microbial Community Under a Rice–Wheat System

A field experiment was conducted in the town of Junchuan in central China to investigate the effects of short-term (less than 2 years) tillage practices and residue return on topsoil (0 to 5 cm), microbial community structure, and microbial diversity (Guo et al., 2013). Treatments were established following a split-plot design of a randomized complete block, with tillage practices (CT and NT) as the main plot and wheat straw returning levels—0 (SR0), 3000 (SR1), 4000 (SR2), and 6000 (SR3) kg ha⁻¹—as the subplot treatment. Soil samples were collected in October 2012 (just after rice harvest). Phospholipid fatty acid (PLFA) analysis has been used as a culture-independent method for assessing the structure of soil microbial communities and determining the gross changes that accompany different agricultural soil disturbances (Hill et al., 2000). The PLFA approach has been shown to be more sensitive than nucleic acid-based techniques (Tscherko et al., 2005; Ramsey et al., 2006). Therefore, the present study observed and determined the composition of soil microbial communities using PLFA analysis. In this section, we discuss the effects of tillage practices on microbial communities in paddy soil.

In our study, no differences were observed in the total soil PLFA, Actinobacteria, bacteria, and Gram-positive bacteria between NT and CT, whereas Gram-negative bacteria and fungi were found to be higher in NT than in CT (Table 9.3). Kay and VandenBygaart (2002) reported that assessments of tillage-induced changes in organic matter are most consistent when measured for a minimum of 15 years following the conversion of CT to NT after soils have had sufficient opportunity to become adapted or equilibrated under NT management. Therefore, the duration (less than 2 years) of short-term conservation tillage is not long enough to result in a significant change in the microbial community in paddy soil, except for fungi and Gram-negative bacteria. The results of the study also indicated that fungi and Gram-negative bacteria had a more sensitive response to shifts in tillage practices in paddy soil compared with other microbial communities. The lack of differences in total PLFA under different tillage practices might be attributed to the fact that tillage practices affecting microbial communities vary with the types of microbial populations and that the shift in the abundance of different microbial communities offset each other (Adl et al., 2006). L. Zhang et al. (2002) suggested that NT practice significantly improves bacterial biomass and fungi biomass and that microbial biomass shows obvious seasonal changes, with the highest points observed in spring and the lowest points observed in early autumn, generally in paddy soil. Seasonal changes in microbial communities could provide another explanation for the effect of tillage practices on microbial communities.

The fungal-to-bacterial ratio, which represents the relative abundance of a microbial population, has been shown to be sensitive to changes in physicochemical conditions and in the efficiency of the carbon use of microbial communities induced by environmental stress (Bossio et al., 1998; J.H. Chen et al., 2013). In our study, the fungal-to-bacterial ratio was significantly lower under NT than under CT (Table 9.3), suggesting that the soil ecosystem buffering capacity was stronger under CT than under NT (Bossio et al., 1998). In addition, in the present study, tillage practices had no significant effect on bacterial biomass, whereas fungal biomass was lower under NT than under CT (Table 9.3). Therefore, the lower fungal-to-bacterial ratio under NT than under CT may be due to the low fungal biomass under NT. A fungal-dominated microbial community has been known to improve carbon stabilization and produce protected and stable carbon storage (Zak et al., 1996; Holland and Coleman, 1987; Parton et al., 1987; Six et al., 2006; M. Liu et al., 2011). Additionally, a fungi-predominant microbial community was not observed under the NT treatments (Table 9.3). A similar result was reported by Mathew et al. (2012). This finding is attributable to the fact that the duration of NT (less than 2 years) might have been insufficient to change soil properties (Sotomayor–Ramírez et al., 2006), which would promote fungus breeding. The results could have also been affected by the special physical and chemical conditions of the paddy soils, which were regularly flooded and intermittently irrigated.

Table 9.3 Phospholipid Fatty Acid (PLFA) Profiles under Different Treatments

Treatment	Total PLFAs (nmol g⁻¹)	Bacteria (nmol g⁻¹)	Fungi (nmol g⁻¹)	Fungi/ Bacteria Ratio	Actinobacteria (nmol g⁻¹)	Gram-Positive Bacteria (G⁺) (nmol g⁻¹)	Gram-Negative Bacteria (G⁻) (nmol g⁻¹)	G⁺/G⁻ Ratio	MUFA/ STFA Ratio
CT+SR0	33.58 ± 3.87	20.34 ± 1.77	0.55 ± 0.06	0.03 ± 0.00	4.21 ± 0.83	6.66 ± 0.81	10.28 ± 1.14	0.65 ± 0.0	1.21 ± 0.32
CT+SR1	31.09 ± 3.95	18.29 ± 2.21	0.44 ± 0.14	0.02 ± 0.00	3.92 ± 0.41	8.73 ± 0.87	6.59 ± 0.81	1.33 ± 0.04	0.49 ± 0.06
CT+SR2	34.05 ± 3.03	21.25 ± 2.24	0.68 ± 0.24	0.03 ± 0.01	3.97 ± 0.21	7.96 ± 0.63	9.91 ± 1.32	0.81 ± 0.07	1.17 ± 0.22
CT+SR3	33.86 ± 2.60	21.76 ± 2.10	0.47 ± 0.05	0.02 ± 0.00	3.72 ± 0.26	8.48 ± 0.17	9.42 ± 1.41	0.91 ± 0.13	1.09 ± 0.06
NT+SR0	25.27 ± 0.93	14.50 ± 0.56	0	0	3.85 ± 0.22	7.86 ± 0.37	5.44 ± 0.30	1.45 ± 0.13	0.26 ± 0.06
NT+SR1	32.92 ± 3.84	20.2 ± 2.09	0.48 ± 0.04	0.02 ± 0.00	3.83 ± 0.75	8.08 ± 0.72	8.32 ± 0.86	0.97 ± 0.08	1.01 ± 0.27
NT+SR2	25.3a ± 3.07	14.79 ± 2.62	0.38 ± 0.01	0.03 ± 0.00	3.52 ± 0.13	7.32 ± 0.18	5.32 ± 0.77	1.47 ± 0.41	0.55 ± 0.39
NT+SR3	43.06 ± 4.64	27.24 ± 3.20	0.72 ± 0.10	0.03 ± 0.00	4.61 ± 0.45	10.82 ± 1.30	11.28 ± 1.31	0.96 ± 0.02	1.02 ± 0.11
T	ns	ns	9.48**	15.24**	ns	ns	9.00**	17.62**	9.54**
SR	9.08**	12.75**	9.59**	13.73**	ns	12.60**	7.83**	ns	ns
T+SR	9.46**	10.56**	15.68**	15.4**	ns	6.14**	14.85**	15.58**	12.44**

Note: MUFA, monounsaturated fatty acids; STFA, saturated fatty acids; CT, conventional tillage; NT, no-tillage; T, tillage practice; SR, straw returning; SR0, no wheat straw returning; SR1, 3000 kg ha⁻¹ wheat straw returning; SR2, 4000 kg ha⁻¹ wheat straw returning; SR3, 6000 kg ha⁻¹ wheat straw returning; **, $p <$ 0.01; ns, not significant.

A high proportion of Gram-negative bacteria is usually interpreted as a shift from oligotrophic to copiotrophic soil conditions (Borga et al., 1994; Saetre and Baath, 2000). Table 9.3 shows that the ratio of Gram-negative bacteria to Gram-positive bacteria under NT is higher than that under CT, thus indicating that NT decreases soil nutritional conditions (Borga et al., 1994; Saetre and Baath, 2000). A partial correlation analysis has indicated that the ratio of Gram-positive bacteria to Gram-negative bacteria is significantly negatively correlated with fungal biomass (Guo et al., 2013); this correlation indicates that low fungal biomass under NT results in a high Gram-negative to Gram-positive bacteria ratio (Table 9.3). Kierkegaard et al. (2000) and Ogram (2000) reported that fungi have great effects on soil fertility. Fungi play an important role in some soil biochemical processes, such as ammonification, nitrification, and cellulose and humus decomposition (Y. Hu et al., 2007). Therefore, low fungal biomass in paddy soil may lower soil fertility. The ratio of monounsaturated fatty acids to saturated fatty acids (MUFA/STFA) reflects aeration of the soil; a greater MUFA/STFA indicates better aeration conditions of the soil. The ratio is commonly used as a sensitive indicator for agricultural management (Bossio et al., 1998). A significantly low MUFA/STFA ratio in NT soils (Table 9.3) indicates a highly anaerobic condition under NT (Spedding et al., 2004; Gouaerts et al., 2007) because of minimal physical disturbance.

In the present study, compared with CT, NT decreases the MUFA/STAFA and increases the ratio of Gram-positive bacteria to Gram-negative bacteria (Table 9.3), all of which are indicative of the poor aeration and nutrition conditions of soil under NT (Bossio et al., 1998; Saetre and Baath, 2000). These conditions may explain the low Shannon–Weiner index under NT. Noll et al. (2005) suggested that the succession of bacterial community structure and diversity in paddy soil can also be affected by oxygen gradient. Our results are inconsistent with other studies conducted under long-term NT (Adl et al., 2006; Ceja-Navarro et al., 2010). Soil microbial diversity is known to be enhanced by long-term NT (Adl et al., 2006; Ceja-Navarro et al., 2010). Ceja-Navarro et al. (2010) claimed that long-term NT (more than 10 years) significantly increases soil microbial diversity and abundance. Similarly, Adl et al. (2006) reported that NT (4 to 25 years) significantly enhances soil microbial diversity in cotton fields. Carpenter-Boggs et al. (2003) pointed out that soil microbial parameters are negatively related to tillage intensity. These inverse results suggest that further research on the effects of tillage practices on microbial communities in paddy soil is needed.

9.4.2 Tillage-Induced Effects on Microbial Community Associated with Soil Aggregates in Paddy Soil

Young and Ritz (2000) reported that the primary effect of tillage on soil microbial communities depends largely on the degree of soil disturbance that destroys soil aggregates at the microscale. Moreover, the chemical and physical properties of soil aggregates contribute to the heterogeneous distribution of microorganisms among aggregates with different sizes (Young et al., 2008). van Gestel et al. (1996) reported that microbial distribution patterns may be influenced by pore size, which is associated with particular aggregates; however, tillage effects on the distribution pattern of microorganisms associated with soil aggregates are not widely reported (Bronick and Lal, 2005; Väisänen et al., 2005).

We conducted a field experiment in the town of Huaqiao in central China to investigate the changes in topsoil (0 to 5 cm) microbial communities within soil aggregates under different tillage treatments. Treatments were established following a split-plot design of a randomized complete block with tillage practices (CT and NT) as the main plot and wheat straw returning levels—0 (NS) and 6000 (SR) kg ha[-1]—as the subplot treatment. Soil samples were collected in October 2012 (just after rice harvest). PLFA analysis was employed to determine the composition of the soil microbial community within the aggregates. In this section, we investigate the effects of tillage practices on microbial communities within different soil aggregates.

Table 9.4 Changes in Microbial Communities within Soil Aggregates under Different Treatments

Treatment	Size	Microbial Biomass Carbon (mg kg^{-1})	Total PLFAs (nmol g^{-1})	Bacteria (nmol g^{-1})	Actinobacteria (nmol g^{-1})
NT	Bulk soil	1133 ± 94	27.95 ± 3.48	25.25 ± 3.35	4.00 ± 0.54
	2–0.25 mm	854 ± 180	28.18 ± 3.53	25.44 ± 3.65	3.81 ± 0.38
	0.25–0.053 mm	1585 ± 67	32.54 ± 0.69	29.81 ± 0.54	4.34 ± 0.05
	<0.053 mm	620 ± 119	15.47 ± 1.00	14.34 ± 1.15	1.79 ± 0.06
CT	Bulk soil	927 ± 79	19.82 ± 2.38	18.05 ± 2.51	2.37 ± 0.22
	2-0.25 mm	622 ± 102	26.93 ± 1.16	24.23 ± 0.72	3.67 ± 0.58
	0.25–0.053 mm	1441 ± 38	31.47 ± 2.38	28.54 ± 2.26	3.98 ± 0.30
	<0.053 mm	823 ± 18	14.59 ± 0.75	12.38 ± 0.64	2.00 ± 0.10
T		*	**	**	**
A		**	**	**	**
T+A		**	*	ns	**

Note: NT, no-tillage; CT, conventional tillage; T, tillage practice; A, soil aggregate; *, $p < 0.05$; **$p < 0.01$; ns, not significant.

In our study, we also observed that, compared with CT, NT significantly increased soil microbial biomass carbon by 9.9%, total PLFAs by 12.2%, bacteria by 14.0%, and Actinobacteria by 16.0% for bulk soil (Table 9.4) but did not affect fungi (data not shown). Numerous studies have also found that the soil microbial community is greatly affected by tillage practices and that microbial biomass is usually higher in soils under NT than under CT (Spedding et al., 2004; Nyamadzawo et al., 2009; González-Chávez et al., 2010). Jiang et al. (2011) investigated the effects of NT on microbial communities within paddy soil aggregates in southwest China and claimed that NT significantly increases microbial biomass carbon and fungal biomass for bulk soil, as well as all fractions of aggregates, relative to CT. Similar stimulating effects of NT on these microbial parameters were found for all aggregate fractions (Table 9.4), possibly because of the high soil organic carbon under NT (X. Jiang et al., 2011) and the high level of residues in surface soil. Tillage decreased soil microbial biomass mainly by decreasing the proportion of soil macroaggregates in the bulk soil (X. Jiang et al., 2011).

Cropping system type is one of the main factors that affect soil microbial communities, and the effects of tillage practices on soil microbial communities have been reported to vary with cropping systems (U. Chen and Zhang, 2004; Xiong et al., 2008). In double rice cropping systems, NT significantly increases soil bacteria, fungi, and actinomycetes compared with CT (B. Chen and Zhang, 2004). In rice–oilseed rape cropping systems, soil bacteria are lower under NT than under CT in both rice and oilseed rape growing seasons, whereas fungi and actinomycete biomass decrease during the rice growing season but increase during the oilseed rape growing season under NT (Xiong et al., 2008). In rice–wheat cropping systems, NT results in higher soil bacteria, fungi, and actinomycete biomass compared with CT in spring and early winter (Xiong et al., 2008). This difference is attributable to the fact that the effect of NT on soil microbial biomass varies based on the crop growing seasons in rice–upland cropping systems (M. Gao et al., 2004; Xiong et al., 2008). Hence, further studies on tillage effects on soil microbial communities must focus on different cropping systems.

9.5 EFFECTS OF NO-TILLAGE ON FERTILIZER NITROGEN DYNAMICS OF PADDY FIELDS

According to ICAM (2012), the world population was 6.79 billion in January 2010. Meeting the basic food needs of this growing population requires the production of large amounts of grains, which in turn requires large amounts of chemical N fertilizer (X.X. Zhang et al., 2014). Canfield et al. (2010) reported that the use of N fertilizers increased by 800% from 1960 to 2000, with the production of wheat, rice, and maize accounting for almost half of current fertilizer use. China is the largest consumer of chemical N in the world, and approximately 18% of the chemical N consumed by China is applied in rice paddies (Peng et al., 2002b; Heffer, 2009). However, several studies have reported that N use efficiency for rice production is typically below 40% in China (Peng et al., 2002a; Qiu, 2009; J.G. Liu et al., 2010). Therefore, future studies should be focused on N losses in paddy fields, as such losses may cause serious environmental problems (Kyaw et al., 2005). Moreover, highly effective management practices should be developed to enhance N fertilizer use efficiency and thus reduce any adverse effects on the environment.

Soil environmental factors, such as redox potential, bulk density, carbon and N substrates, and microorganisms, significantly affect N biogeochemical processes that occur in flooded paddy fields (X.X. Zhang et al., 2014). Tillage practices can change these soil environment parameters, thereby affecting the N dynamics of paddy fields. In addition, conservation tillage, including NT and crop residue incorporation, has been recommended as a favorable management practice because of its beneficial effects on soil properties, such as increases in soil fertility, improvements of aggregation, and decreases in erosion (Malhi et al., 2006). However, few studies have investigated the effect of tillage practices on N cycles in rice fields. Studies on the pathways and rates of input, output, and internal cycle of N under different tillage practices can provide insights into the fundamental issues of low N fertilizer use efficiency in paddy fields (Ishii et al., 2011).

9.5.1 Effects of No-Tillage with Surface Application of Nitrogen Fertilizer on Nitrogen Fertilizer Loss in Paddy Fields

A similar experiment in Zhonggui County of Dafashi Town in Wuxue City (Ahmad et al., 2009; J.S. Zhang et al., 2011) was conducted to discover the effects of NT with a surface application of N fertilizer on N fertilizer loss in paddy fields during the 2008 and 2009 rice growing seasons. NH_3 volatilization flux was measured using a continuous airflow chamber method described in the work of Tian et al. (1998). Ceramic porous cups were installed to collect percolation water, and the different formations of N concentrations (total N, NO_3^-, and NH_4^+) in the percolation water were measured. The volume of percolation water was calculated based on the balance of water in the field system (Xu and Liu, 1999). Therefore, the N leaching in each treatment was the product of the volume of the percolation water and the mean N concentrations of percolation water during the rice growing season.

Figure 9.6 shows that N fertilization significantly increased NH_3 volatilization. For fertilized treatments, NH_3 volatilization fluxes peaked the next day and then dropped rapidly to unfertilized treatment levels within 1 to 2 weeks after each application of N fertilizer. This result may be attributed to the consumption of N fertilizer in N infiltration, nitrification, denitrification, soil N immobilization, and N uptake by plants, all of which decreased the substrates for NH_3 volatilization (Hou et al., 2007). We also observed that the mean NH_3 volatilization fluxes under NT and CT with N fertilizer treatments were 29.3 and 21.8 mg m^{-2} day^{-1}, respectively (Figure 9.6). The cumulative emissions of NH_3 ranged from 4.44 to 36.7 kg N ha^{-1}, and the cumulative NH_3 emissions in NT with N fertilizer treatment were 29 to 52% higher than those in CT with N fertilizer treatment. The high NH_3 volatilization under NT with N fertilizer treatment (NTC) was possibly due to the high urease activities in the NT soil (J.S. Zhang et al., 2011). M. Gao et al. (2004) suggested that higher urease activities in NT than in CT lead to the production of large amounts of NH_4^+ in soil and floodwater from hydrolyzed N fertilizer, thus boosting the volatilization of NH_3 from NT. Moreover, crop residues on the NT soil surface could decrease the contact of N fertilizer granules with the soil, possibly reducing the adsorption of NH_4^+ on soil particles. A portion of N fertilizer granules fall into shallow cracks, thus potentially lowering NH_3 volatilization in the CT soil (Rochette et al., 2009).

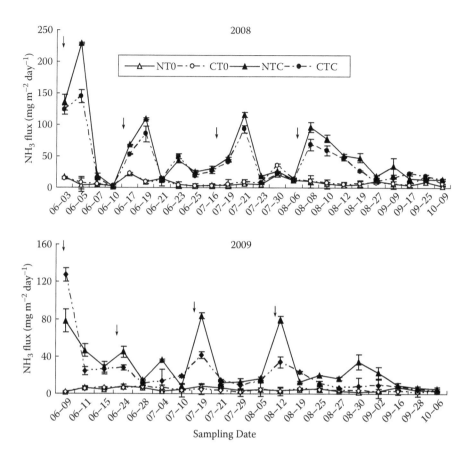

Figure 9.6 Changes in NH_3 volatilization fluxes from different tillage treatments during the 2008 and 2009 rice growing seasons. The vertical bars are standard deviations of the means. The arrows indicate N fertilization in fertilized treatments during the rice growing season. NT0, NT without NPK fertilizers; CT0, CT without NPK fertilizers; NTC, NT with NPK fertilizers; CTC, CT with NPK fertilizers. (From Zhang, J.S. et al., *Agric. Ecosyst. Environ.*, 140, 164–173, 2011.)

In the present study, total N, NO_3^-, and NH_4^+ concentrations in the percolation water were 0.9 to 27.5 mg L^{-1}, 0.2 to 12.2 mg L^{-1}, and 0.6 to 4.5 mg L^{-1}, respectively (J.S. Zhang et al., 2011). Total N and NO_3^- concentrations in the percolation water were significantly lower under NT with N fertilizer treatment than under CT with N fertilizer treatment, possibly because of the decreased N_2O and NH_3 volatilization in CT with N fertilizer treatment. However, the larger percolation water volume under NT with N fertilizer treatment than under CT with N fertilizer treatment did not result in any differences in N losses between NT and CT with N fertilizer treatments; this result is consistent with that observed by Mkhabela et al. (2008). Patni et al. (1998) pointed out that higher percolation water volume under NT than under CT is caused by a preferential flow resulting from macropores, reduced surface runoff resulting from high crop residue on the soil surface, and high infiltration under NT. Our results contradicted those of Kaneki et al. (2000) and Spragye and Triplett (1986).

Table 9.5 shows that the N fertilizer loss in the present study was caused mainly by NH_3 volatilization, which accounted for 64 to 74% of the total N loss, followed by N leaching and N_2O emission. Total N loss under NT with N fertilizer treatment was significantly higher than that under CT with N fertilizer treatment. This result indirectly reflects the low rice yield under NT with N fertilizer treatment. Furthermore, the result suggests that the surface application of N fertilizer

Table 9.5   Fertilizer Nitrogen (N) Loss from NT and CT with Fertilizer N Treatments (kg N ha⁻¹)

Treatment	Cumulative Emissions of N_2O		Cumulative Emissions of NH_3		Nitrogen Leaching		Total Nitrogen Loss	
	2008	2009	2008	2009	2008	2009	2008	2009
NT with fertilizer N	4.28[a]	3.40[a]	36.7[a]	27.6[a]	8.74[a]	8.74[a]	49.72[a]	39.74[a]
CT with fertilizer N	3.24[b]	2.32[b]	28.5[b]	18.2[b]	9.18[a]	8.05[b]	40.92[b]	28.57[b]

Note: NT, no-tillage; CT, conventional tillage. Different letters in a column mean significant difference at the 5% level.

under NT is ineffective in reducing N losses from N fertilizer applied in paddy fields, and methods for reducing N fertilizer loss from NT rice fields should be encouraged to improve N fertilizer use efficiency in the future.

9.5.2 Effects of Nitrogen Application Patterns on NH_3 Volatilization and Nitrogen Utilization Efficiency in No-Tillage Rice Paddy

As mentioned earlier, NT decreases global warming potentials (GWPs) and increases topsoil organic carbon sequestration but enhances N fertilizer loss. Hence, methods must be developed to reduce N fertilizer loss from NT rice fields. We conducted a field experiment to investigate the possibility of decreasing N fertilizer loss and increasing N utilization efficiency through N application patterns in NT paddy fields. The experiment was conducted at an experimental farm located in Dafashi Town, Zhonggui County, in central China during the 2012 rice growing season. For all treatments, the application rate of N fertilizer was kept constant over the rice growing season. All the phosphorus and potassium fertilizers were applied before sowing, and the N fertilizer applied ratios at the seeding, tillering, jointing, and heading stages of rice growth were different for different treatments. The nitrogen ratios of the four treatments (seedling stage:tillering stage:jointing stage:heading stages) were 2:2:3:3, 3:2:2:3, 4:2:2:2, and 4:3:1:2 respectively. Another treatment without any nitrogen was applied, and the five N fertilizer treatments were denoted as N(2:2:3:3), N(3:2:2:3), N(4:2:2:2), N(4:3:1:2), and N(0). The NH_3 volatilization fluxes increased rapidly after each N fertilizer application and peaked 1 to 3 days after N fertilizer application (Figure 9.7). This result is attributable to the hydrolysis of N fertilizer into NH_4^+ by urease, which promotes NH_3 volatilization (X.F. Hu et al., 2009). Then, the NH_3 volatilization fluxes dropped rapidly to N(0) levels 7 to 9 days after N fertilizer application; this result is consistent with those obtained by Griggs et al. (2007), S.H. Yang et al. (2012), and Zhou et al. (2011).

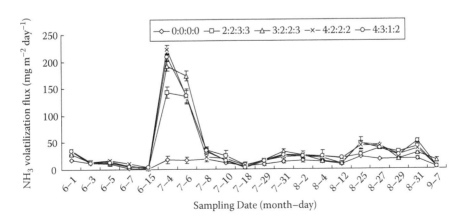

Figure 9.7   Changes in NH_3 volatilization fluxes from different treatments during the rice growing season.

Table 9.6   Changes in Cumulative Emissions of NH_3, Nitrogen Uptake of Rice, and Nitrogen Utilization Efficiency Under Different Treatments

Treatment	Cumulative NH_3 Emissions (kg N ha^{-1})	Nitrogen Uptake of Rice (kg N ha^{-1})	Nitrogen Utilization Efficiency (%)
N(0)	8.52 ± 0.20[c]	114.1 ± 5.47[e]	—
N(2:2:3:3)	19.59 ± 2.30[b]	216.0 ± 3.17[a]	57.41 ± 0.01[a]
N(3:2:2:3)	21.85 ± 0.68[a]	194.6 ± 0.49[b]	44.73 ± 0.03[b]
N(4:2:2:2)	21.98 ± 0.45[a]	180.5 ± 2.97[c]	36.94 ± 0.05[c]
N(4:3:1:2)	23.79 ± 1.15[a]	167.1 ± 5.77[d]	32.83 ± 0.06[c]

Note: Different letters in a column mean significant difference at the 5% level.

During the rice growing season, the application of N fertilizer significantly increased NH_3 volatilization (Figure 9.7 and Table 9.6). The mean NH_3 volatilization fluxes of treatments N(2:2:3:3), N(3:2:2:3), N(4:2:2:2), and N(4:3:1:2) were 13.8 ± 2.0, 15.3 ± 0.2, 15.8 ± 0.1, and 14.2 ± 0.1 mg m^{-2} day^{-1}, respectively. These values were 57.0%, 81.0%, 88.0%, and 69.0% higher than that of N(0). Compared with N(2:2:3:3), the N(3:2:2:3) and N(4:2:2:2) treatments increased NH_3 volatilization by 9.3% and 12.5%, respectively.

The cumulative emissions of NH_3 varied with different rice growing stages, and the cumulative emissions of NH_3 at the tillering stage were significantly higher than those at the seeding, jointing, and heading stages (Table 9.6). The ratios of the cumulative NH_3 emissions to the applied N fertilizer at the seeding, tillering, jointing, and full heading stages were 2.2 to 2.8%, 11.9 to 14.7%, 4.3 to 15.9%, and 4.7 to 6.6%, respectively. The ratio of the cumulative NH_3 emissions to the applied N fertilizer for the entire rice growing stage was 6.2 to 8.5%, which is similar to the result (3.7 to 11.7%) obtained by J.B. Huang et al. (2006) but is lower than the result (13.8%) reported by J. Wang et al. (2013). The difference in these results may be attributed to the different pH levels of the tested soils, N fertilizer sources, and application levels of N fertilizer. In this study, the soil pH was 5.16, which is slightly lower than that of the yellow soil (pH = 5.23) used by J.B. Huang et al. (2006) and that of the Wushan soil (pH = 7.4) in the work of J. Wang et al. (2013). These findings indicate that soil pH is one of the main factors that affect NH_3 volatilization in paddy ecosystems. The present study used chemical compound fertilizers as base fertilizers, whereas J.B. Huang et al. (2006) and J. Wang et al. (2013) used urea as the base fertilizer. The application rate of N fertilizer in the present study was 180 kg N ha^{-1}, which is lower than the N fertilizer applied by J.B. Huang et al. (2006) (100 to 350 kg N ha^{-1}) and J. Wang et al. (2013) (240 kg N ha^{-1}). The result of our study suggests that the selection of proper sources of N fertilizer and the implementation of reasonable N application levels are effective approaches to reduce N fertilizer loss in paddy fields; this result is consistent with those reported by Xiao et al. (2013) and Ye et al. (2011).

In the present study, NH_3 volatilization reached its peak at the tillering stage and gradually reduced at the heading, seeding, and jointing stages (Figure 9.7). This result is similar to those observed by J. Zhang and Wang (2007). The peak of NH_3 volatilization at the tillering stage may be attributed to the high temperature at this stage. P.P. Wu et al. (2009) suggested that urease activity increases with temperature because of the increased proportion of ammonia N in the total amount of ammonia and ammonium N in the liquid phase, thus accelerating NH_3 volatilization. Hao et al. (2012) reported that temperature is one of the important factors that affect NH_3 volatilization. In the early stage of rice growth, rice roots are limited in size and extension, which results in weak N uptake. Hence, large quantities of N fertilizer are lost through surface runoff. Although the temperature was relatively high in the jointing stage in the present study, the demand for rice nutrients was the highest at the jointing stage during the entire rice growth period (Yu et al., 2012). This condition led to the absorption of a large portion of N fertilizers for rice growth and to the subsequent reduction of NH_3 volatilization.

The respective cumulative NH_3 emissions under N(2:2:3:3), N(3:2:2:3), N(4:2:2:2), and N(4:3:1:2) were 2.3, 2.5, 2.6, and 2.8 times higher, respectively, than under N0 (Figure 9.7 and Table 9.6). The lowest cumulative emissions of NH_3 under N(2:2:3:3) may be attributed to the fact that the application of N fertilizer matched the time of rice growth for N demand. The results suggest that postponing the application of N fertilizer significantly decreases NH_3 volatilization, which is consistent with the report of Xia et al. (2010). In this study, the cumulative NH_3 emission under the fertilized NT treatments ranged from 19.6 to 23.8 kg N ha^{-1} (Table 9.6), which is 56.2 to 63.9% lower than the results (54.4 kg N ha^{-1}) reported by J. Huang (2012) and 13.8 to 29.0% lower than the results (27.6 kg N ha^{-1}) reported by H. Zhang et al. (2011). The low NH_3 volatilization found in the present study may be attributed to the difference in the soil pH. The soil pH in our study was 5.16, whereas the NT soil pH in the work of J. Huang (2012) ranged from 6.31 to 6.52 and that in the work of H. Zhang et al. (2011) was 5.39.

The respective N uptakes of rice plants under N(2:2:3:3), N(3:2:2:3), N(4:2:2:2), and N(4:3:1:2) in the present study were 89.3, 70.6, 58.2, and 46.5% higher, respectively, than under N0 (Table 9.6). N(2:2:3:3) showed the highest N utilization efficiency of rice, the value of which was 11.0, 19.7, and 29.3 higher, respectively, than under N(3:2:2:3), N(4:2:2:2), and N(4:3:1:2) (Table 9.6). These results suggest that postponing the application of N fertilizer can increase N fertilizer efficiency, as suggested by S.G. Pan et al. (2012), possibly because the N fertilizer application of N(2:2:3:3) matched the time of rice growth for N demand, thus promoting N uptakes of rice plants. L.G. Jiang et al. (2004) claimed that cumulative N uptakes at different rice growing stages increase with the application rates of N fertilizer. For example, in the early stage of rice growth, tender rice roots commonly result in weak N uptakes, which lead to increased N fertilizer loss and decreased N utilization efficiency if excess N fertilizer is applied at this stage. Therefore, postponing the application of N fertilizer could meet the N demand of rice at the late stage of rice growth and reduce N fertilizer loss through emissions of NH_3 and N_2O. Consequently, the N utilization efficiency of rice increases, which is in agreement with the results reported by P.P. Chen et al. (2011).

Although NT can decrease the GWPs of paddy fields and topsoil organic carbon sequestration (Tables 9.1 and 9.2), the surface application of N fertilizer under NT leads to the stratification of the N fertilizer. Such stratification induces N losses through N leaching and emissions of N_2O and NH_3 (Table 9.5), which in turn decreases N utilization efficiency. In the present study, postponing the application of N fertilizer (N applied at the seedling, tillering, jointing, and heading stages = 2:2:3:3) can significantly decrease NH_3 volatilization from NT rice paddies and increase N fertilizer utilization efficiency (Table 9.6), and can be used as an effective approach to enhance the N utilization efficiency of rice under NT of the middle and lower reaches of the Yangtze River.

9.6 EFFECTS OF NO-TILLAGE ON RICE GRAIN YIELDS

The history of rice cultivation in China goes as far back as over 3000 years ago, but NT has only been applied in paddy fields in China in the last 50 years (Cao and Li, 2014). In the 1960s, NT was developed as a representative of conservation tillage techniques (Cao and Li, 2014). In recent years, NT has become increasingly attractive in China because of its social, economic, and environmental benefits (M. Huang et al., 2011). We investigated the effects of NT on rice grain yields in three experimental sites (Dafashi Town, Wuxue City; Huaqiao Town, Wuxue City; and Junchuan Town, Suizhou City) (Table 9.7). Table 9.7 shows that rice yields in the experimental sites were not significantly affected by tillage practices. This result is consistent with those reported by Jat et al. (2009) (a 3-year wheat–rice rotation experiment) and Sharma (2005) (a 4-year wheat–rice rotation experiment). Similar results have also been reported by other studies (Bhushan et al., 2007; Bhattacharyya et al., 2008; Qin et al., 2010; Xu et al., 2010; Su et al., 2011). In contrast, X.J. Jiang and Xie (2009) and Mishra and Singh (2012) reported higher rice grain yields under NT than under CT because of the improvement of soil

Table 9.7 Differences in Rice Grain Yields between No-Tillage (NT) and Conventional Tillage (CT) at Different Experimental Sites

Rice Regime	Year	Experimental Site	Planting Pattern	Yield (t ha⁻¹) CT	NT
Middle rice	2008	Dafashi Town, Wuxue City	Direct seeding	7.68[a]	7.30[a]
Middle rice	2009	Dafashi Town, Wuxue City	Direct seeding	6.79[a]	7.16[a]
Middle rice	2010	Dafashi Town, Wuxue City	Direct seeding	9.10[a]	9.06[a]
Middle rice	2011	Junchuan Town, Suizhou City	Direct seeding	6.81[a]	6.55[a]
Middle rice	2012	Junchuan Town, Suizhou City	Direct seeding	10.65[a]	10.08[a]
Early rice	2011	Huaqiao Town, Wuxue City	Throwing of seedlings	5.51[a]	5.80[a]
Later rice	2011	Huaqiao Town, Wuxue City	Throwing of seedlings	7.85[a]	8.00[a]
Early rice	2012	Huaqiao Town, Wuxue City	Throwing of seedlings	6.37[a]	6.40[a]
Later rice	2012	Huaqiao Town, Wuxue City	Throwing of seedlings	7.75[a]	8.10[a]

Note: Different letters in a column mean significant difference at the 5% level.

physicochemical properties. In some cases, lower crop yields under NT than under CT were observed (Singh et al., 2001; Sharma et al., 2005; Jat et al., 2009; Kumar and Ladha, 2011). Xie et al. (2007) reviewed the effect of NT on crop yield in China and found that in most cases NT does not reduce crop yield, but in some cases it does. Xie et al. (2007) surmised that the difference in crop yields between tillage systems may depend on different climates, soil fertility, or duration of NT.

In the present study, the difference observed in crop yield was not significant during the first 5 years after conversion from CT to NT (Table 9.7). However, Basamba et al. (2006) indicated that, compared with CT, NT achieves lower maize grain yield in the first year after conversion but higher yield in the second year, and thus indicated that changes in crop yield during the initial years after the conversion from CT to NT might be a result of the differences in the nutrient immobilization/mineralization dynamics of soil. Triplett and Dick (2008) analyzed crop responses to NT in the United States and suggested that 3 years may be necessary for NT systems to become fully functional after conversion from CT. Griffith et al. (1988) reported that corn yields improve with time in continuous NT on a poorly drained soil with low organic matter. In the study, the corn yields under NT were low for the first 3 years but showed a considerable increase by the fourth year of the study.

9.7 EVALUATING THE PRACTICALITY AND EFFECTIVENESS OF NO-TILLAGE

Given the vast territory, complex climate, and cropping system diversity in the middle and lower reaches of the Yangtze River, the dominant factors that influence soil carbon, N content, and GHG emissions vary among several districts of this region. These differences present an urgent need to further investigate the dynamics of regional soil carbon and N and GHG emissions for the evaluation of the effects of NT on soil carbon and N cycles in this region.

Most rice-based cropping system types in the world can be found in the middle and lower reaches of the Yangtze River. The major Chinese rice cropping regions are also found in this area. These cropping systems include the annual rice–upland crop rotation systems in the provinces of Hunan, Hubei, and Anhui and the annual double rice cropping system in Jiangxi Province. The regional differences in natural resources and social and economic conditions restrain and regulate the choice of NT technology by local farmers. Chinese farmers commonly consider NT a lazy agricultural practice (F. Liu et al., 2010). This perspective hinders the adoption of NT by these farmers. Another constraint that affects the success of NT is the growth of weeds (Rao et al., 2007). Estimated losses from weeds in rice are around 10% or more of the total grain yield (Rao et al., 2007). H. Zhang et al. (2011) observed high fertilizer N loss under NT. Hence, low rice grain yields have been reported in China (Lu et al., 2001; Xie et al., 2007). Finally, a shortage in seeding and fertilizing equipment for NT cultivation is common in China (J.G. Liu et al., 2010) and contributes to the low efficiency of rice production. All of these problems constrain the adoption of NT by farmers and its extension in agricultural experiment stations across China.

As mentioned above, wide-scale adoption of NT by farmers faces several challenges, including weed infestation, availability of developed varieties, N availability, and lack of NT machinery. Based on our research and experience, many opportunities exist to treat these issues. Weed infestation is one of the major limiting factors impeding the success of NT through competition for nutrients, light, and resource space throughout the rice growing season. Tilling soil may help to control weeds temporarily by burying weeds to a sufficient soil depth, but doing so increases GHG emissions and GWPs compared with NT (Table 9.1). An integrated method involving NT and crop residue mulching may be useful for good weed control, as demonstrated by reduced weed density and improved or maintained rice yields. Crop residue mulching on the soil surface has been regarded as a part of an integrated weed management program under NT that works by selectively suppressing weeds (Teosdale et al., 1991). Moreover, our previous study indicated that crop residue mulching compared with no residue mulching on NT rice fields could effectively decrease net GWPs and GHG intensity, without decreasing rice grain yield (Cao and Li, 2014). In addition, the development of new and improved herbicides for NT rice is also needed.

For efficient use of N in rice production, it is important to decrease NH_3 volatilization from NT rice fields (Table 9.4). Methods of decreasing NH_3 volatilization from NT rice may include using N efficient varieties, improving timing and application methods of N fertilizer, and incorporating basal N fertilizer application without standing water in the paddy (Ali et al., 2007). In our study, split application of N fertilizer (2:2:3:3 at the seeding, tillering, jointing, and heading stages) has been reported as a good method to decrease NH_3 volatilization and improve N fertilizer use efficiency of NT rice (Table 9.6). The split application is also reported to reduce denitrification and improve plant N uptake (Farooq et al., 2011).

In the middle and lower reaches of the Yangtze River, a large proportion of farmlands are unsuitable for tractors because of either difficult terrain limiting access or small terraces (L.L. Li et al., 2011). Because the large machines developed in the United States and Canada are not directly applicable to use on small farms in China, the development of NT machinery for small-scale operations is necessary for this region. Based on our previous study, we proposed an integrated approach involving NT with ridge culture, seedling throwing, crop residue mulching, deep application of fertilizers, and moist irrigation (Figure 9.8) (Cao and Li, 2014). For this approach, the ridges (1.2 to 1.5 m in width) and furrows (0.33 m in width) keep undisturbed channels for machinery operation during crop growing seasons. Moreover, NPK fertilizers are applied to a depth of 10 cm at the seedling stage, crop residues are mulched on the soil surface of the ridges, and, finally, rice seedlings are thrown on the ridges. During rice growing seasons, moist irrigation is used. We conducted field experiments in two sites (Shayang County and Wuxue City of Hubei Province) to assess the effects of this integrated approach on GWPs and rice grain yield during the 2006 and 2007 rice growing seasons. We found that this integrated approach, compared with CT cultivation, could increase rice grain yield by 6.7% and decrease GPWs by 13% on average (Cao and Li, 2014). In the face of labor scarcity and environmental problems, the future of rice production is under threat. NT rice offers an attractive alternative. A successful transition of rice cultivation from CT to NT demands the breeding of developed varieties and development of appropriate management strategies for controlling weeds and increasing fertilizer N availability. Despite controversies, and if properly managed, NT may produce rice grain yields comparable to CT yields in our study (Cao and Li, 2014).

The popularization and application of NT involves not only the innovation of agricultural technology but also shifts in the relationships between farmers and the government. The government and the agricultural extension agencies should play positive roles in promoting the wide-scale adoption of NT by farmers through administrative, legal, and economic means. First, a system in China

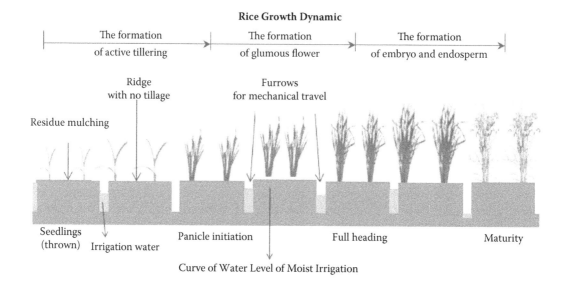

Figure 9.8 No-tillage (NT) rice with ridge culture throughout the rice growing season.

should be established to administer agricultural clean production regulations to advance the development of NT rice to the point where it becomes standardized and institutionalized. Second, the government and agencies should formulate preferential policies for NT rice and provide financial subsidies to encourage farmers to adopt NT. Finally, much publicity should be used to change the attitudes of traditional farmers toward NT. It is important to develop NT through the administrative, legal, and economic means of the government and agricultural extension agencies.

ACKNOWLEDGMENT

This work was funded by the National Natural Science Foundation of China (31100319, 31471454), National Technology Project for High Food Yield of China (2011BAD16B02), Fundamental Research Funds for the Central Universities (2013PY106), and Hubei Collaborative Innovation Center for Grain Industry (Yangtze University)..

REFERENCES

Adl, S.M., Coleman, D.C., and Read, F. 2006. Slow recovery of soil biodiversity in sandy loam soils of Georgia after 25 years of no tillage management. *Agric. Ecosyst. Environ.*, 114: 323–334.

Ahmad, S., Li, C.F., Dai, G.Z., Zhan, M., Wang, J.P., Pan, S.G., and Cao, C.G. 2009. Greenhouse gas emission from direct seeding paddy field under different rice tillage systems in central China. *Soil Tillage Res.*, 106: 54–61.

Ali, M.A., Ladha, J.K., Rickman, J., and Lales, J.S. 2007. Nitrogen dynamics in lowland rice as affected by crop establishment and nitrogen management. *J. Crop Prod.*, 20: 89–105.

Ali, M.A., Lee, C.H., Lee, Y.B., and Kim, P.J. 2009. Silicate fertilization in no-tillage rice farming for mitigation of methane emission and increasing rice productivity. *Agric. Ecosyst. Environ.*, 132: 16–22.

Aslam, T., Choudhary, M.A., and Saggar, S. 2000. Influence of land-use management on CO_2 emissions from a silt loam soil in New Zealand. *Agric. Ecosyst. Environ.*, 77: 257–262.

Attard, E., Poly, F., Commeaux, C., Laurent, F., Terada, A., Smets, B.F., Recous, S., and Roux, X.L. 2010. Shifts between *Nitrospira-* and *Nitrobacter*-like nitrite oxidizers underlie the response of soil potential nitrite oxidation to changes in tillage practices. *Environ. Microbiol.*, 12: 315–326.

Aulakh, M.S., Rennie, D.A., and Paul, E.A. 1984. Gaseous N losses from soils under zero-till as compared to conventional-till management systems. *J. Environ. Qual.*, 13: 130–136.

Bai, X.L., Xu, S.Q., Tang, W.G., Chen, F., Hu, Q., and Zhang, H.L. 2009. Ecosystem service value and carbon cycle of double cropping paddy under different tillage [in Chinese]. *J. Agro-Environ. Sci.*, 28: 2489–2494.

Bai, X.L., Zhang, H.L., Chen, F., Sun, G.F., Hu, Q., and Li, Y. 2010. Tillage effects on CH_4 and N_2O emission from double cropping paddy fields [in Chinese]. *Trans. Chin. Soc. Agric. Eng.*, 26: 282–289.

Baker, J.M., Ochsner, T.E., Venterea, R.T., and Griffis, T.J. 2007. Tillage and soil carbon sequestration—what do we really know? *Agric. Ecosyst. Environ.*, 118: 1–5.

Balesdent, J. 1996. The significance of organic separates to carbon dynamics and its modelling in some cultivated soils. *Eur. J. Soil Sci.*, 47: 485–493.

Ball, B.C., Scott, A., and Parker, J.P. 1999. Field N_2O, CO_2 and CH_4 fluxes in relation to tillage, compaction and soil quality in Scotland. *Soil Tillage Res.*, 53: 29–39.

Basamba, T.A., Barrios, E., Amézquita, E., Rao, I.M., and Singh, B.R. 2006. Tillage effects on maize yield in a Colombian savanna oxisol: soil organic matter and P fractions. *Soil Tillage Res.*, 91: 131–142.

Beheydt, D., Boeckx, P., Ahmed, H.P., and Van Cleemput, O. 2008. N_2O emission from conventional and minimum-tilled soils. *Biol. Fertil. Soils*, 44: 863–873.

Bhattacharyya, R., Kundu, S., Pandey, S.C., Singh, K.P., and Gupta, H.S. 2008. Tillage and irrigation effects on crop yields and soil properties under the rice–wheat system in the Indian Himalayas. *Agric. Water Manage.*, 95: 993–1002.

Bhushan, L., Ladha, J.K., Gupta, R.K., Singh, S., Tirolpadre, A., Saharawat, Y., Gathala, M., and Pathak, H. 2007. Saving of water and labor in a rice–wheat system with no-tillage and direct seeding technologies. *Agron. J.*, 99: 1288–1296.

Blair, G.J., Lefory, R.D.B., and Lise, L. 1995. Soil carbon fractions based on their degree of oxidation and the development of a carbon management index for agricultural systems. *Aust. J. Agric. Res.*, 46: 1459–1466.

Borga, P., Nilsson, M., and Tunlid, A. 1994. Bacterial communities in peat in relation to botanical composition as revealed by phospholipid fatty acid analysis. *Soil Biol. Biochem.*, 7: 841–848.

Bossio, D.A., Scow, K.M., Gunapala, N., and Graham, K.J. 1998. Determination of soil microbial communities: effects of agricultural management, season, and soil type on phospholipid fatty acid profiles. *Microb. Ecol.*, 36: 1–12.

Bronick C.J. and Lal, R. 2005. Soil structure and management: a review. *Geoderma*, 124: 3–22.

Cai, T.Y., Huang, Y.W., Huang, H.J., Jia, Z.K., Li, L.K., Yang, B.P., and Han, S.M. 2011. Soil labile organic carbon and carbon pool management index as affected by different years no-tilling with straw mulching [in Chinese]. *Chin. J. Ecol.*, 30: 1962–1968.

Cai, Z.C., Xing, G.X., Yan, X.Y., Xu, H., Tsuruta, H.R., Yagi, K.Y., and Minami, K.Y. 1997. Methane and nitrous oxide emissions from rice paddy fields as affected by nitrogen fertilisers and water management. *Plant Soil*, 196: 7–14.

Canfield, D.E., Glazer, A.N., and Falkowski, P.G. 2010. The evolution and future of Earth's N cycle. *Science*, 330: 192–196.

Cao, C.G. and Li, C.F. 2014. *The Theory and Practice of Low Carbon Rice Farming* [in Chinese]. Beijing: Science Press.

Carpenter-Boggs, L., Stahl, P.D., Lindstrom, M.J., and Schumache, T.E. 2003. Soil microbial properties under permanent grass, conventional tillage, and no-till management in South Dakota. *Soil Tillage Res.*, 71: 15–23.

Causarano, H.J., Franzluebbers, A.J., Reeves, D.W., and Shaw, J.N. 2006. Soil organic carbon sequestration in cotton production systems of the southeastern United States: a review. *J. Environ. Qual.*, 35: 1374–1383.

Ceja-Navarro, J.A., Rivera-Orduna, F.N., Patino-Zuniga, L., Vila-Sanjurjo, A., Crossa, J., Govaerts, B., and Dendooven, L. 2010. Phylogenetic and multivariate analyses to determine the effects of different tillage and residue management practices on soil bacterial communities. *Appl. Environ. Microbiol.*, 76: 3685–3691.

Chaparro, J.M., Sheflin, A.M., Manter, D.K., and Vivanco, J.M. 2012. Manipulating the soil microbiome to increase soil health and plant fertility. *Biol. Fertil. Soils*, 48: 489–499.

Chatskikh, D. and Olesen, J.E. 2007. Soil tillage enhanced CO_2 and N_2O emissions from loamy sand soil under spring barley. *Soil Tillage Res.*, 97: 5–18.

Chen, B. and Zhang, R.Z. 2004. Effects of no-tillage and mulch on soil microbial quantity and composition [in Chinese]. *J. Gansu Agric. Univ.*, 39: 634–638.

Chen, H.Q., Hou, R.X., Gong, Y.S., Li, H.W., Fan, M.S., and Kuzyakov, Y. 2009. Effects of 11 years of conservation tillage on soil organic matter fractions in wheat mono-culture in Loess Plateau of China. *Soil Tillage Res.*, 106: 85–94.

Chen, J.H., Liu, X.Y., Zheng, J.W., Zhang, B., Lu, H.F., Chi, Z.Z., Pan, G.X., Li, L.Q., Zheng, J.F., Zhang, X.H., Wang, J.F., and Yu, X.Y. 2013. Biochar soil amendment increased bacterial but decreased fungal gene abundance with shifts in community structure in a slightly acid rice paddy from southwest China. *Appl. Soil Ecol.*, 71: 33–44.

Chen, L.Y., Xiao, Y.H., Tang, W.B., and Lei, D.Y. 2007. Practices and prospects of super hybrid rice breeding. *Rice Sci.*, 14: 71–77.

Chen, P.P., Zhang, X.P., Wu, X.J., Yi, Z.X., and Tu, N.M. 2011. Effects of nitrogen fertilizer planning mode on yield formation and nitrogen utilization efficiency of "Luliangyou 996" [in Chinese]. *Chin. Agric. Sci. Bull.*, 27: 238–242.

Chen, S., Xia, G.M., Zhao, W.M., Wu, F.B., and Zhang, G.P. 2007. Characterization of leaf photosynthetic properties for no-tillage rice [in Chinese]. *Rice Sci.*, 14: 283–288.

Cheneby, D., Brauman, A., Rabary, B., and Philippot, L. 2009. Differential responses of nitrate reducer community size, structure, and activity to tillage systems. *Appl. Environ. Microbiol.*, 75: 3180–3318.

Christopher, S., Lal, R., and Mishra, U. 2009. Long-term no-till effects on carbon sequestration in the midwestern U.S. *Soil Sci. Soc. Am. J.*, 73: 207–216.

Chu, H.Y., Lin, X.G., Fujii, T.S., Morimoto, S., Yagi, K.Y., Hu, J.L., and Zhang, J.B. 2007. Soil microbialbiomass dehydrogenase activity, bacterial community structure in response to long-term fertilizer management. *Soil Biol. Biochem.*, 39: 2971–2976.

Culman, S.W., Snapp, S.S., Freeman, M.A., Schipanski, M.E., Beniston, J. et al. 2012. Permanganate oxidizable carbon reflects a processed soil fraction that is sensitive to management. *Soil Sci. Soc. Am. J.*, 76: 494–504.

Davidson, E.A. and Janssens, I.A. 2006. Temperature sensitivity of soil carbon decomposition and feedbacks to climate change. *Nature*, 440: 165–173.

Deen, K. and Kataki, P.K. 2003. Carbon sequestration in a long-term conventional versus conservation tillage experiment. *Soil Tillage Res.*, 74: 143–150.

Delprat, L., Chassin, P., Lineres, M., and Jambert, C. 1997. Characterization of dissolved organic carbon in cleared forest soils converted to maize cultivation. *Dev. Crop Sci.*, 25: 257–266.

Derpsch, R. and Friedrich, T. 2009. Development and current status of no-till adoption in the world. In: *Proceedings of the 18th Triennial Conference of the International Soil Tillage Research Organization (ISTRO)*, Article T1-041. ISTRO: The Netherlands.

Derpsch, R., Friedrich, T., Kassam, A., and Li, H. 2010. Current status of adoption of no-till farming in the world and some of its main benefits. *Int. J. Agric. Biol. Eng.*, 3: 1–25.

Dolan, M.S., Clapp, C.E., Allmaras, R.R., Baker, J.M., and Molina, J.A.E. 2006. Soil organic carbon and nitrogen in a Minnesota soil as related to tillage, residue and nitrogen management. *Soil Tillage Res.*, 89: 221–231.

Dou, F.G., Wright, A.L., and Hons, F.M. 2008. Sensitivity of labile soil organic carbon to tillage in wheat-based cropping systems. *Soil Sci. Soc. Am. J.*, 72: 1445–1453.

Elder, J.W. and Lal, R. 2008. Tillage effects on gaseous emissions from an intensively farmed organic soil in north central Ohio. *Soil Tillage Res.*, 98: 45–55.

Fabrizzi, K.P., Rice, C.W., Amado, T.J.C., Fiorin, J., Barbagelata, P., and Melchiori, R. 2009. Protection of soil organic C and N in temperate and tropical soils: effect of native and agroecosystems. *Biogeochemistry*, 92: 129–143.

FAO. 2014. *World Reference Base for Soil Resources: International Soil Classification System for Naming Soils and Creating Legends for Soil Maps.* Food and Agriculture Organization of the United Nations: Rome (http:://www.fao.org/3/a-i3794e.pdf).

Farooq, M., Siddique, K.H., M.Rehman, H., Aziz, T., Lee, D.J., and Wahid, A. 2011. Rice direct seeding: experiences, challenges and opportunities. *Soil Tillage Res.*, 111: 87–98.

Feiziene, D., Feiza, V., Kadziene, G., Vaideliene, A., Povilaitis, V., and Deveikyte, I. 2012. CO_2 fluxes and drivers as affected by soil type, tillage and fertilization. *Acta Agric. Scand. B*, 62: 311–328.

Filip, Z. 2002. International approach to assessing soil quality by ecologically related biological parameters. *Agric. Ecosyst. Environ.*, 88: 169–174.

Franzluebbers, A.J. and Arshad, M.A. 1997. Particulate organic carbon content and potential mineralization as affected by tillage and texture. *Soil Sci. Soc. Am. J.*, 61: 1382–1386.

Freibauer, A., Rounsevell, M.D.A., Smith, P., and Verhagen, J. 2004. Carbon sequestration in the agricultural soils of Europe. *Geoderma*, 122: 1–23.

Gajda, A.M. 2010. Microbial activity and particulate organic matter content in soils with different tillage system use. *Int. Agrophys.*, 24: 129–137.

Gao, M., Zhou, B.T., Wei, C.F., Xie, D.T., and Zhang, L. 2004. Effect of tillage system on soil animal, microorganism and enzyme activity in paddy field [in Chinese]. *Chin. J. Appl. Ecol.*, 15: 1177–1181.

Gao, Y.J., Zhu, P.L., Huang, D.M., and Wang, Z.M. 2000. Long-term impact of different soil management on organic matter and total nitrogen in rice-based cropping system [in Chinese]. *Soil Environ. Sci.*, 9: 27–30.

Ge, Y., Zhang, J.B., Zhang, L.M., Yang, M., and He, J.H. 2008. Long-term fertilization regimes affect bacterial community structure and diversity of an agricultural soil in Northern China [in Chinese]. *J. Soils Sediment.*, 8: 43–50.

Ghani, A., Dexter, M., and Perrott, W.K. 2003. Hot-water extractable carbon in soils: a sensitive measurement for determining impacts of fertilization, grazing and cultivation. *Soil Biol. Biochem.*, 35: 1231–1243.

González-Chávez, C.A., Aitkenhead-Peterson, J.A., Gentry, T.J., Zuberer, D., Hons, F., and Loeppert, R. 2010. Soil microbial community, C, N, and P responses to long-term tillage and crop rotation. *Soil Tillage Res.*, 106: 285–293.

Gouaerts, B., Mezzalama, M., and Unno, Y. 2007. Influence of tillage, residue management, and crop rotation on soil microbial biomass and catabolic diversity. *Appl. Soil Ecol.*, 37: 18–30.

Grace, P.R., Harrington, L., Jain, M.C., and Robertson, G.P. 2003. Long-term sustainability of tropical and subtropical rice–wheat system: an environmental perspective. In: Ladha, J.K., Ed., *Improving the Productivity and Sustainability of Rice–Wheat Systems: Issues and Impacts*, Special Publ. 65. Crop Science Society of America and Soil Science Society of America: Madison, WI, pp. 27–41.

Griffith, D.R., Kladivko, E.J., Mannering, J.V., West, T.D., and Parsons, S.D. 1988. Long-term tillage and rotation effects on corn growth and yield on high and low organic matter poorly drained soils. *Agron. J.*, 80: 599–605.

Griggs, B.R., Norman, Jr., R.J., Wilson, C.E., and Slaton, N.A. 2007. Ammonia volatilization and nitrogen uptake for conventional and conservation tilled dry-seeded, delayed-flood rice. *Soil Sci. Soc. Am. J.*, 71: 745–751.

Guo, L.J., Cao, C.G., Zhang, Z.S., Liu, T.Q., and Li, C.G. 2013. Short-term effects of tillage practices and wheat-straw returned to rice fields on topsoil microbial community structure and microbial diversity in central China [in Chinese]. *J. Agro-Environ. Sci.*, 32: 1577–1584.

Halvorson, A.D., Wienhold, B.J., and Black, A.L. 2002. Tillage, nitrogen, and cropping system effects on soil carbon sequestration. *Soil Sci. Soc. Am. J.*, 66: 906–912.

Hao, X.Y., Gao, W., Wang, Y.J., Jin, J.Y., Huang, S.W., Tang, J.W., and Zhang, Z.Q. 2012. Effects of combined application of organic manure and chemical fertilizers on ammonia volatilization from greenhouse vegetable soil [in Chinese]. *Sci. Agric. Sin.*, 45: 4403–4414.

Harada, H., Kobayashi, H., and Shindo, H. 2007. Reduction in greenhouse gas emissions by no-tilling rice cultivation in Hachirogata polder, northern Japan: life-cycle inventory analysis. *Soil Sci. Plant Nutr.*, 53: 668–677.

Haynes, R.J. 2005. Labile organic matter fractions as central components of the quality of agricultural soils: an overview. *Adv. Agron.*, 85: 221–268.

He, Y.Y., Zhang, H.L., Sun, G.F., Tang, W.G., Li, Y., and Chen, F. 2010. Effect of different tillage on soil organic carbon and the organic carbon storage in two-crop paddy field [in Chinese]. *J. Agro-Environ. Sci.*, 1: 200–204.

Heffer, P. 2009. *Assessment of Fertilizer Use by Crop at the Global Level: 2006/07–2007/08*. International Fertilizer Industry Association: Paris.

Helgason, B.L., Walley, F.L., and Germida, J.J. 2010. Long-term no-till management affects microbial biomass but not community composition in Canadian prairie agroecosytems. *Soil Biol. Biochem.*, 42: 2192–2202.

Hill, G.T., Mitkowski, N.A., Aldrich-Wolfe, L., Emele, L.R., Jurkonie, D.D., Ficke, A., Maldonado-Ramirez, S., Lynch, S.T., and Nelson, E.B. 2000. Methods for assessing the composition and diversity of soil microbial communities. *Appl. Soil Ecol.*, 15: 25–36.

Holland, E.A. and Coleman, D.C. 1987. Litter placement effects on microbial and organic matter dynamics in an agroecosystem. *Ecology*, 68: 425–433.

Hou, H., Zhou, S., Hosomi, M., Toyota, K., Yosimura, K., Mutou, Y., Nisimura, T., Takayanagi, M., and Motobayashi, T. 2007. Ammonia emissions from anaerobically-digested slurry and chemical fertilizer applied to flooded forage rice. *Water Air Soil Poll.*, 183: 37–48.

Hu, X.F., Wang, Z.Y., Sun, Q.Q., and You, Y. 2009. Characteristics of ammonia volatilization of slow release compound fertilizer in different pH values of purple soils [in Chinese]. *Trans. CSAE*, 25: 100–103.

Hu, Y., Jiang, X.J., Tian, B., and Li, H. 2007. Effect of ridge–no-tillage on distribution of soil fungi in different sizes of soil aggregates [in Chinese]. *Soils*, 39(6): 964–967.

Huang, J. 2012. Nitrogen Behaviors and Soil Quality Effects of No-Tillage with Rice Straw Returning, PhD dissertation, Guangxi University, Nanning.

Huang, J.B., Fan, X.H., and Zhang, S.L. 2006. Ammonia volatilization from nitrogen fertilizer in the rice field of Fe-leachi-Stagnic Anthrosols in the Taihu Lake region [in Chinese]. *Acta Pedol. Sin.*, 43: 786–792.

Huang, M., Ibrahim, M.D., Xia, B., and Zou, Y.B. 2011. Significance, progress and prospects for research in simplified cultivation technologies for rice in China. *J. Agric. Sci.*, 149: 487–496.

Huang, Y. and Sun, W.J. 2006. Farmland topsoil organic carbon content changes of the past 20 years in China [in Chinese]. *Chin. Sci. Bull.*, 51: 750–763.

ICAM. 2012. The demand and supply, price changes of domestic rice in China [in Chinese]. *World Agric.*, 140.

IPCC. 2007. *Climate Change 2007: Climate Change Impacts, Adaptation and Vulnerability*. Geneva: Intergovernmental Panel on Climate Change.

IRRI. 2004. *Rice Statistical Database*. Los Banõs, Philippines: International Rice Research Institute.

Ishii, S., Keda, S., Minamisawa, K., and Senoo, K. 2011. Nitrogen cycling in rice paddy environments: past achievements and future challenges. *Microbes Environ.*, 26: 282–292.

Jacobs, A., Rauber, R., and Ludwing, B. 2009. Impact of reduced tillage on carbon and nitrogen storage of two Haplic Luvisols after 40 years. *Soil Tillage Res.*, 102: 158–164.

Jat, M.L., Gathala, M.K., Ladha, J.K., Saharawat, Y.S., Jat, A.S., Kumar, V., Sharma, S.K., Kumar, V., and Gupta, R.K. 2009. Evaluation of precision land leveling and double zero-tillage systems in the rice-wheat rotation: water use, productivity, profitability and soil physical properties. *Soil Tillage Res.*, 105: 112–121.

Jiang, L.G., Cao, W.X., Gan, X.Q., Wei, S.Q., Xu, J.Y., Dong, D.F., Chen, N.P., Lu, F.Y., and Qin, H.D. 2004. Nitrogen uptake and utilization under different nitrogen management and influence on grain yield and quality in rice [in Chinese]. *Sci. Agric. Sin.*, 37: 490–496.

Jiang, X., Wright, A.L., Wang, X., and Liang, F. 2011. Tillage-induced changes in fungal and bacterial biomass associated with soil aggregates: a long-term field study in a subtropical rice soil in China. *Appl. Soil Ecol.*, 48: 168–173.

Jiang, X.J. and Xie, D.T. 2009. Combining ridge with no-tillage in lowland rice-based cropping system: long-term effect on soil and rice yield. *Pedosphere*, 19: 515–522.

Kaneki, R., Kyuma, K., Inagaki, C., Hiromichi, D., and Miki, S. 2000. Reduction of surface and subsurface effluent loads by no-puddling cultivation in combination with a single application of coated fertilizer in the nursery box. *Jpn. J. Soil Sci. Plant Nutr.*, 71: 502–511.

Kang, X., Huang, J., Jiang, J.C., Lan, L.B., Lu, J.Z., and Liang, H. 2010. Effects of no-tillage and straw covering for sweet potato on paddy soil carbon pool and quantity of soil microorganism [in Chinese]. *Guangxi Agric. Sci.*, 41: 236–239.

Kay, B.D. and VandenBygaart, A.J. 2002. Conservation tillage and depth stratification of porosity and soil organic matter. *Soil Tillage Res.*, 66: 107–118.

Khan, S.A., Mulvaney, R.L., Ellsworth, T.R., and Boast, C.W. 2007. The myth of nitrogen fertilization for soil carbon sequestration. *J. Environ. Qual.*, 36: 1821–1832.

Kennedy, A.C. and Smith, K.L. 1995. Soil microbial diversity and the sustainability of agricultural soils. *Plant Soil*, 170: 75–86.

Kierkegaard, J.A., Stawar, M.P., Wong, T.W., Mead, A., Howe, G., and Newell, M. 2000. Field studies on the biofumigation of take-all by *Brassica* break crops. *Aust. J. Agric. Res.*, 51: 445–456.

Ko, J.Y. and Kang, H.W. 2000. The effects of cultural practices on methane emission from rice fields. *Nutr. Cycl. Agroecosyst.*, 58: 311–314.

Kumar, V. and Ladha, J.K. 2011. Direct-seeding of rice: recent developments and future research needs. *Adv. Agron.*, 111: 297–413.

Kyaw, K.M., Toyota, K., Okazaki, M., Motobayashi, T., and Tanaka, H. 2005. Nitrogen balance in a paddy field planted with whole crop rice (*Oryza sativa* cv. Kusahonami) during two rice-growing seasons. *Biol. Fertil. Soils*, 42: 72–82.

Lal, R. 2004. Soil carbon sequestration impacts on global climate change and food security. *Science*, 304: 1623–1627.

Lashof, D.A. and Ahuja, D.R. 1990. Relative contributions of greenhouse gas emissions to global warming. *Nature*, 344: 529–531.

Leggett, J.A. 2012. China's greenhouse gas emissions and mitigation policies. *Environ. Res. J.*, 6: 451–471.

Li, C.F., Kou, Z.K., Yang, J.H., Cai, M.L., Wang, J.P., and Cao, C.G. 2010. Soil CO_2 fluxes from direct seeding rice fields under two tillage practices in central China. *Atmos. Environ.*, 44: 2696–2704.

Li, C.F., Zhou, D.N., Kou, Z.K., Zhang, Z.S., Wang, J.P., Cai, M.L., and Cao, C.G. 2012. Effect of tillage and N fertilizers on CH_4 and CO_2 emissions and soil organic carbon in paddy fields of central China. *PLoS ONE*, 7: e34642.

Li, C.F., Zhang, Z.S., Guo, L.J., Cai, M.L., and Cao, C.G. 2013. Emissions of CH_4 and CO_2 from double rice cropping systems under varying tillage and seeding methods. *Atmos. Environ.*, 80: 438–444.

Li, D.M., Liu, M.Q., Cheng, Y.H., Wang, D., Qin, J.T., Jiao, J.G., Li, H.X., and Hu, F. 2011. Methane emissions from double-rice cropping system under conventional and no tillage in southeast China. *Soil Tillage Res.*, 113: 77–81.

Li, F.P., Liang, W.J., Zhang, X.O., Jiang, Y., and Wang, J.K. 2008. Changes in soil microbial biomass and bacterial community in a long-term fertilization experiment during the growth of maize. *Adv. Environ. Biol.*, 2: 1–8.

Li, J., Zhao, B.Q., Li, X.Y., Jiang, R.B., and Bing, S.H. 2008. Effects of long-term combined application of organic and mineral fertilizer on microbial biomass, soil enzyme activities and soil fertility [in Chinese]. *Agric. Sci. China*, 7: 336–343.

Li, J.T. and Zhang, B. 2007. Paddy soil stability and mechanical properties as affected by long-term application of chemical fertilizer and animal manure in subtropical China. *Pedosphere*, 17: 568–579.

Li, K.R., Wang, S.Q., and Cao, M.K. 2003. Vegetation and soil carbon storage in China [in Chinese]. *Sci. China (Ser. D)*, 33: 72–80.

Li, L.L., Huang, G.B., Zhang, R.Z., Bellotti, B., Li, G., and Kwong, Y.C. 2011. Benefits of conservation agriculture on soil and water conservation and its progress in China. *Agric. Sci. China*, 10: 850–859.

Liang, W., Shi, Y., Zhang, H., Yue, J., and Huang, G.H. 2007. Greenhouse gas emissions from northeast China rice fields in fallow season. *Pedosphere*, 17: 630–638.

Liu, F., Lei, H.X., Wang, Y., Zhou, G.S., and Wu, J.S. 2010. Development and adoption status of no-tillage in China [in Chinese]. *Hubei Agric. Sci.*, 49: 2557–2562.

Liu, J.G., You, L.Z., Amini, M., Obersteiner, M., Herrero, M., Zehnder, A.J.B., and Yang, H. 2010. A high-resolution assessment on global nitrogen flows in cropland. *PNAS*, 107: 8035–8040.

Liu, M., Yu, W.T., Jiang, Z.H., and Ma, Q. 2006. A research review on soil active organic carbon [in Chinese]. *Chin. J. Ecol.*, 25: 1412–1417.

Liu, M., Ekschmitt, K., Zhang, B., Holzhauer, S.I.J., Li, Z.P., Zhang, T.L., and Rauch, S. 2011. Effect of intensive inorganic fertilizer application on microbial properties in a paddy soil of subtropical China [in Chinese]. *Agric. Sci. China*, 10: 1758–1764.

Liu, M.Q., Hu, F., Chen, X.Y., Huang, Q.R., Jiao, J.G., Zhang, B., and Li, H.X. 2009. Organic amendments with reduced chemical fertilizer promote soil microbial development and nutrient availability in a subtropical paddy field: the influence of quantity, type and application time of organic amendments. *Appl. Soil Ecol.*, 42: 166–175.

Liu, S.W., Zhang, L., Jiang, J.Y., Chen, N.N., Yang, X.M., Xiong, Z.Q., and Zou, J.W. 2012. Methane and nitrous oxide emissions from rice seedling nurseries under flooding and moist irrigation regimes in Southeast China. *Sci. Total Environ.*, 426: 166–171.

Liu, X.J., Mosier, A.R., Halvorson, A.D., and Zhang, F.S. 2006. The impact of nitrogen placement and tillage on NO, N_2O, CH_4 and CO_2 fluxes from a clay loam soil. *Plant Soil*, 280: 177–188.

López-Bellido, R.J., Fontán, J.M., López-Bellido, F.J., and Lopez-Bellido, L. 2010. Carbon sequestration by tillage, rotation, and nitrogen fertilization in a Mediterranean Vertisol. *Agron. J.*, 102: 310–318.

Lou, Y.L., Xu, M.G., Wang, W., Sun, X.L., and Zhao, K. 2011. Return rate of straw residue affects soil organic C sequestration by chemical fertilization. *Soil Tillage Res.*, 113: 70–73.

Lu, F., Wang, X.K., Han, B., Ouyang, Z.Y., Duan, X.N., Zheng, H., and Miao, H. 2009. Soil carbon sequestrations by nitrogen fertilizer application, straw return and no-tillage in China's cropland. *Global Change Biol.*, 15: 281–305.

Ma, E.D., Ji, Y., Ma, J., Xu, H., and Cai, Z.-C. 2010. Effects of soil tillage and rice cultivation pattern on methane emission from paddy fields [in Chinese]. *J. Ecol. Rural Environ.*, 26: 513–518.

Ma, Y.C., Wang, J.Y., Zhou, W., Yan, X.Y., and Xiong, Z.Q. 2012. Greenhouse gas emissions during the seedling stage of rice agriculture as affected by cultivar type and crop density. *Biol. Fertil. Soils*, 48: 589–595.

Malhi, S.S., Lemke, R., Wang, Z.H., and Chhabra, B.S. 2006. Tillage, nitrogen and crop residue effects on crop yield, nutrient uptake, soil quality, and greenhouse gas emissions. *Soil Tillage Res.*, 90: 171–183.

Marschner, P., Kandeler, E., and Marschner, B. 2003. Structure and function of the soil microbial community in a long-term fertilizer experiment. *Soil Biol. Biochem.*, 35: 453–461.

Mathew, R.P., Feng, Y.C., Githinji, L., Ankumah, R.A., and Balkcom, K.S. 2012. Impact of no-tillage and conventional tillage systems on soil microbial communities. *Appl. Environ. Soil Sci.*, 2012: 1–10.

Melero, S., López-Garrido, R., and Murillo, J.M., and Moreno, F. 2009. Conservation tillage: short-and long-term effects on soil carbon fractions and enzymatic activities under Mediterranean conditions. *Soil Tillage Res.*, 104: 292–298.

Ministry of Agriculture of the People's Republic of China. 2012a. *China Rural Statistical Yearbook*. Beijing: China Statistics Press.

Ministry of Agriculture of the People's Republic of China. 2012b. *Hubei Rural Statistical Yearbook*. Beijing: China Statistics Press.

Mishra, J.S. and Singh, V.P. 2012. Tillage and weed control effects on productivity of a dry seeded rice–wheat system on a Vertisol in Central India. *Soil Tillage Res.*, 123: 11–20.

Mishra, U., Lal, R., Slater, B., Calhoun, F., Liu, D.S., and Van Meirvenne, M. 2009. Predicting soil organic carbon stock using profile depth distribution functions and ordinary kriging. *Soil Sci. Soc. Am. J.*, 73: 614–621.

Mkhabela, M.S., Madani, A., Gordon, R., Burton, D., Cudmore, D., Elmi, A., and Hart, W. 2008. Gaseous and leaching nitrogen losses from no-tillage and conventional tillage systems following surface application of cattle manure. *Soil Tillage Res.*, 98: 187–199.

Nayak, P., Patel, D., Ramakrishnan, B., Mishra, A.K., and Samantaray, R.N. 2009. Long-term application effects of chemical fertilizer and compost on soil carbon under intensive rice–rice cultivation. *Nutr. Cycl. Agroecosyst.*, 83: 259–269.

Noll, M., Matthies, D., Frenzel, P., Derakshani, M., and Liesack, W. 2005. Succession of bacterial community structure and diversity in a paddy soil oxygen gradient. *Environ. Microbiol.*, 7: 382–395.

Nyamadzawo, G., Nyamangara, J., Nyamugafata, P., and Muzulu, A. 2009. Soil microbial biomass and mineralization of aggregate protected carbon in fallow-maize systems under conventional and no-tillage in Central Zimbabwe. *Soil Tillage Res.*, 102: 151–157.

Ogram, A. 2000. Soil molecular microbial ecology at age 20 methodological challenges for the future. *Soil Biol. Biochem.*, 32: 1499–1504.

Oorts, K., Garnier, P., Findeling, A., Mary, B., Richard, G., and Nicolardot, B. 2007a. Modeling soil carbon and nitrogen dynamics in no-till and conventional tillage using PASTIS model. *Soil Sci. Soc. Am. J.*, 71: 336–346.

Oorts, K., Merckx, R., Grehan, E., Labreuche, J., and Nicolardot, B. 2007b. Determinants of annual fluxes of CO_2 and N_2O in long-term no-tillage and conventional tillage systems in northern France. *Soil Tillage Res.*, 95: 133–148.

Pan, G.X., Li, L.Q., Wu, L.S., and Zhang, X.H. 2004. Storage and sequestration potential of topsoil organic carbon in China's paddy soils. *Global Change Biol.*, 10: 79–92.

Pan, S.G., Huang, S.Q., Zhai, J., Cai, M.G., Cao, C.G., Zhan, M., and Tang, X.G. 2012. Effects of nitrogen rate and its basal to dressing ratio on uptake, translocation of nitrogen and yield in rice [in Chinese]. *Soils*, 44: 23–29.

Pandey, D., Agrawal, M., and Bohra, J.S. 2012. Greenhouse gas emissions from rice crops with different tillage permutations in rice–wheat system. *Agric. Ecosyst. Environ.*, 159: 133–144.

Parton, W.J., Schimel, D.S., Cole, C.V., and Ojima, D.S. 1987. Analysis of factors controlling soil organic matter levels in Great Plains grasslands. *Soil Sci. Soc. Am. J.*, 51: 1173–1179.

Parton, W.J., Scurlock, J.M.O., Ojima, D.S., Ggilmanov, T., Scholes, R.J., Schimel, D.S., Kirchner, T., Menaut, J.C., Seastedt, T., Moya, E.G., Kamnalrut, A., and Kinyamario, J.L. 1993. Observations and modeling of biomass and soil organic matter dynamics for the grasslands biome worldwide. *Global Biogeochem. Cycles*, 7: 785–809.

Pathak, H., Sankhyan, S., Dubey, D.S., Bhatia, A., and Jain, N. 2013. Dry direct-seeding of rice for mitigating greenhouse gas emission: field experimentation and simulation. *Paddy Water Environ.*, 11: 593–601.

Patni, N.K., Masse, L., and Jui, P.Y. 1998. Groundwater quality under conventional and no-tillage. I. Nitrate, electrical conductivity, and pH. *J. Environ. Qual.*, 27: 869–877.

Paustian, K., Six, J., Elliott, E.T., and Hunt, H.W. 2000. Management options for reducing CO_2 emissions from agricultural soils. *Biogeochemistry*, 48: 147–163.

Pendell, D.L., Williams, J.R., Boyles, S.B., Rice, C.W., and Nelson, R.G. 2007. Soil carbon sequestration strategies with alternative tillage and nitrogen sources under risk. *Appl. Econ. Perspect. Policy*, 29: 247–268.

Peng, S.B., Huang, J.L., Zhong, X.H., Yang, J.C., Wang, G.H. et al. 2002a. Challenge and opportunity in improving fertilizer–nitrogen use efficiency of irrigated rice in China [in Chinese]. *Sci. Agric. Sin.*, 1: 776–785.

Peng, S.B., Huang, J.L., Zhong, X.H., Yang, J.C., Wang, G.H. et al. 2002b. Research strategy in improving fertilizer–nitrogen use efficiency of irrigated rice in China [in Chinese]. *Sci. Agric. Sin.*, 35: 1095–1103.

Puget, P. and Lal, R. 2005. Soil organic carbon and nitrogen in a Mollisol in central Ohio as affected by tillage and land use. *Soil Tillage Res.*, 80: 201–213.

Qin, J., Hu, F., Li, D., Li, H., Lu, J., and Yu, R. 2010. The effect of mulching, tillage and rotation on yield in non-flooded compared to flooded rice production. *J. Agron. Crop Sci.*, 196: 397–406.

Qiu, J. 2009. Nitrogen fertilizer warning for China. *Nature*, 10: 103–105.

Quincke, J.A., Wortmann, C.S., Mamo, M., Franti, T., and Drijber, R.A. 2007. Occasional tillage of no-till systems. *Agron. J.*, 99: 1158–1168.

Ramsey, P.W., Rilling, M.C., Feris, K.P., Holben, W.E., and Gannon, J.E. 2006. Choice of methods for soil microbial community analysis: PLFA maximizes power compared to CLPP and PCR-based approaches. *Pedobiologia*, 50: 275–280.

Rao, A.N., Johnson, D.E., Sivaprasad, B., Ladha, J.K., and Mortimer, A.M. 2007. Weed management in direct-seeded rice. *Adv. Agron.*, 93: 153–255.

Reicosky, D.C., Dugas, W.A., and Torbert, H.A. 1997. Tillage-induced carbon dioxide loss from different cropping systems. *Soil Tillage Res.*, 41: 105–118.

Reicosky, D.C., Reeves, D.W., Prior, S.A., Runion, G.B., Rogers, H.H., and Raper, R.L. 1999. Effects of residue management and controlled traffic on carbon dioxide and water loss. *Soil Tillage Res.*, 52: 153–165.

Reiner, W. and Milkha, S.A. 2000. The role of rice plants in regulating mechanisms of methane missions. *Biol. Fertil. Soils*, 31: 20–29.

Rochette, P., Angers, D.A., Chantigny, M.H., MacDonald, J.D., Bissonnette, N., and Bertrand, N. 2009. Ammonia volatilization following surface application of urea to tilled and no-till soils: a laboratory comparison. *Soil Tillage Res.*, 103: 310–315.

Roper, M.M., Gupta, V.V.S.R., and Murphy, D.V. 2010. Tillage practices altered labile soil organic carbon and microbial function without affecting crop yields. *Aust. J. Soil Res.*, 48: 274–285.

Rudrappa, L., Purakayastha, T.J., Singh, D., and Bhadraray, S. 2006. Long-term manuring and fertilization effects on soil organic carbon pools in a Typic Haplustept of semi-arid sub-tropical India. *Soil Tillage Res.*, 88: 180–192.

Saetre, P. and Baath, E. 2000. Spatial variation and patterns of soil microbial community structure in a mixed spruce-birch stand. *Soil Biol. Biochem.*, 32: 909–917.

Shang, Q.Y., Yang, X.X., Gao, C.M., Wu, P.P., Liu, J.J., Xu, Y.C., Shen, Q.R., Zou, J.W., and Guo, S.W. 2011. Net annual global warming potential and greenhouse gas intensity in Chinese double rice-cropping systems: a 3-year field measurement in long-term fertilizer experiments. *Global Change Biol.*, 17: 2196–2210.

Shao, J.A., Li, Y.B., Wei, C.F., and Xie, D.T. 2009. Effects of land management practices on labile organic carbon fractions in rice cultivation. *Chin. Geograph. Sci.*, 19: 241–248.

Sharma, P., Tripathi, R.P., and Singh, S. 2005. Tillage effects on soil physical properties and performance of rice–wheat-cropping system under shallow water table conditions of Tarai, Northern India. *Eur. J. Agron.*, 23: 327–335.

Shi, S.W., Li, Y., Liu, Y.T., Wan, Y.F., Gao, Q.Z., and Zhang, Z.X. 2010. CH_4 and N_2O emissions from rice fields and mitigation options based on field measurements in China: an integration analysis [in Chinese]. *Sci. Agric. Sin.*, 43: 2923–2936.

Shine, K.P., Derwent, R.G., Wuebbles, D.J., and Morcrette, J.J. 1995. Radiative forcing of climate. In: Houghton, J.T., Ed., *Climate Change: The IPCC Scientific Assessment.* Cambridge University Press: Cambridge, pp. 47–68.

Singh, S., Sharma, S.N., and Prasad, R. 2001. The effect of seeding and tillage methods on productivity of rice–wheat cropping system. *Soil Tillage Res.*, 61: 125–131.

Singh, S.K., Venkatesh, B., Thakur, T.C., Pachauri, S.P., Singh, P.P., and Mishra, A.K. 2009. Influence of crop establishment methods on methane emission from rice fields. *Curr. Sci.*, 97: 84–89.

Six, J., Elliot, E.T., and Paustian, K. 1999. Aggregate and soil organic matter dynamics under conventional and no-tillage systems. *Soil Sci. Soc. Am. J.*, 63: 1350–1358.

Six, J., Elliott, E.T., and Paustian, K. 2000a. Soil macroaggregate turnover and microaggregate formation: a mechanism for C sequestration under no-tillage agriculture. *Soil Biol. Biochem.*, 32: 2099–2103.

Six, J., Paustian, K., Elliott, E.T., and Combrink, C. 2000b. Soil structure and organic matter. I. Distribution of aggregate-size classes and aggregate-associated carbon. *Soil Sci. Soc. Am. J.*, 64: 681–689.

Six, J., Conant, R.T., Paul, E.A., and Paustian, K. 2002a. Stabilization mechanisms of soil organic matter: implications for C-saturation of soils. *Plant Soil*, 241: 155–176.

Six, J., Feller, C., Denef, K., Ogle, S.M., Sa, J.C.D., and Albrech, A. 2002b. Soil organic matter, biota and aggregation in temperate and tropical soils: effects of no-tillage. *Agronomie*, 22: 755–775.

Six, J., Ogle, S.M., Breidt, F.J., Conant, R., Mosier, A.R., and Paustian, K. 2004. The potential to mitigate global warming with no-tillage management is only realized when practised in the long term. *Global Change Biol.*, 10: 155–160.

Six, J., Frey, S.D., Thiet, R.K., and Batten, K.M. 2006. Bacterial and fungal contributions to carbon sequestration in agroecosystems. *Soil Sci. Soc. Am. J.*, 70: 555–569.

Smith, P., Martino, D., Cai, Z., Gwary, D., Janzen, H., Kumar, P., McCarl, B., Ogle, S., O'Mara, F., Rice, C., Scholes, B., and Sirotenko, O. 2007. Agriculture. In: Metz, B., Davidson, O.R., Bosch, P.R., Dave, R., and Meyer, L.A., Eds., *Climate Change 2007: Mitigation.* Cambridge University Press: Cambridge.

Song, G.H., Li, L.Q., Pan, G.X., and Zhang, Q. 2005. Topsoil organic carbon storage of China and its loss by cultivation. *Biogeochemistry*, 74: 47–62.

Sotomayor–Ramírez, D., Espinoza, Y., and Rámos–Santana, R. 2006. Short-term tillage practices on soil organic matter pools in a tropical Ultisol. *Aust. J. Soil Res.*, 44: 687–693.

Spedding, T.A., Hamel, C., Mehuys, G.R., and Madramootoo, C.A. 2004. Soil microbial dynamics in maize-growing soil under different tillage and residue management systems. *Soil Biol. Biochem.*, 36: 499–512.

Spragye, M.A. and Triplett, G.B. 1986. *No-Tillage and Surface-Tillage Agriculture: The Tillage Revolution.* Wiley Interscience: New York, pp. 467.

Su, W., Lu, J.W., Zhou, G.S., Li, X.K., Han, Z.H., and Lei, H.X. 2011. Effect of no-tillage and direct sowing density on growth, nutrient uptake and yield of rapeseed (*Brassica napus* L.) [in Chinese]. *Sci. Agric. Sin.*, 44: 1519–1526.

Sundermeier, A.P., Islam, K.R., Raut, Y., Reeder, R.C., and Dick, W.A. 2011. Continuous no-till impacts on soil biophysical carbon sequestration. *Soil Sci. Soc. Am. J.*, 75: 1779–1788.

Tang, K. and Zhang, C.E. 1996. Research on minimum tillage, no-tillage and mulching systems and its effects in China. *Theor. Appl. Climatol.*, 54: 61–67.

Tang, X.H., Shao, J.G., Gao, M., Wei, C.F., Xie, D.T., and Pan, G.X. 2007. Effects of conservational tillage on aggregate composition and organic carbon storage in purple paddy soil [in Chinese]. *Chin. J. Appl. Ecol.*, 18: 1027–1032.

Teosdale, J.R., Beste, C.E., and Potts, W.E. 1991. Response of weeds to tillage and cover crop residues. *Weed Sci.*, 39: 95–199.

Tian, G.M., Cao, J.L., Cai, Z.C., and Ren, L.T. 1998. Ammonia volatilization from winter wheat field top-dress with urea. *Pedosphere*, 8: 331–336.

Tian, K., Zhao, Y.C., Xing, Z., Sun, W.X., Huang, B.A., and Hu, W.Y. 2013. A meta-analysis of long-term experiment data for characterizing the topsoil organic carbon changes under different conservation tillage in cropland of China [in Chinese]. *Acta Pedol. Sin.*, 50: 433–440.

Tilman, D., Cassman, K.G., Matson, P.A., Naylor, R., and Poolasky, S. 2002. Agricultural sustainability and intensive production practices. *Nature*, 418: 671–677.

Tirol-Padre, A. and Ladha, J.K. 2004. Assessing the reliability of permanganate-oxidizable carbon as an index of soil labile carbon. *Soil Sci. Soc. Am. J.*, 68: 969–978.

Triplett, Jr., G.B. and Dick, W.A. 2008. No-tillage crop production: a revolution in agriculture! *Agron. J.*, 100: S-153–S-165.

Tscherko, D., Hammesfahr, U., Zeltner, G., Kandeler, E., and Böcker, R. 2005. Plant succession and rhizosphere microbial communities in a recently deglaciated alpine terrain. *Basic Appl. Ecol.*, 6: 367–383.

Ussiri, D.A.N., Lal, R., and Jarecki, M.K. 2009. Nitrous oxide and methane emissions from long-term tillage under a continuous corn cropping system in Ohio. *Soil Tillage Res.*, 104: 247–255.

Väisänen, R.K., Robers, M.S., Garland, J.L., Frey, S.D., and Dawson, L.A. 2005. Physiological and molecular characterization of microbial communities associated with different water-stable aggregate size classes. *Soil Biol. Biochem.*, 37: 2007–2016.

van Gestel, M., Merckx, R., and Vlassek, K. 1996. Spatial distribution of microorganisms to soil drying. *Soil Biol. Biochem.*, 28: 503–510.

Vargas, G.S., Becker, A., Oddino, C., Zuza, M., Marinelli, A., and March, G. 2009. Field trial assessment of biological, chemical, and physical responses of soil to tillage intensity, fertilization and grazing. *Environ. Manage.*, 44: 378–386.

Wang, C.J., Pan, G.X., Tian, Y.G., Li, L.Q., Zhang, X.H., and Han, X.J. 2010. Changes in cropland topsoil organic carbon with different fertilizations under long-term agro-ecosystem experiments across mainland China. *Sci. China Life Sci.*, 53: 858–867.

Wang, J., Wang, D.J., Zhang, G., and Wang, Y. 2013. Comparing the ammonia volatilization characteristic of two typical paddy soil with total wheat straw returning in Taihu Lake region [in Chinese]. *Environ. Sci.*, 34: 27–33.

Wang, M.X. and Li, J. 2002. CH_4 emission and oxidation in Chinese rice paddies. *Nutr. Cycl. Agroecosyst.*, 64: 43–55.

Wang, Y.F. and Chen, Z.Z. 1998. Distribution of soil organic carbon in the major grasslands of Xilinguole, inner Mongolia, China. *Acta Phytoecol. Sin.*, 22: 545–551.

West, T.O. and Marland, G. 2002. A synthesis of carbon sequestration, carbon emissions, and net carbon flux in agriculture: comparing tillage practices in the United States. *Agric. Ecosyst. Environ.*, 91: 217–232.

Wright, A.L. and Hons, F.M. 2005a. Carbon and nitrogen sequestration and soil aggregation under sorghum cropping sequences. *Biol. Fertil. Soils*, 41: 95–100.

Wright, A.L. and Hons, F.M. 2005b. Soil carbon and nitrogen storage in aggregates from different tillage and crop regimes. *Soil Sci. Soc. Am. J.*, 69: 141–147.

Wright, A.L., Hons, F.M., Lemon, R.G., McFarland, M.L., and Nichols, R.L. 2008. Microbial activity and soil C sequestration for reduced and conventional tillage cotton. *Appl. Soil Ecol.*, 38: 168–173.

Wu, F.L., Li, L., Zhang, H.L., and Chen, F. 2007. Effects of conservation tillage on net carbon flux from farmland ecosystems [in Chinese]. *Chin. J. Ecol.*, 26: 2035–2039.

Wu, F.L., Zhang, H.L., Li, L., Chen, F., Huang, F.Q., and Xiao, X.P. 2008. Characteristics of CH_4 emission and greenhouse effects in double paddy soil with conservation tillage [in Chinese]. *Sci. Agric. Sin.*, 41: 2703–2709.

Wu, P.P., Liu, J.J., Yang, X.X., Shang, Q.Y., Zhou, Y. et al. 2009. Effects of different fertilization systems on ammonia volatilization from double-rice cropping field in red soil region [in Chinese]. *Chin. J. Rice Sci.*, 23: 85–93.

Wu, X.Y., Liao, H.P., and Yang, W. 2013. Effect of different tillage systems on distribution of organic carbon and readily oxidation carbon in purple paddy soil [in Chinese]. *J. Agric. Mech. Res.*, 35: 184–188.

Xia, W.J., Zhou, W., Liang, G.Q., Wang, X.B., Sun, J.W., Li, S.L., and Hu, C. 2010. Effect of optimized nitrogen application on ammonia volatilization from paddy field under wheat–rice rotation system [in Chinese]. *Plant Nutr. Fertil. Sci.*, 16: 6–13.

Xiao, X., Yang, L.L., Deng, Y.P., and Wang, J.F. 2012. Effects of irrigation and nitrogen fertilization on ammonia volatilization in paddy field [in Chinese]. *J. Agro-Environ. Sci.*, 31: 2066–2071.

Xiao, X.P., Wu, F.L., Huang, F.Q., Li, Y., Sun, G.F., Hu, Q., He, Y.Y., Chen, F., and Yang, G.L. 2007. Greenhouse air emission under different pattern of rice–straw returned to field in double rice area [in Chinese]. *Res. Agric. Modern.*, 28: 629–632.

Xie, R.Z., Li, S.K., Li, X.J., Jin, Y.Z., Wang, K.R., Chu, Z.D., and Gao, S.J. 2007. The analysis of conservation tillage in China—conservation tillage and crop production: reviewing the evidence [in Chinese]. *Sci. Agric. Sin.*, 40: 1914–1924.

Xiong, H.Y., Li, T.X., and Yu, H.Y. 2008. Amount of soil microorganism and influencing factor of different no-tillage years in "paddy-upland" rotation systems [in Chinese]. *J. Wuhan Univ.*, 54: 244–248.

Xu, X.Y. and Liu, J. 1999. Simulation of organic matter loss in the area around Taihu Lake [in Chinese]. *J. Lake Sci.*, 11: 81–85.

Xu, Y.Z., Nie, L.X., Buresh, R.J., Huang, J.L., Cui, K.H., Xu, B., Gong, W.H., and Peng, S.B. 2010. Agronomic performance of late-season rice under different tillage, straw, and nitrogen management. *Field Crops Res.*, 115: 79–84.

Xue, J.F., Zhao, X., Dikgwatlhe, S.B., Chen, F., and Zhang, H.L. 2013. Advances in effects of conservation tillage on soil organic carbon and nitrogen [in Chinese]. *Acta Ecol. Sin.*, 33: 6006–6013.

Yan, C.R., Liu, E.K., He, W.Q., Liu, S.A., and Liu, Q. 2010. Effect of different tillage on soil organic carbon and its fractions in the Loess Plateau of China [in Chinese]. *Soil Fertil. Sci. China*, 6: 58–63.

Yan, L.D., Deng, Y.J., and Qu, Z.G. 2010. Research on development of low-carbon agriculture from the angle of ecology [in Chinese]. *China J. Pop. Res. Environ.*, 20: 40–45.

Yan, X.Y., Akiyama, H.K., Yagi, K.Y., and Akimoto, H.J. 2009. Global estimations of the inventory and mitigation potential of methane emissions from rice cultivation conducted using the 2006 Intergovernmental Panel on Climate Change Guidelines. *Global Biogeochem. Cycles*, 23: GB2002.

Yang, C.M., Yang, L.H., and Zhu, O.Y. 2005. Organic carbon and its fractions in paddy soil as affected by different nutrient and water regimes. *Geoderma*, 124: 133–142.

Yang, S.H., Peng, S.Z., Xu, J.Z., Yao, J.Q., Jin, X.P., and Song, J. 2012. Characteristics and simulation of ammonia volatilization from paddy fields under different water and nitrogen management [in Chinese]. *Trans. CSAE*, 28: 99–104.

Yao, X., Liao, D.X., Tang, Y.Q., Jiang, G., Zhang, X.W., Li, J.Y., and Wang, L.C. 2011. Research progress of ecological efficiency and developing strategies for paddy-field conservation tillage [in Chinese]. *Ecol. Environ. Sci.*, 20: 372–378.

Ye, S.C., Lin, Z.C., Dai, Q.G., Jia, Y.S., Gu, H.Y. et al. 2011. Effects of nitrogen application rate on ammonia volatilization and nitrogen utilization in rice growing season [in Chinese]. *Chin. J. Rice Sci.*, 25: 71–78.

Young, I.M. and Ritz, K. 2000. Tillage, habitat space and function of soil microbes. *Soil Tillage Res.*, 53: 201–213.

Young, I.M., Crawford, J.W., Nunan, N., Otten, W., and Spiers, A. 2008. Microbial distribution in soils: physics and scaling. *Adv. Agron.*, 100: 81–121.

Yu, Q.G., Ye, J., Yang, S.N., Fu, J.R., Ma, J.W., Sun, W.C., Jiang, L.N., and Wang, Q. 2012. Effects of different nitrogen application levels on rice nutrient uptake and ammonia volatilization [in Chinese]. *Chin. J. Rice Sci.*, 26: 487–494.

Zak, D.R., Ringelberg, D.B., Pregitzer, K.S., Randlett, D.L., White, D.C., and Curtis, P.S. 1996. Soil microbial communities beneath *Populus grandidentata* grown under elevated atmospheric CO_2. *Ecol. Appl.*, 6: 257–262.

Zhan, M., Wang, J.P., Yue, L.X., Jiang, Y., Yu, J., and Pan, S.G. 2009. Effects of short-term no-tillage on soil carbon pool in paddy fields [in Chinese]. *Hubei Agric. Sci.*, 4: 834–837.

Zhan, M., Cao, C.G., Jiang, Y., Wang, J.P., Yue, L.X., and Cai, M.L. 2010. Dynamics of active organic carbon in a paddy soil under different rice farming modes [in Chinese]. *Chin. J. Appl. Ecol.*, 21: 2010–2016.

Zhang, F.S., Wang, J.Q., Zhang, W.F., Cui, Z.L., Ma, W.Q., Chen, X.P., and Jiang, R.F. 2008. Nutrient use efficiency of major cereal crops in China and measures for improvement [in Chinese]. *Acta Pedol. Sin.*, 45: 915–924.

Zhang, H., Yang, Z.G., Luo, L.G., Zhang, Q.W., Yi, J. et al. 2011. Study on the ammonia volatilization from paddy field in irrigation area of the Yellow River [in Chinese]. *Plant Nutr. Fertil. Sci.*, 17: 1131–1139.

Zhang, H.L., Bai, X.L., Xue, J.F., Chen, Z.D., Tang, H.M., and Chen, F. 2013. Emissions of CH_4 and N_2O under different tillage systems from double-cropped paddy fields in southern China. *PLoS ONE*, 8: e65277.

Zhang, J. and Wang, D.J. 2007. Ammonia volatilization in gleyed paddy field soils of Taihu Lake region [in Chinese]. *Chin. J. Eco-Agric.*, 15(6): 84–87.

Zhang, J.B., Song, C.C., and Yang, W.Y. 2005. Influence of land-use type on soil dissolved organic carbon in the Sanjiang Plain. *China Environ. Sci.*, 25: 343–347.

Zhang, J.K., Jiang, C.S., Hao, Q.J., Tang, Q.W., Cheng, B.H., Li, H., and Chen, L.H. 2012. Effects of tillage-cropping systems on methane and nitrous oxide emissions from agro-ecosystems in a purple paddy soil [in Chinese]. *Environ. Sci.*, 33: 1979–1986.

Zhang, J.S., Zhang, F.P., Yang, J.H., Wang, J.P., Cai, M.L., Li, C.F., and Cao, C.G. 2011. Emissions of N_2O and NH_3, and nitrogen leaching from direct seeded rice under different tillage practices in central China. *Agric. Ecosyst. Environ.*, 140: 164–173.

Zhang, L., Xiao, J.Y., Xie, D.T., and Wei, C.F. 2002. Study on microbial characteristics in paddy soil under long-term no-tillage and ridge culture. *J. Soil Water Conserv.*, 16: 111–114.

Zhang, X.X., Yin, S., Li, Y.S., Zhuang, H.L., Li, C.S., and Liu, C.J. 2014. Comparison of greenhouse gas emissions from rice paddy fields under different nitrogen fertilization loads in Chongming Island, Eastern China. *Sci. Total Environ.*, 472: 381–388.

Zhao, J.S., Zhang, X.D., Yuan, X., and Wang, J. 2003. Characteristics and environmental significance of soil dissolved organic matter [in Chinese]. *Chin. J. Appl. Ecol.*, 14: 126–130.

Zheng, X.H., Han, S.H., Huang, Y., Wang, Y.S., and Wang, M.X. 2004. Requantifying the emission factors based on field measurements and estimating the direct N_2O emission from Chinese croplands. *Global Biogeochem. Cycles*, 18: GB2018.

Zhou, W., Tian, Y.H., and Yin, B. 2011. Ammonia volatilization and nitrogen balance after topdressing fertilization in paddy fields of Taihu Lake region [in Chinese]. *Chin. J. Eco-Agric.*, 19: 32–36.

Zou, J.W., Huang, Y., Jiang, J.Y., Zheng, X.H., and Sass, R.L. 2005. A 3-year field measurement of methane and nitrous oxide emissions from rice paddies in China: effects of water regime, crop residue, and fertilizer application. *Global Biogeochem. Cycles*, 19: GB2021.

Zou, J.W., Huang, Y., Zheng, X.H., and Wang, Y.E. 2007. Quantifying direct N_2O emissions in paddy fields during rice growing season in mainland China: dependence on water regime. *Atmos. Environ.*, 41: 8030–8042.

CHAPTER 10

Framework for Conversion
Ecological Mechanisms and Regulation of Greenhouse Organic Vegetable Production in North China

Li Ji, Xu Ting, Hui Han, Ding Guoying, Li Yufei,
Wang Xi, Yang Hefa, and Li Shengnan

CONTENTS

10.1 INTRODUCTION TO PRODUCTION AND RESEARCH ON ORGANIC VEGETABLES

10.1.1 Introduction to Organic Farming

Organic farming originated in Europe and America in the early 20th century and has become an important alternative form of agriculture. The total area of organically managed farmland grew rapidly at a rate of 10.4% per annum worldwide from 1999 to 2012. By the end of 2012, approximately 2.3% of arable land in Europe had been converted to organic farmland, accounting for 30% of the total organic farmland worldwide (Willer and Lernoud, 2014).

The total area of certified organic farmland reached 37.5 million hectares worldwide in 2012; however, only 245,000 hectares (about 3.3% of total worldwide organic farmland) were used for vegetable production, representing 1% of the total worldwide area used for horticulture. Nevertheless, the area of organic vegetable production still increased from 106,000 hectares in 2004 to 245,000 hectares in 2012, as shown in Figure 10.1 (Willer and Lernoud, 2014).

In 2012, the top 10 leading countries with the largest area of organic vegetable production were the United States, Mexico, Italy, France, United Kingdom, Germany, Spain, Poland, Ukraine, and the Netherlands, as shown in Figure 10.2 (Willer and Lernoud, 2014). Interestingly, Austria, Switzerland, Germany, and the United Kingdom, also major producers of organic food in Europe, had the highest proportion of organic farmland for vegetable production in comparison to their total arable land.

Organic farming in China started as late as the early 1990s; however, the total area of organic farmland quickly reached 2 million hectares in 2012, representing 0.3% of the total arable land in China. By June 2012, 23 organic certification agencies had been established, issuing 10,478 certificates for 7266 companies in China. Both the Chinese domestic and international markets for organic products have grown swiftly, reaching 12.8 billion and 400 million dollars, respectively, in 2012 (Organic Food Development and Certification Center, 2013).

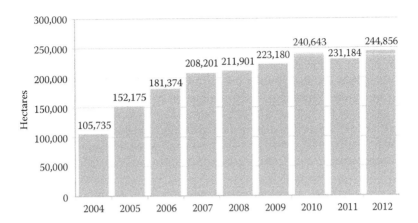

Figure 10.1   Growth of organic vegetable land worldwide (2004 to 2012). (Data from Willer, H. and Lernoud, J., Eds., *The World of Organic Agriculture: Statistics and Emerging Trends 2014*, FiBL and IFOAM, Bonn, 2014.)

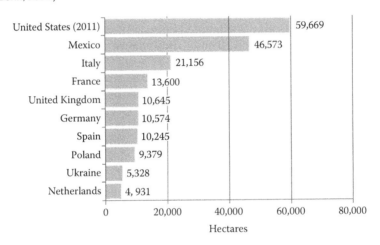

Figure 10.2   Ten leading countries with the largest organic vegetable growing area in 2012. (Data from Willer, H. and Lernoud, J., Eds., *The World of Organic Agriculture: Statistics and Emerging Trends 2014*, FiBL and IFOAM, Bonn, 2014.)

Twenty years of development in organic farming in China have resulted in expanded markets for diverse organic products and numerous indigenous organic farms, as well as recognition of their importance by local governments. Currently, organic farming in China is in a transition from scattered small businesses to an industry that will not heavily rely on international markets but instead focus on Chinese domestic needs. Thus, the domestic market in China for organic food, especially for vegetables, is predicted to increase dramatically over the next 5 to 10 years.

10.1.2 Organic Farming Research

The strongest and most active organic farming research has been conducted in Europe. Approximately 100 different projects have been sponsored since 2000, targeting such concerns in organic farming systems as selecting proper plants, suppressing phytopathogens or weeds, improving rotation and tillage systems, optimizing soil management and maintaining soil fertility, and the health benefits of organic products. The International Centre for Research in Organic Food Systems (ICROFS) established a project titled "Organic Cropping Systems for Vegetable Production," which compared the effects of conventional or organic cropping systems on the quality of products, as well as broader environmental impacts. A couple of practices were proven effective in suppressing pests and maintaining soil fertility. To assess and reduce the risks of pathogen contamination in organic vegetables, the Coordination of European Transnational Research in Organic Food and Farming Systems (CORE) sponsored a project named "PathOrganic: Assessing and Reducing Risks of Pathogen Contamination in Organic Vegetables," which studied the effects of environmental factors, plant genotypes, fertilizer application techniques, and soil buffering on the spread and persistence of pathogens in organic vegetable products. Beyond these advancements described above, several other achievements on breeding, innovative bioregulation or rotation systems, increasing soil fertility, and reducing the emission of greenhouse gases were presented in detail at the 16th International Federation of Organic Agriculture Movements (IFOAM) meeting in Moderna, Italy, in 2008.

Organic farming research in China has suffered from a lack of financial support since its late birth. Approximately 85% of related papers recorded in the China National Knowledge Infrastructure only reviewed the progress on techniques for organic vegetable production. Reports on innovations for fertilizer application, rotation, and pest management are very rare. The Beijing Agriculture Commission and Beijing Municipal Science and Technology Commission established projects in 2003 and 2004, respectively, both of which aimed to study techniques used in suppressing agricultural pests in organic farming systems. Except for these government grants, a few organic-food-export-oriented enterprises (e.g., Taian Taishan Asia Food Co., Ltd.) also supported research on improving techniques for growing organic vegetables.

Despite this limited support, some achievements have still come about. In the study by Lu et al. (2005), the content of heavy metals in soil was found to be lower on organic farms located in the middle of eastern China than in their corresponding surroundings. It also suggested that compost contaminated with heavy metals could be a potential threat to organic products. From 2006 to 2007, scientists at the Qingdao Academy of Agricultural Science studied different vegetable varieties and rotation systems in order to improve the adoption rate of organic farming systems there. Notably, researchers from China Agricultural University and Nanjing Agricultural University have established long-term experiments to optimize systems that could be used to sustainably produce organic vegetables. The effects of their cropping systems on the environment were also comprehensively evaluated. Despite the advancements made on techniques, systematic research on organic vegetable production is urgently needed to innovate and improve current organic farming systems for a growing organic industry in China.

10.2 LONG-TERM EXPERIMENT ON AN ORGANIC VEGETABLE PRODUCTION SYSTEM IN QUZHOU

Instability is one of the characteristics of an agricultural ecosystem. As agricultural environmental conditions change constantly, the developing rules and trends of agriculture are generally difficult to observe during a short-term field study. However, the long-term field study adopted as an effective research method in agricultural production worldwide could overcome those shortcomings (Zhao, 2012). Many researchers have paid great attention to long-term field studies, and the data from these studies have been employed to solve some agricultural problems both theoretically and practically. The long-term experiment reported here is focused on organic vegetables and was started in June 2002 at the Agricultural Experimental Station (36°52′N, 115°01′E) located at Quzhou, Hebei, China. The average annual precipitation in this area is about 600 mm

Table 10.1   Content of Soil Nutrients before Experiment

Type of Greenhouse	Soil Depth (cm)	Organic Matter (g kg⁻¹)	Total N (g kg⁻¹)	Total P (g kg⁻¹)	Available K (mg kg⁻¹)	Alkaline Hydrolytic Nitrogen (mg kg⁻¹)	Available Phosphorus (mg kg⁻¹)
CON	0–20	18.93	1.36	2.22	212.83	128.38	163.05
	20–40	8.75	0.74	1.08	135.28	47.66	48.75
LOW	0–20	15.25	1.19	1.24	364.28	95.35	81.68
	20–40	7.13	0.68	0.79	131.18	34.95	39.42
ORG	0–20	16.63	1.17	1.38	257.30	101.28	139.13
	20–40	9.60	0.77	1.04	129.30	40.43	33.03

Note: CON, conventional farming; LOW, low-input farming; and ORG, organic farming.

Table 10.2   Average Amount of Nutrient Inputs in Each Type of Greenhouse (2002–2012)

Nutrient	ORG	LOW	CON
N (kg ha⁻²)	934	757	836
P_2O_5 (kg ha⁻²)	424	316	306
K_2O (kg ha⁻²)	1378	943	912
Animal dung and compost (t ha⁻²)	60	30	16

Note: ORG, organic farming; LOW, low-input farming; CON, conventional farming.

with a primary peak in July and August that accounts for 60% of the rainfall for the entire year. The maximum yearly evaporation was recorded as 1841 mm. Spring is the most serious drought season for the year. The climate in this area is warm and semi-humid with abundant sunlight and heat. The temperature ranges from –2.9°C in January to 26.8°C in July, with an average of 13°C. The annual frost-free period is around 200 days. Based on the climate, it is possible to grow two crops a year.

The experiment was conducted in three side-by-side greenhouses consisting of three different systems: organic farming (ORG), low-input farming (LOW), and conventional farming (CON). The greenhouses were built up in a typical format with a semi-round arch that is widely used in North China. Each one is 52 m long east–west and 7 m wide south–north, with an area of 0.04 ha. The soil nutrient contents in samples collected in March 2002 were measured (Table 10.1).

The three farming systems differed mainly in fertilizer application and pest control. The organic system was conducted in accordance with IFOAM basic standards by using only organic fertilizers and biological control for pests. The low-input system was conducted with lower inputs of agrochemicals and organic fertilizer. The conventional system was conducted according to the local greenhouse vegetable growing style. In the low-input and conventional systems, the most used pesticides were lambda-cyhalothrin, imidacloprid, cartap, dimethomorph, and carbendazim. The average amounts of fertilizer nutrients used in each greenhouse were recorded from 2002 to 2012 (Table 10.2). Details of the cropping patterns are shown in Table 10.3. The rotation system mainly consists of tomatoes or eggplants in March and cucumbers in September. The frequency of flood irrigation was adapted to crops and planting seasons for all farming systems. The effects of the different farming systems on vegetable growth, yields and fruit quality, key soil nutrient elements, soil organisms, severity of important plant diseases, and nitrogen balance were studied systematically from 2002 to 2012. One-way ANOVA in conjunction with the Duncan test was used to test for significant difference (adjusted $p < 0.05$) among the farming systems.

Table 10.3 Cropping Patterns in Greenhouse Experiment (2002–2012)

Crop Code	Crop	Growing Period	Crop Code	Crop	Growing Period
1	Cucumber	9/1/2002–12/9/2002	11	Cucumber	9/21/2007–1/22/2008
2	Tomato	3/8/2003–7/8/2003	12	Tomato	3/14/2008–6/25/2008
3	Celery	10/15/2003–1/18/2004	13	Fennel	10/21/2008–1/7/2009
4	Tomato	2/5/2004–6/23/2004	14	Tomato	3/5/2009–6/27/2009
5	Cucumber	8/25/2004–12/5/2004	15	Fennel	10/8/2010–1/11/2011
6	Tomato	3/1/2005–6/20/2005	16	Eggplant	2/21/2011– 7/16/2011
7	Cucumber	9/5/2005–12/26/2005	17	Fennel	11/18/2011–3/8/2012
8	Tomato	3/2/2006–6/20/2006	18	Tomato	3/28/2012–7/14/2012
9	Celery	11/23/2006–2/12/2007	19	Cucumber	8/26/2012–12/11/2012
10	Tomato	3/21/2007–6/30/2007			

10.3 PLANT GROWTH, YIELD, AND QUALITY OF ORGANIC VEGETABLES

In recent years, many studies have been carried out worldwide comparing the yield and quality of vegetables in organic and conventional systems. Martini et al. (2004) showed that tomato yields were, surprisingly, 150% higher in the organic system than in the conventional treatment, while the yield of organic corn production was lower than the yields in the conventional system. X.J. Zhu et al. (2010) compared the effects of organic, integrated, and conventional farming systems on kidney bean production. No significant difference in yield was observed among these three systems, despite the yield in the organic system being slightly higher than in the two other systems.

Vegetable quality could be affected by several factors, including variety, growth period, temperature, light, fertilizer, and cropping system. In Shen et al. (2010), the quality of celery, broccoli, radishes, and lettuce in organic farming systems was better than in the conventional farming system: Higher contents of vitamin C, dry matter, and soluble sugar and lower contents of soluble protein and nitrate were found in the organic farming system. Cai et al. (2008) found similar results by comparing the quality of cucumbers in organic, integrated, and conventional farming systems. Hoefkens et al. (2009) found that the vitamin C concentration of tomato was significantly higher in the organic system than in the conventional system, whereas the opposite result was obtained for carrots. Significantly higher concentrations of beta-carotene were observed in the carrots, tomatoes, and spinach in the organic system when compared to the conventional system. Worthington (2001) showed that organic crops contained significantly higher concentrations of vitamin C, iron, magnesium, and phosphorus and significantly lower nitrate concentrations than in conventional crops.

In conclusion, the influence of organic and conventional systems on yield depended on the crop species. Generally, crop yields in organic systems were lower than in conventional systems during the transition, but they tended to be higher after a few years of organic fertilizer application. Organic systems are superior to conventional systems in terms of vegetable quality, including higher vitamin C content and lower nitrate content. It is possible that organic fertilizer contains many types of nutrients and trace elements and that organic fertilizer in general is stable enough to provide nutrients to crops for quite a long time.

In our experiment, vegetable growth, yields, and characteristics of fruit quality (vitamin C, soluble sugar, soluble solids, and nitrate content) were measured for tomatoes, celery, and eggplants from 2002 to 2012 in order to evaluate the influence of different farming systems on the quality and yield of products.

10.3.1 Methods

Ten plants were randomly selected from each plot in each greenhouse to measure plant height. Two rows of plants from each plot were selected to measure yield, which was used to estimate the gross yield in each greenhouse. Around 1 kg of randomly selected fruit was used to measure vitamin C content, soluble sugar, and soluble substances in tomatoes or nitrate content in celery and cucumbers. Plant height was recorded at different time points during each growing season. Levels of vitamin C, soluble sugar, soluble substances, and nitrates were determined via liquid chromatography, anthrone colorimetry, refractometry, and cadmium reduction methods, respectively.

10.3.2 Results and Discussion

10.3.2.1 Plant Height

Heights were recorded of tomatoes grown in 2003, 2004, 2005, 2006, and 2012. In general, average tomato height was higher in the organic farming system than in the other two systems (Figure 10.3). A significant difference between organic and conventional farming systems was only observed in 2012 (Figure 10.3). In 2006, the average height in the organic system was significantly higher than in the low-input system (Figure 10.3).

10.3.2.2 Vegetable Yield

Tomato yields from 2003 to 2008 and in 2012; cucumber yields in 2002, 2004, 2005, and 2007; fennel yields in 2008; and celery yields in 2003 and 2006 were compared among three different farming systems. As shown in Tables 10.4 and 10.5 (W.J. Liang et al., 2009), tomato yields in 2005 and 2006 in the organic farming system were significantly higher than those in the conventional farming system. The same result was found for cucumbers in 2005 and celery in 2003 and 2006. In all other years besides 2007, tomato yields in the organic farming system were also the highest, but no significant difference was found among the three systems. The lowest cucumber yields were observed in the organic farming system and the highest yields in the low-input system (Table 10.5).

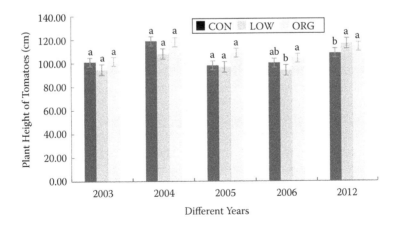

Figure 10.3   Plant height of tomatoes in different farming systems from 2003 to 2012. CON, conventional farming; LOW, low-input farming; ORG, organic farming.

Table 10.4 Yield of Tomatoes Grown in Spring in Different Farming Systems (2003–2008)

Year	CON (kg fresh weight ha⁻¹)	LOW (kg fresh weight ha⁻¹)	ORG (kg fresh weight ha⁻¹)
2003	81,361.7[a]	78,672.8[a]	88,479.8[a]
2004	83,020.8[a]	80,124.7[a]	88,881.3[a]
2005	60,273.4[b]	58,315.3[b]	68,506.9[a]
2006	45,985.6[b]	52,070.0[a]	53,203.7[a]
2007	86,325.9[a]	82,697.6[a]	86,161.1[a]
2008	85,728.7[a]	85,576.1[a]	86,398.0[a]

Source: Liang, L.N. et al., *Trans. CSAE*, 25(8), 186–191, 2009.

Note: CON, conventional farming; LOW, low-input farming; ORG, organic farming. Same letter in a row indicates no significant difference ($p < 0.05$), and different letters indicate significant difference ($p < 0.05$).

Table 10.5 Yield of Vegetables Grown in Autumn in Different Farming Systems (2002–2008)

Vegetable	Year	CON (kg fresh weight ha⁻¹)	LOW (kg fresh weight ha⁻¹)	ORG (kg fresh weight ha⁻¹)
Cucumber	2002	7812.0[b]	9100.0[a]	3584.0[c]
Cucumber	2004	17,258.9[a]	18,431.2[a]	16547.0[a]
Cucumber	2005	25,134.3[b]	28,384.6[a]	27,862.2[a]
Cucumber	2007	29,631.6[b]	32,608.6[a]	31,533.8[ab]
Fennel	2008	43,535.5[a]	44,002.2[a]	46,602.3[a]
Celery	2003	30,315.0[b]	29,085.0[c]	32,148.0[a]
Celery	2006	83,983.2[b]	85,508.0[b]	89,620.6[a]

Source: Liang, L.N. et al., *Trans. CSAE*, 25(8), 186–191, 2009.

Note: CON, conventional farming; LOW, low-input farming; ORG, organic farming. Same letter in a row indicates no significant difference ($p < 0.05$), and different letters indicate significant difference ($p < 0.05$).

In the present study, organic fertilizer was applied at a rate of 60 tons per hectare in the organic farming system—higher than European input levels but still comparable to those applied in organic systems in South Korea and Japan. In our study, different crop yields in organic farming systems were higher than those in the other two farming systems for most of the observed years. No significant decrease in crop yield was detected during the transition period in which low yields were frequently reported in other studies; thus, we assumed that high organic fertilizer inputs could still supply enough nutrients for plant growth. However, we hypothesized that a combination of organic and inorganic NPK fertilizer would be more efficient than the application of only organic NPK fertilizer, in terms of increasing crop yield (C.M. Ren et al., 2005; Y. Liu et al., 2008). In general, our results are consistent with this hypothesis, as the yield in the low-input system in which both organic and inorganic NPK fertilizers were applied was similar to yields in the organic farming system in which the total input of NPK was over 23% higher. However, recent results tend to reflect a significantly higher yield in the organic farming system than yields in the low-input system. These results suggested that the effect of such organic and inorganic fertilizer combinations on increasing crop yield was less sustainable than that of long-term organic fertilizer application.

10.3.2.3 Quality of Harvested Vegetables

In the present study, the contents of vitamin C, soluble sugars, soluble substances, and nitrates in different harvested vegetables were used as indicators to evaluate their quality. The harvested parts with high levels of vitamin C, soluble sugars, and soluble substances and low nitrate levels

Table 10.6 Vitamin C Content of Tomatoes in Different Farming Systems (2003–2005)

Year	CON (mg/kg)	LOW (mg/kg)	ORG (mg/kg)
2003	166.0[b]	167.0[b]	185.8[a]
2004	107.0[b]	109.0[b]	122.0[a]
2005	93.4[c]	128.0[a]	106.0[b]

Note: CON, conventional farming; LOW, low-input farming; ORG, organic farming. Same letter in a row indicates no significant difference ($p < 0.05$), and different letters indicate significant difference ($p < 0.05$).

Table 10.7 Soluble Sugar Content of Tomatoes in Different Farming Systems (2003–2005)

Year	CON (% w/w)	LOW (% w/w)	ORG (% w/w)
2003	3.26[a]	2.86[a]	3.72[a]
2004	5.07[b]	6.08[a]	5.36[ab]
2005	3.50[ab]	4.00[a]	2.90[b]

Note: CON, conventional farming; LOW, low-input farming; ORG, organic farming. Same letter in a row indicates no significant difference ($p < 0.05$), and different letters indicate significant difference ($p < 0.05$).

are regarded as high-quality vegetable products. To study the effect of different farming systems on the quality of vegetable products, the vitamin C content, soluble sugars, and soluble substances in tomatoes from 2003 to 2005 and 2012; the nitrate content in celery from 2003; and the nitrate content in cucumbers from 2012 were compared.

Vitamin C from daily edible vegetables is indispensable to humans and other animals (X.R. Liu et al., 2003). Our results suggested that the influence of farming systems on vitamin C levels in fruits is quite complicated (Table 10.6) (Xie et al., 2007). The content of vitamin C in the organic farming system was 11.9% higher in 2003 than that in the conventional system, and in 2004 the levels in the organic farming and low-input farming systems were significantly higher than those in the conventional systems. The highest levels in 2005 were detected in the low-input system, which were 2.08% and 37.04% higher than those in the organic and conventional farming systems, respectively. In 2012, the highest levels were again observed in the low-input system, which were 39.8% and 50.2% higher than levels in the organic and conventional, respectively. It seems that procedures in organic and low-input systems might help to improve the vitamin C content in tomatoes. For each farming system, dramatic differences among fruits from different years were also observed.

Soluble sugars are one of the important factors affecting the taste of fruits. The influence of farming systems on the content of soluble sugar in tomatoes in 2003, 2004, 2005, and 2012 was also investigated (Table 10.7) (Xie et al., 2007). The effect of organic farming systems varied among different years. In 2003, the content in the organic farming system was 30% and 14% higher than those in the low-input and conventional farming systems, respectively, but the difference was not significant. In 2012, a significant difference in soluble sugar content was found among the different farming systems, with the highest levels being found in the organic farming system, followed by the low-input and conventional systems. However, in 2005 the lowest content was found in the organic system. For each farming system, different soluble sugar content levels were also detected among samples collected from different years.

Soluble solids in fruit—consisting mainly of soluble sugars, acids, vitamins, and minerals—are indicators of the maturity and quality of many fruits. As shown in Table 10.8, the levels of soluble solids in tomatoes were comparable among different farming systems in 2003 and 2004 (Xie et al., 2007). However, in 2005 the organic farm system had a significantly lower soluble solid content compared to other two systems. In contrast, the highest levels were detected in 2012 in the organic

Table 10.8   Soluble Solids Content of Tomatoes in Different Cropping Systems (2003–2005)

Year	CON (% w/w)	LOW (% w/w)	ORG (% w/w)
2003	5.07[a]	5.07[a]	5.27[a]
2004	4.80[a]	5.60[a]	5.20[a]
2005	5.40[a]	5.40[a]	4.40[b]

Note: CON, conventional farming; LOW, low-input farming; ORG, organic farming. Same letter in a row indicates no significant difference ($p < 0.05$), and different letters indicate significant difference ($p < 0.05$).

farming system. Compared to vitamin C and soluble sugars, the variation of soluble solids levels among different years was smaller for all farming systems. Vegetables are the main source of dietary nitrates, which pose a potential threat to human health, and the content of nitrates is regarded as an important indicator for vegetable quality (J.H. Wang et al., 2004). The nitrate content in celery and cucumbers from 2003 and 2012 was compared among the different farming systems. In both years, the content of nitrates was significantly higher in the conventional systems than in the other two. No significant difference was detected between organic and low-input farming systems. In general, the organic and low-input systems mostly showed higher improvement in terms of vegetable quality and yield when compared with conventional system.

10.4 SOIL PHYSICAL AND CHEMICAL PROPERTIES IN ORGANIC VEGETABLE SYSTEMS

In terms of the influence of organic farming on soil physical and chemical properties, most researchers agree that organic farming is better than conventional farming in improving soil nutrients and soil quality. A study on the effects of onion–lettuce organic and conventional production systems on the soil environment was carried out in field trials (Xi et al., 1999). The results showed that soil organic matter and microbe levels in organic production systems were higher than those in conventional production systems. Cai et al. (2008) compared soil nutrients among organic, integrated, and conventional culture systems; the results showed that the soil total nitrogen in the organic culture was higher than that found in the integrated and conventional cultures, with an increase of 8.78% compared with levels before the study. Another study on the effects of traditional farming and organic farming on vegetable productivity, soil conditions, and soil environment has been carried out since the spring of 1992 at the Gosford Institute of Horticulture located in New South Wales, Australia. (Ryan et al., 2004). The results showed that total nitrogen, total phosphorus, water-holding capacity, microbial biomass, and aggregate stability of soils in the organic farming system were higher than those in the conventional farming system. The available phosphorus increased in both systems. This indicated that organic agriculture could improve soil health. The effects of different greenhouse farming systems on soil physical and chemical properties were evaluated for the long-term experiment reported in this chapter, which started in 2002 at the Quzhou Agricultural Experimental Station in Handan, Hebei Province.

10.4.1 Methods

Each greenhouse space was divided into three plots. Five soil cores were collected from the surface layer (0 to 20 cm) on an "S" curve of each plot and then mixed. The soil in the deeper layer (20 to 40 cm) was collected in the same way. The soil samples were air dried, sieved (0.25 mm and 1 mm), and then stored at room temperature until chemical analysis. The methods of chemical property analysis are listed as follows: total organic matter (TOM) by dichromate oxidation; total organic nitrogen (TON) by semi-micro Kjeldahl; total phosphorus (TOP) by $HClO_4$–H_2SO_4 digestion and molybdenum–antimony anti-colorimetric method; alkaline hydrolysis by nitrogen-diffusion method; available P by the Olsen method; and available K by NH_4OAc flame photometer.

10.4.2 Results and Discussion

10.4.2.1 Soil Bulk Density Dynamics in Three Different Farming Systems

As shown in Figure 10.4 (A. Zhang et al., 2013), in the 0- to 20-cm soil layer the order of soil bulk density from low to high was organic < low-input < conventional, with significant differences observed. The soil bulk densities of the conventional and low-input systems were 44.2% and 22.6% higher than the organic system, respectively. In the 20- to 40-cm layer, the highest soil bulk density was shown in the conventional system, and no significant difference was observed between the organic system and the low-input system. The bulk densities of soil in the 0- to 20-cm layer in the conventional, low-input, and organic systems were 10.1%, 21.7%, and 1.47% higher, respectively, than those in the 20- to 40-cm layer. In the organic system, no significant difference was observed between the 0- to 20-cm and 20- to 40-cm layers. Soil bulk density in the organic system was less than that of the conventional system.

10.4.2.2 Dynamic of Soil Nitrogen in Three Different Farming Systems

As shown in Figure 10.5 (A. Zhang et al., 2013), a significant difference in soil total nitrogen was observed among the conventional, low-input, and organic systems in 2012. In the 0- to 20-cm layer, the soil total nitrogen in the conventional, low-input, and organic systems in 2012 increased 23.2%, 78.7%, and 166.9%, respectively, compared to 2002 measurements. The soil total nitrogen levels of the low-input and organic systems were 29.5% and 88.1% higher, respectively, than in the conventional system. In the 20- to 40-cm layer, the soil total nitrogen of the conventional, low-input, and organic systems in 2012 increased 8.69%, 58.20%, and 94.44%, respectively, compared to 2002 measurements. The soil total nitrogen levels of the low-input and organic systems were 27.2% and 68.0% higher, respectively, than in the conventional system.

In both soil layers, there was a significant difference in levels of alkaline hydrolysis nitrogen among the three different systems in 2012. In the 0- to 20-cm layer, alkaline hydrolysis nitrogen in the conventional system in 2012 decreased 6.47%, while levels in the low-input and organic systems increased 80.52% and 108.40%, respectively, compared to those in 2002. In the 20- to 40-cm layer,

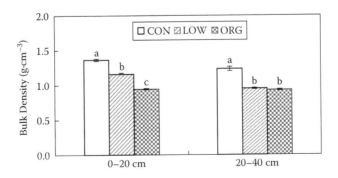

Figure 10.4 Dynamic of soil bulk density in different farming systems. (From Zhang, A. et al., *Jiangsu J. Agric. Sci.*, 29(6), 1345–1351, 2013.)

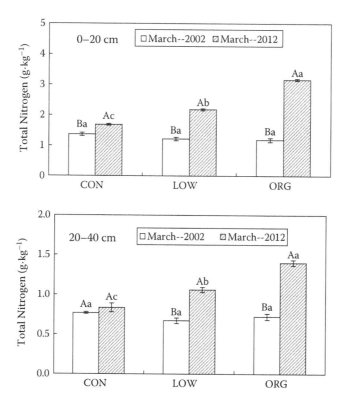

Figure 10.5 Dynamic of soil total nitrogen in different farming systems. (From Zhang, A. et al., *Jiangsu J. Agric. Sci.*, 29(6), 1345–1351, 2013.)

alkaline hydrolysis nitrogen in the conventional system in 2012 had decreased 7.33%, while levels in the low-input and organic systems increased 88.27% and 169.82%, respectively, compared to those in 2002. Most alkaline hydrolysis nitrogen in the conventional system was from chemical fertilizer. The possible reason is that the samples were collected in March 2012, a term between the harvest of the previous crop and the planting of the next crop; there were no fertilizer inputs, and little alkaline hydrolysis nitrogen remained in the soil of the conventional system during that time. When a large amount of organic fertilizer is added to the low-input and organic systems, mineralization of organic matter may increase the alkaline hydrolysis nitrogen and provide enough nutrients during the later period of crop growth. In the conventional system, a topdressing fertilizer applied later in the season may meet the nitrogen demand of plants.

10.4.2.3 Soil Phosphorus Dynamics in Three Different Farming Systems

No significant difference in soil total phosphorus was observed among the three farming systems at both soil layers in 2012 (Figure 10.6) (A. Zhang et al., 2013). In the 0- to 20-cm layer, soil total phosphorus in the conventional, low-input, and organic systems in March 2012 increased 69.6%, 63.5%, and 62.1%, respectively, compared with those in March 2002. In the 20- to 40-cm layer, soil total phosphorus in the conventional, low-input, and organic systems increased 30.9%, 76.8%, and 57.1%, respectively. All soil total phosphorus levels increased in each of the three farming systems. In the conventional system, the application of calcium superphosphate led to the increase of the soil available phosphorus content. Phosphorus mobility in the soil is weak, and phosphorus residue would remain in the soil for a long time after its application.

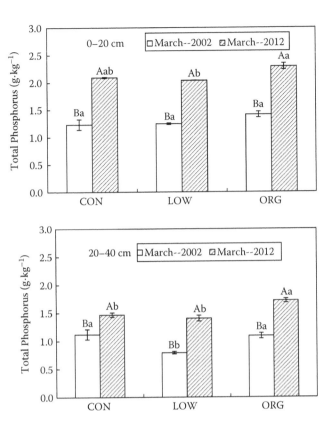

Figure 10.6 Dynamic of soil total phosphorus in different farming systems. (From Zhang, A. et al., *Jiangsu J. Agric. Sci.*, 29(6), 1345–1351, 2013.)

A significant difference was observed in available phosphorus content in the 0- to 20-cm and 20- to 40-cm layers among the three farming systems. In the 0- to 20-cm layer, available phosphorus content in the conventional, low-input, and organic systems in 2012 increased 42.3%, 70.8%, and 154.0%, respectively, compared to those in 2002. In the 20- to 40-cm layer, the available phosphorus content in the conventional, low-input, and organic systems increased 117.9%, 356.2%, and 671.2%, respectively. The available phosphorus content in the 20- to 40-cm layer was significantly lower than that in the 0- to 20-cm layer. The largest increase of soil available phosphorus was measured in the organic system. The high levels of organic fertilizer inputs increased the available phosphorus content in the subsoil, obviously, which may also increase the risk of groundwater contamination via soil phosphorus leaching.

10.4.2.4 Soil Available Potassium Dynamics

The soil available potassium contents all increased in the 0- to 20-cm and 20- to 40-cm layers in 2012 (Figure 10.7) (A. Zhang et al., 2013). In the 0- to 20-cm layer, soil available potassium levels in the conventional, low-input, and organic systems were 1.36, 0.56, and 1.38 times higher, respectively, than those before the experiment; in the 20- to 40-cm layer, the values were 1.94, 2.89, and 3.68, respectively. A significant difference in soil available potassium content was observed in the 20- to 40-cm layer. In the 0- to 20-cm layer, higher soil available potassium content was observed in the organic system than in the conventional and low-input systems, indicating that the application of organic fertilizer obviously increased levels of soil available potassium.

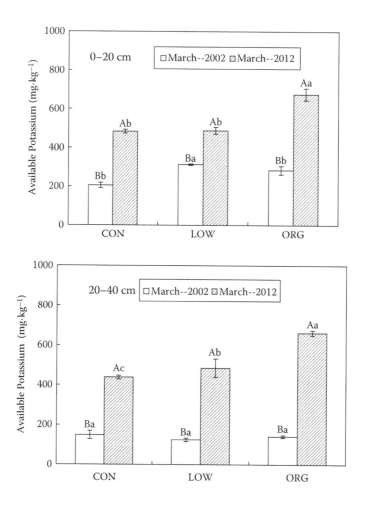

Figure 10.7 Dynamic of soil available potassium in different farming systems. (From Zhang, A. et al., *Jiangsu J. Agric. Sci.*, 29(6), 1345–1351, 2013.)

10.4.2.5 Dynamics of Soil Organic Matter in Three Different Farming Systems

The soil organic matter content increased in the 0- to 20-cm and 20- to 40-cm layers in 2012 (Figure 10.8) (A. Zhang et al., 2013). A significant difference in organic matter content was observed in the 0- to 20-cm and 20- to 40-cm layers among the three farming systems. In the 0- to 20-cm layer, the organic matter content in the conventional, low-input, and organic systems in 2012 was 0.23, 1.17, and 1.90 times higher, respectively, than that in 2002. In the 20- to 40-cm layer, the values were 0.26, 0.91, and 1.57 higher, respectively, than in 2002. Organic matter content in the three farming systems increased with higher levels of organic fertilizer inputs. Organic fertilizer showed a greater capacity than chemical fertilizer to improve soil organic matter content.

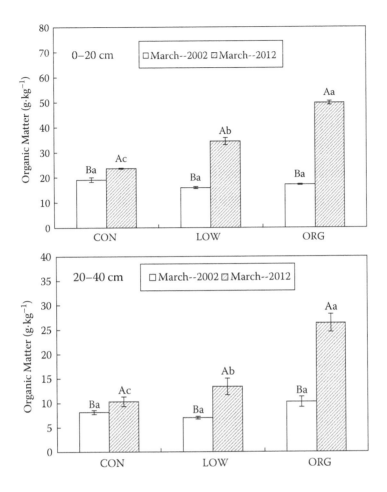

Figure 10.8 Dynamic of soil organic matter in different farming systems. (From Zhang, A. et al., *Jiangsu J. Agric. Sci.*, 29(6), 1345–1351, 2013.)

10.5 STATUS OF SOIL BIOLOGY IN ORGANIC VEGETABLE PRODUCTION SYSTEMS

Soil organisms play a key role in maintaining the material cycling and energy flow of soil ecological systems. The soil organism activity helps plants to mineralize nutrients and resist soil disease infections (Wardle et al., 2004). In a natural system, various soil functions can be achieved through the soil food web in order to maintain high productivity (B.G. Zhang, 1995). However, in an agricultural system, soil organisms are significantly disturbed due to intensive management and high chemical inputs. Agricultural activities, such as fertilizer application, tillage, and pest control, could change the soil environment and thus affect the biomass and community structure of soil organisms. Meanwhile, soil biological properties could reflect soil conditions. In the present study, soil organisms were surveyed in different periods.

10.5.1 Methods

The survey of protozoa, nematodes, and mites was carried out identically. Samples were collected five times monthly from August to December 2012, in both 0- to 10-cm and 10- to 20-cm soil layers. Nematodes had been surveyed in the same way, but the sampling time was from March to July 2012. In this section, we compare protozoa, nematodes and mites in the three different farming systems and try to provide scientific support for the reasonable management of vegetable production in organic agricultural systems and optimization of the soil biochemical environment.

10.5.2 Results and Discussion

10.5.2.1 Soil Protozoa

Soil protozoa exist between decomposers and predators in the soil food web. They feed on bacteria and are also preyed on by larger protozoa and small soil animals (Foissner, 1999). They are involved in soil substance cycling and the energy flow regulated by soil microbes. Protozoan abundance could be affected by agricultural activities such as fertilization, cultivation, and pesticide spraying (Foissner, 1997; Pankhurst et al., 2002; Spedding et al., 2004; Z.P. Cao et al., 2005).

Flagellate levels were the highest in all three systems, followed by amoeba and ciliate levels, with proportions of 85.2%, 13.1%, and 1.7%, respectively. The total number of protozoa, flagellates, and amoebas in the organic system was higher than in the low-input and conventional systems (Figure 10.9), and no significant difference in abundance was observed between the low-input and conventional systems. Protozoan abundance was obviously enhanced by organic fertilizer inputs,

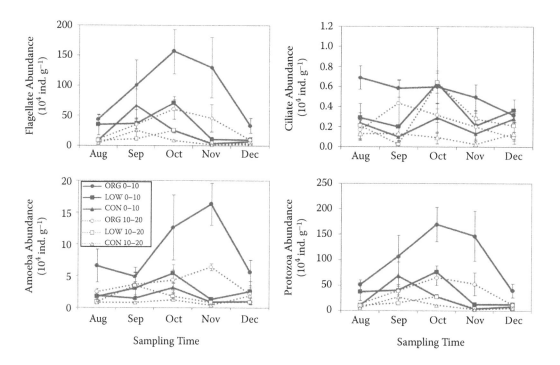

Figure 10.9 Dynamic of the abundance of protozoa and functional groups in different farming systems. (From Wan, Q. et al., *J. China Agric. Univ.*, 19(1), 107–117, 2014.)

which would stimulate the growth of flagellates and amoebas feeding on bacteria (Z.P. Cao et al., 2005). On the other hand, chemical fertilizer, insecticide, and fungicide inputs in the low-input and conventional systems might suppress protozoa to some extent (Foissner, 1997; Z.P. Cao et al., 2005).

The highest abundances of flagellates in the organic and low-input systems were observed in October, while those in the conventional system were observed in September (Figure 10.9). The peak of amoeba abundance in the organic system was observed in November and was observed in low-input and conventional systems in October. The variation of ciliate abundance in the three systems was complicated. In the 0- to 20-cm soil layer, ciliate abundance declined in the order of ORG, LOW, and CON. In the low-input and conventional systems, fluctuations in ciliate abundance were obvious, unlike in the organic system. As for the vertical distribution, more protozoa were observed in the upper soil layer than in the lower layer; however, some reversed distribution patterns occurred in the low-input and conventional systems.

10.5.2.2 Nematodes

Nematode feeding behaviors are broad and occupy a central position in different nutrition levels of the food web (Moore and de Ruiter, 1991), playing a critical role in soil organic matter decomposition and nutrient element cycling (Verhoef and Brussaard, 1990). In addition, soil nematodes can serve as bioindicators of soil conditions because of their sensitivity to environmental changes (Neher, 2001; Fiscus and Neher, 2002; Yeates, 2003).

In both soil layers, total nematode abundance in the organic system was higher in comparison with the low-input and conventional systems (Figure 10.10) (Wan et al., 2014). Nematode amounts in the low-input and conventional systems were close. Nematode abundance peaked in the organic system in both April and June, while abundance in the other two systems only peaked in April. The abundance and variation of nematodes were closely related to soil organic matter, moisture, and disturbances (Sohlenius, 1985; W.J. Liang et al., 2009). They were also affected by some biotic factors such as the levels of predators or food resources.

Bacterivore was the dominant trophic nematode in all three systems, with a mean proportion of 73.3%, which was in accordance with previous studies (Y.J. Liu et al., 2006; Y.F. Chen et al., 2008; Dong et al., 2008). Bacterivore abundance was related to manure application, which leads to high bacteria abundance (Bulluck and Ristaino, 2002); however, fungivore abundance was the lowest among all the trophic groups. In the 10- to 20-cm soil layer, the proportions of nematodes feeding on plants declined in the organic and low-input systems, especially in the organic system, indicating that the organic system was superior to the low-input system in suppressing nematodes that feed on plants.

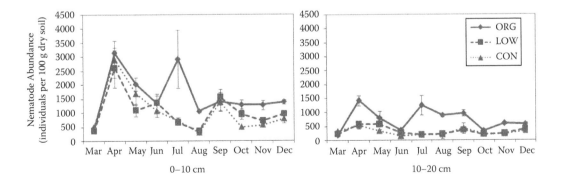

Figure 10.10 Dynamic of nematode abundance in different farming systems. (From Wan, Q. et al., *J. China Agric. Univ.*, 19(1), 107–117, 2014.)

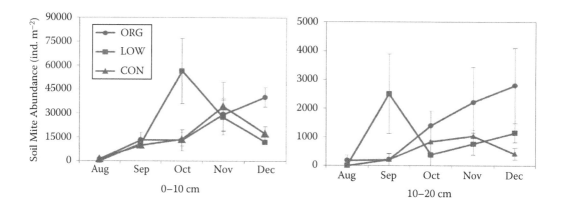

Figure 10.11   Dynamic of soil mite abundance in different farming systems.

The Shannon index for nematode abundance (H') did not increase in the organic system compared to the other systems in this study. The lowest H' values were observed in the organic system in August, November, and December. No significant difference in H' has been observed between the organic and conventional farming systems in other studies (Ferris et al., 1996; Neher, 1999; Van Diepeningen et al., 2006; García-Ruiz et al., 2009). The inverse proportion was obtained between the λ index and the H' index, suggesting that the organic system lowered the nematode diversity but increased certain nematode dominance.

10.5.2.3 Soil Mites

Mites belong to the mesofauna group, which maintains normal soil ecosystem functions by promoting the production and decomposition of substances in the soil, accelerating the cycling and flowing of energy, and increasing the rate of soil aggregate formation (Moore et al., 1990; Beare et al., 1997; Renker et al., 2005). In agricultural ecosystems, the distribution and abundance of soil mites have been correlated with land use, tillage, fertilization, and so on (Behan-Pelletier, 1999; Kae et al., 2002; Badejo et al., 2004). The mean abundance of mites in the 0- to 10-cm soil layer was 20 times higher than that in the 10- to 20-cm soil layer, with a value of 18,707 and 937 individuals (ind.) m^{-2}, respectively (Figure 10.11). Mite densities in the organic, low-input, and conventional systems were 18,707, 11,180, and 7918 ind. m^{-2}, respectively. A significant difference was observed in mite abundance among the three systems. The variation of mite density in the low-input system was obvious, with a peak in the 0- to 10-cm soil layer in October and in the 10- to 20-cm soil layer in September. A small fluctuation of mite abundance was observed in the conventional system, with a peak in November.

The relative abundance of mite trophic groups and ecology indices are shown in Table 10.9. Astigmatid and Mesostigmata dominated in the 0- to 10-cm and 10- to 20-cm soil layers, with a mean abundance of 60.3% and 47.7%, respectively. The results of one-way ANOVA indicated that the abundance of mite trophic groups was not affected significantly by the farming system ($p > 0.05$), except for Mesostigmata in the 0- to 10-cm soil layer ($p < 0.05$). No significant influence of farming systems was observed on the abundance index (H') and dominance index (D) of soil mites in either soil layer ($p < 0.05$). The results suggest that mite communities and diversity were slightly affected by the different farming systems.

Table 10.9 Relative Abundance of Mite Suborders and Mite Ecology Indices in Different Farming Systems

	0–10 cm			10–20 cm		
	ORG	LOW	CON	ORG	LOW	CON
Prostigmata	3.8[a]	11.4[a]	8.8[a]	27.5[a]	37.7[a]	4.2[a]
Mesostigmata	30.2[a]	20.0[ab]	11.1[b]	46.7[a]	40.1[a]	56.3[a]
Astigmatid	53.5[a]	63.7[a]	63.7[a]	25.8[a]	13.6[a]	35.4[a]
Oribatida	12.4[a]	4.8[a]	16.5[a]	0.0[a]	8.6[a]	4.2[a]
Shannon index	1.05[a]	1.01[a]	0.91[a]	0.51[a]	0.47[a]	0.22[a]
Dominance index	0.51[a]	0.53[a]	0.56[a]	0.68[a]	0.72[a]	0.85[a]

Note: ORG, organic farming; LOW, low-input farming; CON, conventional farming. Same letter in a row indicates no significant difference($p < 0.05$), and different letters indicate significant difference ($p < 0.05$).

10.6 PHYTOPATHOGEN THREATS TO ORGANIC GREENHOUSE VEGETABLE PRODUCTION

Greenhouses are continually being built for vegetable production in China. Heating systems even allow vegetable production year-round, which also contributes to continuous pest accumulation. Numerous phytopathogens that do not seriously impact open field vegetables pose additional threats to greenhouse vegetable production. Disease control is more complicated in organic greenhouses where chemical pesticide applications are not allowed. According to our 10-year survey, over 400 diseases were found to cause damage to greenhouse vegetables. Members of the Peronosporaceae and Sphaerotheca often cause serious disease in plants of Cucurbitaceae. Frequently, losses in vegetables belonging to the family of Solanaceae can be attributed to pathogens such as *Phytophthora capsici*, *Alternaria solani*, *Botrytis cinerea*, and *Fulvia fulva*.

10.6.1 Method

The experiment was performed in solar-style greenhouses in Quzhou from 2002 to 2007. Details on the setup of the experiment were described in Section 10.2, above. Briefly, the experiment consisted of organic, low-input, and conventional farming systems. The influence of different farming systems on the severity of common diseases in cucumbers (downy mildew, bacterial angular leaf spot, powdery mildew, and gray mold) and tomatoes (gray mold, late blight, and early blight) was evaluated with five replicates per treatment. Each replicate contained five individual plants. The numbers of diseased plants, leaves, and fruits were recorded every three to four days during the whole growing season.

The severity index for each disease observed on leaves was recorded as described by H.L. Li and Xu (2000). Disease severity and percentage of diseased fruit were calculated according to the following formulas, respectively:

$$\text{Disease severity} = \sum (\text{No. of diseased plants in this index} \times \text{Disease index})$$
$$\div (\text{Total no. of plants investigated} \times \text{Highest disease index}) \times 100\%$$

$$\text{Percentage of diseased fruit} = (\text{No. of diseased fruits} \div \text{Total no. of investigated fruits}) \times 100\%$$

10.6.2 Results and Discussion

10.6.2.1 Reduced Severity of Four Cucumber Diseases in Organic Farming Systems

Downy mildew (*Pseudoperonospora cubensis*), bacterial angular leaf spot (*Pseudomonas syringae* pv. Lachrymans), powdery mildew (*Podosphaera fusca*), and gray mold (*Botrytis cinerea*) are major threats to cucumber production in greenhouses in Northern China. To study the influence of different greenhouse farming systems on the severity of each of these diseases in cucumber plants or fruit, off-season winter cucumber was repeatedly cultivated in 2002, 2004, 2005, and 2007.

For all four years investigated, the severity of each foliar disease initially increased with plant age and reached a peak in the late middle of each growing season, followed by a decrease in the remainder of the growing season. Most strikingly, the lowest peaks of severity in all diseases investigated were consistently found in the organic farming system, while the highest peaks were always detected in the conventional farming system. In general, the peak of disease severity increased progressively with continuous cucumber cultivation, except for 2007, when the lowest peak was observed for each disease and farming system. More abundant sunlight and higher temperatures in November and December 2007 possibly impeded the development of these diseases. It is also worth noting that the greenhouses were heat-treated more often using solar energy in the summer of 2007, and celery was planted instead of cucumber in 2006.

Downy mildew in cucumber plants, caused by *Pseudoperonospora cubensis* (Berkeley & Curtis), is a potentially catastrophic disease in most parts of China that can result in significant yield loss. The disease affects cucumber plants in both fields and greenhouses. For greenhouse cucumber cultivation, downy mildew normally appears in late autumn in Northern China, where a large difference in temperature between day and night frequently causes increased moisture content in the plant canopy. For each of the years studied, the highest disease severities of downy mildew in the organic farming system ranged from 5.1 to 8.1 (Figure 10.12), over 44% lower than those values detected in the conventional farming system (Figure 10.12). Peaks of disease severity in the low-input farming system were also lower than those in the conventional farming system (Figure 10.12) (Yang et al., 2009).

Pseudomonas syringae pv. Lachrymans, the causal agent for bacterial angular leaf spot, is responsible for enormous cucumber losses in China, especially greenhouse cucumbers. Compared to downy mildew, the highest disease severity of bacterial leaf spot was much lower in each of the studied farming systems. In the organic farming system, the peaks of disease severity were 2.3, 2.7, 4.4, and 3.1 for 2002, 2004, 2005, and 2007, respectively (Figure 10.13). In contrast, the peaks were higher in both the conventional and low-input farming systems (Figure 10.13) (Yang et al., 2009).

Cucumber powdery mildew caused by *Podosphaera fusca* is a common disease in greenhouses in Northern China. The disease severity in the conventional farming system increased faster than in the low-input and organic farming systems. For all three systems, disease severity peaked in the middle term of cucumber growth and then declined in the later term of cucumber growth. The disease index peaks of the organic farming system in 2002, 2004, 2005, and 2007 were 3.5, 4.7, 4.7, and 2.9, respectively; the low-input farming system, 4.1, 6.7, 5.2, and 3.9, respectively; and the conventional farming system, 6.3, 7.5, 7.7, and 5.6, respectively. In each farming system, the highest disease severity of cucumber powdery mildew ranged between those of downy mildew and bacterial angular leaf spot for all 4 years.

Gray mold caused by *Botrytis cinerea* on cucumber fruits appears frequently in China. In the present study, the percentage of diseased fruits increased after the growth stage of bearing fruit. The highest 20% of diseased fruits was detected in the conventional farming system in 2005. In the organic farming system, percentages of diseased fruits were lower than 9%, except for 2005, in contrast to more than 10% of diseased fruits in other two farming systems.

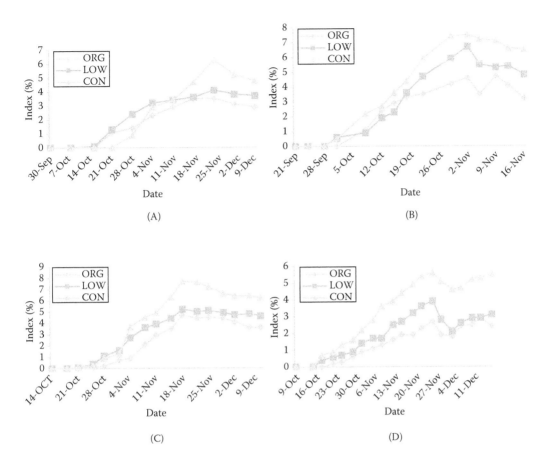

Figure 10.12 Change of disease severity of downy mildew in cucumber in (A) 2002, (B) 2004, (C) 2005, and (D) 2007. (From Yang, H.F. et al., *Acta Agric. Boreali-Sin.*, 24, 240–245, 2009.)

10.6.2.2 Reduced Severity of Three Common Diseases of Tomato in Organic Farming Systems

Botrytis cinerea, *Phytopthora infestans*, and *Alternaria solani* are causal agents of gray mold, late blight, and early blight in tomatoes, respectively. In Northern China, these diseases create challenges for greenhouse tomato production. The effect of different greenhouse farming systems on the severity of these diseases was investigated from 2003 to 2005. Again, the lowest disease severities of gray mold, late blight, and early blight were observed in the organic farming system, and the highest was found in the conventional farming system. The severities of both early blight and late blight reached peaks in the middle late of each growing season. (See Figure 10.14.)

For gray mold, the foliar disease severity in all studied farming systems increased annually from 2003 to 2005, suggesting that continual tomato cultivation possibly hastened the development of the disease caused by *Botrytis cinerea*. The percentages of diseased fruit in the organic farming system were 10.4%, 14.5%, and 11.1% for 2003, 2004, and 2005, respectively, more than 37% lower than those in the conventional farming system (Figure 10.15) (Yang et al., 2009).

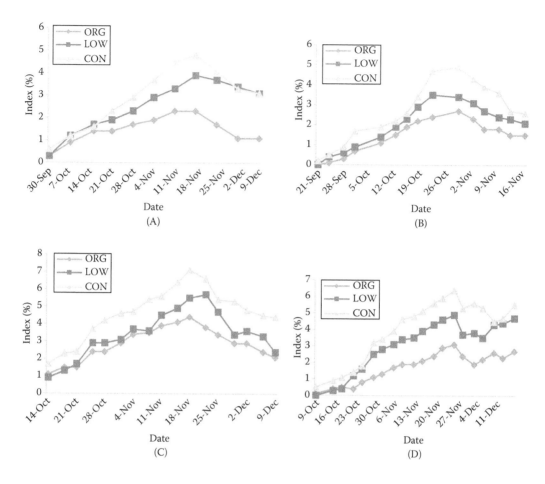

Figure 10.13 Change of disease severity of bacterial angular leaf spot in cucumber in (A) 2002, (B) 2004, (C) 2005, and (D) 2007. (From Yang, H.F. et al., *Acta Agric. Boreali-Sin.*, 24, 240–245, 2009.)

Interestingly, increased disease severity of late blight with continual tomato cultivation was observed in conventional and low-input farming systems but not in the organic farming system, suggesting that some organic procedures possibly compensated for the adverse effects caused by continual cultivation. In the organic farming system, the disease severity of late blight was only about 4%, which was over 47% lower than that in the conventional farming system and over 26% lower than that in the low-input farming system.

In the present study, the peaks of early blight disease were the lowest among peaks of the diseases detected in each farming system. Most of the peaks of the disease index were less than 1. The highest peak, of only 1.6, was detected in the conventional farming system in 2005. Nevertheless, the peaks in the organic farming system were still lower than those in other farming systems.

10.7 NITROGEN BALANCE IN ORGANIC CROPPING SYSTEMS

Nitrogen balance is defined as the balance of nitrogen inputs and outputs. In vegetable crop–soil systems, nitrogen inputs include nitrogen fertilizer, animal excrements, rainwater, irrigation water, and biological nitrogen fixation; nitrogen output refers to the nitrogen in drainage/seepage water and the nitrogen concentration of aboveground plants, which is removed from the field during the

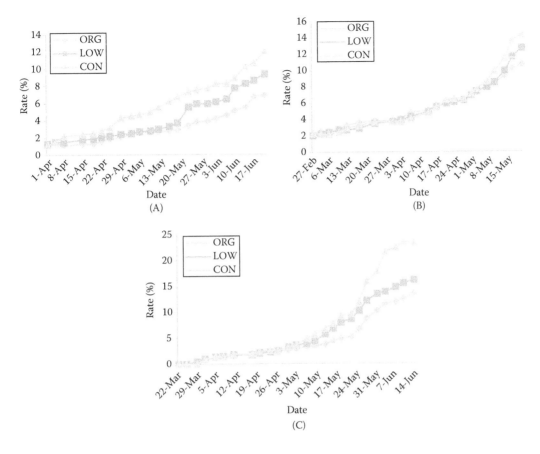

Figure 10.14 Change of disease severity of gray mold on tomato leaf in (A) 2003, (B) 2004, and (C) 2005. (From Yang, H.F. et al., *Chin. J. Eco-Agric.*, 17(5), 933–937, 2009.)

harvest period. Previous studies suggested that in vegetable production systems large amounts of nitrogen fertilizer were used with very low efficiency, resulting in a large nitrogen surplus (C.P. Li et al., 2005; J.H. Zhu et al., 2005; B. Cao et al., 2007; He et al., 2008). C.P. Li et al. (2005) found that the N surplus was as high as 2051 N kg ha^{-1} for celery on a farm. B. Cao et al. (2007) reported that nitrogen use efficiency was in the range of 18.1 to 24.6% in Chinese cabbage fields. Whereas 33.4 to 33.5% of total N still remained in the soil, 41.9 to 48.5% was lost during the Chinese cabbage growing season. In the study by He et al. (2008), the nitrogen use efficiency was about 70% higher in manure treatments than in conventional nitrogen treatments, suggesting organic fertilizer such as manure is more environmental friendly because it reduces the severity of nitrogen surplus. Here, nitrogen balance was studied in organic, low-input, and conventional farming systems in 2011.

10.7.1 Methods

Details on the long-term experiment were described in Section 10.2. To recap, the three farming systems mainly differed in fertilization application. The organic system was conducted in accordance with IFOAM basic standards by using organic fertilizers exclusively. In the low-input and conventional systems, both organic and inorganic fertilizers were used but the total amount of nitrogen applied was reduced in the low-input system. Compared to the organic system, less organic fertilizer was also applied in the low-input system. The total yearly nitrogen inputs were 1868, 1514, and 1672 kg ha^{-1} for the organic, low-input, and conventional systems, respectively.

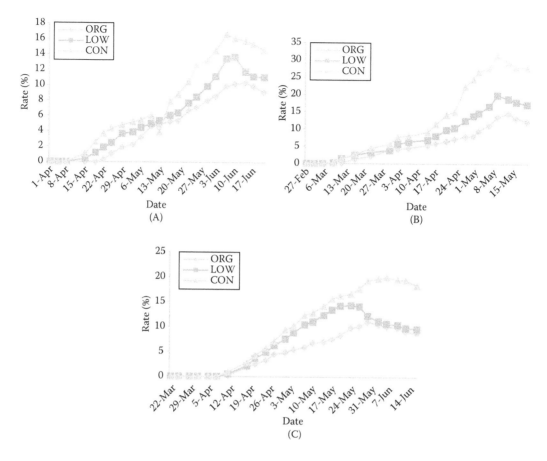

Figure 10.15 Change of percent of diseased tomato fruit caused by *Botrytis cinerea* in (A) 2003, (B) 2004, and (C) 2005. (From Yang, H.F. et al., *Chin. J. Eco-Agric.*, 17(5), 933–937, 2009.)

Eggplants were transplanted into each greenhouse on March 12, 2011. For each farming system, whole plants were collected on April 3, May 12, May 23, June 1, June 11, and July 15, as well as seedlings analyzed with three replicates. Root, stem, leaf, and fruit were analyzed separately. The dry weight of each plant organ was determined, and all dry material was crushed and passed through a 0.5-mm mesh for further analysis. Total nitrogen in plant material was measured by the kjeldahl method (Bao, 2000).

About 8 kg of compost fertilizer were collected from four points on different sides of the compost pile and mixed thoroughly. Two kilograms of sample were used to determine the total nitrogen content. The sample was dried and sieved through a 0.149-mm mesh. Total nitrogen was measured using the semi-micro Kjeldahl method (Younie and Watson, 1992). Total nitrogen content in irrigation water from a nearby well was measured by the alkaline potassium super sulfate digestion–ultraviolet spectrophotometric method. Soil nitrogen balance was calculated using the following formula (Van Eerdt and Fong, 1998; Oenema et al., 2003):

$$\text{N surplus} = \text{Input components (Fertilizer + Manure + N from transplant + N from irrigation)} - \text{Output components (N in plant)}$$

The N surplus represented the N that was lost by ammonia volatilization, denitrification, or leaching, or stored in various soil fractions.

10.7.2 Results and Discussion

10.7.2.1 Nitrogen Input

Some studies indicated that nitrogen fertilizer could significantly affect residual nitrogen in different soil layers (Tang et al., 2002). Studies also indicate that the residual nitrogen exists in many forms, such as nitrate, exchangeable ammonium nitrogen, microbial biomass nitrogen, non-exchangeable ammonium in mineral clay, nitrogen in organic components that are easy to mineralize, and organic nitrogen in stable soil compositions. In general, residual nitrogen exists in organic forms, but the proportion of nitrate nitrogen was elevated when nitrogen applications were high (Ju et al., 2000). Total nitrogen inputs to the organic, low-input, and conventional systems were 1150, 1182, and 1433 kg ha^{-1}, respectively. Inorganic fertilizer, compost, and animal manure for this crop were the main source of N, contributing more than 96% of total nitrogen input.

10.7.2.2 Nitrogen Output

According to the equation for nitrogen balance, nitrogen output refers to nitrogen in plants, including fruits and other organs. For all three farming systems, nitrogen uptake and utilization by eggplants were lower during the vegetative stage and higher during the reproductive stage, especially after the fruiting stage. Total nitrogen outputs in the organic, low-input, and conventional systems were 178, 135, and 116 kg ha^{-1}, respectively. Differences in total nitrogen output among the different farming systems, possibly caused by varied nitrogen content as the yield of eggplant varied between different farming systems, was comparable, with less than 4% difference.

10.7.2.3 Nitrogen Balance

Generally, the net nitrogen surplus increased with plant age in the three cropping systems, possibly due the continual application of fertilizer during the growing seasons (Figure 10.16) (Guo et al., 2014). A decrease of net nitrogen surplus was observed for organic and low-input systems in the late growing season, but not for the conventional system. It was possibly due to a higher frequency and amount of fertilizer application in the conventional system. The nitrogen uptake by eggplants increased greatly at the fruiting stage, which could lead to a decline of nitrogen surplus in organic and low-input systems when no additional fertilizer is applied. The net nitrogen surpluses in the organic, low-input, and conventional systems were 971, 1046, and 1317 kg ha^{-1}, respectively. In summary, lower nitrogen surpluses and higher eggplant yields

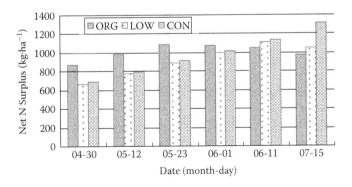

Figure 10.16 Net nitrogen surplus in different periods in different farming systems. (From Guo, R.H. et al., *Chin. J. Eco-Agric.*, 22(1), 10–15, 2014.)

were obtained from the organic farming system, in comparison to the low-input and conventional farming systems. The results strongly indicated that compost fertilization application in organic farming system is superior to that in conventional and low-input systems because it reduces the intensity of nitrogen surplus.

10.8 CONCLUSIONS AND PERSPECTIVES

The long-term study of organic agriculture includes a broad range of fields, such as organic agriculture productivity, soil physicochemical properties, soil biology, pest and disease control, plant roots, crop yield, and the quality of agroproducts. Some valuable results were obtained from the present long-term study on different farming systems.

10.8.1 Conclusions

10.8.1.1 Effects of Farming Systems on Plant Growth and Yield

During the earlier transition period (about 3 years) after the experiment started, no significant difference in plant growth or yield was observed among the three different farming systems. Organic system production then began to stabilize and surpassed that of the conventional system in the fourth year.

10.8.1.2 Effects of Farming Systems on Vegetable Quality

Much higher levels of vitamin C and soluble sugar and substances—and lower nitrate concentration—were detected in vegetables from the low-input and organic systems compared to those from the conventional system. No significant differences of these chemical levels were observed between the organic system and the low-input system.

10.8.1.3 Effects of Farming Systems on Soil Physical Properties

In the 0- to 20-cm layer, a significant difference was shown of soil bulk density in the three systems, with a minimum in the organic system after 10 years of cultivation. In the 20- to 40-cm layer, the highest soil bulk density was observed in the conventional system; meanwhile, the soil bulk density of the organic system was similar to that of the low-input system, and there was no significant difference among the three systems.

10.8.1.4 Effects of Farming Systems on Soil Nutrients

After 10 years of the experiment, the concentrations of soil total nitrogen, total phosphorus, available phosphorus, available potassium, and organic matter in the three systems in both 0- to 20-cm and 20- to 40-cm soil layers clearly increased. In the 0- to 20-cm layer, the soil organic matter in the low-input system and the organic system increased by 116.8% and 190.1%, respectively, compared to the initial data. In the 20- to 40-cm layer, the soil available phosphorus in the low-input system and the organic system increased by 356.2% and 368.1%, respectively, compared to initial data. By 2012, the levels of soil organic matter, total N, total P, and available NPK in the 0- to 20-cm and 20- to 40-cm layers in the organic system were all higher than those in the conventional and low-input systems. This indicates that more plant nutrients increased in the organic system than in the low-input and conventional systems.

10.8.1.5 Effects of Farming Systems on Soil Bioactivity

The total numbers of protozoa, flagellates, and amoebas in the organic system were higher than those in the other systems, and no significant difference was observed between the low-input and the conventional systems. In the 0- to 20-cm soil layer, the amount of nematodes in the organic system was the highest among the three systems. Meanwhile, the amount of nematodes in the low-input system was close to the conventional system. During the sampling time, mites tended to live in the upper soil layer, as the abundance of mites in the 0- to 10-cm layer was 20 times greater than that in the 10- to 20-cm layer (18,707 m^{-2} vs 937 m^{-2}). The average mite abundances in the organic system, low-input system, and conventional system were 10,368, 11,180, and 7918 ind. m^{-2}, respectively.

10.8.1.6 Effects of Farming Systems on Vegetable Diseases

The occurrence of downy mildew, bacterial angular leaf spot, powdery mildew, and gray mold on cucumbers and the occurrence of gray mold, late blight, and early blight on tomatoes showed the following trend in the three systems: conventional system > low-input system > organic system.

10.8.1.7 Effects of Farming Systems on Nitrogen Balance

The total nitrogen inputs in the organic, low-input, and conventional systems in spring were 1150, 1182, and 1433 kg ha^{-1}, respectively; the outputs were 178, 135, and 116 kg ha^{-1}, respectively; and the final nitrogen surpluses were 971, 1046, and 1317 kg ha^{-1}, respectively. At similar yield levels, the amount of total nitrogen input and the nitrogen surplus in the organic system were the lowest among the three systems.

10.8.2 Perspectives

1. This study was carried out on intensive greenhouse vegetable production in the north of China. The data could be employed to support organic farming in intensive agricultural production regions worldwide, both theoretically and practically.
2. During the earlier period of the study, nitrogen inputs in the different systems were around 800 kg ha^{-1} to ensure vegetable yield. Further research will be focused on nitrate content in groundwater and the emission of gaseous nitrogen in order to assess the risks of nitrogen contamination in organic systems.
3. The results of the present study show that soil biodiversity in the organic system was significantly improved, whereas the occurrence of vegetable diseases tended to be lower. Further study will be focused on the infection mechanism of soil diseases in organic systems, which will help to establish indicators of soil health and improve the sustainability of these agricultural production systems.

REFERENCES

Badejo, M.A., De Aquino, A.M., De-Polli, H., and Correia, M.E.F. 2004. Response of soil mites to organic cultivation in an Ultisol in southeast Brazil. *Exp. Appl. Acarol.*, 34: 345–364.

Bao, S.D. 2000. *Soil and Agricultural Chemistry Analysis*. China Agriculture Press: Beijing.

Beare, M.H., Reddy, M.V., Tian, G., and Srivastava, S.C. 1997. Agricultural intensification, soil biodiversity and agroecosystem function in the tropics: the role of decomposer biota. *Appl. Soil Ecol.*, 6: 87–108.

Behan-Pelletier, V.M. 1999. Oribatid mite biodiversity in agroecosystems: role for bioindication. *Agric. Ecosyst. Environ.*, 74: 411–423.

Bulluck, L.R. and Ristaino, J.B. 2002. Effect of synthetic and organic soil fertility amendments on southern blight, soil microbial communities, and yield of processing tomatoes. *Phytopathology*, 92: 181–189.

Cai, L.L., Yan, S.H., Sun, L.F., and He, W.L. 2008. Study on cucumber yield and quality and soil nutrients under three different cultural practices. *Acta Agric. Shanghai*, 24(4): 51–54.

Cao, B., He, F.Y., Xu, Q.M., and Cai, G.X. 2007. Nitrogen use efficiency and N losses from Chinese cabbage grown in an open field. *Plant Nutr. Fert. Sci.*, 13(6): 1116–1122.

Cao, Z.P., Chen, G.K., Zhang, K., and Wu, W.L. 2005. Impact of soil fertility maintaining practices on protozoa abundance in high production agro-ecosystem in northern China. *Acta Ecol. Sin.*, 25: 2992–2996.

Chen, Y.F., Steinberger, Y., and Cao, Z.P. 2008. Effects of alternatives to methyl bromide on soil free-living nematode community dynamics in a greenhouse study. *J. Sustain. Agric.*, 31: 95–113.

Dong, D.F., Chen, Y.F., Steinberger, Y., and Cao, Z.P. 2008. Effects of different soil management practices on soil free-living nematode community structure, Eastern China. *Can. J. Soil Sci.*, 88: 115–127.

Ferris, H., Venette, R.C., and Lau, S.S. 1996. Dynamics of nematode communities in tomatoes grown in conventional and organic farming systems, and their impact on soil fertility. *Appl. Soil Ecol.*, 3: 161–175.

Fiscus, D.A. and Neher, D.A. 2002. Distinguishing sensitivity of free-living soil nematode genera to physical and chemical disturbances. *Ecol. Appl.*, 12: 565–575.

Foissner, W. 1997. Protozoa as bioindicators in agroecosystems, with emphasis on farming practices, biocides, and biodiversity. *Agric. Ecosyst. Environ.*, 62: 93–103.

Foissner, W. 1999. Soil protozoa as bioindicators: pros and cons, methods, diversity, representative examples. *Agric. Ecosyst. Environ.*, 74: 95–112.

García-Ruiz, R., Ochoa, V., Vinegla, B., Hinojosa, M.B., Pena-Santiago, R., Liébanas, G., Linares, J.C., and Carreira, J.A. 2009. Soil enzymes, nematode community and selected physico-chemical properties as soil quality indicators in organic and conventional olive oil farming: influence of seasonality and site features. *Appl. Soil Ecol.*, 41: 305–314.

Guo, R.H., Yang, Y.B., and Li, J. 2014. Comparative study of nitrogen budget in three different vegetable planting patterns under greenhouse condition. *Chin. J. Eco-Agric.*, 22(1): 10–15.

He, F.F., Ren, T., Chen, Q., Jiang, R.F., and Zhang, F.S. 2008. Nitrogen balance and optimized potential of integrated nitrogen management in greenhouse vegetable system. *Plant Nutr. Fert. Sci.*, 14(4): 692–699.

Hoefkens, C., Vandekinderen, I., De Meulenaer, B., Devlieghere, F., Baert, K. et al., 2009. A literature-based comparison of nutrient and contaminant contents between organic and conventional vegetables and potatoes. *Br. Food J.*, 111(10): 1078–1097.

Ju, X.T. 2000. Transformation and Fate of Soil-Fertilizer Nitrogen in Winter Wheat/Summer Maize Rotation System, PhD thesis, China Agricultural University, Beijing.

Kae, M., Hiroyuki, T., Makoto, Y., Hiroshi, N., and Tomomi, N. 2002. The effects of cropping systems and fallow managements on microarthropod populations. *Plant Prod. Sci.*, 5: 257–265.

Li, C.P., Xu, Y.B., Li, Y.M., Zheng, Y., Zhang, W.L., and Liu, H.B. 2005. The nutrient balance in the protected fields of vegetable and flower cultivation in Dian Lake front. *J. Yunnan Agric. Univ.*, 20(6): 804–809.

Li, H.L. and Xu, J.Y. 2000. *Instructor for Experiment and Exercitation of Agricultural Plant Pathology.* China Agricultural University Press: Beijing, pp. 148–150.

Liang, L.N., Li, J., Yang, H.F., and Xie, Y.L. 2009. Effect of organic, low-input and conventional production model on soil quality in solar greenhouse vegetable growing system. *Trans. CSAE*, 25(8): 186–191.

Liang, W.J., Lou, Y.L., Li, Q., Zhong, S., Zhang, X.K., and Wang, J.K. 2009. Nematode faunal response to long-term application of nitrogen fertilizer and organic manure in Northeast China. *Soil Biol. Biochem.*, 41: 883–890.

Liu, X.R., Ren, J.Q., and Zhen, L. 2003. Advance in the study of nitrate accumulation in vegetables and influence factors. *Chin. J. Soil Sci.*, 34(4): 356–358.

Liu, Y., Tang, L.S., and Li, Y. 2008. The effect of different fertilization treatments on soil nutrient and crop yield in oasis farmland. *Agric. Res. Arid Areas*, 26(3): 151–156.

Liu, Y.J., Hua, J.F., Jiang, Y., Li, Q., and Wen, D.H. 2006. Nematode communities in greenhouse soil of different ages from Shenyang suburb. *Helminthologia*, 43: 51–55.

Lu, D., Zong, L.G., and Xiao, X.J. 2005. A comparison of heavy metals concentrations in soils of organic and conventional farming in typical regions of eastern China. *J. Agric. Environ. Sci.*, 24(1): 143–147.

Martini, E.A., Buyer, J.S., Bryant, D.C., Hartz, T.K., and Denison, R.F. 2004. Yield increases during the organic transition: improving soil quality or increasing experience. *Field Crops Res.*, 86: 255–266.

Moore J.C. and De Ruiter, P.C. 1991. Temporal and spatial heterogeneity of trophic interactions within belowground food webs. *Agric. Ecosyst. Environ.*, 34: 371–397.

Moore J.C., Zwetsloot, H.J.C., and De Ruiter, P.C. 1990. Statistical analysis and simulation modelling of the belowground food webs of two winter wheat management practices. *Netherlands J. Agric. Sci.*, 38: 303–316.

Neher, D.A. 2001. Role of nematodes in soil health and their use as indicators. *J. Nematol.*, 33: 161–168.

Neher, D.A. 1999. Nematode communities in organically and conventionally managed agricultural soils. *J. Nematol.*, 31: 142–154.

Oenema, O., Kros, H., and de Vries, W. 2003. Approaches and uncertainties in nutrient budgets: implications for nutrient management and environmental policies. *Eur. J. Agron.*, 20: 3–16.

Organic Food Development and Certification Center. 2013. Develop organic agriculture and promote ecological civilization. *Organic Food Times*, 67: 1–4.

Pankhurst, C.E., Kirkby, C.A., Hawke, B.G., and Harch, B.D. 2002. Impact of a change in tillage and crop residue management practice on soil chemical and microbiological properties in a cereal-producing red duplex soil in NSW, Australia. *Biol. Fert. Soils*, 35: 189–196.

Ren, C.M., Hu, X.L., Xie, J.Y., and Sun, D. 2005. Fertilize in the location long time had the effect on soil nutrient and rice yield. *Reclaim. Rice Cult.*, 4: 37–40.

Renker C., Otto, P., Schneider, K., Zimdars, B., Maraun, M., and Buscot, F. 2005. Oribatid mites as potential vectors for soil microfungi: study of mite-associated fungal species. *Microb. Ecol.*, 50: 518–528.

Ryan, M.H., Derrick, J.W., and Dann, P.R. 2004. Grain mineral concentrations and yield of wheat grown under organic and conventional management. *J. Sci. Food Agric.*, 84: 207–216.

Shen, M.H., He, W.L., Yan, S.H., and Cheng, H.L. 2010. Effects of organic farming and special farming on the yield and quality of four kinds of vegetables. *Jiangsu J. Agric. Sci.*, 26(40): 729–734.

Sohlenius, B. 1985. Influence of climatic conditions on nematode coexistence: a laboratory experiment with a coniferous forest soil. *Oikos*, 44: 430–438.

Spedding, T.A., Hamel, C., Mehuys, G.R., and Madramootoo, C.A. 2004. Soil microbial dynamics in maize-growing soil under different tillage and residue management systems. *Soil Biol. Biochem.*, 36: 499–512.

Tang, L.L., Chen, Q., Zhang, H.Y., Zhang, X.S., Li, X.L., and Liebig, H.P. 2002. Effects of different irrigation and fertilization strategies on soil inorganic N residues in open field of vegetable rotation system. *Plant Nutr. Fert. Sci.*, 8(3): 282–287.

Van Diepeningen, A.D., de Vos, O.J., Korthals, G.W., and van Bruggen, A.H.C. 2006. Effects of organic versus conventional management on chemical and biological parameters in agricultural soils. *Appl. Soil Ecol.*, 31: 120–135.

Van Eerdt, M.M. and Fong, P.K.N. 1998. The monitoring of nitrogen surpluses from agriculture. *Environ. Pollut.*, 102: 227–233.

Verhoef, H.A. and Brussaard, L. 1990. Decomposition and nitrogen mineralization in natural and agroecosystems: the contribution of soil animals. *Biogeochemistry*, 11: 175–211.

Wan, Q., Li, J., Cao, Z.P., and Li, Y.F. 2014. Effects of greenhouse planting modes on soil nematode community structure. *J. China Agric. Univ.*, 19(1): 107–117.

Wang, J.H., Liu, J.S., Yu, J.B., and Wang, J.D. 2004. Effect of fertilizing N and P on soil microbial biomass carbon and nitrogen of black soil corn agroecosystem. *J. Soil Water Conserv.*, 18: 35–38.

Wardle, D.A., Bardgett, R.D., Klironomos, J.N., Setälä, H., Van Der Putten, W.H., and Wall, D.H. 2004. Ecological linkages between aboveground and belowground biota. *Science*, 304: 1629–1633.

Willer, H. and Lernoud, J., Eds. 2014. *The World of Organic Agriculture: Statistics and Emerging Trends 2014*. Bonn: FiBL and IFOAM.

Worthington, V. 2001. Nutritional quality of organic versus conventional fruits, vegetables, and grains. *J. Alt. Comp. Med.*, 7(2): 161–173.

Xi, Y.G., Li, Z.F., Tai, C.M., Wang, Q.H., and Ding, W. 1999. A comparison of energy flow and economy flow between organic and inorganic vegetable production systems. *Eco-Agric. Res.*, 7(2): 39–42.

Xie, X.G., Li, J., and Yang, H.F. 2007. Quality of vegetables in different cultivation patterns in greenhouse. *Chin. J. Soil Sci.*, 38(4): 718–721.

Yang, H.F., Fan, J.F., Ge, Z.Q., Shen, G.C., Lv, R.H., and Li, J. 2009a. Main diseases and control effects of organic, integrated and conventional cultivation patterns of greenhouse tomato. *Chin. J. Eco-Agric.*, 17(5): 933–937.

Yang, H.F., Fan, J.F., Liang, L., Meng, Y.L., Zhang, S.K., and Li, J. 2009b. Studies on the main diseases and control effects under organic, integrated and conventional cultivation patterns of cucumber in greenhouse. *Acta Agric. Boreali-Sin.*, 24: 240–245.

Yeates, G.W. 2003. Nematodes as soil indicators: functional and biodiversity aspects. *Biol. Fert. Soils*, 37: 199–210.

Younie, D. and Watson, C.A. 1992. Soil nitrate-N levels in organically and intensively managed grass and systems. *Aspects Appl. Biol.*, 30: 235–238.

Zhang, A., Han, H., Yang, H.F., and Li, J. 2013. Effect of conventional, low-input and organic vegetable cropping systems on soil properties. *Jiangsu J. Agric. Sci.*, 29(6): 1345–1351.

Zhang, B.G. 1995. The role of soil invertebrate in soil fertility. In: Zhang, F.S. et al., Eds., *New Trends of Soil and Plant Nutrition Research*. China Agricultural University Press: Beijing, pp. 82–97.

Zhao, F.J. 2012. Long-term experiments at Rothamsted Experimental Station: introduction and experience. *J. Nanjing Agric. Univ.*, 35(5): 147–153.

Zhu, J.H., Li, X.L., Christie, P., and Li, J.L. 2005. Environmental implications of low nitrogen use efficiency in excessively fertilizer hot pepper (*Capsicum frutescens* L.) cropping systems. *Agric. Ecosyst. Environ.*, 111: 70–80.

Zhu, X.J., Wu, B., Feng, X., and He, W.L. 2010. Impacts of different cultivation methods on yield and quality of kidney bean and soil nutrient. *Hunan Agric. Sci.*, 13: 32–34.

Figure 1.8 Traditional Yuanyang terrace with large preserved forest above in Yunnan Province.

Figure 2.7 Visualization of rhizosphere acidification of fava bean (left) and maize (right). The roots were imbedded for 6 hr in agar gel containing a pH indicator (bromocresol purple) without P supply. Yellow indicates acidification, and purple indicates alkalization. (From Li, L. et al., *Proc. Natl. Acad. Sci. U.S.A.*, 104(27), 11192–11196, 2007. With permission.)

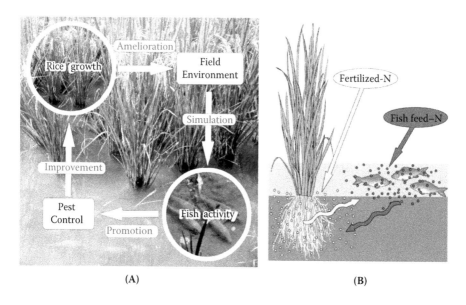

(A)   (B)

Figure 4.8 Positive interactions and complementary use of nitrogen (N) between rice and fish explain why the rice–fish co-culture system can remain productive for long periods with low input of chemicals. (A) *Positive interactions between rice and fish:* Fish remove pests from rice through feeding activity, and rice plants moderate the field environment for fish which in turn promotes fish activity and pest removal. (B) *Complementary use of N by rice and fish*: Unused fertilizer N promotes plankton in paddy fields that is consumed by fish. The unconsumed fish feed acts as an organic fertilizer, and the N in the unconsumed feed can be gradually used by rice. Thus, rice and fish use different forms of N, resulting in a high efficiency of N utilization in RF.

Figure 5.1 Rice–duck co-culture in the field.

Figure 14.2 Degraded mountains of the Yuanmou dry–hot valley area before eco-agriculture development.

Figure 14.3 Gully control model integrated with *Leucaena leucocephala* (*Leucaena leucocephala* + *Albizia kalokora* + *Jatropha curcas* + *Azadirachta indica* + *Pennisetum purpureum*, etc.) in eco-agricultural development.

Figure 16.2 Author Wang Shuqing standing before a newly developed agroecosystem with terraces and shelterbelt in Baiquan County.

Figure 16.3 Well-developed agroecosystem with terraces, shelterbelt, and pond in Baiquan County. Author Wang Shuqing is standing in front.

PART III

Regional Eco-Agricultural Practice and Management

CHAPTER **11**

Status quo and Development Policies of Organic Agriculture in China

Wu Wenliang, Qiao Yuhui, Meng Fanqiao, Li Ji, Guo Yanbin, and Li Huafen

CONTENTS

Organic agriculture has been under development for more than 20 years in China. This chapter introduces the background and developmental stages of organic agriculture, the *status quo* of organic production and marketing, developmental policies, strategies, and measures for the continued development of organic farming in China.

11.1 INTRODUCTION

Conventional farming has contributed greatly to worldwide food security, including that of China; however, it also poses tremendous challenges, including the depletion of natural resources, the deterioration of the ecological environment, and some social and economic problems (Jiang, 2011). Organic agriculture, an environmentally friendly farming system, has been promoted to resolve the above problems. During China's 5000-year-long history of agricultural production, the Chinese people have developed much scientific skill and theory for traditional agriculture and have established diverse cropping and animal husbandry systems utilizing intensive cultivation, rational utilization of natural resources, soil fertility improvement, and many ecologically sound practices. These can be thought of collectively as an important heritage, now intensively utilized in modern organic agriculture in China.

11.1.1 Initiation Stage (1990–1994)

Modern organic agriculture in China began in the early 1990s. Organic tea was produced first and exported to European countries; SKAL (a Dutch certifier now known as Control Union) inspected and certified the first two tea farms in Zhejiang and Anhui provinces. The initial development of organic agriculture in China was mainly oriented toward exports. Before 1999, 95% of Chinese organic products were marketed to the European Union, North America, and Japan. More recently, the domestic market is becoming the main force driving organic agriculture in China.

11.1.2 Growing Stage (1994–2003)

Research on organic agriculture in China began in the late 1970s. In addition to research on environmentally friendly farming technologies at agricultural universities and institutes, the Nanjing Institute of Environmental Science collaborated with the University of California, Santa Cruz, to compare the energy flow, logistics, and economic flow of organic and conventional production systems in the Pacific Rim region and to perform a comparative study of organic and conventional wheat, vegetables, and rice production, a program launched by the Rockefeller Brothers Fund. In 1989, the Institute became an International Federation of Organic Agriculture Movements (IFOAM) member; in 1994, the Organic Food Development Center (OFDC) was established within the institute. The OFDC worked with the China branch of the Organic Crop Improvement Association (OCIA) on inspection and certification activities in order to promote the exportation of organic food. In the late 1990s, some other foreign certification bodies, including BCS Öko-Garantie GmbH, Ecocert, Japan Organic and Natural Foods Association (JONA), and IMO, jointly worked with Chinese organizations to certify Chinese organic products. Local certifiers were also established to control the quality of organic products. The first ministerial organic standard was developed by the State Environmental Protection Administration (SEPA) on June 19, 2001 (Du and Dong, 2007). Since the standard was still at the ministerial level, organic agriculture had not yet been promoted at the national level.

11.1.3 National Standards Development and Implementation Stage (2003–2012)

In 2003, authority for the certification management of organic products was transferred to the Certification and Accreditation Administration (CNCA) of the People's Republic of China, which is part of the General Administration of Quality Supervision, Inspection and Quarantine, and Standardization Administration (AQSIQ). The China National Accreditation Service for Conformity Assessment (CNAS) was also authorized to begin accreditation for certification bodies. National standards for organic products were officially developed and implemented in 2005. The organic standards of China were developed based on the basic standards of IFOAM, the *Codex Alimentarius* of the United Nations' Food and Agriculture Organization, the European Union EU-2092/91 standards and regulations, and the U.S. National Organic Program (NOP) (Wang and Lv, 2012). The new standards consisted of four parts: production, processing, labeling and marketing, and management systems. The promulgation and implementation of organic standards in China meant that all marketing, quality control, and labeling of organic products must be in accordance with these standards. In addition to the unified organic product certification and management system, since 2005 AQSIQ has issued the publications *Administrative Measures on Organic Product Certification* (AQSIQ Decree No. 155) and *Implementation Rules for Organic Product Certification*. It was during this period that the production and consumption of organic products in China began to develop rapidly.

11.1.4 Fast Growth and Quality Control Stage (2012–Present)

Since 2005, the production and consumption of organic products have experienced rapid growth; however, with the weak social trust system and low cost of illegal activities, organic products have not always been produced and marketed with respect for the national standards and rules. In order to protect the interests of consumers and maintain organic integrity, national standards and rules on organic product management have been undergoing revision since 2012 with the goal of establishing stricter controls for nonconformity, misrepresentation, labeling errors, and other unacceptable operations during the production, processing, and marketing of organic products. These revised standards and rules will further enhance the quality of products produced by organic farming in China.

11.2 ORGANIC PRODUCTION AND MARKETING IN CHINA

We collected organic agriculture data through the end of December 31, 2013, from the CNCA, which has a database of acreage, yield, and marketing.

11.2.1 Organic Production and Processing

11.2.1.1 Production Enterprises

By December 31, 2013, 10,000 organic certifications had been issued for 7894 production enterprises in China, covering 6628 organic farms and 3910 organic processing plants. Compared with 2010, the number of organic enterprises increased 44%, and certificates increased by 32%. The certified enterprises were mainly distributed in the eastern regions or developed regions, including Zhejiang, Shandong, Heilongjiang, Jilin, Beijing, Jiangsu, Liaoning, Guangdong, and Anhui provinces.

11.2.1.2 Organic Production Acreage

In 2013, the total area of organic production was 2.72 million ha, including 1.28 million ha of organic planting and 1.44 million ha of wild collection. Total yields of organic crops were 7.67 million tons, with 7.07 million tons of organic products and 0.6 million tons of wild collection products. The top three largest areas of organic production were grains, beans, and oil crops.

11.2.1.3 Organic Livestock and Poultry Production

About 830,000 organic beef cattle, 6.28 million sheep, and 198,000 pigs were organically certified in China in 2013. The main types of poultry under organic production were broilers and laying hens (14.6 million birds), ducks (0.11 million birds), and geese (65,000 birds). Over 421,000 tons of organic milk were produced.

11.2.1.4 Organic Aquatic Production

In 2013, the 195,000 tons of kelp and seaweed products accounted for 61% of the total organically certified aquatic products. The others included freshwater fish, with a yield of 87,000 tons (27%), and invertebrate animal products, with a yield of 27,000 tons (7%). The share of turtles, crabs, shrimp, and saltwater fish was less than 5%.

11.2.1.5 Organic Processed Products

A total of 4.34 million tons of organically processed products was produced, including 4.14 million tons of organic products and 1.95 million tons of organic conversion products. In processed products, the milled grain yield was 2.33 million tons (54% of total organically processed products). The yield of fruit juice and vegetable juice was 0.79 million tons (18.1%), followed by processed dairy products (0.39 million tons).

11.2.2 Organic Products Marketing

11.2.2.1 Domestic Organic Products Marketing

The total output value of organic products was 11.7, 9.6, and US$13.2 billion in 2010, 2012, and 2013, respectively (here, US$1 = 6.2 RMB). The output value of organic food in 2012 was 18% lower than in 2010, mainly due to the implementation of stricter organic product certification rules on March 1, 2012. However, the output value in 2013 increased again, 37% higher than in 2012. In 2013, the output value of organically processed products was the largest, with a value of US$8.97 billion (82% of total organic products). Organic plant products were valued at US$0.35 billion (13%); the others were aquatic products (1%) and animal products (4%).

11.2.2.2 Marketing Channels

At present, there are six sales channels for organic products: supermarkets; grocery stores; the Internet; group consumption, including hotels; farm direct sales and ecotourism; and home delivery (Zhang, 2011).

11.2.2.3 *Consumer Consumption Attitudes*

We undertook a survey on the organic product consumption of about 400 Chinese consumers in some large cities such as Shanghai, Nanjing, Xi'an, and Shenzhen (Zou and Jia, 2009). The survey revealed the following:

1. Organic food consumers are mainly 20 to 40 years old (69% of the surveyed consumers). The next largest group of consumers are 40 to 50 years old (17% of those surveyed), followed by consumers under 20 years of age (8%) and those older than 50 years (6%).
2. Male consumers accounted for 53% of the total organic food consumers surveyed, compared with 47% of female consumers.
3. Among the organic food consumers, 25% are white-collar workers, and 58% are civil servants, non-governmental organization (NGO) staff, or teachers. The monthly income of all organic food consumers is between US$484 and US$806.
4. Of organic food consumers, 68% have an undergraduate degree or higher, 23% have a college degree, and the remaining have secondary degrees.

11.2.2.4 *Exportation of Organic Products*

In 2013, the export trade value of Chinese organic products was US$456 million, twice the value of the export trade in 2012. Most organic products were exported to the European Union, with a trade value of US$363 million (90,252 tons), accounting for 80% of the total amount traded. The second best market was North America, with a trade value of US$74 million (65,681 tons), accounting for 16% of the total. The third most valuable market was the Asian market (mainly Japan, Hong Kong, Taiwan, and Macao), with a value of US$17 million, accounting for 4% of the total.

11.3 IMPLEMENTED SUPPORT POLICIES FOR ORGANIC AGRICULTURE DEVELOPMENT IN CHINA

11.3.1 Central Government

11.3.1.1 *Constructing an Organic Production Base/Area for Demonstration*

Constructing an organic production base for demonstration is the most important measure needed to promote organic agriculture in China. The Ministry of Environmental Protection (MEP) was the first ministry to set up a national organic food production demonstration base as a means of protecting the natural environment, beginning in 2003. By 2013, 138 national demonstration bases had been established throughout the country. Since 2010, the Ministry of Agriculture (MOA) has set up 13 organic production demonstration areas, and the Certification and Accreditation Administration (CNCA) has also set up 23 organic certification regions in China. This policy mainly promotes assessments of local governments and seldom directly assists organic producers, especially individual small households.

11.3.1.2 *Promotion of Organic Trade and Marketing*

Because the development of an organic industry in China is very much driven by the market, central government departments have issued some policies to promote both domestic and international marketing.

11.3.1.2.1 Attendance at Organic Product Expositions

Organization and participation in organic product expositions by central ministries are important to the support of organic industry development. Participation in an exposition can promote both exports and domestic market consumption of organic products as well as increase consumer understanding of the organic industry. The Trade Development Council Affairs of the Ministry of Commerce held the first China international organic agricultural products fair in Beijing in 2005; based on their experience there, they prepared a business guide for the export of organic agricultural products and set up an information exchange platform. In February 2005, the Chinese Green Food Development Center organized enterprises to participate in the 2005 Nuremberg International Organic Products Expo. Since then, the MOA has organized enterprises to participate in Biological Fachöffentlichkeit (BIOFACH), the world's leading trade fair for organic food, every year and promotes the export trade of organic products. During BIOFACH 2014, 80% of Chinese enterprises made organic deals, worth a total revenue of nearly US$50 million. Organized by Nürnberg Messe and the China Green Food Development Center, the first BIOFACH China was held from May 31 to June 2, 2007, in Shanghai. BIOFACH China has become the most influential domestic and professional expo for the organic industry. It is an important avenue for publicizing and developing the organic food industry in China.

11.3.1.2.2 Market Link

China's Ministry of Commerce issued guidelines to set up large supermarket chains with enterprises in order to implement the "agriculture super docking" project. This measure supports certified organic food production, improves the competitiveness of agricultural products, and promotes the benefits for farmers.

11.3.1.2.3 Support to Export Enterprises

Since 2006, the Ministry of Commerce and CNCA have jointly organized organic agricultural product export training and have provided financial support and funds for export enterprises to promote agricultural produce and textile product trade and export. These measures mainly concentrate on both large enterprises and export-oriented enterprises and are less related to organic cooperatives and individual farmers.

11.3.1.3 Research and Technical Support for Organic Agriculture

The Ministry of Science and Technology and the Ministry of Finance have implemented a special action program to improve farmer livelihoods using science and technology. This program focuses on promoting employment and farmers' incomes and enhancing local economic strength and the ability of science and technology to serve farmers. Many local governments implemented what are referred to as local pillar industries, which serve as examples of organic production, in addition to approving various projects and providing funding to the organic sector. From 2004 to 2006, the Ministry of Science and Technology funded research projects on the development of organic agriculture systems. A science and technology support program known as the Green/Organic Production and Fresh-Storage of Key Technology Development and Demonstration, was also funded from 2007 to 2010. In 2013, in order to address certification and accreditation technology problems in China's food safety and organic industry development, CNCA undertook a national technology support project with the long title of "Regional Advantages and Characteristics of Organic Product Certification—Key Technology Research and Demonstration," funded by the Ministry of Science and Technology. Implementation of the project will further promote the implementation and certification of organic products and promote both exports and domestic markets.

11.3.2 Local Government

Local governments have issued policies and taken many measures to support organic industry development at all levels. Since 2000, about two thirds of the provinces (municipalities, autonomous regions) have issued provincial support policies related to organic agriculture, including Beijing, Guangxi, Anhui, Hebei, Heilongjiang, Jiangsu, Jiangxi, Ningxia, Shandong, Xinjiang, Yunnan, and Zhejiang, in addition to other organic agriculture development areas. Within the provinces, cities or counties have also issued relevant policies to support organic agriculture. During the period from 2008 to 2010, about 10 local policies were created each year, which greatly promoted the regional development of organic agriculture. According to the policy content and project support, the policies can basically be divided into four types, described in the following text.

11.3.2.1 Capital Subsidies

Local governments give special funds to enterprises, organizations, and farmers both directly as organic production inputs and as certification fee subsidies. Of these, 109 support policies are considered capital subsidies, accounting for 73% of total policies. Direct subsidies to enterprises for organic certification accounted for most (70 to 80%) of the capital subsidies. Funds allocated to each enterprise are generally in the amount of US$1613 to US$8064 and basically cover the cost for organic certification.

11.3.2.2 Technical Support

Local government has taken the lead with special funding for regional organic agricultural planning, training, consulting, and technology promotion and for establishing organic production bases with high standards. They have also supported some research projects designed to demonstrate organic agriculture technology as well as the relationship between organic agriculture and ecological protection.

11.3.2.3 Financial Policy

Since 2009, local governments have supported enterprises, organizations, and individuals in carrying out organic production with the preferential and diversified policies of infrastructure construction, loans, and taxes. In 2010, Chengdu City in Sichuan Province introduced support policies for the overall construction of organic agriculture, including construction of production bases, establishment of an organization management system, implementation of supportive technology for organic production and deep processing, and development of organic markets. Funding and support policies have been diversified, including priority enterprise bank loans to the organic sector, development and promotion of organic production techniques, export tax rebates to organic enterprises, and the cultivation, application, market expansion, and protection, among other policies, of organic product brands.

11.3.2.4 Evaluation Indicator for Local Government Performances

Some local governments have integrated organic industry development with performance evaluations for local officials. In Beipiao City of Liaoning Province, agricultural standardization and organic certification are included in the evaluation index for village and town officials. In Jiangsu Province, the development of organic agriculture falls under the county environmental protection responsibility.

11.4 STRATEGIES, POLICIES, AND MEASURES FOR ORGANIC AGRICULTURE DEVELOPMENT IN CHINA

In general, Chinese organic farming is still in the early stage of development compared with other developed countries. Domestic sales are rising gradually, and the formulation of standards, laws, and regulations of organic agriculture has just been completed. Industry policy and management systems are far from being developed thoroughly enough. It is necessary to promote the sustained and healthy development of organic agriculture at nationwide levels of strategy and policy making (Wu, 2008).

11.4.1 Strategies

"Making people richer, strengthening the enterprises, rejuvenating the country, promoting public health, and creating a harmonious development" can be considered to be the core idea of the organic industry development strategy in China, based on ecological and resource advantages guided by scientific principles, sustained and healthy development, and market demand.

11.4.2 Key Policies

Construction of organic agriculture is a process of engaging contemporary society with nature in a harmonious way, with the participation of the whole society and to the benefit of future generations. By the end of 2020, key strategies include research and development of key technologies for organic production and processing; improvement of the organic industry technology support and promotion system; training personnel for local management, marketing, technology, processing, and organic farm work; and continuous transformation of scientific and technological achievements into organic farming, demonstrations, and promotion of standardized production.

11.4.3 Measures

With regard to promoting production efficiency, harmonious development, and environment protection, the concepts of "quality" and "efficiency" are key to establishing an organic industry base in China, to creating and nurturing China's organic brands, and to improving the marketing and competitiveness of organic products. The following steps must be taken:

- Integrate and optimize all of the resources in the management system, policy development, project support, scientific and technological support, standards, personnel training, information exchange, and pilot demonstrations necessary to build capacity.
- Fully mobilize the enthusiasm of all sectors in society. The government needs to establish social participation and investment guarantee mechanisms, encourage the establishment of various forms of organic product development and sales models, and fully respect enterprises, farmers, and other practitioners of practical experience and pioneering spirit. In addition, the government should further strengthen supervision and services, guaranteeing the organic agriculture industry's sustainable and healthy development.
- Under the guidance, coordination, and support of the government, the various enterprises, technical personnel, and farmers/producers should follow the natural law of economics; make efforts to promote the development of organic production, processing, and preservation technology; and establish organic industry standards in China. They should strengthen the application and demonstration of technical standards, quality management, internal inspection, and management of organic products, with full traceability.

REFERENCES

Chinese Green Food Development Center. 2005. *The German Nuremberg International Organic Fair Successfully Held with Chinese Delegation Organized by the Ministry of Agriculture.* China Agricultural Information Network: Beijing (http://www.moa.gov.cn/sydw/lssp/zt/zhxxi/jwblh/201203/t20120320_2512738.htm)

Du, X.G. and Dong, M. 2007. Organic agriculture development in China: situation, advantages and counter-measures. *Agric. Qual. Stand.*, (1): 4–7.

Jiang, G.M. 2011. *Chinese Ecological Environment Crisis.* Hainan University Press: Haikou.

Wang, M.H. and Lv, Yan. 2012. The overview of organic product certification management. *Cert. Technol.*, 5: 35–36, 39.

Wu, W.L. 2008. The challenges and countermeasures for the development of organic agriculture in China. *Beijing Agric.*, 6: 1–3.

Zhang, X.M. 2011. Development mode of the organic products' supply chain, bottleneck and countermeasures. *Hebei Agric. Sci.*, 15(3): 125–127.

Zou, W.H. and Jia, J.R. 2009. Research on obstacle factors of organic food marketing—an empirical analysis based on the domestic organic food market. *Shaanxi Agric. Sci.*, 6: 178–182.

CHAPTER 12

Highly Efficient Utilization of Rainfall Promotes Dryland Agroecosystem Sustainable Development on the Loess Plateau

Li Fengmin and Guan Yu

CONTENTS

12.1 REGIONAL BACKGROUND

The Loess Plateau is located in the north central region of China, at latitude 34° to 40°, longitude 103° to 114°. It stretches over 1000 km from east to west and about 700 km from north to south, including the areas west of the Taihang Mountains, east of the Riyue Mountains in Qinghai Province, north of the Qinling Mountains, and south of the Yinshan Mountains (Figure 12.1A). It covers many provinces, including Shanxi, Shaanxi, Gansu, Qinghai, Ningxia, Inner Mongolia, and Henan, and has a total area of about 640,000 km² and an altitude ranging from 800 to 2400 m. Apart from a few rocky mountains, most parts of the Loess Plateau are covered with thick loess which is a deep silt loam blown in from central Asian desert area and deposited over many millennia. The average thickness of loess is about 50 to 80 m and sometimes may even be up to 150 to 180 m. The loess has a delicate and uniform texture with a particle size of only 0.005 to 0.05 mm.

The average annual precipitation in the Loess Plateau is 434 mm. This value varies from 200 to 700 mm in different areas, with a general trend of more precipitation in the southeast than in the northwest (Figure 12.1B). The precipitation also has many seasonal variations. Most rainfall occurs in the summer (June to August), accounting for 54.8% of the annual precipitation. Of the total annual precipitation, 25.8% occurs in the autumn (August to November) and 7.7% in the spring (March to May). In the winter (December to February), rainfall accounts for only 0.06% of total

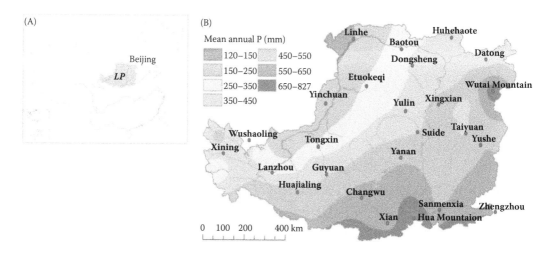

Figure 12.1   (A) Location and coverage of Loess Plateau, and (B) mean annual precipitation distributions in the Loess Plateau. (From Wan, L. et al., *Hydrol. Proc.*, 28(18), 4971–4983, 2013.)

precipitation. About 78 to 92% of the annual precipitation occurs during the rainy season (June to September). The annual average temperature is 8.8°C (spring, 10.0°C; summer, 20.9°C; autumn, 8.8°C; and winter, –4.6°C). The temperature also shows a higher value in the southeast than in the northwest (Y.R. Wang et al., 2004). Over the past 50 years, the average temperature of the Loess Plateau has significantly increased, at a higher rate than that of the Northern Hemisphere and even the world average (Q.X. Wang et al., 2012). However, the annual precipitation has decreased by 49 mm, with an average decrease of 1.5 mm per year, which equates to 0.26% of the average annual precipitation in the Loess Plateau. The average decreases in precipitation were –0.09 mm/a, –0.57 mm/a, and –0.19 mm/a in spring, summer, and autumn, respectively, and the average increase in winter was 0.065 mm/a (Wan, 2011; Wan et al., 2013).

The Loess Plateau is one of the world's most critical soil erosion areas, where the soil erosion is closely linked to the development of dryland agriculture. Thousands of years ago, the Loess Plateau was composed of many large, flat plateaus with very small gullies, and the forest coverage reached 53% (N.H. Shi, 1988, 2001; H. Shi and Shao, 2000; X.Z. Xu et al., 2004; Chen et al., 2007). With population growth, large areas of natural vegetation (forest, shrub, and grassland) were reclaimed for farmland, and the erosion of sloping farmland greatly increased. The forest coverage was 50% about 2000 years ago, fell to 33% about 1500 years ago, and then to 6.1% in 1949 (N.H. Shi, 1988, 2001). The decrease of vegetation cover accelerated soil erosion. Throughout the Loess Plateau, with a total area 642,000 km², soil erosion areas account for 60% of the 454,000 km², with 337,000 km² of water erosion areas and 117,000 km² of wind erosion areas.

Due to severe land degradation, agricultural production in the Loess Plateau was weak and unstable for a long period, which resulted in food shortages and impoverishment of the local people. In the mid-1990s, the State Council approved a national poverty alleviation plan that included 592 counties and 80 million people; of these, 115 counties and 23 million people belonged to the Loess Plateau and accounted for 21.3% of the total counties in poverty and 28.8% of the total national poor population (CSC, 1994).

The Loess Plateau, with a long history of development and serious land degradation, is one of the origins of the world's dryland agriculture. The area of arable land in this region is 1548 × 10⁴ ha. Data from 2008 show that farmland with a slope of >5° accounted for 69% of the total arable area (farmland with a slope of 5 to 15° accounted for 42%, 15 to 25° for 20%, and >25° for 7%). The area of irrigated farming accounted for only 30% of the total arable area, mainly distributed in the flat zone along the rivers of Hetao Plain and Fenwei Plain. The area of rainfed farming was 1020.6×10^4 ha, accounting for 70% of the total arable area, most of which was sloping land and mainly distributed in semiarid hilly regions with an average annual precipitation of 250 to 550 mm (Wan, 2011).

12.2 DEVELOPMENT OF DRYLAND AGRICULTURE

Since the founding of new China in 1949, the development of dryland agriculture in the Loess Plateau has undergone several important changes (R.Y. Guo and Li, 2014). Before the 1970s, people learned from their traditional dry farming experience and built terraces; however, most terraces with gentle slopes had little effectiveness on soil and water conservation. Between the 1980s and 1990s, horizontal terraces and soil-retaining dams were constructed to manage the small watersheds more comprehensively. Farmers focused on the reasonable use of fertilizer and adjustment of soil moisture, and biological resources were also used to improve overall agricultural efficiency. In the late 1990s, rainwater catchment agriculture was developed which offered more efficient use of rainfall, and the use of rainwater harvesting for supplemental irrigation and ridge–furrow mulching techniques also developed rapidly. Since the beginning of the 21st century, field rain-harvesting cultivation technology has developed more fully, and the technical system continues to mature and stabilize. Growing potatoes and maize with this technology can produce high yields, and the economic benefit has been improved due to rapidly increasing market demand. The planting areas of maize and potatoes have increased quickly, and the cultivation of traditional low-yielding grain crops, such as buckwheat, millet, oats, lentils, broad beans, and mung beans, has been steadily decreasing. Farmers now adjust crop structure according to market demands.

During the 30 years of reform and opening up, the grain yield of dry land in Gansu Province in the Loess Plateau increased from 1200 kg ha^{-1} in 1978 to 2700 kg ha^{-1} in 2008, and the per capita share of grain increased from 250 kg to 350 kg (Figure 12.2) (Yang et al., 2009). During this period, the contract responsibility system of the 1980s brought an almost linear improvement to grain yields. In the 1990s, grain yields fluctuated tremendously, with almost no increase overall. Grain yields increased from 1800 kg ha^{-1} at the end of the 1990s to 2400 kg ha^{-1} in 2010, and the per capita share of grain increased from 300 kg to 350 kg.

With the support of micro-field rain-harvesting technology, especially ridge–furrow mulching, potatoes and maize yields increased rapidly. In 2010, the government of Gansu Province formulated and published a construction and development plan (GPPG, 2010) that established the central semiarid regions as a hub to produce 2.5 billion kg of grain. The wheat planting areas were required to remain unchanged from 2009 to 2015, but the average yield increased from 2730 kg ha^{-1} (2000 to 2008 data) to 3000 kg ha^{-1}. The maize planting areas increased from 500,000 ha to 733,000 ha, and the average yield increased from 4650 kg ha^{-1} (2000 to 2008 data) to 5490 kg ha^{-1}. The potato planting areas increased from 643,500 ha to 666,700 ha, and the average yield increased from 3105 kg ha^{-1} (2000 to 2008 data) to 4500 kg ha^{-1}. The planting areas of other crops were planned to decrease from 476,000 ha to 333,000 ha, and the average yield remained at about 2310 kg ha^{-1}.

12.3 YIELD EFFECT OF FARMLAND MULCHED WITH PLASTIC FILM

Plastic film mulch is a field crop production method in semiarid areas. The ridge of a field is covered with plastic film to prevent heavy evaporation and preserve water in the soil. At the same time, the mulch can increase soil temperature in early spring (Zhou et al., 2009, 2012). Research into the effect of plastic film mulching on field crops began in the 1980s in China. As the most important food crop

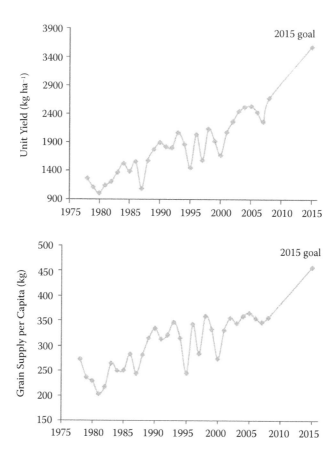

Figure 12.2 Dynamics of grain yield per hectare and grain supply per capita in the Loess Plateau of Gansu Province. (Data partially from Yang, X.M. et al., *The Great Changes in Gansu Rural Areas—Data Collection of Gansu Rural Reform and Open for 30 Years*, Gansu People Press, Lanzhou, 2009.)

in northern China, wheat is often a pressing concern of the government, residents, and researchers. Film-mulching technology increased wheat yield by 50% compared with normal cultivation, as demonstrated in the irrigation district of the North China Plain (CDATES, 1983), dry lands of Loess Plateau (F.M. Li et al., 1999), and irrigation areas of drought oases (S.Q. Li and Lan, 1995). Due to the large fluctuation of annual precipitation in dryland agricultural areas in the Loess Plateau, the yield fluctuation is also large. Film-mulching technology may cause a decrease of grain yield under severe drought (F.M. Li et al., 2005). After the development of a seeder machine designed for planting wheat with the film-mulching technology, the practice of film mulching wheat increased rapidly in the 1990s. In order to save labor when wheat seeding using film-mulching technology, Xue et al. (2006) proposed film side planting, which involves planting wheat in the furrow by the side of the mulching film. Film side planting reduced many field management activities and increased wheat yield by 40% compared with traditional cultivation, by 31% compared with no-tillage high stubble mulching, and by 20% compared with stubble mulching and deep loosening of the soil The grain yield of wheat reached 7311 kg ha^{-1} with film side planting; the annual precipitation was 685 mm during the period the experiment was carried out (Xue et al., 2006). Mulching a layer of film on soil before wheat planting could significantly reduce weed growth and save labor. The land requires no tillage for three to four years, and the grain yield can increase by 35% compared with conventional cultivation (Han, 2013), providing significant economic benefits compared to other farming methods.

Potatoes with high yields are traditionally dominant crops in the Loess Plateau. Potatoes grow in the same period as temperature increases and rainfall occurs; however, farmers had little initiative to plant potatoes due to the low level of market demand. Since the beginning of the 21st century, potatoes have gained acceptance in the market in this region. Because of the growing market demand, farmers increased their production of potatoes. The ridge–furrow mulching technologies have played an increasingly important role in potato planting. X.L. Wang et al. (2005) achieved the highest yields and efficiency of water use in potatoes when the widths of the ridges and furrows were 45 cm and 60 cm, respectively. The highest yield was 50% greater than that for traditional cultivation. In an area with an average annual precipitation of less than 400 mm and free water evaporation of 1500 mm, potatoes were planted by the side of V-shaped ridges. The yield of potatoes with a fully mulched treatment reached 4311 to 5300 kg ha^{-1}, which was 57 to 78% greater than that for flat-plot sowing without ridge–furrow mulching (H. Zhao et al., 2014). Taking off the film in the middle of the potato growing season (65 days after sowing) could not only lower the risk of plastic residue pollution but also maintain the potato yield at the same level as yields from fields mulched throughout the entire growing season (H. Zhao et al., 2012).

Maize is a thermophilic crop characterized by high consumption of water and fertilizer. It is often planted on flat land with more suitable growing conditions in the Loess Plateau. Over the past several decades, low temperatures and drought have occurred often in rainfed areas, and, in addition to its lower selling prices, maize was not a main food for the local people, so farmers had little incentive to plant maize. In the past decade, however, the onset of a global food crisis and the strong demand for livestock feeding have caused the price of maize to increase up to nearly the same level as wheat, which has a saturated supply in the domestic market. As a high-yielding crop, maize can produce more economic benefits than wheat. The stalks of maize can not only be used as forage and biogas but can also increase soil fertility when applied back into the field. With so many advantages, maize has gained the attention of the government, farmers, and scientists.

The yield of conventional cultivated maize is often higher than that of wheat. The Loess Plateau is located in a zone with southeast monsoons and northwest cold waves, which results in low temperatures in spring and autumn and drought, both of which have a negative impact on maize production. After the application of ridge–furrow mulching, both the production and the planting area of maize greatly expanded.

Many mulching and planting arrangements have been developed, including (1) flat plots mulched with plastic film, (2) alternating ridges and furrows with only the ridges mulched with plastic film, and (3) ridge–furrows coupled with plastic mulching (Zhou et al., 2009, 2012). The third arrangement is considered easy and efficient. In Xiaokangyin Village, Yuzhong County, Gansu Province, the average annual precipitation is 388 mm and average temperature is 6.7°C. Maize yields with ridge–furrow mulching technology reached 8 t ha^{-1} in 2009 and 12 t ha^{-1} in 2010, 105% and 89% higher, respectively, than the yields of flat plots mulched without plastic film (Wu et al., 2012). Ridge–furrow mulching technology can effectively improve topsoil temperature and humidity, which enhances the range of suitable areas for maize planting. In the Northern Mountain regions (2400 m above sea level), Yuzhong County, Gansu Province, the average annual precipitation is only 300 mm, and the average annual temperature is 6.5°C. This region was not considered a place to grow maize in the past, but the use of ridge–furrow film mulching technology could produce a maize yield of 6 to 7 ton ha^{-1} (C.A. Liu et al., 2009; Zhou et al., 2009). Despite a slight decline in maize production, the use of plastic film once every two years, coupled with the no-tillage practice, was shown to reduce inputs and produce greater economic benefits compared to the existing practice of plastic film mulching with conventional tillage (C.A. Liu et al., 2009). Our results from a 4-year field experiment indicate that the lower limit of average annual precipitation for the use of the ridge–furrow film mulching technology was 273 mm (X.F. Liu et al., 2013). Therefore, the entire region of the Loess Plateau is well within the suitable range for maize production. In the Loess Plateau, maize production utilizing the ridge–furrow mulching technology can reach

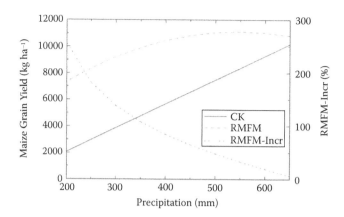

Figure 12.3   Increase in maize grain yield for ridge–furrow with plastic film mulching. (Data from Ye, J.S. and Liu, C.A., *J. Agric. Sci. (Can.)*, 4(10), 182–190, 2012.)

6 to 9 ton ha^{-1} in areas where the average annual precipitation is 300 to 400 mm, and 9 to 12 ton ha^{-1} in areas where the average annual precipitation is 400 to 600 mm (L.X. Li et al., 2009). Ye and Liu (2012) performed a meta-analysis of the published data for ridge–furrows with plastic film mulching and found that this technology is suitable for areas of annual precipitation less than 600 mm. The percentage of grain yield increase was higher in regions with less precipitation than in regions with more precipitation, which indicates a significant opportunity for semiarid areas to increase crop productivity (Figure 12.3).

Large-scale research on rainwater-harvesting agriculture practices started in the early 1990s (S.L. Zhao et al., 1995; S.L. Zhao and Li, 1995). In the late 1990s, the development of water-harvesting agriculture was rapid due to the support of the government. Micro-field water-harvesting systems developed for the production of field crops and rainwater collecting led to more efficient, higher value farming (F.M. Li et al., 1995, 1999; J. Wang et al., 2001; F.M. Li and Xu, 2002). These systems have played an important role in raising grain yields to new levels and enhancing the production and scale of economic crops.

Film mulching is the core of micro-field rainwater-harvesting systems. Over the past 10 years, ridge–furrow mulching technology has continued to improve and develop, stimulating a significant increase in the production capacity of potatoes (Tian et al., 2003; X.L. Wang et al., 2005; H. Zhao et al., 2012) and maize (L.X. Li et al., 2009; C.A. Liu et al., 2009, 2014; Zhou et al., 2009), as well as increasing the scale of production. The planting area of other traditional crops has decreased due to low yields and low economic returns. Even the planting area of wheat (the main food crop of the region) has decreased gradually. The adjustment of farming structures, oriented by market demand, has pushed the continuous improvement of agricultural production capacity in the region.

12.4 POSSIBLE ENVIRONMENTAL RISK OF FILM MULCHING

Increased yields with film mulching should be attributed to the improvement of topsoil moisture and temperature, which could significantly increase the soil respiration rate (F.M. Li et al., 1999, 2004b, 2005). In Dingxi City, Gansu Province, we mulched films on a field without planting crops and measured the soil respiration rate during the growing season. The results showed that the soil respiration rate with film mulching was 1.9 to 2.96 times higher than that with no film mulching (F.M. Li et al., 2004a). We concluded that film mulching accelerated mineralization of soil organic matter, which could potentially lead to a worsening of soil quality, especially for loess soil. In wheat farmland

Table 12.1 Correlation among Soil Microbial Biomass and Soil Organic C, Total N, Available N, and Olsen P Content

Year	Organic C (g kg⁻¹)	Total N (g kg⁻¹)	Available N (mg kg⁻¹)	Olsen P (mg kg⁻¹)
1	−0.949**	−0.732**	−0.702*	−0.822**
2	−0.795**	−0.519*	−0.829***	−0.678**
Both	−0.873***	−0.632***	0.140	−0.634***

Source: Adapted from Li, F.M. et al., *Soil Biol. Biochem.*, 36(11), 1893–1902, 2004.

Note: $p < 0.05$, $p < 0.01$, $p < 0.001$.

with film mulching, there were significant negative correlations between microbial biomass carbon and some soil nutritional factors, including soil organic matter, soil total nitrogen, mineral nitrogen, and available phosphorus (see Table 12.1) (F.M. Li et al., 2004a), suggesting that these important parameters were sensitive to a change of topsoil moisture and temperature, which decreased with the increase of soil microbial activity.

These phenomena indicated a positive-feedback mechanism in farmland soil ecosystems in that film mulching improved soil water and temperature, and the soil respiration rate increased, which strengthened the mineralization of organic matter. Also, a decrease in the soil C/N ratio and increase of C/P ratio disturbed the balance of N and P supply. The result was an increase of nitrogen loss and a decrease of phosphorus availability, leading to further degradation of soil quality.

In a 2-year experiment with maize, film mulching significantly increased the soil microbial biomass carbon (MBC). Significant negative correlations were found between MBC and soil organic carbon, light fraction organic carbon, total nitrogen, and mineral nitrogen (C.A. Liu et al., 2014), findings similar to those for film-mulching wheat. In another 2-year experiment with maize, MBC was negatively correlated with soil organic carbon and mineral nitrogen but was positively correlated with light fraction organic carbon and available phosphorus. Film mulching improved the light fraction organic carbon compared with no-mulching treatment, indicating that soil organic matter might also be increased by film mulching (L.M. Zhou et al., 2012). In a 5-year experiment with maize, although no significant difference was observed in soil light fraction organic carbon with or without film mulching, total soil organic carbon (SOC) was significantly lower with film mulching than with no mulching at the initial stage. However, with increased planting duration, SOC and light soil organic carbon increased gradually and significantly with film mulching (X.E. Liu et al., 2014).

The above results remind us that wheat production, which uses less organic fertilizer input and produces lower biomass than corn, can lead to a continuous decline in soil organic matter with film mulching; however, long-term maize planting with film mulching may actually increase soil organic matter inputs due to the high biomass of maize. Although film mulching will accelerate mineralization of soil organic matter, the input of organic matter by maize planting is much larger than the enhanced consumption through mineralization by film mulching. Therefore, the total soil organic matter content should continue to increase, which benefits dryland agroecosystems on the Loess Plateau. Further research is required to learn more about the mechanisms of organic matter dynamics in film-mulched farmland.

12.5 LIMITATIONS OF PRODUCTION FUNCTION AND ADVANTAGES OF ECOLOGICAL FUNCTION FOR ALFALFA PLANTATIONS

Since the 1980s, the government and scientists have reached a consensus that, in order to control excessive reclamation of land for agriculture, reduce soil erosion, increase people's income, and promote poverty alleviation, farmers in the Loess Plateau should reduce the planting area of sloping farmland and develop animal husbandry by increasing the area of artificial pasture/grassland. At the beginning of the 1980s, the entire society answered the call of the central government and participated in the introduction of vegetation to the Loess Plateau, with an emphasis on planting trees and grass to encourage animal husbandry (S. Shi, 1986). From 1994 to 2004, the World Bank Loan Project carried out an ecological restoration project in the Loess Plateau, including the provinces of Shaanxi, Gansu, Ningxia, and Inner Mongolia, among others. The project planned to build terraced fields to improve food production capacity, in addition to planting trees and grass to increase vegetation coverage and develop animal husbandry. The project achieved significant success in promoting sustainable development in the Loess Plateau (Qi et al., 2008); however, planting grass to develop animal husbandry did not go smoothly. As Shan and Xu (2009) pointed out, the positive effects would decrease once the project finished, and the quick retreat from planting grass became an indisputable fact. Upon entering the 21st century, the importance of returning cultivated land to its original forest and grass habitat was again highlighted, but the government and scientists advocated for the use of barn feeding instead. Over the past 10 years, the practical impacts of planting alfalfa to develop animal husbandry have been limited and much lower than expected (Ren and Chang, 2009; Ren et al., 2009; Shan and Xu, 2009; Shan, 2011).

In many countries, such as Australia and New Zealand, the rotation of legume pastures and crops is an important measure taken to maintain farmland sustainability. In the semiarid Loess Plateau, alfalfa is the most important legume pasture crop, but the biomass production of alfalfa is often lower than that of wheat, potato, and maize in the same areas (Jia and Li, 2011). In areas with an average annual precipitation of 300 mm, biomass production of alfalfa was improved by ridge–furrow film mulching technology and was 25% higher than the aboveground biomass of wheat, but still 30% lower than that of maize (Jia et al., 2006, 2009). As forage to meet the requirements of animal husbandry, alfalfa has a lower competitive advantage than maize and wheat and offers fewer economic benefits than potatoes, so farmers are not willing to plant alfalfa.

Similar results were found in a survey of various regions in the Loess Plateau (W.S. Zhang et al., 2012; W.S. Zhang, 2013). In Zhonglianchuan Village, Yuzhong County, Gansu Province (which has an average annual precipitation of 300 mm), the scale of barn-feeding sheep and interest in developing barn feeding were negatively correlated with the sales volume of potatoes, positively correlated with total grain production (mainly maize), and negatively correlated with the biomass production of alfalfa (but not significantly). This indicates that planting alfalfa is not enough to support the development of barn-feeding sheep, and the production capacity of maize determines the scale of barn feeding. Because potato stalks and tubers cannot be used directly to feed the sheep, barn-feeding sheep on a smaller scale resulted in larger potato-planting areas (W.S. Zhang et al., 2012).

Mowing is a common method of utilizing alfalfa pasture. We conducted a study on whether or not mowing can restore the soil productivity of alfalfa pasture. We found that soil organic carbon (SOC), soil total nitrogen (STN), and soil SOC/STN ratios declined slowly in an alfalfa pasture with mowing and conventional cultivation methods (Figure 12.4), suggesting that the soil quality did not improve. For this reason, it is not possible to restore farmland soil productivity by crop rotation with this type of alfalfa pasture (H.M. Jiang et al., 2006).

Ecological vegetation restoration is an important step in improving the ecological environment of semiarid areas. Returning farmland to forests and pasture/grassland is one of the main policies supported by the Chinese government. In a 6-year experiment, we found that the biomass on fallow land was always approximately 1000 kg ha^{-1}. On sweet clover pasture, it was approximately 2000 kg ha^{-1}, except for a significant decrease in the third year. On the alfalfa pasture, it was over 2000 kg ha^{-1} after the third year and over 6000 kg ha^{-1} in the sixth year (Z.B. Guo, 2011). The result means that the biomass of introduced vegetation restoration is higher than for natural recovery on abandoned land. Obviously, the alfalfa pasture has clear advantages in increasing the carbon pool and reducing soil erosion (Figure 12.5).

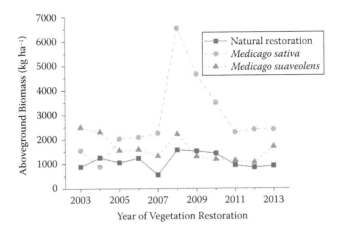

Figure 12.4 Aboveground biomass dynamics of alfalfa and sweet-clover vegetation restoration after cropland fallow (unpublished data).

Alfalfa pasture is often criticized for its large consumption of deep soil moisture. We used a 9-year-old alfalfa pasture field for annual crop (millet) production. After 4 years of rotation with different crops, the soil moisture content within the 0- to 5-m layer recovered by over 80% (X.L. Wang et al., 2008, 2009). With regard to alfalfa vegetation for ecological protection, the decrease of soil moisture should not be so serious due to low water consumption rate.

We can conclude that alfalfa is not ideal as the main source of forage for developing animal husbandry in the Loess Plateau, due to lower productivity than main crops such as maize and potatoes. Fallow vegetation with an extremely low production rate cannot support animal husbandry on grasslands. The practice of barn-feeding sheep has demonstrated that alfalfa is an inferior type of forage, and maize is becoming the main source of forage for developing animal husbandry. Alfalfa can only be considered to be a secondary forage in support of animal husbandry that uses both grain

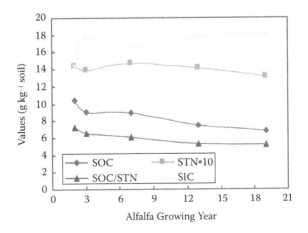

Figure 12.5 Soil organic carbon (SOC), soil total nitrogen (STN), ratio of SOC to STN, and soil inorganic carbon (SIC) in the 0- to 20-cm topsoil layer of alfalfa pasture over various growing years. (Data from Jiang, H.M. et al., *Soil Biol. Biochem.*, 38(8), 2350–2358, 2006.)

crops and grassland/pasture. On the other hand, alfalfa offers much greater productivity compared with natural recovered vegetation or sweet clover and is a better option than natural restoration or sweet clover pasture for the restoration of vegetation cover in Loess Plateau.

12.6 CONCLUSIONS AND RECOMMENDATIONS

The Loess Plateau is one of the birthplaces of Chinese culture. The population keeps increasing, and the climate tends to be warming up and drying out. In order to solve the problem of food shortages and impoverishment, the local people have continued to reclaim the land to grow crops and to increase grazing intensity on the natural grassland. Despite all of these efforts, those living in this area have only been able to maintain a basic livelihood. In a severe drought year, many are forced to beg for a living. Excess reclamation and grazing caused by low agricultural productivity lead to more severe soil erosion and land degradation. Severe poverty and ecological degradation have plagued the local people. Since the founding of new China in 1949, the government, together with scientists and local residents, has endeavored to explore how to further develop agricultural production and restore ecosystems in the region (R.Y. Guo and Li, 2014). In recent years, ecological restoration and reconstruction on the Loess Plateau have made significant progress. Through ridge–furrow mulching technology, maize yields increased by 138% in an area with annual precipitation of 300 mm and increased by 33% in an area with average annual precipitation of 550 mm (Ye and Liu, 2012). In areas or years with deficient water and low temperatures, yield increase potential is greater than in areas or years with better water and temperature conditions. This technology can greatly increase farming production capacity in semiarid areas. Similar results were observed with potatoes (H. Zhao et al., 2012, 2014). Crop production systems are concentrating on maize and potatoes as market demand increases. Meanwhile, the planting areas of other traditional crops with short growth periods and low yields are gradually being reduced. The natural adjustment of planting structures is promoting the continual increase of crop production capacity in this region.

In the Loess Plateau, legume pasture combined with grassland animal husbandry with mostly cattle and sheep has been considered the main strategy for coordinating agricultural production with environmental improvement in the region. People have made great efforts to promote the development of pasture/grassland animal husbandry, but the result has been far from ideal. Over the past 30 years, people hoped to develop cattle and sheep husbandry with introduced alfalfa pasture and other grassland, but, due to lower economic benefits than production crops, farmers lacked the incentives to plant alfalfa/grassland. With the support of new dryland farming technologies, crop production capacity is greatly improved. The improvement of maize production offered an unprecedented opportunity to develop cattle and sheep husbandry. Maize stalks and grain can provide rich sources of forage for the development of animal husbandry. With the support of high-efficiency dryland crop production, animal husbandry will be an important foundation in increasing the income of local farmers.

Although alfalfa pasture is difficult to use as the primary forage plant in support of the development of pasture animal husbandry, it is still indispensable as an important forage protein source in cattle and sheep husbandry. At the same time, alfalfa plants have deep root systems with a strong ability to use soil water, which plays an important role in the regulation of deep soil water for plant production. Compared with natural restoration on fallow land, planting alfalfa on fallow land can rapidly improve biomass and coverage, better control soil erosion, increase soil carbon and nitrogen storage, and improve soil quality, all of which contribute to the restoration and sustainable development of degraded ecosystems. Therefore, planting alfalfa on hillside fallow fields is a simple, quick, and efficient ecological engineering measure to speed up ecological restoration that should be extended.

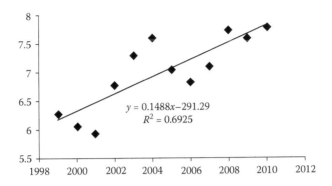

$$y = 0.1488x - 291.29$$
$$R^2 = 0.6925$$

Figure 12.6 Interannual change trend of the growing season normalized difference vegetation index (NDVI) in the Loess Plateau. (Adapted from Yi, L. et al., *Resour. Sci.*, 36(1), 166–174, 2014.)

Over the past decade, the per capita possession of grain and economic income continued to increase with the rapid development of dryland agriculture in the Loess Plateau (Y.L. Zhang et al., 2010; B.Q. Xu and Shi, 201). The rapid urbanization in China prompted more and more local farmers to work in cities, resulting in a decrease in the resident population, and the pressure on local ecosystems lessened. The project of returning cultivated cropland to original forest or grassland, started by the central government in 2000, also provided huge support to vegetation restoration. Although the climate tends to be warming up and drying out in the Loess Plateau, the vegetation coverage has tended to increase since the start of the 21st century (X.F. Liu et al., 2013; Yi et al., 2014). This phenomenon is obviously both important evidence that degraded ecosystems have been recovering (Figure 12.6) and an important milestone of dryland agriculture development and ecological construction during the past few decades.

The restoration of degraded dryland ecosystems in the Loess Plateau required over 60 years of hard work. A variety of measures and policies for developing dryland agriculture and restoring ecological sustainability were tried; yet, the state of the ecological environment only showed improvement in some areas and appeared to generally deteriorate. That situation had not changed until the end of the last century (X.C. Liu et al., 2001); however, today the degraded ecosystem is shifting from deterioration to restoration with increasing vegetation coverage (X.F. Liu et al., 2013) and is moving in a sustainable direction. This transformation has not come easily, and more research is necessary to promote ecological restoration and dryland agriculture development in the semiarid Loess Plateau.

REFERENCES

Cangzhou District Agricultural Technological Extension Station (CDATES). 1983. Cultivation technology of plastic film mulching once for two years. *Hebei Agric. Technol.*, 6: 12–13.

Chen, L.D., Wei, W., Fu, B.J., and Lü, Y.H. 2007. Soil and water conservation on the Loess Plateau in China: review and perspective. *Prog. Phys. Geogr.*, 31(4): 389–403.

Chinese State Council (CSC). 1994. China's National Targeted Poverty Reduction Program, http://www.people.com.cn/item/flfgk/gwyfg/1994/112103199402.html.

Gansu Provincial People's Government (GPPG). 2010. Plans for increasing by 2.5 million tons the grain production capacity of dryland farming in Gansu Province, China (2009–2015), http://www.gansu.gov.cn/art/2010/10/18/art_801_188770.html.

Guo, R.Y. and Li, F.M. 2014. Agroecosystem management in arid areas under climate change: experiences from the semiarid Loess Plateau, China. *World Agric.*, 4(2): 19–29.

Guo, Z.B. 2011. Dynamics of the Aboveground Biomass and Soil Physical Properties in the Abandoned Land with Different Treatments in the Semi-Arid Loess Plateau of China, PhD thesis, Lanzhou University, Lanzhou.

Han, X.J. 2013. Thin soil covering plastic film mulching cultivation and hole sowing for dryland winter wheat. *Agric. Technol. Equip.*, 2: 8–10.

Jia, Y. and Li, F.M. 2011. Production and ecological benefits of alfalfa fields in the Loess Plateau. In: Wu, J.G. and Li, F.M., Eds., *Lectures in Modern Ecology*. V. *Large-Scale Ecology and Sustainability Science*. Higher Education Press: Beijing, pp. 262–288.

Jia, Y., Li, F.M., Wang, X.L., and Yang, S.M. 2006. Soil water and alfalfa yields as affected by alternating ridges and furrows in rainfall harvest in a semiarid environment. *Field Crops Res.*, 97: 167–175.

Jia, Y., Li, F.M., Zhang, Z.H., Wang, X.L., Guo, R.Y., and Siddique, K.H.M. 2009. Productivity and water use of alfalfa and subsequent crops in the semiarid Loess Plateau with different stand ages of alfalfa and crop sequences. *Field Crops Res.*, 114: 58–65.

Jiang, H.M., Jiang, J.P., Jia, Y., Li, F.M., and Xu, J.H. 2006. Soil carbon pool and effects of soil fertility in seeded alfalfa fields on the semi-arid Loess Plateau in China. *Soil Biol. Biochem.*, 38(8): 2350–2358.

Li, F.M. and Xu, J.H. 2002. Rainwater-collecting eco-agriculture in semi-arid region of Loess Plateau. *Chin. J. Eco-Agric.*, 10(1): 101–103.

Li, F.M., Zhao, S.L., Duan, S.S., Gao, S.M., and Feng, B. 1995. Preliminary study on limited irrigation for spring wheat field in semi-arid region of Loess Plateau. *Chin. J. Appl. Ecol.*, 6(3): 259–264.

Li, F.M., Guo, A.H., and Wei, H. 1999. Effects of plastic film mulch on the yield of spring wheat. *Field Crops Res.*, 63: 79–86.

Li, F.M., Song, Q.H., Jjemba, P.K., and Shi, Y.C. 2004a. Dynamics of soil microbial biomass C and soil fertility in cropland mulched with plastic film in a semiarid agro-ecosystem. *Soil Biol. Biochem.*, 36(11): 1893–1902.

Li, F.M., Wang, J., Xu, J.H., and Xu, H.L. 2004b. Productivity and soil responses to plastic film mulching durations for spring wheat on entisols in the semiarid Loess Plateau of China. *Soil Tillage Res.*, 78(1): 9–20.

Li, F.M., Wang, J., and Xu, J.H. 2005. Plastic film mulch effect on spring wheat in a semiarid region. *J. Sustain. Agric.*, 25(4): 5–17.

Li, L.X., Liu, G.C., Yang, Q.F., Zhao, X.W., and Zhu, Y.Y. 2009. Research and application development for the techniques of whole plastic film mulching on double ridges and planting in catchment furrows in dry land. *Agric. Res. Arid Areas*, 27(1): 114–118.

Li, S.Q. and Lan, N.J. 1995. The research results and progresses of plastic film mulching wheat. *Gansu Agric. Technol.*, 5: 1–3.

Liu, C.A., Zhou, L.M., Li, F.M., Jin, S.L., Jia, Y., Xiong, Y.C., and Li, X.G. 2009. Effects of plastic film mulch and tillage on maize productivity and soil parameters. *Eur. J. Agron.*, 31(4): 241–249.

Liu, C.A., Zhou, L.M., Jia, J.J, Wang, L.J., Si, J.T., Li, X., Pan, C.C., Siddique, K.H.M., and Li, F.M. 2014. Maize yield and water balance is affected by nitrogen application in a film-mulching ridge–furrow system in a semiarid region of China. *Eur. J. Agron.*, 52(B): 103–111.

Liu, X.C., Yang, H.J., and Yin, H.T. 2001. Effect comment on comprehensive control and reconstructing ecosystem frame of Loess Plateau. *Bull. Soil Water Conserv.*, 21(6): 1–6.

Liu, X.E., Li, X.G., Hai, L., Wang, Y.P., Fu, T.T., Turner, N.C., and Li, F.M. 2014. Film-mulched ridge–furrow management increases maize productivity and sustains soil organic carbon in a dryland cropping system. *Soil Sci. Soc. Am. J.*, 78(4): 1434–1441.

Liu, X.F., Yang, Y., Ren, Z.Y., and Lin, Z.H. 2013. Changes of vegetation coverage in the Loess Plateau in 2000–2009. *J. Desert Res.*, 33(4): 1244–1249.

Qi, Y.X., Liu, Z.R., and Wang, X.Z. 2008. Effect analysis of soil and water conservation from the Loess Plateau watershed rehabilitation project (phase II). *Res. Soil Water Conserv.*, 15(5): 204–207.

Ren, J.H. and Chang, S.H. 2009. Using grassland agricultural systems to ensure the food security. *Chin. J. Grassland*, 31(5): 3–6.

Ren, J.H., Lin, H.L., and Wei, L. 2009. Grassland farming is an important approach for the sustainable development of agriculture in Gansu Province. *Acta Grestia Sin.*, 17(4): 405–412.

Shan, L. 2011. Does it alternate forbidding and then grazing or forbid grazing forever on the Loess Plateau? *China Awards Sci. Technol.*, 6: 6–7.

Shan, L. and Xu, B.C. 2009. Exploration of constructing sustainable artificial grasslands on the semiarid Loess Plateau. *Acta Prataculturae Sin.*, 18(2): 1–2.

Shi, H. and Shao, M.G. 2000. Soil and water loss from the Loess Plateau in China. *J. Arid Environ.*, 45: 9–20.

Shi, N.H. 1988. The research on the changing of forest in history. *Hist. Geogr. Works China*, 1: 1–17.

Shi, N.H. 2001. *Study on Historical Geography of Chinese Loess Plateau*. Zhenzhou: Yellow River Water Conservancy Press.

Shi, S. 1986. The Loess Plateau is starting to green treasury. *Bull. Soil Water Conserv.*, 1: 8–13.

Tian, Y., Li, F.M., and Liu, P.H.. 2003. Economics analysis of rainwater harvesting and irrigation methods, with an example from China. *Agric. Water Manage.*, 60: 217–226.

Wan, L. 2011. Spatial and Temporal Trend of Precipitation on the Loess Plateau During the Past 53 Years, PhD thesis, Northwest Agriculture and Forestry University, Yanglin.

Wan, L., Zhang, X.P., Ma, Q., Zhang, J.J., Ma, T.Y., and Sun, Y.P. 2013. Spatiotemporal characteristics of precipitation and extreme events on the Loess Plateau of China between 1957 and 2009. *Hydrol. Proc.*, 28(18): 4971–4983.

Wang, J., Li, Z.X., Wang, Z.H., Liu, H.S., and Nan, D.C. 2001. Combined construction of rainwater catchment techniques with methane pool and greenhouse. *Chin. J. Appl. Ecol.*, 12(1): 51–54.

Wang, Q.X., Fan, X.H., Qin, Z.D., and Wang, M.B. 2012. Change trends of temperature and precipitation in the Loess Plateau region of China, 1961–2010. *Global Planet. Change*, 92–93: 138–147.

Wang, X.L., Li, F.M., Jia, Y., and Shi, W.Q. 2005. Increasing potato yields with additional water and increased soil temperature. *Agric. Water Manage.*, 78: 181–194.

Wang, X.L., Sun, G.J., Jia, Y., Li, F.M., and Xu, J.H. 2008. Crop yield and soil water restoration on 9-year-old alfalfa pasture in the semiarid Loess Plateau of China. *Agric. Water Manage.*, 95: 190–198.

Wang, X.L., Jia, Y., Li, X.G., Long, R.J., Ma, Q.F., Li, F.M., and Song, Y.J. 2009. Effects of land use on soil total and light fraction organic, and microbial biomass C and N in a semi-arid ecosystem of northwest China. *Geoderma*, 153: 560–565.

Wang, Y.R., Yin, X.Z., and Yuan, Z.P. 2004. Main characteristics of climate system in Loess Plateau of China. *J. Catastrophol.*, 10(S1): 39–44.

Wu, R.M., Wang, Y.P., Li, F.M., and Li, X.G. 2012. Effects of coupling film-mulched furrow-ridge cropping with maize straw soil-incorporation on maize yields and soil organic carbon pool at a semiarid Loess site of China. *Acta Ecol. Sin.*, 32(9): 2855–2862.

Xu, B.Q. and Shi, W.Q. 2011. The spatial relationship analysis of rural per capital revenue based on GIS in Zulihe River basin, Gansu Province. *Acta Ecol. Sin.*, 31(9): 2585–2592.

Xu, X.Z., Zhang, H.W., and Zhang, O.Y. 2004. Development of check-dam systems in gullies on the Loess Plateau, China. *Environ. Sci. Policy*, 7(2): 79–86.

Xue, S.P., Zhu, R.X., Yang, Q., Han, S.M., and Han, W.T. 2006. Rainfall mechanical utilization techniques for winter wheat in Loess Plateau dry land. *J. Northwest A&F Univ. (Nat. Sci.)*, 34(1): 1–8.

Yang, X.M., Lu, M.S., and Xie, D. 2009. *The Great Changes in Gansu Rural Areas—Data Collection of Gansu Rural Reform and Open for 30 Years*. Gansu People Press: Lanzhou.

Ye, J.S. and Liu, C.A. 2012. Suitability of mulch and ridge-furrow techniques for maize across the precipitation gradient on the Chinese Loess Plateau. *J. Agric. Sci. (Can.)*, 4(10): 182–190.

Yi, L., Ren, Z.Y., Zhang, C., and Liu, W. 2014. Vegetation cover, climate and human activities on the Loess Plateau. *Resour. Sci.*, 36(1): 166–174.

Zhang, W.S. 2013. Analysis of the Social and Economic Factors Affecting the Adoption of Raising Sheep in Folds in the Semiarid Loess Plateau, PhD thesis, Lanzhou University, Lanzhou.

Zhang, W.S., Li, F.M., Xiong, Y.C., and Xia, Q. 2012. Econometric analysis of the determinants of adoption of raising sheep in folds by farmers in the semiarid Loess Plateau of China. *Ecol. Econ.*, 74: 145–152.

Zhang, Y.L., Xie, Y.S., Li, X., Jiang, Q.L., and Zhang, Y. 2010. Stability and cost benefit analysis of main grain crops production in gully area of the Loess Plateau. *Bull. Soil Water Conserv.*, 30(4): 201–204.

Zhao, H., Xiong, Y.C., Li, F.M., Wang, R.Y., Qiang, S.C., Yao, T.F., and Mo, F. 2012. Plastic film mulch for half growing-season maximized WUE and yield of potato via moisture–temperature improvement in a semi-arid agroecosystem. *Agric. Water Manage.*, 104: 68–78.

Zhao, H., Wang, R.Y., Ma, B.L., Xiong, Y.C., Qiang, S.C., Wang, C.L., Liu, C.A., and Li, F.M. 2014. Ridge-furrow with full plastic film mulching improves water use efficiency and tuber yields of potato in a semiarid rainfed ecosystem. *Field Crops Res.*, 161: 137–148.

Zhao, S.L. and Li, F.M. 1995. Discussion on development of water-harvested agriculture in semi-arid region Northwest China. *Acta Bot. Boreali-Occid. Sin.*, 15(8): 9–12.

Zhao, S.L., Wang, J., and Li, F.M. 1995. A study on the limitation of agriculture development by conserving soil and water in semi-arid regions of Loess Plateau. *Acta Bot. Boreali-Occid. Sin.*, 15(8): 13–18.

Zhou, L.M., Li, F.M., Jin, S.L., and Song, Y.J. 2009. How double ridges and furrows mulched with plastic film affect soil water, soil temperature and yield of maize on the semiarid Loess Plateau of China. *Field Crops Res.*, 113: 41–47.

Zhou, L.M., Jin, S.L., Liu, C.A., Xiong, Y.C., Si, J.T., Li, X.G., Gan, Y.T., and Li, F.M. 2012. Ridge-furrow and plastic-mulching tillage enhances maize-soil interactions: opportunities and challenges in a semiarid agroecosystem. *Field Crops Res.*, 126: 181–188.

CHAPTER 13

Agroecology Research and Practice in the Oasis Region, Northwest China

Su Peixi and Xie Tingting

CONTENTS

13.1 INTRODUCTION

Oases lie in desert regions and are surrounded by desert, mountains, hills, and lakes, with desert as the main body. Arid desert in China covers about 1.25×10^6 km², accounting for 13% of China's total land area, and mainly occurs in northwest China (H.J. Gao, 1987). An oasis in arid desert regions is a naturally and economically integrated system with water resources, inhabitable environmental conditions, and various socioeconomic activities (Fang, 1994). It is also a unique ecosystem with an annual precipitation of <200 mm and ideal natural resources assembling around a center in arid regions.

Oases in China are distributed in the arid desert regions that are north of a line extending from Kunlun Mountain to Altun Mountain, Qilian Mountain, and Wushaoling Mountain, and west of a line extending from Yanchi to Ertuoke, Bailingmiao, and Wenduermiao, including the desert oases in Qaidam Basin (Figure 13.1) (H.J. Gao, 1987). They include the north Xinjiang, south Xinjiang, Hexi Corridor, Qaidam Basin, Alxa Plateau, and Hetao Plain oases (Shen et al., 2002). Oases in China cover an estimated total area of 8.6×10^4 km², among which the Hexi Corridor oases cover about 1.1×10^4 km² (Shen et al., 2002). The second largest river in China, the Yellow River, flows about 1178 km through desert regions and nourishes the Hetao oases. The largest inland river in China, the Tarim River, traverses 1321 km and nourishes the south Xinjiang oases. The second largest inland river in China, the Heihe River, 821 km in length, nourishes oases in the middle section of the Hexi Corridor (Figure 13.1). Oases are the source of life in arid desert regions.

High summer temperatures, strong solar radiation, and low precipitation are key characteristics of an oasis. Total solar radiation in oases is over 5.04×10^5 J cm^{-2} yr^{-1} on average and can even reach 6.49×10^5 J cm^{-2} yr^{-1}. This intensity is second only to the Qinghai–Tibet Plateau. Heat resources in oases are also abundant. For example, $\geq 10°C$ annual cumulative temperature in the Tarim Basin adds up to an excess of 4000°C, and the corresponding figure in the Hexi Corridor region varies between 2500 and 3500°C (Guan et al., 2002). Sunlight and heat conditions in oases are quite favorable to the growth and photosynthesis of crops. The light and temperature potential productivity in oases is 37 to 75% higher than that in North China, but only under abundant irrigation conditions in oases can light and temperature fully play their roles. Therefore, with proper management, crop yields in oases are 12 to 16% higher than those in the rest of North China (Shen, 2000).

Oasis agriculture, also known as irrigation agriculture, is an agricultural production approach using irrigation facilities in accordance with crop growing periods and agricultural production levels in arid desert regions. Unlike agriculture in semiarid regions, its agricultural productivity is much higher than that of natural precipitation-dependent, rain-fed agriculture. Because oasis agriculture depends heavily on irrigation in desert regions (Su et al., 2004), it is very important to coordinate the following three aspects: (1) water-saving agriculture, which refers to how to optimize the efficiency of canal-system water use, field water use, and crop water use (Su et al., 2002); (2) economic benefits, with regard to how to enhance the income of oasis agriculture; and (3) ecological

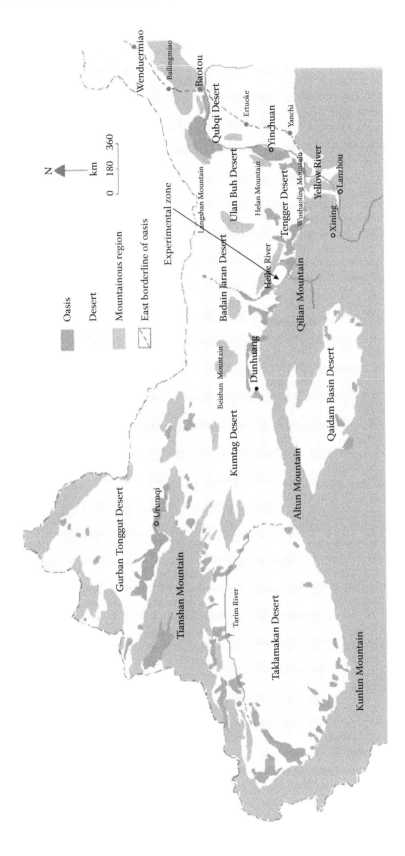

Figure 13.1 Distribution of the oases in China.

conditions, referring to how to produce green food, prevent sand building up *in situ* on farmland, and reduce dust storms. These three aspects can either inhibit or amplify each other. Any contradiction among them could produce a bottleneck effect.

The shortage of irrigation water resources is becoming more serious in the northwestern arid region of China. The competition for water resources among agriculture, industry, and ecology is very intensive. Due to agriculture and industrial water use having appropriated the share of ecological water, the ecological environment of inland river basins is worsening, and the problem of desertification and salinization is becoming more serious. Agriculture is the major water consumer in oasis agricultural regions. The problems associated with agricultural water use are as follows:

1. The ability for the water supply to be temporally and spatially regulated should be improved. The ability for the timely diversion of water from the water supply to meet the evaporation and transpiration needs of crops is quite weak. It is especially weak in the capacity to coordinate the use of surface water with groundwater. It becomes the limiting factor for the development of water-saving agriculture in oasis regions.

2. The improvement of water use efficiency is restricted by the lack of a good technological package. A single water-saving technique could not successfully reach the target yield without both a sound coupling of water-saving irrigation with irrigation quotas and knowledge of the crop biological characteristics and environmental conditions. The irrigation quota is the amount of water set for a specific crop for optimum irrigation based on field research.

3. Just focusing on the soil water environment and ignoring atmospheric humidity conditions is another problem. Attention has usually focused on water-saving irrigation and the reduction of soil evaporation only; the effect of suitable air humidity on crop growth and development has usually been ignored. Under low air humidity, high temperatures, and strong solar radiation, the vapor pressure deficit between leaves and air increases, triggering a phenomenon of serious photoinhibition. This results in growth inhibition and yield reduction.

4. Focusing on saving water during crop growth periods has resulted in a lack of countermeasures for the salinity caused by irrigation. As much as 1.6 t salt ha^{-2} could actually be introduced into farmland, even when 500 mm of high-quality water with electrical conductivity (EC) of 0.5 dSm^{-1} and total dissolved solids (TDS) of 320 mg L^{-1} is run into a field through irrigation (Loomis and Connor, 2002). The TDS of the Heihe River is 456.5 mg L^{-1} in the middle of the Hexi Corridor, and the TDS of well water is 559.6 mg L^{-1} in the irrigation region. This region mainly produces maize and wheat. A simple calculation shows that 2.7 t salt ha^{-2} yr^{-1} is brought into farmland through river water irrigation, and 3.4 t salt ha^{-2} yr^{-1} is brought into farmland through well water irrigation, when the average irrigation reaches 600 mm. Without a leaching operation, this salt could gather around root systems and be harmful to crop productivity.

5. Groundwater is overexploited, and inadequate attention has been paid to ecological water use. The imbalance of the groundwater supply has resulted in overexploitation of water resources. The decreasing groundwater table affects the stabilization of the shelterbelt system, and the cooling and high humidity effects within the oasis are undermined.

6. There is a lack of guidance from the circular economy and ecology for water-saving agriculture. The "white pollution" caused by large amounts of mulching film, plastic film, drip irrigation belts, and other materials used for water-saving operations is a serious issue. There is a lack of value-increasing paths for recycling wastes from crop production, animal production, processing industries, and farmers' households.

Improvement of the ecological environment in oases is the foundation of oasis sustainable development. Establishing agroforestry systems that save water and have a high water use efficiency is important for sustainable development of oases. Agroforestry systems and their interspecies interactions are affected by climate, trees, cropping patterns, and other factors with distinguishable regional characteristics. Agriculture in the oasis regions with inland rivers in Northwest China totally relies on glacier and snowmelt, as well as precipitation from the upper reaches of mountains in the catchments. Along the Heihe River in the middle part of the Hexi Corridor region, for example, precipitation from catchments accounts for about 95% of the water resources, and glacial meltwater accounts for about 5%. In this article, research on oasis agroecology in China and the construction of water-saving agroecology practices are presented. "Eco-agriculture" is the term used in China for "agroecology practice." The main topics include the regulatory effects of shelter-belts on crops in oasis agroforestry systems, irrigation methods for fruit tree–crop systems in oases, yield and water use efficiency of economic crops and energy crops in a cluster planting pattern, and other eco-agricultural models for oases and their economic benefit analyses.

13.2 MAIN ECO-AGRICULTURAL MODES AND THEIR WATER USE EFFICIENCY IN OASES

13.2.1 Specialized Agriculture Production in the Oasis Region

Special agricultural products such as cotton, hops, processing tomatoes, grapes, Chinese wolf-berry, melons, and fruits produced in the oasis region are very important in China, and even to the world. As much as 95% of processing tomatoes in China come from the Xinjiang and Hexi Corridor oases, and tomato ketchup produced in the oasis region and exported abroad accounts for 20% of the world's trade volume for this product. All green raisins in China are produced in the oasis region. The Chinese wolfberries grown in the Hetao oasis of Ningxia are sold to China and Southeast Asia regions.

13.2.1.1 Cotton Production

China is the number one cotton producer in the world, producing a third of the world's total cotton. The cotton-producing oasis regions are important for high-quality cotton in China. The cotton area in Xinjiang produces 25% of the cotton in China and 34% of the high-quality cotton, fine-staple cotton, long-staple cotton, and colored cotton in China.

13.2.1.2 Hops Production

Hops were planted in Hexi Corridor, northern Xinjiang, and other oasis regions in the early 1960s, and oasis regions gradually became the major hops production regions in China. Almost all hops exported from China come from oasis regions.

13.2.1.3 Beet Production

Beet production began in northwestern oases in the 1950s. Now, the average yield is more than 22.5 t ha^{-2}, varying from 4.5 to 52.5 t ha^{-2} in the Hexi Corridor of Gansu Province due to different site conditions. The average sugar content reaches over 18%.

13.2.1.4 Grassland Production

Natural grassland on the outskirts of oases is an abundant resource. The ecological environment of oases can be improved through the development of grassland. Compared to grain crop produc-tion, the production of forage such as alfalfa requires fewer external inputs for field management and less water and fertilizer, yet it increases income to about 3000 RMB yuan per hectare. The total cost of planting alfalfa is only 87% that of wheat and 84% of maize in the Jiuquan region of the Hexi Corridor, but the net income per hectare is 2.18 times that of the income from wheat and 1.51 times that of the income from maize.

13.2.1.5 Chinese Medicinal Herb Production

The main cultivated species of Chinese medicinal herbs are *Carehamus tinctorious* L., *Lycium barbarum* L., *Gastrodia elata* Bl., *Eucommia ulmoides* Oliver, *Angelica sinensis* (Oliv.) Diels, and *Codonopsis pilosula* (Franch.) Nannf., among others. The agrestic species are *Glycyrrhiza uralensis* Fisch., *Sophora alopecuroides* Linn., *Ephedra sinica* Stapf, *Rheum officinale* Baill., *Gentiana macrophylla* Pall., *Cistanche deserticola* Ma, and *Cynomorium songaricum* Rupr., among others. Rational harvests of wild medicinal resources and the expansion of artificially planted medicinal herbs are both important in the oasis regions.

13.2.1.6 Seed Production

Seed production of maize began in the Hexi Corridor of Gansu in the 1980s, and seed production of maize and oil sunflowers began in the Shihezi and Yili of Xinjiang in the 1990s. Now seed production of cotton, cucurbitaceous vegetables, and others is developing quickly. The Hexi Corridor is widely recognized as an ideal seed production region both at home and abroad, and it is the second biggest seed production base in China (Lai, 2005). The production of seeds includes about 1000 varieties of 20 different crops, including wheat, maize, cotton, oil sunflowers, beans, vegetables, melons, and flowers. Tomato, watermelon, and muskmelon seeds have been sold to the United States, Holland, Japan, Korea, and the Taiwan Province of China. Agricultural areas in oases are ideal for the production of parental seed, protospecies, and hybrid seeds due to their low air humidity, low occurrence of pests and disease outbreaks, good irrigation conditions, and geographic isolation by the Gobi desert and sand dunes. The main period of seed maize production that requires water is June to August, when precipitation and snow melt are both at their peaks. The irrigation for seed production fields is guaranteed. Seed yields are stable and seldom affected by drought. Compared to income from food production, income from seed fields is 3 to 16 times that for other crop productions. The maximum income from seed production can reach 45×10^4 RMB yuan ha^{-2}.

13.2.2 Regional Characteristics and Main Agricultural Utilization Models of the Oasis

A good agricultural system in the oasis regions should have reasonable structures, with high economic returns and sustainable functions. Depending on the maturity of oasis development and position in the oasis, three regions in oases can be recognized: center oasis, inner oasis, and edge oasis (Figure 13.2). Suitable models of eco-agriculture are determined based on convenience of transportation, level of economic development and ecological environment, ambient environmental effects, and potential for sustainable development.

Center oasis is also known as *mature oasis*. Covered with irrigated-farming soil, center oasis is developed from gray–brown desert soil through irrigation and cultivation. This soil is formally named *irrigated desert soil* according to the technological rule of the second soil survey in China (Planning Commission of Gansu Province, 1992). It is also called *thick irrigated desert soil* (Soil Survey Office of Gansu Province, 1993). This soil has a good water-retention capacity. Center oasis regions have the converging characteristics of convenient transportation, secure water resources, and predictable development. The characteristic agriculture in this region has become one of mainly producing fruits and vegetables with water-saving irrigation technology, greenhouse technology, and other advanced technologies. Water use efficiency is significantly improved by using modern water-saving engineering technology.

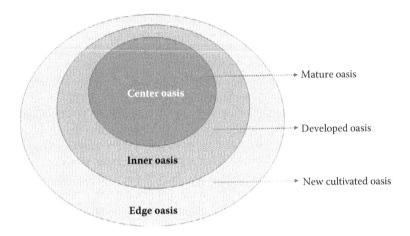

Figure 13.2 Diagram showing various oasis regions.

Inner oasis is also known as *developing oasis* and is covered with irrigated desert soil. The thickness and organic content in the cultivation layer are lower than those of the mature irrigated desert soil in the center oasis. It is formally named *shallow irrigated desert soil* (Soil Survey Office of Gansu Province, 1993). In order to distinguish soil in the inner oasis vs. soil from the center oasis, this soil is also called *irrigation-silting soil*. Its agricultural production level is stable and relatively high due to the protection provided by the oasis edge and shelter forest system. The main crops used in this area must be more resilient due to unpredictable natural and market conditions. Water saving and the economic return of water used are both emphasized in this part of the oasis. Seed production for corn, tomatoes, and radishes are also very important. In addition to seed production, there are large areas in the Hexi Corridor of intercropping jujubes with grain crops and intercropping almonds with other crops.

Edge oasis is also known as *newly cultivated oasis* and is covered with irrigated aeolian sandy soil (formally named *cultivated aeolian sandy soil*) (Soil Survey Office of Gansu Province, 1993). This region is found on the edge of oases. The water retention capacity of this soil is very poor. Due to insecure water resources, agricultural production in this newly reclaimed area is often interrupted by the lack of irrigation water. Both input and output levels of agriculture are quite low. Major agricultural production includes livestock, medicinal plants, hops, and wine grapes. In light of the instability in market conditions and water resources, intercropping of forage crops and grain crops or just returning the land to natural forest and grassland should be considered.

Soil evaporation is strong after 2 to 3 days of irrigation or after precipitation in oasis regions. Because evaporation in the edge oasis is greatly accelerated by nearby desert and crop transpiration in the edge oasis is higher than in other parts of the oasis, the irrigation consumption necessary for crop growth is also higher. The results of optimum field irrigation show that the irrigation water consumption of the inner oasis is significantly lower than that of the edge oasis. The irrigation quantity needed for one-time irrigation in a 0.03-ha^2 plot in the inner oasis is 930 m^3 ha^{-2}; whereas, it is 1160 m^3 ha^{-2} in the edge oasis, an increase of 24.7%. The quantity needed for one-time irrigation in a 0.07-ha^2 plot in the inner oasis is 1230 m^3 ha^{-2}; whereas, it is 1950 m^3 ha^{-2} in the edge oasis, an increase of 58.5%. The quantity needed for one-time irrigation for a 0.10-ha^2 plot in the inner oasis is 1950 m^3 ha^{-2}; whereas, it is 3100 m^3 ha^{-2} in the edge oasis, an increase of 59%. This indicates that the irrigation water required decreases with the reduction of plot size. The irrigation water consumption in a 0.03-ha^2 irrigation plot in the edge oasis was close to that of a 0.07-ha^2 irrigation plot in the inner oasis. Most irrigation plots in oasis farmland are 0.03 ha^2.

Farmland protected by a shelterbelt system formed by trees is very important in the oasis region. Another method to protect the oasis from wind erosion and to reduce evaporation is plastic sheeting for soil cover or greenhouses covered with plastic sheeting; however, the waste plastic sheeting can cause "white pollution" if it is not collected and recycled.

Compared with grain crop monoculture, the development of grain crop intercropping with forage crops can significantly reduce soil erosion in oases (Su et al., 2004). Hence, intercropping can reduce the rate of dust transportation contributing to sandstorms. The intercropping systems used can be grain crop intercropping with forage grass, economic crop intercropping with forage grass, or forage crop intercropping with leguminous forage. The systems can be very flexible in response to market conditions. For example, it is easy to convert intercropping systems into grain monoculture systems when food prices increase, into larger forage grass areas when the economic return of grassland husbandry improves, or into seed crop production when seed prices increase. No matter how the system changes, the balance of soil fertility depletion and fertility enrichment must be maintained.

13.2.3 Water Use Efficiency and Benefits of Main Crops and Different Intercropping Systems

Cultivated crop species in oases are numerous and vary with different regions. The main grain crops in oases include wheat, maize, rice, millet, and soybeans. The main economic crops are cotton, rape, sunflowers, benne or sesame, barley, beets, hemp, and hops. The forage crops are silage corn, sorghum, and oats. The energy crops are sweet sorghum and Jerusalem artichokes. The vegetable crops include tomatoes, caraway, radishes, cumin, and melons. The forages include alfalfa, shamrock, and perennial ryegrass. The fruits include apples, pears, grapes, peaches, apricots, and jujubes. Crop water use efficiency can be tested at leaf level, canopy level, and yield level. Water use efficiency at the leaf level is water physiological use efficiency, or transpiration rate. It is important in the formation of dry matter. Canopy level efficiency is based on groups of leaves or individual plants (S. Gao et al., 2010). Canopy water use efficiency is the ratio between canopy photosynthetic rate and canopy transpiration rate. It can more accurately reflect the water use conditions of a plant community. Water use efficiency at the yield level is an important index of agricultural production that can be compared between crops under monoculture or in intercropping systems. However, the economic return of units of water used is mostly used to compare different crops or different intercropping systems formed with entirely different crops.

13.2.3.1 Wheat Production

The water use benefit is indicated by production value per cubic meter of water used, shown in Table 13.1 and calculated using 2009 prices. Comparing spring wheat monoculture with jujube–spring wheat intercropping in the inner oasis, the total irrigation time is the same, but the irrigation quota decreased from 7200 m^3 ha^{-2} to 6300 m^3 ha^{-2} due to irrigation optimization in jujube–wheat intercropping. Although the spring wheat yield decreased slightly in the same area, water use efficiency was significantly increased in the jujube–spring wheat complex system. Together with the yield and income from jujube trees, production value per cubic meter of water increased by 74%. The quantity of irrigation in the edge oasis for nine times reached 1.26 $\times$ 10^4 m^3 ha^{-2} in the spring wheat monoculture and was 1.11 $\times$ 10^4 m^3 ha^{-2} in the jujube–wheat intercropping system. The production value per cubic meter of water increased by 87% in the intercropping system. Overall, water use efficiency and production value per cubic meter of water in the inner oasis in monoculture are two times those in the edge oasis. Water use efficiency and production value per cubic meter of water were significantly increased in the jujube–spring wheat intercropping system compared to the spring wheat monoculture system in both the inner oasis and edge oasis; however, these indexes increased nearly twice as much in the inner oasis than in the edge oasis.

13.2.3.2 Maize Production

A comparison of maize production in different regions (Table 13.1) shows that the number of times that irrigation was needed was 7 in the inner oasis but up to 11 in the edge oasis due to differences in soil texture between the irrigated-farming soil in the inner oasis and irrigated aeolian sandy soil in the edge oasis. Differences in soil evaporation also played a role. Soil evaporation in the edge oasis was significantly higher than in the inner oasis. Irrigation levels were 9300 m^3 ha^{-2} for maize monoculture and 8700 m^3 ha^{-2} for jujube–maize intercropping in the inner oasis. The production value per cubic meter of water for the jujube–maize intercropping system increased by 37% compared to maize monoculture. Irrigation levels were 1.45×10^4 m^3 ha^{-2} for maize monoculture in the edge oasis and 1.28×10^4 m^3 ha^{-2} for the jujube–maize intercropping system; the production value per cubic meter of water in the complex system increased by 54%. When comparing maize monocultures between the edge oasis and the inner oasis, the production value per cubic meter of water in the inner oasis was 92% higher than that in the edge oasis. Results showed that the rate of water use efficiency increased faster in the edge oasis than in the inner oasis when jujube–grain crop intercropping replaced grain monoculture.

13.2.3.3 Seed Corn Production

Table 13.1 shows that seed corn monoculture in the inner oasis needed to be irrigated seven times, and the total irrigation volume was 9300 m^3 ha^{-2}, with a production value per cubic meter of water of 2.29 RMB yuan. When jujube–seed corn intercropping was implemented, total irrigation amounts declined to 8700 m^3 ha^{-2} with no changes in irrigation times. The irrigation quota each time was 1200 m^3 ha^{-2} during the growth stages, and the overwintering irrigation quota was 1500 m^3 ha^{-2}. The yield of fresh jujube was 3200 kg ha^{-2}, and the yield of maize seed was 6700 kg ha^{-2}. The production value was 2.65×10^4 RMB yuan ha^{-2}, with 2.0 RMB yuan kg^{-1} for fresh jujube and 3.0 RMB yuan kg^{-1} for seed maize. The water use efficiency was 1.14 kg m^{-3}, and the production value per cubic meter of water was 3.05 RMB yuan.

13.2.3.4 Cotton Production

Cotton production needed to be irrigated five times, and the total irrigation volume was 7000 m^3 ha^{-2}. The production value per cubic meter of water was 2.36 RMB yuan for cotton monoculture in the inner oasis. The total irrigation volume declined to 6300 m^3 ha^{-2}, and the production value per cubic meter of water increased to 3.59 RMB yuan when jujube–cotton intercropping was implemented. Compared to monoculture, jujube–cotton intercropping showed an increase in production value per cubic meter of water of 52%. The irrigation volume significantly increased to 9400 m^3 ha^{-2} for cotton monoculture in the edge oasis. It was two times that of the inner oasis irrigation volume. The production value per cubic meter of water was 49 RMB yuan m^{-3}, less than that in the inner oasis. However, the number of irrigation times required decreased from seven to six, and the total irrigation volume was 7500 m^3 ha^{-2} due to optimization of the irrigating unit in jujube–cotton intercropping in the edge oasis. The irrigation quota in each time was 1200 m^3 ha^{-2} during growth stages, and the overwintering irrigation quota was 1500 m^3 ha^{-2}. The yield of fresh jujube was 3300 kg ha^{-2}, and the yield of seed cotton was 3100 kg ha^{-2}. The production value was 2.21×10^4 RMB yuan ha^{-2}. The production value per cubic meter of water was 2.95 RMB yuan, 98% higher than that of monoculture cotton in the edge oasis area (Table 13.1).

Table 13.1 Water Use Efficiency and Production Value of Crops and Their Intercropping Systems in the Inner and Edge Oases

Crop and Complex Pattern	Location in Oasis	Total Irrigation Times	Total Irrigation Quota (×10⁴ m³ ha⁻²)	Yield (×10⁴ kg ha⁻²)	Unit Price (yuan kg⁻¹)	Total Production Value (×10⁴ yuan ha⁻²)	Water Use Efficiency (kg m⁻³)	Production Value per Cubic Meter of Water (yuan m⁻³)
Spring wheat monoculture	Inner	5	0.72	0.68	1.6	1.09	0.94	1.51
	Edge	9	1.26	0.56	1.6	0.90	0.44	0.71
Spring wheat–soybean intercropping	Inner	7	0.84	0.60 (wheat); 0.35 (soybean)	1.6; 3.2	0.96; 1.12	1.13	2.48
Jujube–spring wheat intercropping	Inner	5	0.63	0.62 (wheat); 0.33 (jujube)	1.6; 2.0	0.99; 0.66	1.51	2.62
	Edge	9	1.11	0.54 (wheat); 0.31 (jujube)	1.6; 2.0	0.86; 0.62	0.77	1.33
Soybean monoculture	Inner	6	0.72	0.38	3.2	1.22	0.53	1.69
Benne monoculture	Inner	5	0.60	0.23	6.0	1.38	0.38	2.30
Benne–soybean intercropping	Inner	7	0.84	0.21 (benne); 0.30 (soybean)	6.0; 3.2	1.26; 0.96	0.61	2.64
General maize monoculture	Inner	7	0.93	1.18	1.5	1.77	1.27	1.90
	Edge	11	1.45	0.95	1.5	1.43	0.66	0.99
Jujube–general maize intercropping	Inner	7	0.87	1.11 (maize); 0.30 (jujube)	1.5; 2.0	1.67; 0.60	1.62	2.61
	Edge	11	1.28	0.91 (maize); 0.29 (jujube)	1.5; 2.0	1.37; 0.58	0.94	1.52
Producing seed maize monoculture	Inner	7	0.93	0.71	3.0	2.13	0.76	2.29
Jujube–producing seed maize intercropping	Inner	7	0.87	0.67 (maize); 0.32 (jujube)	3.0; 2.0	2.01; 0.64	1.14	3.05
Alfalfa–producing seed maize intercropping	Inner	7	0.93	0.57 (maize); 1.11 (alfalfa)[a]	3.0; 1.0	1.71; 1.11	1.81	3.03
Producing seed caraway monoculture	Inner	5	0.63	0.44	5.0	2.20	0.70	3.49
Producing seed radish monoculture	Inner	5	0.63	0.30	7.0	2.10	0.48	3.33
Cumin monoculture	Inner	4	0.47	0.15	10.0	1.50	0.32	3.19

Continued

Table 13.1 (Continued) Water Use Efficiency and Production Value of Crops and Their Intercropping Systems in the Inner and Edge Oases

Crop and Complex Pattern	Location in Oasis	Total Irrigation Times	Total Irrigation Quota ($\times 10^4$ m^3 ha^{-2})	Yield ($\times 10^4$ kg ha^{-2})	Unit Price (yuan kg^{-1})	Total Production Value ($\times 10^4$ yuan ha^{-2})	Water Use Efficiency (kg m^{-3})	Production Value per Cubic Meter of Water (yuan m^{-3})
Cotton monoculture	Inner	5	0.70	0.33[b]	5.0	1.65	0.47	2.36
	Edge	7	0.94	0.28	5.0	1.40	0.30	1.49
Jujube–cotton intercropping	Inner	5	0.63	0.32 (cotton); 0.33 (jujube)	5.0; 2.0	1.60; 0.66	1.03	3.59
	Edge	6	0.75	0.31 (cotton); 0.33 (jujube)	5.0; 2.0	1.55; 0.66	0.85	2.95
Hop	Edge	7	0.67	1.50	1.5	2.25	2.24	3.36
Grape	Edge	7	0.67	1.65	1.4	2.31	2.46	3.45
Processing tomato monoculture	Inner	5	0.63	9.65	0.2	1.93	15.32	3.06
	Edge	6	0.75	9.23	0.2	1.85	12.31	2.47
Jujube–processing tomato intercropping	Inner	5	0.63	9.05 (tomato)[c]; 0.34 (jujube)	0.2; 2.0	1.81; 0.68	14.90	3.95
	Edge	6	0.71	8.55 (tomato); 0.30 (jujube)	0.2; 2.0	1.71; 0.60	12.46	3.25
Sweet sorghum monoculture	Inner	5	0.61	8.85[d]	0.22	1.95	14.51	3.20
	Edge	10	0.97	7.20	0.22	1.58	7.42	1.63
General sorghum monoculture	Inner	6	0.72	0.84	1.3	1.09	1.17	1.51
	Edge	10	0.97	0.75	1.3	0.98	0.77	1.01

Note: The yield of crop is grain dry weight; jujube yield is fresh jujube weight. Production value per cubic meter of water is calculated by the price in 2009. The irrigation time is the number of irrigations for the crop field.

[a] Alfalfa yield is dry weight.
[b] Cotton yield is seed cotton yield.
[c] Tomato yield is fresh weight.
[d] Sweet sorghum yield is aboveground fresh biomass weight.

13.2.3.5 Tomato Production

When processing tomatoes were monocultured in the inner oasis, irrigation was required five times and total irrigation volume was 6300 m³ ha⁻². The production value per cubic meter of water was 3.06 RMB yuan. When jujube–processing tomato intercropping was implemented, the production value per cubic meter of water increased to 3.95 RMB yuan, an increase of 29%. The number of irrigation times required for processing tomato monoculture was six times in the edge oasis, the total irrigation volume was 7500 m³ ha⁻², and the production value per cubic meter of water was 2.47 RMB yuan. It increased to 3.25 RMB yuan m⁻³ for jujube–processing tomato intercropping, an increase of 32% (Table 13.1). The production value per cubic meter of water increased 22% for jujube–processing tomato intercropping compared to monocultured tomatoes.

13.2.3.6 General Conclusions

Comparing production values per cubic meter of water in jujube–spring wheat intercropping, jujube–maize intercropping, jujube–cotton intercropping, and jujube–processing tomato intercropping with their respective monocultures show the following: Production values in intercrops increased 74%, 37%, 52%, and 29%, respectively, in the inner oasis with an average increase of 48%. They increased by 87%, 54%, 98%, and 32%, respectively, in the edge oasis, with an average increase of 68%. Generally, the production value per cubic meter of water of economic crops was higher than that of grain crops, jujube–crop intercropping was higher than monoculture, and the degree of increase in production value per cubic meter of water for jujube–crop intercropping in the edge oasis was higher than that in the inner oasis. The production value per cubic meter of water for economic crops and seed crop production was significantly higher than for grain crops; however, the economic risk of processing tomatoes is high due to its short storage time. The risk for seed crop production is also high due to unstable market prices. Because of the high population density and limited water resources in the Hexi Corridor oasis region, if only grain crops are considered the water use efficiency is low and farmer income is difficult to increase. Intensive economic crop and seed crop production can beneficially increase farmers' incomes, but intercropping systems should be adopted and contract production should be considered in order to reduce market risk and improve water use efficiency and economic return per unit of irrigation water.

13.3 CHARACTERISTICS OF THE OASIS AGROFORESTRY SYSTEM AND CROP PRODUCTION

Oases are frequently affected by various natural disasters such as sandstorms, dry–hot wind, frost damage, low temperatures, and strong winds. The dry–hot wind is a major threat to high and stable grain yields. In the Hexi Corridor region, dry–hot winds often occur from June to July and may cause a 20 to 40% decrease in wheat yields at the edge of oases. The average annual number of days with wind velocities exceeding 17 m s⁻¹ in this region is 68.5 days. When a strong wind arrives, sand particles are lifted into the air, causing a dust storm, sandstorm, haze, or blowing sand weather. Croplands and canals can even be buried under these weather conditions. An oasis agroforestry system is a production system with an artificially established forest system within farmland to protect against natural disasters in the desert. Optimization of stand structures is beneficial in improving natural environmental conditions and creating a favorable climate for crop growth. Developing an economic agroforestry system with disaster prevention and mitigation effects, higher economic value, and water-saving potential is essential. Therefore, tree species selection and configuring agroforestry on farmland are very important in order to maximize the beneficial effects and minimize the negative effects.

13.3.1 Composition of the Protective System in Oasis Areas and Environmental Effects of Different Planting Patterns on Agriculture

13.3.1.1 Composition of the Protective System

A protective system is essential to the sustainable development of oases. This kind of system includes a grass belt for blocking sand, a forest belt dominated by psammophilous shrubs for stabilizing sand, a forest belt in the oasis edge for blocking wind and sand, and a shelterbelt system within the oasis for farmland protection. The main model calls for constructing grass, shrubs, and forest on the outside part of the oasis; establishing a farmland protection forest with a narrow forest belt and small grid in the inside part of the oasis; and building an economic forest, timber forest, and firewood forest patches on the wasteland in the inner part of oasis. In regions with the potential for serious sand damage, a sand-stabilizing forest belt 20 to 50 m wide and dominated by trees can be established at the margin of the oasis perpendicular to the prevailing wind direction. The protective system in the oasis plays an important role in improving farmland microclimates, mitigating natural disasters, and enhancing crop yield. Farmland protective forests can be established in the form of a network to protect each cultivated field against spring gales and the dry–hot winds in summer. The size of the farmland surrounded by shelterbelt trees differs: 10 to 15 hm^2 in the inner part of oasis, 5 to 10 hm^2 at the edge of oasis, and 3 to 5 hm^2 in the wind-gap section. Generally, the crop yield in the protected field can be increased by 20 to 30% or even 50 to 70% compared to adjacent unprotected fields.

13.3.1.2 Wind Erosion Effect of Different Planting Patterns

Oasis agriculture can suffer from soil erosion caused by strong winds. It is an important factor in oasis land degradation, seedling damage, and yield reduction. The occurrence of sandstorms and the increased dust content are related to farmland tillage and the extent of exposed surface. Some methods, such as stubble mulch tillage, zero tillage, plastic film-covered planting, and changing spring sowing into winter seeding, can enhance the spring mulch in farmland to prevent soil erosion (Table 13.2), thus improving the environmental condition of oases (Su et al., 2004).

As Table 13.2 shows, the spring-plowed bare cropland, sparse shrub forest, and mobile desert had high sand transportation rates. Although the suspended dust-to-sand ratio in desert areas is the lowest, the sand transport rate in the desert was significantly higher than other agricultural planting patterns due to its strong suspended dust flux. The next highest was the spring-plowed bare cropland; the sand transport rate of the spring-plowed bare cropland in the edge oasis was 71% higher than that in the inner oasis. The sand transport rate of the spring-plowed bare cropland in the inner oasis was similar to the sparse shrub forest, with 10 to 15% coverage rates at the edge of the oasis. The ratios of suspended dust content to sand in other agricultural planting patterns were greater than 60%. The amounts of suspended dust and sand in sandstorms were lowered due to the reduction of soil erosion. The distribution ratio of dust to sand particles in the alfalfa field was the lowest. Next to it was the grain crop and grass intercropping with alfalfa belt. In addition, plastic-film mulching planting in the spring could significantly mitigate soil erosion.

13.3.2 Effects of Shelterbelts on Farmland Environment and Crop Production

13.3.2.1 Effects of Shelterbelts on Crop Yield and the Mechanisms

Shelterbelts play an important role in maintaining higher production levels, high efficiency, and sustainable development in oasis agriculture. The width and length of shelter belt grids in the inner oasis and center oasis range from 300 × 350 m to 350 × 400 m, respectively, and from 200 × 250 m to 250 × 300 m, respectively, at the edge of the oasis. In order to mitigate strong sandy winds, areas

Table 13.2   Sand Transportation Rate and Dust Content in Relation to Different Use and Cultivation Practices in Oasis Agricultural Area and Desert (2001–2002) in the Middle Reaches of the Heihe River in the Hexi Corridor Region

Land Use and Cultivation Pattern	Sand Transport Rate[a] (g min^{-1} cm^{-2})	Surface Roughness[b] (cm)	Suspended Dust Content Rate[c] (g min^{-1} cm^{-2})	Ratio of Suspended Dust Content to Sand Transport (%)
Bare mobile desert around oasis	1.25	0	0.192	15.4
Sparse shrub forest at the edge of oasis	0.17	0.01	0.028	16.5
Spring-plowed bare cropland in edge oasis	0.29	0.01	0.083	28.7
Spring-plowed bare cropland in inner oasis	0.17	0.04	0.051	30.5
Wheat field[d]	1.8×10^{-3}	1.04	1.24×10^{-3}	68.9
Plastic-film mulching planted row middle	1.6×10^{-3}	0.01	1.06×10^{-3}	66.3
Winter-irrigated, zero-tillage cropland	1.8×10^{-3}	0.86	1.23×10^{-3}	68.3
Crop and grass intercropping with alfalfa	1.4×10^{-3}	2.15	1.03×10^{-3}	72.1
Crop and grass intercropping with bare belt	3.5×10^{-3}	0.95	2.15×10^{-3}	61.4
Alfalfa field	1.2×10^{-3}	5.25	0.92×10^{-3}	80.7

[a] Sand transport rate is an index that reflects the degree of surface sand raising and farmland wind erosion; it was measured in late April with a wind velocity of 8.5 to 9.0 m s^{-1} 2 m above the surface. The sand transport rate is the mean value at 0- to 20-cm height.

[b] Surface roughness is an index that reflects anti-wind erosion; it is proportional to the anti-wind erosion capability. As wind blows over the surface, different heights have different wind velocities; the closer to the surface, the smaller the wind velocity. There may be zero wind velocity at a certain height; this height is referred to as the surface roughness. It can be calculated from the wind velocities at two points in the same observation plot by the following formula (Wu, 1987): $\lg Z_0 = (\lg Z_2 - A \lg Z_1)/(1 - A)$, where Z_0 is the surface roughness (cm); Z_2 and Z_1 are different observation heights (cm); and $A = V_2/V_1$ is the wind velocity ratio of different heights.

[c] In the sand transport rate, the dust may be formed by suspended dust with particle size < 0.063 mm; the suspended dust level can be acquired through the sieve method, including silt and clay (Su et al., 2004).

[d] Wheat height in the single seeding field was 5 cm, alfalfa height in the grain crop and grass intercropping field was 4 cm, and alfalfa height in the single seeding field was 8 cm.

ranging from 150×200 m to 200×250 m are utilized in the wind-gap section of the oasis. The main tree for farmland shelterbelts is poplar (*Populus gansensis* C. Wang et H.L. Yang) (Figure 13.3). The experimental site was set in the inner oasis. In the poplar–maize compound system, the plant spacing of poplars is 1.2 to 1.5 m, the row spacing is 1.5 to 2.0 m, tree age is 14 years, average height is 13.2 m, and average diameter at breast height is 15.1 cm (Figure 13.3B). The yield variations of different regions in the poplar–maize compound system were studied using a randomized block experimental design. In Figure 13.3, sites 0 to 13 m from the shelterbelt tree edge to the field on the west side of the shelterbelt (W–S) and 0 to 13 m within the east side of the shelterbelt (E–S) were the shaded zones in the morning and in the afternoon, respectively. The field zones that were 13 to 20 m from the west side of the shelterbelt (W–NS), 13 to 20 m from the east side of the shelterbelt (E–NS), and in the middle position between tree rows (M) were non-shaded zones. The results showed that grain yields of maize exhibited significant differences between sites in the poplar–maize system (Figure 13.4). The mean crop yields in shaded zones were 27% and 22% lower than for the non-shaded zones for the western and eastern sides, respectively, and the mean crop yields for the western side were 23% lower than for the eastern side ($p < 0.0001$) (Figure 13.4A) (Ding and Su, 2010). Compared with the middle position (M, 7.3 t ha^{-2}), crop yields at the three sites (W–S, W–NS, and E–S) were significantly reduced, whereas the value at the E–NS site was a little higher (7.6 t ha^{-2}). As a whole, the negatively affected

Figure 13.3 (A) Poplar–wheat and (B) poplar–maize complex systems in oases.

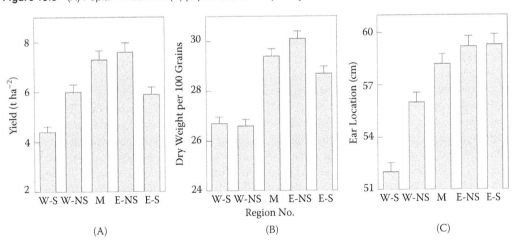

Figure 13.4 (A) Maize yield, (B) dry weight per 100 grains, and (C) ear location at different sites within the poplar–maize system. The error bar represents one standard error of the mean. Regions W–S and E–S were shaded regions in the morning and afternoon, respectively; regions W–NS and E–NS were non-shaded regions; and M represents the middle position between tree rows, also a non-shaded region.

area of crop yield was larger on the western side of shelterbelts; crop yield was seriously reduced within 13 m in the shading zones when the shelterbelt tree height was around 13 m. The affected area of crop yields on the eastern side was less than that on the western side, with only a moderate decrease within 9 m on the eastern side and some increase outside of 9 m. The order of maize yields in the poplar–maize compound system at the different sites was E–NS > M > W–NS > E–S > W–S.

We also studied two other parameters related to yield. Regarding dry weight per 100 grains, Figure 13.4B shows that the mean value of 100-grains weight from the western sites (W–S and W–NS) was 26.7 g, which was lower than the means from both the eastern sites (E–S and E–NS) and the middle site (M). Shaded and non-shaded regions for both the western and eastern site did not exhibit any significant differences in dry weight per 100 grains. For the height of the ear-bearing location, there was a similar pattern with the previous parameter, which showed that the mean value for the western site was lower than both the eastern site and the middle site (Figure 13.4C). For example, the ear locations at W–S were 6 cm and 7 cm lower than those at the M and E–S sites, respectively. In addition, differences between shaded and non-shaded regions were significant for the western site but not for the eastern site.

Further, we examined how the photosynthetic physiological parameters affected the maize yield and which parameter played the predominant role in impacting the variation in crop yield at different locations within the system. Results show that the gas exchange process of maize in different directions and different distances away from trees can vary significantly within agroforestry systems. The middle location of farmland within shelterbelts showed a greater maize net photosynthetic rate (P_n). The values were smaller at the locations close to trees. Morning, midday, and daily mean P_n values of maize on the eastern side of the shelterbelts were greater than at the same distance on the western side; the opposite pattern was observed in the afternoon. For example, P_n values of W–S, M, and E–S sites in mid-August were 1.7, 24.1, and 24.7 µmol m^{-2} s^{-1} in the morning and 19.5, 21.8, and 18.9 µmol m^{-2} s^{-1} in the afternoon, respectively.

We analyzed the relationships between corn yield and gas exchange parameters (Table 13.3). Maize yields in the shelterbelt system were significantly affected by the average P_n. There was a positive linear relationship between maize yield and noon value of P_n, transpiration rate (E) at the seeding stage, and average value of P_n ($p < 0.05$). At the flowering stage, the average value of P_n, E, and stomatal conductance (G_s) were closely related to maize yield ($p < 0.05$). The linear relationship between the average value of P_n, E, and maize yield was very significant ($p < 0.01$). In the filling stage, the noon value of G_s, the average value of P_n, and E were significantly correlated with maize yield ($p < 0.05$). In general, the maize yield was higher within the zones that had higher P_n values in all growth stages; it was also higher within the zones that had higher average E values. G_s was also correlated with the yield, but with different correlation levels at different stages.

Both photosynthesis and transpiration are driven in part by environmental conditions, particularly within the leaf and canopy boundary layers of the plant. In order to understand the main environmental factors causing the variation in maize yields, the linear relationship between main gas exchange parameters and microclimate factors was analyzed. The results showed that maize yields were affected by photosynthetically active radiation (PAR) in all growth stages. The stronger the radiated region, the higher the maize yield was. The filling stage is a critical period for dry matter production; the average value of air temperature (T_a) has a positive function with the maize yield during this time. The higher the average T_a in the site, the higher the yield was. In addition, the vapor pressure deficit (VPD) during the seeding stage and the average of atmospheric relative humidity (RH) during the flowering stage also had important effects on maize yields. Whereas the VPD was negatively correlated with yield, RH was positively correlated with yield. The maize yield in shelterbelts is the direct combined result of all microclimate factors in the field.

Negative effects on the farmland of shelterbelts were observed at both aboveground and belowground levels. The negative effects on roots or the canopy varied within different positions in the shelterbelt cell. Many studies have concluded that shade is a major factor causing crop yield

Table 13.3 Correlation Coefficients for Yield and Gas Exchange Parameters in Maize During Seeding, Flowering, Filling Stages within Poplar–Maize System

Parameter	Time							
	Morning (8:00–11:00)		Noon (12:00–14:00)		Afternoon (15:00–18:00)		Average (8:00–18:00)	
	R	p	R	p	R	p	R	p
Seeding Stage								
P_n	0.62	0.263	0.91	0.031˙	0.83	0.082	0.92	0.028˙
E	0.44	0.460	0.89	0.045˙	0.65	0.232	0.85	0.066
WUE	0.79	0.110	−0.72	0.171	0.55	0.336	0.51	0.382
G_s	0.58	0.301	0.67	0.219	0.84	0.077	0.81	0.097
Flowering Stage								
P_n	0.84	0.074	0.94	0.018˙	0.02	0.980	0.97	0.007˙˙
E	0.72	0.168	0.89	0.045˙	0.23	0.715	0.97	0.005˙˙
WUE	0.38	0.522	−0.75	0.147	−0.18	0.778	−0.16	0.803
G_s	0.91	0.034	0.61	0.273	0.31	0.612	0.88	0.046˙
Filling Stage								
P_n	0.87	0.057	0.88	0.051	0.69	0.195	0.92	0.026˙
E	0.72	0.169	0.82	0.088	0.60	0.282	0.90	0.040˙
WUE	0.88	0.052	−0.48	0.412	−0.09	0.884	0.86	0.060
G_s	0.84	0.074	0.92	0.027˙	0.30	0.628	0.85	0.065

Note: R, correlation coefficient; P_n, net photosynthetic rate; E, transpiration rate; WUE, water use efficiency; G_s, stomatal conductance. ˙$p < 0.05$, ˙˙$p < 0.01$.

reductions in tree–crop systems (e.g., Lawson and Kang, 1990). Maize, with the C_4 photosynthetic pathway, is sensitive to shading; earlier investigations in different agroforestry systems attributed declines in crop yield to competition for light. A study by Miller and Pallardy (2001) on maize–silver maple temperate alley cropping stands in Missouri also showed that decreases in sunlight near the tree row limited maize production. In our poplar–maize system in the Hexi Corridor desert oasis of China, morning shade had a more significant negative effect on maize. Afternoon shade was in some cases beneficial to crop growth. The trend of negative effects on crop yields within shelterbelt farms in North and Northeast China is that more serious negative effects occurred on the northern side than on the southern side of the field within the shelterbelt, and more serious negative effects occurred on the eastern side than on the western side (Sun, 1982; Liu et al., 2004). However, the opposite was observed in the Hexi Corridor desert oasis in Northwest China. More serious negative effects occurred on the southern side than on the northern side, and more serious negative effects occurred on the western side than on the eastern side of the field within the shelterbelt. Crop yields within the tree heights' distance from the shelterbelt are usually negatively affected; however, the distance on the eastern side was less than tree height, and the yield decreased by 4 to 19%. The distance on the western side was more than tree height and the yield decreased by 18 to 40%.

13.3.2.2 *Measures to Reduce Negative Effects on Farmland*

The construction of a buffering belt using pasture or forage crops between trees and crop fields can help to prevent or reduce the negative effects of the shelterbelt on the crops. The species or varieties selected for the pasture or forage in the buffering belt should be water saving and adaptive to low soil fertility. In order to reduce negative effects of shelterbelts, the water and fertilizer management schedule should be carefully coordinated. Cutting off tree roots penetrating the crop field is an effective way to reduce negative effects on crops. In addition, pruning the tree canopy can increase the light permeability of the tree crown, which promotes crop growth, eventually increasing crop yield. Replacing highly water-demanding tree species, such as poplars, in the shelterbelt system with low water-consuming tree species, such as jujube trees, with wider row spacing (12 to 15 m) can help reduce water consumption rates in an agroforestry system.

13.3.3 Effects of Jujube Tree–Crop Intercropping Systems on Crops and Their Environment

Jujube–crop intercropping systems are an important type of agroforestry system (Figure 13.5). Understanding the trend of negative effects on farmland caused by jujube trees can provide an important theoretical foundation for increasing jujube and crop yields with high water use efficiency. In the Hexi Corridor desert oasis in Northwest China, the planting pattern of jujubes is mainly north–south rows. Although some east–west rows are also used in the region, the duration of sunlight between north–south rows is more consistent and the total solar energy accepted by crops is higher than with east–west rows.

A jujube–maize intercropping system with jujube trees planted in the north–south direction was used for this research. The row spacing of jujube trees was 12 m, and the plant spacing was 3 m. A double row strip pattern was adopted for maize with 60-cm strip spacing, 40-cm row spacing, and 28-cm plant spacing. A random block experimental design was adopted. Measurements were taken in three random lines. Each sampling line was extended from west to east, starting from one middle point in a field. The sampling line then crossed a jujube tree row and ended at the middle point in the field located on the other side of the jujube tree row. There were 12 sampling points along each line from west to east. The distance between two sampling points was 1 m. The positions of the sampling points were denoted from west to east as W5, W4, W3, W2, W1, jujube tree, E1, E2, E3, E4, and E5. The variation of maize yield in this jujube–maize intercropping system is shown in Figure 13.6. First, it can be seen that the closer the distance to the jujube tree, the lower the maize grain yield. The maize yield was 9.6 t ha^{-2} in W5. Compared to the yield in this position, the maize yield decreased by 25.0% in W4, 33.3% in W3, 44.8% in W2, and 56.3% in W1. The maize yield was 9.8 t ha^{-2} in E5. Compared to the yield in this position, the maize yield decreased by 5.1% in E4, 13.3% in E3, 28.6% in E2, and 42.9% in E1. Second, the maize yield east of the tree was higher than west of the tree. For example, the yields for E1 and W1 were 5.6 and 4.2 t ha^{-2}, respectively. The former was 1.4 t ha^{-2} higher than that of the latter. The yield of W4 was 7.2 t ha^{-2}, which was close to 7.0 t ha^{-2} in E2.

In order to understand the reason for the variations in maize yields in jujube–maize intercropping systems, the relationships between maize yields and the main gas exchange parameters were analyzed by linear regression (Table 13.4). The results showed that the net photosynthetic rate (P_n) was the most significant parameter affecting maize yields during filling stages. Morning, noon, and average P_n values all had a significant linear correlation with maize yield. The higher the P_n was in a position in the field, the higher the maize yield in that position. The transpiration rate (E) at noon, average E value, and stomatal conductance (G_s) at noon also had significant positive relationships with maize yields.

The results of the relationship between main gas exchange parameters and field microclimate factors in filling stages showed that the temperature (T_a) at noon presented a significant positive correlation with P_n ($R = 0.66$, $p = 0.036$). The average value of T_a from 8:00 to 18:00 was significantly correlated with P_n and E ($R = 0.74$ and 0.70; $p = 0.014$ and 0.025, respectively). The position in the maize field with higher T_a values at noon and higher average T_a values also had higher P_n and E values. The air relative humidity (RH) at noon showed a significant positive correlation with E and

(A)

(B)

Figure 13.5   Jujube–maize and jujube–soybean intercropping systems in oases. (A) Jujube–maize intercropping system with more development time; the row spacing of jujube trees is 10 × 15 m, and the plant spacing is 3 × 4 m. The crop strip should become more narrow with the growth of the jujube trees; the intercrop strip of crop should be fixed when the height of the jujube trees is limited to 4 m. As shown in the figure, the fresh fruit yield is 3200 kg ha^{-2}, and the yield of producing seed maize is 6700 kg ha^{-2}. (B) Jujube–soybean intercropping system with less development time; the row spacing of the jujube trees is 4 × 6 m, and the plant spacing is 3 × 4 m. The crop that is intercropped with jujube should be gradually removed with growth of the jujube trees to increase the yield of the jujube orchard. The fresh fruit yield was 10,130 kg ha^{-2}, and the soybean yield was 3150 kg ha^{-2}.

G_s ($R = 0.75$ and 0.85; $p = 0.013$ and 0.002, respectively). The photosynthetically active radiation (PAR) in the morning had a significant linear correlation with P_n ($R = 0.71, p = 0.022$). The position with higher PAR also had a higher P_n value. Compared to the relationship observed in the morning, the correlation of PAR with P_n at noon was higher and reached a significant level ($R = 0.82, p = 0.003$). Thus, the spatial variation of maize yields was due to the spatial differences of gas exchange parameters in a jujube–maize system that is influenced by environmental factors.

The trend of negative effects of jujube trees on crops is similar to the effects of poplar trees in a shelterbelt system: More negative effects appeared on the southern side of the tree row than on the northern side, and more negative effects appeared on the western side of the tree row than on

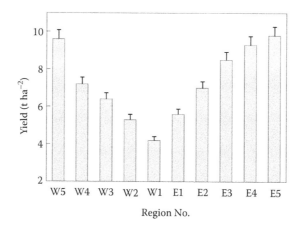

Figure 13.6   Maize yields at different locations within the jujube–maize system. The error bar represents one standard error of the mean. Regions W and E were located west and east, respectively, of the jujube row; the number represents the distance to the jujube trees (m).

Table 13.4   Correlation Coefficients for Yield and Gas Exchange Parameters for Filling Stages within the Jujube–Maize System

Parameter	Time							
	Morning (8:00–11:00)		Noon (12:00–14:00)		Afternoon (15:00–18:00)		Average (8:00–18:00)	
	R	p	R	p	R	p	R	p
P_n	0.69	0.027*	0.77	0.009**	0.57	0.084	0.82	0.004**
E	0.58	0.078	0.73	0.017*	0.35	0.328	0.74	0.015*
WUE	0.30	0.400	−0.51	0.134	0.08	0.823	−0.05	0.885
G_s	0.38	0.275	0.64	0.048*	0.39	0.271	0.62	0.057

Note: R, correlation coefficient; P_n, net photosynthetic rate; E, transpiration rate; WUE, water use efficiency; G_s, stomatal conductance. *$p < 0.05$, **$p < 0.01$.

the eastern side. The crop yield is usually negatively affected within the distance of one tree height from the tree edge. In our case, the negatively affected distance on the eastern side was less than one tree height, and on the western side was greater than one tree height. The width of the field negatively affected by the jujube tree belt was about one third of that affected by the poplar tree belt. The percentage of yield reduction caused by the jujube tree belt was only 25% of that caused by the poplar tree belt.

13.4 WATER DEMAND IN A JUJUBE TREE–CROP COMPLEX SYSTEM

Fruit tree–field crop complex agroforestry systems are an important model for oasis ecological agriculture. Every oasis region has its unique fruit tree resources, such as seedless grapes (*Vitis vinifera* L. cv. *Thompson seedless*), bergamot pears (*Pyrus sinkiangensis* Yü), almonds (*Amygdalus communis* L.), pistachios (*Pistacia vera* L.), figs (*Ficus carica* L. var. *hortensis* Shinn.), pomegranates (*Punica granatum* L.), thin-shell walnuts, kernel-apricots, dried apricots, gold flat peaches, cherries, peaches, pears, apples, and Hami jujubes, among others, in Xinjiang and table grapes, wine grapes, plums, apple-pears, Linze jujubes, and Liguang apricots, among others, in the Hexi Corridor of Gansu. These fruit trees can be used to establish monoculture orchards and can also be used to form a complex agroforestry system by intercropping with field crops.

Three additional measures should be considered in order to improve the rate of water resource use and water use efficiency in oasis agricultural regions. The first measure is to prevent leakage and water evaporation in water canal systems. This mainly depends on engineering methods, including canal lining or piped water supply. The second measure is to increase field water utilization rates by preventing field leakage and decreasing soil surface evaporation in order to meet the water requirements for crop growth. This can be realized using water-saving irrigation facilities and a suitable irrigation scheme. It is very important to understand the water requirement rules of different crops under different cropping patterns. The third measure is to enhance plant water utilization rates. This mainly relies on breeding or introducing crop varieties with short growth durations and either high drought resistance or high drought tolerance. Water-saving cultivation methods are also important (Su et al., 2002). The results of our research exploring the water demands of fruit–crop complex systems in an effort to enhance water use efficiency are presented in the following section.

13.4.1 Phonological Phases of Linze Jujube and Crops

Phonological phase refers to the annual rhythm of morphological phases of plant growth and development, which adapt to the seasonal variation of light, temperature, and rainfall. Different morphology characteristics also appear in different crop development stages. These morphological characters can be used as indicators to guide fertilization, irrigation, and other field operations.

Jujubes (*Zizyphus jujuba* Mill. var. *inermis* Rehd.) are native to China and have a cultivation history of over 2500 years; consequently, the various endemic varieties have unique characteristics. Linze jujube (*Zizyphus jujuba* Mill. var. *inermis* Rehd. cv. *Linze jujube*) is a unique native selection that produces high-quality fruits and grows endemically in arid northwest China and in the Linze Oasis of Gansu Province. It has adapted to high temperatures, high irradiance, and a dry environment. Compared with more than 20 other jujube accessions, Linze jujube had the highest fruit-setting rate and the highest vitamin C content, reaching an average of 579.3 mg vitamin C per 100 g fresh fruit. A comparative study of desert oasis plant species showed that the stable carbon isotope ratio ($\delta^{13}C$) in the leaves of Linze jujube was –26.6‰, while those of the desert plants *Caragana korshinskii*, *Nitraria sphaerocarpa*, and *Hedysarum scoparium* were –25.8‰, –25.8‰, and –26.4‰, respectively, with no significant difference among them (Su and Liu, 2005). Foliar $\delta^{13}C$ values can reflect water use efficiency associated with plant photosynthesis and transpiration and can be used to indicate the long-term water use efficiency of plants. To a certain extent, $\delta^{13}C$ values were positively correlated with water use efficiency. Thus, it is evident that the water use efficiency of Linze jujubes was similar to that of some wild desert plants (Su and Liu, 2005).

The root of the Linze jujube begins its activity around April 25 to 30 each year, when the average temperature of the main root distribution layer reaches 15.9°C (Table 13.5). The annual growth period of Linze jujubes from the start of root activity to leaf drop is 163 days. Linze jujubes can tolerate high temperatures and strong solar radiation. They can grow normally under extreme conditions, such as high temperatures of 40.5°C and strong illumination intensity at 2016 µmol m^{-2} s^{-1} (Su and Liu, 2005).

By analyzing weather records and Linze jujube growth, we found that the critical temperature above which the Linze jujube could survive through winter is –20°C. When average daily temperatures are lower than –20°C continuously for 2 days, it suffers from freezing damage; however, temporary low temperatures will not cause freezing damage. The shorter the duration of cold, the lower the temperature it can tolerate. The lowest temperature that it could tolerate was –27°C (Su et al., 2009a). The main measures available to prevent freezing damage include the construction and preservation of oasis shelterbelts, the development of jujube trees around villages and houses and along roads and canals, and more urgent measures such as establishing windbreaks, smoke, and spraying antifreeze and high-lipid film according to weather forecasts (Su et al., 2009a).

Table 13.5 Phenophases and the Related Temperatures of Linze Jujube

Phenophase	Date (month-day)	Air Temperature (°C)	Average Air Temperature (°C)	Ground Temperature (°C)	Average Ground Temperature (°C)	Temperature of Root Main Distribution Layer (°C)	Average Temperature of Root Main Distribution Layer (°C)
Preliminary activating stage of root	04-25–04-30	9–15	11.9	13–26	21.5	13–20	15.9
Root sprouting stage	05-25–05-30	12–23	17.3	19–33	27.6	19–26	22.1
Germination stage	05-03–05-09	8–20	14.1	12–30	23.8	15–25	19.1
Leaf expansion stage	05-06–05-13	8–21	16.4	14–32	24.2	15–25	19.2
Earlier flowering stage	06-11–06-15	19–24	22.0	21–37	32.0	23–29	25.8
Full flowering stage	06-11–08-20	17–29	21.9	19–40	30.9	20–32	26.9
Current shoot growing stage	05-08–07-10	13–27	19.7	12–39	28.9	14–31	23.9
Fruit shoot growing stage	05-05–07-10	10–27	19.4	12–39	28.7	14–31	23.6
Fruit growth stage	06-20–09-23	10–28	20.3	13–40	29.0	17–33	26.1
Fruit coloring stage	08-26–09-20	10–22	16.9	13–32	25.1	17–29	24.2
Fruit harvesting stage	09-10–09-28	10–20	14.1	13–27	22.4	17–26	22.2
Leaves shedding stage	10-05–10-26	1–14	6.9	4–18	11.8	7–19	14.2
Dormancy stage	12-01–02-28	−20–4	−6.0	−16–10	−4.8	−10–6	−1.8

Note: The root mainly distributed in the 20- to 60-cm soil layer, and the temperature of the root main distribution layer is the average temperature of those in the 20-cm, 40-cm, and 60-cm soil layers.

The emergence period for spring wheat generally begins 10 days after sowing. The wheat emerges around April 10. The tillering stage begins around April 23, the jointing stage around May 10, booting stage around May 22, heading stage around May 30, flowering stage around June 7, filling stage around June 22, and milk stage around July 2. The optimum temperature for alfalfa growth is around 25°C. Root growth is best at 15°C, and plants can tolerate higher temperatures with irrigation. Alfalfa has a high resistance to cold. It can geminate at 5 to 6°C and tolerate temperatures as cold as –5°C. Fully grown plants can tolerate temperatures as low as –30°C.

13.4.2 Water Demand

The amounts and variation of plant water requirements are determined by meteorological conditions, plant characteristics, soil properties, and agrotechnical measures. Agricultural research shows that soil moisture content should be maintained at around 70 to 80% of field capacity to meet the crop's normal growth and development needs (Su et al., 2002; Masinde et al., 2005). The water requirements of pure Linze jujube communities and intercropping systems of Linze jujube with both spring wheat and alfalfa were studied using a water balance method. The total water needed by Linze jujubes, from the beginning of root activity in the spring to leaf fall in deep fall, was 497.2 mm (Table 13.6). At the beginning of root activity from April 25 to 30, the major water consumption in the jujube plantation was mainly soil evaporation. The water requirement rates in May, June, July, August, and September were 2.5, 4.3, 4.8, 2.4, and 1.9 mm d^{-1}, respectively. The month requiring the most water was July. The next highest month was June. At the beginning of October, soil evaporation became the major water requirement again, just as in late April. Due to seasonal variation, the water requirement of Linze jujubes was 84.6 mm in spring (April 25 to May 31), 351.1 mm in summer (June 1 to August 31), and 61.3 mm in autumn (September 1 to October 5). The average water requirement rate over the whole growth period was 3.1 mm d^{-1} (Su et al., 2010).

The water requirement of spring wheat for the entire growth duration was 447.9 mm (Table 13.6). Water requirement rates in April, May, and June were 2.6, 4.3, and 5.2 mm d^{-1}, respectively. The average water requirement from sowing to seedling appearance stage was 1.3 mm d^{-1} and was 5.5 mm d^{-1} in the filling stage. The average water requirement rate during the entire growth period was 4.0 mm d^{-1}.

The water requirement of alfalfa during the growth period was 583.7 mm (Table 13.6). Water requirement rates in April, May, June, July, August, and September were 2.4, 3.0, 3.3, 4.3, 3.5, and 2.1 mm d^{-1}, respectively. The maximum water requirement rate was 4.8 mm d^{-1}. The annual average water requirement rate was 3.0 mm d^{-1}. The annual yield of fresh grass was 70.7 t ha^{-2}. The dry yield was 17.2 t ha^{-2}. The water consumption coefficient of the alfalfa dry yield was 340.

Table 13.6 Water Requirements (mm) of Linze Jujube, Spring Wheat, Alfalfa, and Their Intercropping Systems

Date (month-day)	Jujube	Spring Wheat	Alfalfa	Jujube– Spring Wheat Intercropping	Jujube– Alfalfa Intercropping
03-26–03-31	—	3.9	7.2	3.5	7.4
04-01–04-30	5.7	78.4	72.3	75.2	76.0
05-01–05-31	78.9	131.8	93.5	132.3	160.7
06-01–06-30	130.1	156.7	98.7	159.2	189.7
07-01–07-31	147.4	77.0	132.5	142.8	248.6
08-01–08-31	73.8	—	109.1	75.7	170.1
09-01–09-30	55.5	—	63.6	52.8	117.9
10-01-10-05	5.8	—	6.8	6.2	9.9
Total	497.2	447.9	583.7	647.7	980.3

The water requirement of the jujube–spring wheat intercropping system was 647.7 mm (Table 13.6). The daily water requirement rates in April, May, June, July, August, and September were 2.5, 4.3, 5.3, 4.6, 2.4, and 1.8 mm d^{-1}. The water requirement was maximal in June. The annual average water requirement rate was 3.4 mm d^{-1}.

The water requirement of jujube–alfalfa intercropping was 980.3 mm (Table 13.6). The daily water requirement rates in April, May, June, July, August, and September were 2.5, 5.2, 6.3, 8.0, 5.5, and 3.9 mm d^{-1}, respectively. The water requirement reached its maximum level of 8 mm d^{-1} in July. It was significantly higher than the water requirements of jujube or alfalfa monoculture. The annual average water requirement rate was 5.1 mm d^{-1} for jujube–alfalfa intercropping systems.

13.4.3 Irrigation Dates and Quotas

An irrigation schedule with suitable dates and quotas is essential in oasis agriculture and is the key for improving irrigation water efficiency and its benefits. Some crops, such as cotton, can benefit from promoting vegetative growth at a higher water content than during the reproductive growth stage. Controlled irrigation or limited irrigation can increase the yields of some crops. Different irrigation schemes used for fruits during their vegetative stage and their reproductive stage could help to significantly improve their yield and quality.

According to the characteristics of climates in oases, as well as crop biology, irrigating before the winter is important for jujubes, alfalfa, and spring wheat. If water resources are not allowed to be used for irrigation before winter, irrigation is needed either in early spring or before sowing. The irrigation data and quotas of different plant patterns were determined based on phenological phases and water requirement regularities (Table 13.7). The four irrigation times most critical for jujube trees are during the following growth stages: before flowering, fruit swelling, before fruit ripening, and overwintering. Irrigation at these times can increase fruit quality and yield. The plot irrigation method is suitable for jujube monoculture. In a jujube–crop complex system, either furrow irrigation or plot irrigation is suitable for crops.

Irrigation before winter is very important in oases. On the one hand, it is beneficial to early spring crop growth and the fruit trees' ability to overwinter and can increase soil water content. On the other hand, it can prevent secondary salinization. Under natural conditions, flooding during the rainy season plays an important role in maintaining a soil salinity balance in the oasis. Accumulated salts can be carried out from the oasis by flood water to the lower reach of the inland river basin where the inland river finally ends. This is an important self-preservation mechanism of oases. Flood control by humans reduces the ability of natural mechanisms to maintain a salinity balance inside the oasis. Also, the use of excess surface irrigation water to obtain a greater economic return introduces more salt into the oasis. Because the self-preservation mechanism of the oasis has been altered, salts carried from the mountains to the oasis by rivers are entirely deposited in oasis soils and eventually lead to secondary salinization. Irrigating over the winter and higher levels of leaching water are very effective ways to resolve this issue.

Irrigation methods for spring wheat include four irrigation times that can meet the water requirement of wheat during its germination, jointing, booting, and filling stages (Table 13.7). If sprinkler irrigation and drip irrigation methods are used to replace surface irrigation, earlier and more frequent irrigation should be adopted, with less water used. Ensuring a water supply during the critical periods of crop growth will enable optimum allocation of limited water resources. Irrigation methods for alfalfa include four irrigation times during its growth stage, and additional irrigation before winter if there is no ground water supply. The total irrigation quota is 5840 m^3 ha^{-2} (Table 13.7). Irrigation methods for jujube–spring wheat intercropping must consider both the phenological phase of jujube trees and the growing period of wheat. Irrigation should occur six times in a year (Table 13.7), including overwintering irrigation. This type of irrigation can satisfy

Table 13.7 Irrigation Date (month-day) and Quota (m³ ha⁻²) of Jujube and Intercropping Systems

Linze Jujube

Irrigation Type	Before Flowering	Fruit Swelling	Before Fruit Ripening	Overwintering
Irrigation date	06-05–06-07	07-05–07-09	08-07–08-10	10-20–10-30
Irrigation quota	1210	1280	1200	1280

Spring Wheat

Irrigation Type	Before Seeding	Jointing	Booting	Filling
Irrigation date	03-23–03-25	05-09–05-11	05-22–05-24	06-20–06-22
Irrigation quota	1200	1040	1200	1040

Alfalfa

Irrigation Type	Accelerating Growth	Rejuvenation	Producing	Increasing	Overwintering
Irrigation date	04-15–04-20	05-18–05-22	06-15–06-19	07-21–07-25	10-20–10-30
Irrigation quota	1040	1200	1200	1200	1200

Jujube–Spring Wheat Intercropping

Irrigation Type	Wheat Joining	Wheat Booting	Wheat Filling	Jujube Fruit Swelling	Before Jujube Fruit Ripening	Overwintering
Irrigation date	05-09–05-11	05-22–05-24	06-20–06-22	07-05–07-09	08-07–08-10	10-20–10-30
Irrigation quota	1200	1280	1220	750	750	1280

Jujube–Alfalfa Intercropping

Irrigation Type	Alfalfa Accelerated Growth	Alfalfa Rejuvenation	Before Jujube Flowering	Alfalfa Production	Jujube Fruit Swelling	Alfalfa Yield Increasing	Before Jujube Fruit Ripening	Overwintering
Irrigation date	04-15–04-20	05-18–05-22	06-05–06-07	06-20–06-24	07-05–07-09	07-21–07-25	08-07–08-10	10-20–10-30
Irrigation quota	1100	1200	1200	1300	1200	1300	1200	1300

the water requirements of this complex system. In the early spring, water requirements of wheat should be prioritized and the plot irrigation method for wheat should be adopted. In later stages, when the jujube is in its critical stage of water requirement, furrow irrigation along the jujube strip is adopted to specifically irrigate jujube trees only. Irrigation methods for jujube–alfalfa complex systems should consider the large water quantity requirement. Irrigation eight times a year, including overwintering irrigation (Table 13.7), can secure the water requirements for both jujube trees and alfalfa in this system.

The water demand of jujube–alfalfa intercropping systems is much greater than that of jujube–wheat systems, due to differences in the root distribution patterns of the plants. As much as 70% of the root system of wheat is distributed within the 0- to 20-cm soil layer, whereas the major root system of jujubes is within the 20- to 60-cm soil layer. It is beneficial to make good use of soil water resources throughout the whole soil profile; however, alfalfa is a deep-rooting perennial forage crop. Considering the water demand, the development of a jujube–alfalfa system should not be too large. In order to reduce competition for water, alfalfa should not be planted too close to the jujube tree belt or even within the jujube tree belt. There should be at least a 2-m-wide belt purely for jujube trees in a jujube–alfalfa system, and a 1-m-wide belt for jujube trees in a jujube–wheat system should be sufficient. The row spacing between jujube tree belts ranges from 4 to 15 m in a jujube–wheat system, depending on the desired proportion of wheat and jujube products. The distance between jujube trees within tree rows is around 3 to 4 m. The height of jujube trees should be controlled at 4 m or lower. The determinate bole height should be kept at 0.8 to 1.0 m.

Intercropping is an important technique for increasing yields and income by making full use of land and climate resources. The plants selected for the intercropping system should be matched reasonably for their different growth characteristics, such as tall crops with short crops, deep-rooted plants with shallow-rooted plants, early-maturing crops with late-maturing crops, and high light-demanding crops with shade-tolerant crops. This can help to maximize use of water, nutrition, light, and heat resources, as well as obtain higher yields and water use efficiencies in agricultural production.

13.5 YIELD AND WATER USE EFFICIENCY OF CROPS IN CLUSTER PLANTING

Crops in oases include grain crops, economic crops, forage crops, and energy crops. The spatial distribution pattern of plant communities can be random distribution, uniform distribution, or cluster distribution. Cluster distribution, also called *aggregated distribution*, is the way most desert plants grow in order to adapt to severe conditions. We make good use of the principle of cluster distribution and positive correlations among plants to change the traditional uniform crop cultivation pattern into a cluster pattern. The yields and water use efficiency of crop production in arid areas can be improved using cluster patterns that concentrate irrigation and fertilization more than do uniform plantings (Su et al., 2009b).

13.5.1 Yield and Water Use Efficiency of Cotton in Cluster Planting

An experiment was designed in order to test the effect of cluster planting patterns on cotton. There were two rows of cotton in each strip. Cotton was planted in each hole within the row. Each hole had either two or three cotton plants in a cluster planting pattern or one cotton plant per hole in a uniform planting pattern. Details regarding the nine treatments are provided in Table 13.8. The experiment used a completely randomized plot design with three replications. Different planting patterns of cotton are shown in Figure 13.7. The results in Table 13.9 show that seed cotton yields and lint yields in treatment 6 were the highest, and those in treatment 1 were the lowest. Treatments 1, 2, and 3 had similar plant numbers per unit area. Seed cotton yields and lint yields were the

Table 13.8   Plot Experiment Design to Compare Cotton Cluster Planting and Uniform Planting (2006, Northern Linze Oasis, Middle of Hexi Corridor Region)

Treatment No.	Plants per Hole	Distance between Strips (cm)	Distance between Rows (cm)	Distance between Holes (cm)	Holes per hm² (×10⁴)	Plants per hm² (×10⁴)
1	1	40	20	16	20.83	20.83
2	2	40	30	27	10.58	21.16
3	3	40	40	35	7.14	21.42
4	1	50	30	15	16.67	16.67
5	2	50	30	18	13.89	27.78
6	3	50	30	25	10.00	30.00
7	1	50	30	20	12.50	12.50
8	2	50	30	20	12.50	25.00
9	3	50	30	20	12.50	37.50

highest in treatment 3, which had a cluster pattern with three plants per hole. Treatments 4, 5, and 6 had the same 30-cm row spacing and 50-cm strip distance; the yield was the highest in treatment 6 when the distance between holes was properly adjusted. Treatments 7, 8, and 9 also had the same row spacing and same distance between holes. The yield decreased in the treatment 9 due to very low spacing between holes.

The average lint yield in all treatments with one plant per hole was 1132.0 kg ha^{-2}. It was 1217.4 kg ha^{-2} in all treatments with two plants per hole, and 1322.1 kg ha^{-2} in all treatments with three plants per hole. The lint yield of treatment 6 was the highest, with 1498.0 kg ha^{-2}. The lint yield of the cluster pattern with three plants per hole was 32.3% greater than that from treatments with one plant per hole and 23.0% more than that from treatments with two plants per hole. When plant density was too high, yields decreased. The optimum planting pattern was a cluster pattern with a strip/row/hole spacing of 50/30/25 to 30 cm. The field had 8.33 to 10.00 × 10⁴ holes ha^{-2} and 25.00 to 30.00 × 10⁴ cotton plants ha^{-2}.

The biomass results from the experiment (Table 13.9) show an average of 4.42 t ha^{-2} biomass from treatments 1, 4, and 7 with one plant per hole; an average of 9.92 t ha^{-2} from treatments 2, 5, and 8 with two plants per hole; and an average of 14.28 t ha^{-2} from treatments 3, 6, and 9 with three

(A)   (B)   (C)

Figure 13.7   Different planting patterns of cotton in sandy soil. (A) Uniform planting with one plant per hole; (B) cluster planting with two plants per hole; and (C) cluster planting with three plants per hole.

Table 13.9　Yield and Water Use Efficiency of Cotton under Different Planting Patterns in Plot Experiment[a]

Index	Treatment No.				
	1	2	3	4	5
Lint percentage (%)	35	39	38	37	38
Seed cotton yield (kg ha⁻²)	2566.8 ± 312.8	2865.0 ± 148.3	3625.1 ± 18.3	3513.0 ± 239.8	3656.1 ± 757.4
Lint yield (kg ha⁻²)	898.4 ± 109.5	1117.4 ± 57.9	1377.5 ± 7.0	1299.8 ± 88.7	1389.3 ± 287.8
Total biomass yield (t ha⁻²)	4.86 ± 0.59	9.24 ± 0.48	13.07 ± 0.07	4.15 ± 0.28	11.14 ± 2.31
Aboveground biomass yield (t ha⁻²)	4.16 ± 0.58	7.95 ± 0.44	11.46 ± 0.09	3.49 ± 0.28	9.12 ± 1.82
Belowground biomass yield (t ha⁻²)	0.70 ± 0.03	1.29 ± 0.08	1.61 ± 0.12	0.66 ± 0.04	2.02 ± 0.49
Seed cotton yield/water consumption (kg m⁻³)	0.32 ± 0.04	0.35 ± 0.02	0.45 ± 0.00	0.43 ± 0.03	0.45 ± 0.09
Lint yield/water consumption (kg m⁻³)	0.11 ± 0.01	0.14 ± 0.01	0.17 ± 0.00	0.16 ± 0.01	0.17 ± 0.04
	6	7	8	9	
Lint percentage (%)	39	36	34	36	
Seed cotton yield (kg ha⁻²)	3841.0 ± 73.7	3327.2 ± 450.5	3368.8 ± 335.8	3029.9 ± 365.6	
Lint yield (kg ha⁻²)	1498.0 ± 28.7	1197.8 ± 162.2	1145.4 ± 114.2	1090.8 ± 131.6	
Total biomass yield (t ha⁻²)	13.97 ± 0.27	4.25 ± 0.58	9.38 ± 0.94	15.80 ± 1.91	
Aboveground biomass yield (t ha⁻²)	11.58 ± 0.23	3.54 ± 0.48	7.65 ± 0.76	13.42 ± 1.86	
Belowground biomass yield (t ha⁻²)	2.39 ± 0.11	0.71 ± 0.09	1.73 ± 0.19	2.38 ± 0.08	
Seed cotton yield/water consumption (kg m⁻³)	0.47 ± 0.01	0.41 ± 0.06	0.42 ± 0.04	0.37 ± 0.05	
Lint yield/water consumption (kg m⁻³)	0.18 ± 0.00	0.15 ± 0.02	0.14 ± 0.01	0.13 ± 0.02	

[a] The values are mean ± standard error (SE).

plants per hole. Although the total biomass yield was highest in treatment 9, the lint yield was not. The average aboveground biomass yield was 3.73 t ha^{-2} from treatments with one plant per hole, 8.24 t ha^{-2} from treatments with two plants per hole, and 12.15 t ha^{-2} from treatments with three plants per hole. Comparing the aboveground and belowground biomass of treatment 6 to that of treatment 9, the aboveground biomass increased in treatment 9, but root biomass remained stable. In treatments 1, 4, and 7, with one plant per hole, the changes in aboveground biomass were significantly correlated with the change in planting density, but the change in root biomass was much less. In general, with the same irrigation quotas, the biomass increased significantly in cluster planting patterns with three plants per hole. Better groundcover could reduce direct evaporation from the soil surface, increase plant transpiration, and create a better microenvironment with higher moisture content.

The results for water use efficiency (WUE) (Table 13.9) show that the average WUE was 0.39 kg m^{-3} for treatments 1, 4, and 7 with one plant; 0.41 kg m^{-3} for treatment 2, 5, and 8 with two plants; and 0.43 kg m^{-3} for treatment 3, 6, and 9 with three plants per hole. The average WUE of lint yields was 0.14 kg m^{-3} in treatments with one plant, 0.15 kg m^{-3} in treatments with two plants, and 0.16 kg m^{-3} in treatments with three plants. It was the highest in treatment 6 with an average of 0.18 kg m^{-3}. The WUE in cluster planting patterns with three plants per hole was higher than that in patterns with one plant and two plants per hole. In terms of lint WUE, the optimum planting pattern was a cluster pattern using three plants per hole with a strip/row/hole spacing of 50/30/25 cm with 10×10^4 holes ha^{-2} and 30×10^4 plants ha^{-2}. The WUE in this pattern was 28.6% higher than the traditional planting pattern of one plant per hole and 20.0% higher than the planting pattern of two plants per hole. The lint WUE in this pattern was 30.0% higher than one plant per hole and 18.2% higher than two plants per hole.

The statistical results of this field experiment (Table 13.10) show that seed cotton yields had significant differences under different planting patterns ($F = 15.0 > F_{0.05}(2,6) = 5.14$). Meanwhile, the results of multiple comparisons show that seed cotton yields in treatments with three plants per hole were significantly higher than in the other two planting patterns: 34.0% higher than one plant per hole and 19.3% higher than two plants per hole. Lint percentages for one plant per hole, two plants per hole, and three plants per hole were 37.2%, 37.6%, and 37.9%, respectively. Lint percentage in a cluster planting pattern increased moderately. The lint yield in treatments of three plants in a hole was 36.5% higher than in one plant per hole and 20.2% higher than in two plants per hole.

The analysis of aboveground biomass showed that there was no significant difference under different planting patterns ($p > 0.05$). The belowground biomass showed a significant difference ($F = 8.6 > F_{0.05}(2,6) = 5.14$). Multiple comparison results (LSR$_{0.05}(3,6) = 0.43$) show that there was a significant difference between one plant per hole and three plants per hole, but there were no significant differences between one plant per hole and two plants per hole nor between two plants per hole and three plants per hole. The total biomass yield showed a significant difference ($F = 5.8 > F_{0.05}(2,6) = 5.14$). Multiple comparison results (LSR$_{0.05}(3,6) = 2.51$) show that there was significant difference only between one plant per hole and three plants per hole. This was similar to the results of belowground biomass.

The reason why lint yields and water use efficiency were significantly higher in cluster planting patterns with three plants per hole is that the cluster planting pattern can be closer to an optimum community structure in arid environments, which best utilizes existing resources (Li et al., 1993). A multilayer topping operation for three times is beneficial for high cotton yields under a suitable density and cluster structure. Even with only topping one time, the cotton yields in a cluster planting pattern with three plants per hole are still higher than one plant per hole or two plants per hole. The effect of increasing yields is especially obvious in the edge areas of the oasis.

Table 13.10 Yield and Water Use Efficiency of Cotton under Different Planting Patterns in Field Experiment

No. of Plants per Hole	Seed Cotton Yield (kg ha^{-2})	Lint Yield (kg ha^{-2})	Total Biomass Yield (t ha^{-2})	Aboveground Biomass Yield (t ha^{-2})	Belowground Biomass Yield (t ha^{-2})	Seed Cotton Yield/Water Consumption (kg m^{-3})	Lint Yield/Water Consumption (kg m^{-3})
1	2279.3 ± 93.8[a]	847.9 ± 34.9[a]	3.32 ± 0.14[a]	2.85 ± 0.16[a]	0.47 ± 0.02[a]	0.26 ± 0.01[a]	0.10 ± 0.00[a]
2	2560.1 ± 98.2[a]	962.6 ± 36.9[a]	4.92 ± 0.47[ab]	4.11 ± 0.41[a]	0.81 ± 0.07[ab]	0.29 ± 0.01[a]	0.11 ± 0.00[a]
3	3053.9 ± 110.9[b]	1157.4 ± 42.0[b]	6.10 ± 0.87[b]	5.04 ± 0.72[a]	1.06 ± 0.16[b]	0.35 ± 0.01[b]	0.13 ± 0.01[b]

Note: The values are mean ± standard error (SE). In the same column, the different lowercase letters means significantly different at the $p < 0.05$ level. For one plant per hole, the distances between strips, rows, and holes were 50, 30, and 18 cm, respectively (13.89 × 10⁴ holes ha^{-2}, 13.89 × 10⁴ plants ha^{-2}). For two plants per hole, the distances between strips, rows, and holes were 50, 30, and 20 cm, respectively (12.50 × 10⁴ holes ha^{-2}, 25.00 × 10⁴ plants ha^{-2}). For three plants per hole, the distances between strips, rows, and holes were 50, 30, and 25 cm, respectively (10.00 × 10⁴ holes ha^{-2}, 30.00 × 10⁴ plants ha^{-2}).

13.5.2 Yield and Water Use Efficiency of Sweet Sorghum as an Energy Crop with a Cluster Planting Method

In the northwestern region of China, a large area is comprised of arid hillside land, poor sandy land, or low-lying saline–alkaline land. The saline–alkaline land area is 1.8×10^6 hm^2 and is found in the Hexi Corridor of Gansu. In this land, the low-production saline–alkaline land area accounts for 9.0×10^4 hm^2. In Linze County alone, 4100 hm^2 of less-productive saline–alkaline land is located in the middle of the Hexi Corridor. The development of energy crops with high photosynthetic efficiency, high water use efficiency, high biomass yield, and strong stress resistance on this marginal land can make good use of land, sunlight, and heat resources, as well as increase biomass production to promote biomass energy development, in order to improve the livelihoods of farmers and the rural economy.

13.5.2.1 Field Experiment of Traditional Cultivation Under Different Soil Types

Sweet sorghum (*Sorghum bicolor* (L.) Moench) is a crop with a high productive capacity and is referred to as a *high energy crop*. An experiment was conducted in the middle part of the Hexi Corridor in order to test the yield and water use efficiency of sweet sorghum in poor sandy soil and saline–alkaline soil, with loam soil as a control. The water-soluble salts in these three soil types are presented in Table 13.11. Ion contents of CO_3^{2-}, Cl^-, SO_4^{2-}, Ca^{2+}, Mg^{2+}, Na^+, and K^+ in the plow layer (0 to 20 cm) were the highest in saline–alkaline soil. These ion concentrations were at the second highest levels in loam soil, except for Na^+. Their concentrations were the lowest in sandy soil (Xie et al., 2012).

The soil particle size and composition in different soil types (Table 13.12) showed that sand content was the highest in sandy soil, over 80% in the plow layer. It was even higher in the 20- to 40-cm layer. In saline–alkaline soil, sand content was over 60% in the plow layer. The sand content decreased in the deeper layer; however, the silt and clay content increased with depth. The sand content in loam soil was only 46.5% in the plow layer and also decreased in the lower layer.

The result of a field experiment using a traditional planting pattern with one plant per hole showed that the stem and aboveground fresh biomass of sweet sorghum were the highest in the loam soil, reaching 75.7 and 87.8 t ha^{-2}, respectively. The stem biomass accounted for 86.2% of aboveground biomass. The stem biomasses from sandy soil and saline–alkaline soil were 61.2 and 61.0 t ha^{-2}, respectively, and the aboveground biomasses were 71.3 and 74.3 t ha^{-2}, respectively. The leaf biomass and panicle yield were the highest in saline–alkaline soil, reaching 10.6 and 2.7 t ha^{-2}, respectively. Compared to loam soil, the fresh mass of aboveground biomass from sandy soil and saline–alkaline soil decreased by 18.8% and 15.4%, respectively. The analysis of variance showed that the fresh mass of stem and aboveground biomass of sweet sorghum in loam soil were significantly higher than those in saline–alkaline soil and sandy soil ($p < 0.05$), but there were no significant differences between saline–alkaline soil and sandy soil. The fresh mass of aboveground biomass of sweet sorghum is about 75 t ha^{-2} in other regions. The fresh stem mass of sweet sorghum in oasis farmland in the Hexi Corridor is usually higher than 75 t ha^{-2}. The brix of stem juice

Table 13.11 Ion Content of Water-Soluble Salt (mg kg^{-1}) for Different Soil Types (2009–2010, Linze Oasis, Middle of Hexi Corridor Region)

Soil Type	Depth (cm)	CO_3^{2-}	HCO_3^-	Cl^-	SO_4^{2-}	Ca^{2+}	Mg^{2+}	Na^+	K^+
Sandy soil	0–20	20	240	10	20	40	40	180	20
	20–40	20	240	10	20	40	40	110	20
Saline–alkaline soil	0–20	70	380	330	500	200	170	1610	340
	20–40	60	500	120	20	60	80	1290	60
Loam soil	0–20	30	400	50	20	90	50	110	110
	20–40	20	300	50	20	80	40	130	70

Table 13.12   Soil Particle Size Composition (%) in Different Soil Types (2009–2010, Linze Oasis, Middle of Hexi Corridor Region)

Soil Type	Depth (cm)	Soil Particle Size			
		Sand (2–0.05 mm)	Coarse Silt (0.05–0.02 mm)	Fine Silt (0.02–0.002 mm)	Clay (<0.002 mm)
Sandy soil	0–20	81.30	10.40	3.60	4.70
	20–40	99.87	0.13	0	0
Saline–alkaline soil	0–20	65.90	18.55	7.50	8.05
	20–40	32.45	28.95	15.50	23.10
Loam soil	0–20	46.50	20.00	25.45	8.05
	20–40	34.15	35.45	20.05	10.35

of sweet sorghum produced in surrounding areas is 17 to 21%; it reached 20 to 23% in the Hexi Corridor, with maximum brix being measured from sorghum grown in sandy soil. The total soluble sugar content in the stems grown in sandy and saline–alkaline soil were significantly higher than in those grown in loam soil ($p < 0.05$).

The dry mass of plant materials was measured after being dried at 80°C. The stem dry mass and aboveground dry mass of sweet sorghum grown in loam soil were the largest with values of 18.9 and 24.2 t ha^{-2}, respectively. The stem dry mass in sandy soil was similar to that in loam soil, with a value of 18.5 t ha^{-2}, but the dry mass of the aboveground parts was lower than that grown in loam soil, with a value of 22.9 t ha^{-2}. The smallest values of both stem dry mass and aboveground dry mass of sweet sorghum occurred in saline–alkaline soil (16.9 and 22.1 t ha^{-2}, respectively). Compared to loam soil, the stem dry mass in saline–alkaline soil and sandy soil decreased by 2.1% and 10.6%, respectively, and the dry mass of the aboveground parts decreased by 5.4% and 8.7%, respectively. The analysis of variance showed that the stem dry mass in saline–alkaline soil was significantly lower than that in loam soil and sandy soil ($p < 0.05$), but no significant difference was found among the dry mass of aboveground parts from the three soil types.

The belowground dry and fresh biomass from loam soil reached maximums of 5.0 and 11.8 t ha^{-2}, respectively. Both the dry and fresh mass from sandy soil were the lowest, at 3.9 and 8.2 t ha^{-2}, respectively. Compared to loam soil, the dry and fresh mass from sandy soil decreased by 22.0% and 30.5% respectively, but they only decreased by 4.6% and 3.4%, respectively, in saline–alkaline soil. The analysis of variance showed that the belowground fresh mass from sandy soil was significantly lower than that from loam soil ($p < 0.05$), but there was no significant difference between belowground fresh mass from saline–alkaline soil and sandy soil. The dry mass of belowground parts was not significantly different among the three soil types.

The measure of water use efficiency of the aboveground dry biomass yields showed that sweet sorghum grown in loam soil exhibited the highest water use efficiency, with a value of 3.48 kg m^{-3}. Next was saline–alkaline soil with a value of 3.21 kg m^{-3}. The lowest water use efficiency occurred in sandy soil with a value of 2.17 kg m^{-3}. Water use efficiency in loam soil increased by 8.4% over saline–alkaline soil and by 60.4% over sandy soil. The analysis of variance showed that the water use efficiencies of aboveground biomass in loam soil and saline–alkaline soil were both significantly higher than that in sandy soil ($p < 0.05$).

Compared to loam soil, water use efficiencies were lower in saline–alkaline and sandy soils; however, higher sugar contents and ethanol yields of sweet sorghum can be achieved in saline–alkaline and sandy soils. Therefore, considering the competition between food crops and energy crops, it would be more beneficial to cultivate sweet sorghum in saline–alkaline and sandy soils in arid regions in northwestern China. Moreover, soil conditions can be improved and different cultivation methods can be adopted in both saline–alkaline and sandy soils to increase biomass yields and improve the water use efficiency of sweet sorghum.

13.5.2.2 *Plot Experiment of Cluster Planting Patterns in Sandy Soil*

It is very important to learn how to increase biomass yields and water use efficiency of sweet sorghum in sandy soil without increasing the irrigation quota. A plot experiment for the cluster planting method of *Sorghum bicolor* (L.) Moench cv. BJ0601 was carried out in irrigated aeolian sandy soil from the edge of the oasis. A completely randomized plot design was adopted, with three replications and using a double-row strip pattern. The three treatments for the planting patterns (Figure 13.8) were as follows: (1) one plant per hole (local traditional planting method), with 60-cm strip spacing, 40-cm row spacing, and 15 cm between plants in the same row; (2) two plants per hole, with the same strip and row spacing as treatment 1 but with 30 cm between plants in the same row; and (3) three plants per hole (cluster planting), with the same strip and row spacing as in treatment 1 but with 45 cm between plants in the same row. The plant densities in all treatments were the same: 13.34×10^4 plants ha^{-2} (Xie et al., 2014).

Biomass measurements showed that stem dry mass increased significantly with the increase in the number of plants per hole (Table 13.13). The highest leaf dry mass was recorded in the treatment with two plants per hole, and the lowest was in treatment with one plant per hole. On the other hand, panicle dry mass was the highest in the treatment with one plant per hole and the lowest in the treatment with two plants per hole. The trend for the total aboveground dry mass was similar to that observed for the stem dry mass (Xie et al., 2014).

The total soluble sugar content was determined using the anthrone method. Ethanol yields from sugar are calculated as follows: Ethanol yields from sugar (L ha^{-2}) = Total sugar content (%) in dry matter $\times$ Dry biomass (t ha^{-2}) $\times$ 0.51 (conversion factor of ethanol from sugar) $\times$ 0.85 (process efficiency of ethanol from sugar) $\times$ 1000/0.79 (specific gravity of ethanol, g mL^{-1}) (Shi and Hua, 2007). Total soluble sugar content was highest in the treatment with three plants per hole. The values of total soluble sugar between the two other planting patterns did not differ significantly. Total soluble

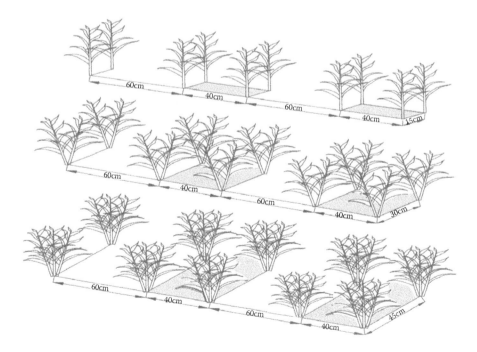

Figure 13.8 Diagram showing three planting patterns of sweet sorghum. (Upper) Single plant per hole with spacing of 60 × 40 × 15 cm (strip spacing × row spacing × hole spacing); (middle) two plants per hole with 60 × 40 × 30-cm spacing; (lower) three plants per hole with 60 × 40 × 45-cm spacing.

Table 13.13   Biomass, Sugar, and Ethanol Yield, and Water Use Efficiency of
Sweet Sorghum under Different Planting Patterns in Plot Experiment

Parameter	One Plant per Hole	Two Plants per Hole	Three Plants per Hole
Stem (g plant^{-1})	20.54[a]	22.97[b]	24.05[c]
Leaf (g plant^{-1})	4.40[a]	5.40[c]	5.07[b]
Panicle (g plant^{-1})	1.68[c]	1.35[a]	1.57[b]
Above ground dry weight (t ha^{-2})	26.62[a]	29.72[b]	30.69[c]
Total soluble sugar content (g kg^{-1})	499.50[a]	504.08[a]	521.26[b]
Total soluble sugar yield (t ha^{-2})	9.14[a]	10.65[b]	11.20[c]
Ethanol yield (L ha^{-2})	5075.34[a]	5844.79[b]	6235.86[c]
Water use efficiency (kg m^{-3})	3.82[a]	4.26[b]	4.40[b]

Note: In the same row, the different lowercase letters means significantly different using
LSD at the $p < 0.05$ level.

sugar yield and ethanol yield increased significantly as the number of plants per hole increased. The treatment with three plants per hole showed the highest water use efficiency, although this value did not differ significantly from that of two plants per hole.

Under the three planting patterns, total soluble sugar and ethanol yields were significantly related to stem biomass and leaf biomass. Furthermore, the correlation coefficients between total soluble sugar and stem biomass under all planting patterns were highly significant. The same trend was observed between ethanol yields and stem biomass, which demonstrates the significant contribution of stem biomass to the accumulation of total soluble sugar and ethanol yield. Thus, a cluster planting pattern of sweet sorghum has great potential in terms of biomass yield and quality, as well as water use efficiency.

13.6 PROSPECTS FOR AGROECOLOGY RESEARCH AND AGROECOLOGY PRACTICE IN OASES

In an arid area, the oasis, mountains, and desert are interdependent and mutually interact with each other as a system. The amount of precipitation in the mountains affects the oasis agricultural production below; the environmental conditions of the desert affect the oasis environment; and the material and energy flow of the inner oasis affects the distribution of water resources from mountain glaciers, snow melt, and rainfall, and hence the stability of the desert oasis transition zone. The protective system is the ecological barrier for oasis development. Water use efficiency and production values per cubic meter of water of a fruit tree–field crop complex agroforestry system are significantly higher than those in crop monoculture. The main negative effects of shelterbelts on nearby crops in the oasis occur through soil and air moisture rather than from nutrition and light intensity; therefore, the coupling of all kinds of resources with water to improve the efficiency of resource utilization is the key to sustainable development in the oasis region.

A thorough understanding of the crop water requirements for plant transpiration and soil evaporation and the exploration of regulatory methods to meet basic crop physiological water demands are both very important to increasing water use efficiency in crop production. For example, the cluster planting pattern of cotton and sweet sorghum can significantly increase crop yields and water use efficiency. The aggregated distribution is also the way most desert plants adapt to the severe dry environment. Research on the eco-physiological mechanisms of desert plants for their high water use efficiency should be strengthened in the future in order to further develop biological water-saving technologies in the oasis.

In order to make good use of the advantages and characteristics of local conditions in oasis regions for agricultural production, it is worth considering setting up linkages among different locations and among crops and animals. A well-arranged linkage will increase the primary and secondary productivities of the system, as well as increase the efficiency of solar energy utilization and water use efficiency, thus promoting harmonious and sustainable development in the entire oasis region.

Based on our findings, the following aspects of future research efforts and development trends of agricultural production in oasis regions should be strengthened:

1. The relationships between crops and a dry environment, water in soil and water in air, water saving and land productivity, and the oasis and surrounding hills and mountains should be better understood in order to set up agricultural production systems in oases that are highly water efficient.
2. The use of digital technology and precision irrigation methods should be more greatly utliized to provide opportunities to further improve irrigation methods and irrigation quotas based on crop water consumption needs.
3. It is important to further consider the linkages among crops, forestry, horticulture, animal husbandry, aquaculture, rural industry, and other types of production together as an integrated oasis agroecosystem that can ensure the provision of food and sustainable development through coordination among each production type.

ACKNOWLEDGMENTS

This work was financially supported by Linze Inland River Basin Research Station, the National Key Basic Research Program of China (2013CB956604), and the National Natural Science Foundation of China (31300323, 91325104).

REFERENCES

Ding, S.S. and Su, P.X. 2010. Effects of tree shading on maize crop within a poplar–maize compound system in Hexi Corridor oasis, northwestern China. *Agroforest Syst.*, 80: 117–129.

Fang, C.L. 1994. Operation mechanism of desert oases ecosystem and its degeneration monitoring. *Chin. J. Ecol.*, 13(5): 50–55.

Gao, H.J. 1987. The distribution and types of oases in China. *Arid Land Geogr.*, 10(4): 23–29.

Gao, S., Su, P.X., Yan, Q.D., and Ding, S.S. 2010. Canopy and leaf gas exchange of *Haloxylon ammodendron* under different soil moisture regimes. *Sci. China Life Sci.*, 53(6): 718–728.

Guan, X., Zhang, F.R., and Li, Q.Y. 2002. Sustainable utilization and exploitation of land resources in Xinjiang. *Agric. Res. Arid Areas*, 20(1): 95–101.

Lai, X.Q. 2005. *China Oasis Agronomy.* China Agriculture Press: Beijing, pp. 130–133.

Lawson, T.L. and Kang, B.T. 1990. Yield of maize and cowpea in an alley cropping system in relation to available light. *Agric. Forest Meteorol.*, 52: 347–357.

Li, Z.Z., Zhao, S.L., and Zhang, P.Y. 1993. The niche-fitness theory and its application to the systems of crop growth. *J. Lanzhou Univ. (Nat. Sci.)*, 29(4): 219–224.

Liu, B.Y., Pang, J.H., Wang, T., Bai, X.J., and Zhang, W.L. 2006. Laws and control countermeasures of negative effect of shelter forest for farmland in Qiqihar region. *Prot. Forest Sci. Technol.*, 2: 66–67.

Loomis, R.S. and Connor, D.J. 2002. *Crop Ecology: Productivity and Management in Agricultural Systems* [Chinese translation by Li, Y.M. et al.]. China Agriculture Press: Beijing, pp. 350–354.

Masinde, P.W., Stützel, H., Agong, S.G., and Fricke, A. 2005. Plant growth, water relations, and transpiration of spiderplant [*Gynandropsisgynandra* (L.) Briq.] under water-limited conditions. *J. Am. Soc. Hortic. Sci.*, 130(3): 469–477.

Miller, A.W. and Pallardy, S.G. 2001. Resource competition across the crop-tree interface in a maize–silver maple temperate alley cropping stand in Missouri. *Agroforest Syst.*, 53: 247–259.

Planning Commission of Gansu Province. 1992. *Land and Natural Resources of Gansu Province*. Gansu Science and Technology Press: Lanzhou, pp. 24–48.

Shen, Y.C. 2000. Development of oases in the 21st century: challenge, direction and prospect. *J. Arid Land Resour. Environ.*, 14(1): 1–11.

Shen, Y.C., Wang, J.W., and Wu, G.H. 2002. Oases as well as their sustainable development and constructions in China. *J. Arid Land Resour. Environ.*, 16(1): 1–8.

Shi, Z.P. and Hua, Z.Z., Eds. 2007. *Biomass and Bioenergy Handbook*. Chemistry Industry Press: Beijing, pp. 166–167.

Soil Survey Office of Gansu Province. 1993. *Soil in Gansu*. Agricultural Press: Beijing, pp. 110–120.

Su, P.X. and Liu, X.M. 2005. Photosynthetic characteristics of Linze jujube in conditions of high temperature and irradiation. *Sci. Hortic.*, 104: 339–350.

Su, P.X., Du, M.W., Zhao, A.F., and Zhang, X.J. 2002. Study on water requirement law of some crops and different planting mode in oasis. *Agric. Res. Arid Areas*, 20(2): 79–85.

Su, P.X., Zhao, A.F., and Du, M.W. 2004. Functions of different cultivation modes in oasis agriculture on soil wind erosion control and soil moisture conservation. *Chin. J. Appl. Ecol.*, 15(9): 1536–1540.

Su, P.X., Ding, S.S., and Xie, T.T. 2009a. Cause of freezing damage and prevention strategies to Linze jujube in Hexi Corridor region of Gansu Province. *J. Natural Disasters*, 18(6): 1–8.

Su, P.X., Xie, T.T., and Ding, S.S. 2009b. Experimental studies on high-yield cluster cultivation of cotton in the Hexi Corridor oases of northwestern China. *Agric. Res. Arid Areas*, 27(6): 108–113.

Su, P.X., Xie, T.T., and Ding, S.S. 2010. Water requirement regularity in Linze jujube (*Zizyphusjujuba* Mill. var. *inermis* Rehd. cv. *Linze jujube*) and jujube/crop complex systems in Linze oasis. *Chin. J. Eco-Agric.*, 18(2): 334–341.

Sun, B.H. 1982. More serious negative effect in the northern side than in the southern side, more serious negative effect in the eastern side than in the western side. *J. Inner Mongolia Forest.*, 6: 34.

Wu, Z. 1987. *Aeolian Sand Geomorphology*. Beijing: Science Press, pp. 24–26.

Xie, T.T., Su, P.X., Shan, L.S., and Ma, J.B. 2012. Yield, quality and irrigation water use efficiency of sweet sorghum [*Sorghum bicolor* (Linn.) Moench] under different land types in arid regions. *Aust. J. Crop Sci.*, 6(1): 10–16.

Xie, T.T., Su, P.X., Zhou, Z.J., Zhang, H., and Li, S.J. 2014. Biomass, sugar and ethanol yield, and water use efficiency of sweet sorghum [*Sorghum bicolor* (L.) Moench.] under different planting patterns. *Philipp. Agric. Sci.*, 97(1): 86–91.

CHAPTER **14**

Ecological Agriculture on Arid, Sloped Land in Yunnan Dry–Hot Valley

Ji Zhonghua and Tan Fengxiao

CONTENTS

In the Yunnan dry–hot valley (DHV), the mountainous area and arid slope area account for 94% and over 90% of the total area, respectively. The region is characterized by a deteriorated natural environment, drought, water scarcity, low vegetation coverage, low land use efficiency, and low rural economic incomes, but it does have abundant light and thermal resources. This chapter introduces agroecology practices in the region that coordinate ecological, economic, and social benefits by applying systematic engineering methods along with modern science and technology to construct an eco-agricultural system with a virtuous eco-economic cycle, multilevels, and multifunctions, guided by the principles of ecology and ecological economics. These eco-agricultural models must be constructed according to local conditions. Three integrated management models have been recognized based on the water conditions: (1) *rain-fed system*, a water preservation forest and grass eco-agricultural model; (2) *supplemental irrigation system*, an eco-agricultural model composed of a vertical community structure with plants and animals internally cycling on arid sloping land; and (3) *irrigation system*, an integrated management model of special, renowned, and high-quality fruit production. Improved techniques used in all of these models are also presented in this chapter. Research results indicate that the efficient use of solar energy, thermal energy, and water resources has been improved as a result of comprehensive evaluations of each model, and the living organisms, working as a whole, maximize their economic, ecological, and social benefits. This experience provides a good example of protecting the natural environment while promoting sustainable development of agriculture in a degraded mountainous region.

Figure 14.1 Jinsha River Valley, a typical dry–hot valley.

14.1 INTRODUCTION

The dry–hot valleys (DHVs) in China cover an area of 32,000 km², which is mainly distributed along the Jinsha River that passes through Yunnan, northwestern Taiwan, southwestern Hainan, and southwestern Sichuan Province. The largest DHV is situated in Yunnan Province and covers the Jinsha River, Yuanjian River, Nujiang River, and Nanpan River (Zhang, 1992). The hot, dry climate creates obvious conflicts around water shortages and the abundant heat in the region. In addition, DHVs are characterized by low vegetation coverage, serious soil erosion, intensive degradation of the natural environment, and challenges to vegetation restoration. In the DHV of the Jinsha River (Figure 14.1), for instance, the average annual temperature ranges from 20.7 to 24.1°C, and its accumulated temperature is 7800 to 8800 degrees, with an average annual precipitation of 610 to 817 mm. However, the rainy season, which often lasts from June to early September, accounts for 80 to 90% of the total annual precipitation. The average annual evaporation levels vary from 2600 to 3700 mm, four to six times the average annual rainfall. In the 1950s, the forest coverage rate was 12.8%, whereas in the late 1980s it dropped to 5.2%; it is currently decreasing by 0.32% each year. Average soil erosion in the region is 2547 t km⁻² (Zhang et al., 2003).

14.1.1 Problems Facing the Development of Agriculture and Economy in Dry–Hot Valleys

14.1.1.1 Difficulties Caused by the Dry–Hot Environment

The limited rainfall, excessive evaporation, and uneven seasonal rainfall allocation in DHVs have resulted in a very challenging environment, particularly in the spring and early summer during the crop growth seasons. Drought and water scarcity have a serious impact on taking advantage of the local warm growing conditions and the rational utilization of land resources and have constrained the sustainable development of agriculture and the local economy in DHV regions, especially in the hilly areas. The degree of water shortages in agricultural production is more than 100 million m³ in counties in DHV regions such as Yongsheng, Binchuan, and Yuanmou. Although some water diversion projects have been finished, the utilization efficiency of water resources is still low. According to a survey, the average irrigation quota of farmland in the Yuanmou DHV (Figure 14.2) is 14,025 $m^3 ha^{-2} a^{-1}$, far more than in the medium arid region of China (i.e., around 10,000 $m^3 ha^{-2} a^{-1}$ for paddy fields and 400 $m^3 ha^{-2} a^{-1}$ for irrigable uplands). In the DHV region of Yunnan Province, the irrigable upland area per capita in hilly areas is only 0.03 ha, whereas the non-irrigable land area per capita on arid slopes is 0.13 ha (Chen et al., 2001).

Figure 14.2 (See color insert.) Degraded mountains of the Yuanmou dry–hot valley area before eco-agri-culture development.

14.1.1.2 Environmental Deterioration and Frequent Natural Disasters

In DHVs, mountainous forest vegetation areas have been greatly reduced due to deforestation caused by firewood collection and wood cutting, as well as grassland degradation caused by overgrazing. In addition, the hot climate and the imbalance of water and heat make it difficult to restore vegetation, especially in the hilly areas where vegetation coverage is relatively low. The forest coverage of the Jinsha River Basin where Youngsheng County, Dongchuan County, and Yuanmou County are located has decreased to only 3.4 to 8.0%, and the total vegetation coverage is less than 50%. The destruction of vegetation has caused environmental degradation and has greatly reduced the ability of ecosystems to self-regulate and buffer disasters; thus, drought and floods happen quite often. The impact of extensive environmental degradation on local resources and the challenge to sustainably develop agriculture have become more serious in DHV regions than in other semiarid regions of China (Chen et al., 2001).

14.1.1.3 Intensive Water and Soil Loss, Soil Degradation, High Proportion of Medium- and Low-Yield Farmland, and Low Land Production Efficiency

Soil erosion is serious, and gullies are well formed in DHVs. Soil A and B horizons on most slope lands have been eroded away, and the parent material layer is exposed. The area of water and soil loss in Yuanmou County accounts for 53.5% of the total county land area and even reaches 62.3% in Dongchuan City. Table 14.1 summarizes the soil erosion conditions in the Yuanmou DHV. The loss of an active surface soil layer lowers both the water infiltration capacity and the adjusting capacity of soil on water, fertilizer, air, and heat. On the arid slopes of the hilly areas without irrigation facilities, it is extremely dry in the dry season; yet, large quantities of soil nutrients are eroded in the rainy season. In addition, high temperatures, high organic matter decomposition rates, a high cropping index, low crop residue returning rates, and low organic fertilizer input have caused severe soil degradation and a significant decrease in soil fertility and land productivity. The ratio of grain output to fertilizer and labor input has also been significantly reduced. In other words, severe soil degradation has greatly restrained the sustainable development of agriculture and economic return in DHV regions (Chen et al., 2001).

Table 14.1   Soil Erosion Conditions in Yuanmou County

Soil Erosion Intensity	Soil Erosion Area			Soil Erosion Amount	
	Area (km²)	Percent of Total Area (%)	Percent of Total Eroded Area (%)	Soil Erosion Modulus (t km^{-2} a^{-1})	Annual Average Erosion Amount (10^4 t)
Slight	568.45	28.1	52.6	2000	113.7
Medium	316.37	15.7	29.3	4000	126.5
High	189.12	9.4	17.5	7000	132.4
Ultra-high	6.85	0.03	0.6	10,000	6.9
Total	1080.79	53.5	100.0	—	379.6

Source: Data from Water and Soil Conservation Office, Yuanmou County, Yunnan Province, China.

14.1.1.4 Crop Germplasm Resource Scarcity

Dry–hot valleys are in a special ecological zone separated from other ecological zones in mountainous areas of Southwest China, thus making it difficult for adapted species to spread into this area naturally. As a result, plant germplasm resources are scarce. It is of particular importance for comprehensive improvement of the degraded environment and development of eco-agriculture in DHVs to introduce plant germplasm from areas with similar climate conditions either at home or abroad (Chen et al., 2001).

14.1.2 Characteristics of This Study

In the pilot area, the average relative water-holding capacity in the plow layer (0 to 60 cm deep) is less than 70% of the capacity in dry season. The duration of soil water content levels that are lower than wilting point is 6 to 7 months, a period of time when no available water for plants is present in the soil. It is so dry that not many plants can survive there. For this reason, eco-agriculture system structures, which we will call "models," should address the specific conditions of local water resources. These models can be classified as *rain-fed systems*, *supplemental irrigation systems*, and *irrigation systems* (Ji et al., 2006; Ji, 2009):

- *Rain-fed systems*—This system has no human-constructed water-conserving facility. All water used is derived from rainfall. Implementing rainwater collection during the rainy season and soil water conservation during the dry season could ensure normal plant growth and certain economic benefits. Rational selection of moderate drought-resistant varieties is important for this model.
- *Supplemental irrigation systems*—This system has some water storage and water collection facilities, but the water is not sufficient and can only partly meet the water demand of plants during the driest season. The system could obtain relatively high economic returns through wise use of the limited water supply. Material cycling and the recycling of water resources are key techniques for this model.
- *Irrigation systems*—The good water-conserving facilities and abundant water supply present in this system can meet the water demand of plants during their critical water-demand period. The production of special cash fruits is a key component of this model.

The research presented in this chapter focuses on integrated upland slope management through the construction of eco-agriculture models in order to provide social, ecological, and economic benefits. The principles of ecology and eco-economics guided our project and ensured that the old development strategy transitioned to a sustainable development strategy. The long-term and short-term benefits could be coordinated through production regulation, ecology protection, resources proliferation, and pollution prevention. Integrated planning for the use of labor, land, and biological and other natural resources could improve the function of the whole system. In this project, we paid great attention to strengthening the integration and innovation of various techniques and providing examples and technical packages to support sustainable development in DHV regions.

14.2 CONSTRUCTION OF ECO-AGRICULTURE MODELS ON ARID SLOPED LAND OF DHVS

Water is the primary limiting factor in the development of agriculture in DHVs; therefore, the construction of eco-agriculture models should be based on water resource conditions.

14.2.1 Supplemental Irrigation System

The goal of constructing a supplemental irrigation system was to take the greatest advantage of local natural resources. Mimicking the forest ecosystem's utilization of multiple levels of sunlight by coupling trees, shrubs, and grasses allows for full use of solar energy, water, and mineral nutrients. Establishing an agroecosystem structure with multiple layers in space and multiple sequences in time could produce high economic and ecological benefits. Building vertical structures using the integrated management models of agroforestry and grazing systems can maximize improvements in land productivity and economic benefits of agriculture as a whole, as well as achieve a virtuous circle of animal feed being provided by plant production and organic fertilizer by animal husbandry (Ji et al., 2005; Ji, 2009). The major eco-agriculture model studied was a vertical (tridimensional) planting community coupled with animal husbandry through cycling.

14.2.1.1 Vertical (Tridimensional) Planting and Breeding Eco-Agriculture Model

This is an integrated agroecosystem referred to as an *arboreal tree–shrub–grass–livestock–poultry–biogas system*. The arboreal trees are mainly composed of economic forest species that are drought resistant and adapted to barren environments, such as *Tamarindus indica*. The shrubs are species dominant in the shrub layer (e.g., *Cajanus cajan*). The grasses are introduced planted forage grasses and forbs selected from among 295 forage germplasm resources with optimum adaptations and maximum yield, such as *Stylosanthes guianensis*, the perennial *Lolium perenne*, and *Pennisetum purpureum*. Chicken–pig (–goat) is a common vertical structural system for animal husbandry.

14.2.2 Rain-Fed System

Water and soil loss is a serious issue in DHVs during the rainy season and is an important factor in eco-agriculture construction. Addressing the problem requires an integration of short-term and long-term efficiency; coordination of economic, social, and ecological benefits; combining biological measures with engineering measures; and coupling large, medium, and small engineering projects to preserve water and soil and to promote economic development (Ji et al., 2005; Ji, 2009). The water-preserving forest–grass model was the main rain-fed agroecosystem studied.

14.2.2.1 Water Preserving Forest–Grass Eco-Agriculture Model

This is referred to as the *forest–grass-grazing–biogas–engineering model*. The forests are mainly composed of timber and firewood species, such as *Albizia kalokora*, *Leucaena leucocephala*, *Tephrosia candida*, *Cajanus cajan*, *Bauhinia purpurea*, *Phyllanthi fructus*, *Acacia farnesiana*,

Figure 14.3   (See color insert.) Gully control model integrated with *Leucaena leucocephala* (*Leucaena leucocephala* + *Albizia kalokora* + *Jatropha curcas* + *Azadirachta indica* + *Pennisetum purpureum*, etc.) in eco-agricultural development.

Jatropha curcas, and *Azadirachta indica*. The grasses are a mixture of Leguminosae and Gramineae species (e.g., *Pennisetum purpureum*, natural grass). Engineering measures such as dams, terraces, trenches, and pits should be used according to local conditions to maximize rainwater utilization (Figure 14.3).

14.2.3 Irrigation System

Dry–hot valleys have been called "natural greenhouses" for their abundant sunlight and heat resources, which make them good bases for the development of crop farming, particularly the farming of special tropical crops. In addition, agriculture is the dominant local industry. Because severe industrial pollution is not present, most of the products produced in this area are pollution free. In recent years, DHVs have been listed by the state as the main pollution-free vegetable production bases. The off-season vegetables and fruits produced there are rare and high quality; thus, the development of special, high-quality fruit products is an important way to promote the local economy. The main system studied in this model was an integrated-management eco-agriculture model for the production of special, high-quality fruit (Ji et al., 2005; Ji, 2009).

14.2.3.1 Integrated-Management Model of Special, High-Quality Fruit Production

This system is referred to as the *fruit–off-season vegetable (medicinal or spice plant)–pig–biogas model*. The fruits are tropical fruit trees (e.g., longan, mango, Taiwan green jujube) selected from 66 new tropical fruit species with high yields and high quality. Off-season vegetable production plays an important supporting role for Yuanmou agriculture. The approximately 20 major varieties of off-season vegetables include fruiting vegetables and leafy vegetables. In addition, high-value medicinal and spice species are planted to enhance benefits (Ji et al., 2005; Ji, 2009).

Table 14.2   Solar Energy Utilization Efficiencies (%) in Different Mixtures of Intercropping Systems

| Ecosystem Pattern | Species | | | | | Efficiency of Total Solar Energy Utilization |
	Tamarind	Pigeon Pea	Elephant Grass	Stylo	Natural Grass	
Tamarind + pigeon pea	1.38	0.39	—	—	—	1.61
Tamarind + elephant grass	1.38	—	1.43	—	—	1.47
Tamarind + pigeon pea + stylo	1.38	0.28	—	0.61	—	0.91
Tamarind + natural grass	1.38	—	—	—	0.12	1.28
Monoculture	—	0.26	—	0.58	—	—

Source: Yang, Y.X. et al., *J. Soil Water Conserv.*, 20(3), 70–73, 2006.

Note: Due to the difficulty of determining the biological yield of fruit trees, the solar energy utilization efficiency (SEUE) of tamarind was replaced by the economic yield. The total utilization efficiency of the entire ecosystem was calculated by weighted averaging of the area of different components in the systems.

14.3 ASSESSMENT OF COMPREHENSIVE BENEFITS OF ECO-AGRICULTURE MODELS

14.3.1 Assessment of Vertical Planting and Breeding Eco-Agriculture Model Used in the Supplemental Irrigation System

14.3.1.1 Solar Energy Use Efficiency

Only 0.5-3.5% of total incoming solar energy is utilized for photosynthesis in most ecosystems. The solar energy use efficiency (SEUE) is calculated as follows:

$$\text{SEUE (\%)} = MT/\sum Q = 0.10 \times 10^{-4} T \times Y/\sum Q \tag{14.1}$$

where SEUE is the solar energy use efficiency, M is the weight of plant dry matter (kg), T is the energy released by burning 1 kg of dry matter, Q is the total radiation during the growing season (kcal cm^{-2}), and Y is the yield of plants in one growing season (kg ha^{-1}). The average T for crop plants is 4.25 kcal g^{-1} and for trees is 4.38 kcal g^{-1} (Pan and Dong, 1984).

The SEUE in a mixed tamarind (*Tamarindus indica*)–pigeon pea (*Cajanus cajan*) system reached 1.61% (Table 14.2), because the tamarind density was relatively low (5 m × 6 m) and pigeon pea (a small shrub intercropped in the understory) increased the area of photosynthesis, enhanced the leaf area index, and made full use of solar radiation, thus improving the total SEUE of the entire ecosystem. In a mixed tamarind–pigeon pea–stylo (*Stylosanthes guianensis*) or tamarind–elephant grass (*Pennisetum purpureum*) intercropping system, stylo, pigeon pea, and elephant grass increased the area of photosynthesis of the entire ecosystem. The total SEUE was 0.90% for the tamarind–pigeon pea–stylo system and 1.47% for the tamarind–elephant grass system. The total SEUE in the mixed tamarind–elephant grass system was significantly higher than in the tamarind–pigeon pea–stylo system, because the SEUE of elephant grass is relatively high, thus enhancing the total SEUE. In the tamarind and natural grass ecosystem, the density of tamarind was 6 m × 8 m, which was lower than that in the mixed tamarind–pigeon pea system; the SEUE of natural grass was only 0.12% (Yang et al., 2006).

In the mixed tamarind–pigeon pea–stylo system, the yields of pigeon pea and stylo were little influenced by the existence of tamarind due to the relatively low tamarind density (Table 14.3). Although a result of the microclimate effect in the intercropping ecosystem, the forage yields of pigeon pea in the mixed systems (13,749.1 and 9634.2 kg ha^{-1}, respectively, in Table 14.3) were higher than in monoculture systems (9022.9 kg ha^{-1}) (Yang et al., 2006). The stylo yield in the

Table 14.3   Comparison of Yields (kg ha⁻¹) among Different Mixtures of Intercropping Systems

	Species				
Model	Tamarind	Pigeon Pea	Elephant Grass	Stylo	Natural Grass
Tamarind + pigeon pea	7845.1	13,749.1	—	—	—
Tamarind + elephant grass	7845.1	—	51,419.9	—	—
Tamarind + pigeon pea + stylo	7845.1	9634.2	—	22,345.2	—
Tamarind + natural grass	7845.1	—	—	—	4260.3
Monoculture	—	9022.9	—	20,592.0	—

Source: Yang, Y.X. et al., *J. Soil Water Conserv.*, 20(3), 70–73, 2006.

Note: Above are cut biomass yields except for the yield of pigeon pea, which was calculated from its fodder utilization.

mixed system (22,345.2 kg ha⁻¹) was also higher than in its monoculture system (20,592.0 kg ha⁻¹). Overall, the order of total SEUE in different ecosystems was as follows: tamarind–pigeon pea > tamarind–elephant grass > tamarind–natural grass > tamarind–pigeon pea–stylo > stylo monoculture > pigeon pea monoculture. The intercropped pigeon pea systems linked with animal production and biogas production could increase the total SEUE up to 519.2% and 250%, respectively, compared to the pigeon pea monoculture system without biogas. The intercropped stylo system enhanced the total SEUE by 56.9% compared to the stylo monoculture system.

The total SEUE in all complex tridimensional cropping systems was higher than in the corresponding monoculture systems. The ecosystems with larger proportions of high-SEUE crops and more complex structures had higher total SEUE values. Even when the proportion of the high-SEUE crops was the same, the ecosystems with more complex structures would have higher total SEUE values as well. Thus, the total SEUE in complex tridimensional cropping systems is related to ecosystem structure and the SEUE of the intercropped plants. Complex tridimensional cropping systems can facilitate the full use of solar energy.

14.3.1.2 Soil Physical Fertility Restoration

In the mixed tamarind–pigeon pea system (Table 14.4), direct seeding of pigeon pea was conducted in pits in the rainy season with a density of 1 m × 0.8 m and a land surface coverage of 85% (Yang et al., 2006). Pigeon peas could be harvested four times a year, and the yearly yield of dried bean was up to 4748.1 kg ha⁻¹ without fertilizer application. In the tamarind–stylo system (T+S, treatment II), the land surface coverage reached 100% (including weeds), and tamarind could be harvested twice a year with an average urea application rate of 180 kg ha⁻¹ a year. In the tamarind–elephant grass system (T+E, treatment III), furrow land preparation was conducted, and the land

Table 14.4   Soil Physical Factors in Pilot Areas

Treatment	Soil Depth (cm)	Soil Bulk Density (g cm⁻³)	Total Porosity (%)	Soil Aggregate Composition (%)			
				>2 mm	2–1 mm	1–0.25 mm	<0.25 mm
Tamarind + pigeon pea	0–20	1.59	40.00	10.23	14.47	32.10	43.20
	20–40	1.65	37.74	7.54	9.66	37.80	45.00
Tamarind + stylo	0–20	1.47	44.71	9.56	17.30	52.18	21.07
	20–40	1.64	39.29	8.54	17.21	58.75	15.04
Tamarind + elephant grass	0–20	1.66	37.36	12.40	10.8	35.10	46.60
	20–40	1.72	35.09	11.23	9.67	36.70	48.00
Natural grass (CK)	0–20	1.78	32.82	9.33	10.00	28.18	52.51
	20–40	1.76	33.58	10.54	8.08	22.65	58.73

Source: Yang, Y.X. et al., *J. Soil Water Conserv.*, 20(3), 70–73, 2006.

surface coverage was up to 100%. Elephant grass could be harvested three times a year with an average urea fertilization rate of 937.5 kg ha^{-1} a year. The control area (CK) was enclosed and no entrance was allowed in order to exclude any human disturbance. The control area eroded naturally, and its vegetation cover was mainly *Heteropogon contortus* (L.) Beauv. and a small number of secondary xerophytic shrubs, with land surface coverage of 45%.

14.3.1.2.1 Soil Bulk Density and Porosity Analysis

Porosity is an important index of soil structure. As shown in Table 14.4, soil bulk density was rather high in the experimental area, with a maximum in treatment II, followed by treatment I, treatment III, and CK (all were above 1.5 g cm^{-3}). Total porosities significantly decreased in the same order. Physical properties such as soil bulk density and total porosity in the 0- to 20-cm soil layer were better than in the 20- to 40-cm layer in all treatments, except CK. In the control area, due to severe surface soil erosion and destruction of soil structure, a large amount of small particles were dispersed and filled the pore spaces. Additionally, due to difficulty in forming macroaggregates as a result of a shortage of organic matter, the tight soil layer and low porosity indicated that the physical properties in the control area are worse in the surface soil than in the deep soil. Thus, rational use of the degraded red soil can improve soil bulk density, enhance soil infiltration, and improve physical conditions.

14.3.1.2.2 Aggregate Analysis

After 3 years of plantation production, the contents of macroaggregates (>1 mm) in the 0- to 20-cm layer under the systems of T+S, T+P, and T+E were 26.83%, 24.7%, and 23.2%, respectively—all higher than that of the control system (19.33%); in the 20- to 40-cm layer, the amounts were 25.75%, 21.2%, and 20.9%, respectively—also higher than in the control (18.62%) (Table 14.4). In the same soil profile, the content of >1-mm aggregates in the surface layer was significantly higher than in the deeper soil layer. The content of <0.25-mm aggregates was lowest in treatment II, followed by treatment I, treatment III, and CK. The contents of <0.25-mm aggregates in the surface layer of treatment I, treatment III, and CK were significantly lower than in the deeper soil layer. This result indicates a serious disintegration of macroaggregates and extensive soil degradation in this area. The soil fertility was improved in all treatments as a result of healthy soil structure formed by returning plant residues and litter from aboveground parts as well as the exudation of underground roots. In the T+P system (treatment I), although the land surface coverage was lower than in treatments II and III due to the low planting density and the system could not stop erosion caused by rainwater and runoff, the physical properties were similar to those of treatment II, indicating its potential ability to improve soil (Yang et al., 2006).

14.3.1.2.3 Water Storage Capability Analysis

In this experiment, water storage and retention capacity were enhanced with the increase of organic matter, improvement of soil physical properties, and formation of a litter layer; thus, whole systems have better soil and water conservation capacities (Table 14.5). The T+P system (treatment I) in particular revealed a high water storage and retention capacity.

14.3.1.2.4 Soil Chemical Fertility Recovery

In this study, organic matter in the surface soil layer in the T+P, T+S, and T+E systems increased by 43.28%, 42.42%, and 8.22%, respectively, compared to the control (0.38%). The nitrogen storage increased by 38.89%, 44.10%, and 8.33%, respectively, indicating that biological

Table 14.5 Soil Water Retention Capacity and Soil Water Storage Capacity for Different Treatments

	Treatment I (Tamarind + Pigeon Pea)	Treatment II (Tamarind + Stylo)	Treatment III (Tamarind + Elephant Grass)	CK (Natural Grass)
Soil water retention capacity (t ha^{-1})	2695.7	1642.7	1229.3	—
Soil water storage capacity (t ha^{-1})	2096.3	925.3	545.2	131.3

Source: Yang, Y.X. et al., *J. Soil Water Conserv.*, 20(3), 70–73, 2006.

Table 14.6 Soil Fertility for Different Treatments

Treatment	Soil Depth (cm)	Organic Matter (%)	Total N (%)	Total P (%)	Total K (%)	Hydrolyzed N (mg kg^{-1})	Available P (mg kg^{-1})	Available K (mg kg^{-1})
Tamarind + pigeon pea	0–30	0.67	0.054	0.022	0.626	19.47	5.36	66.81
	30–60	0.50	0.199	0.018	0.795	11.70	2.96	56.15
Tamarind + stylo	0–30	0.66	0.059	0.017	1.073	25.92	2.13	92.04
	30–60	0.35	0.033	0.016	0.982	15.87	1.05	55.56
Tamarind + elephant grass	0–30	0.414	0.036	0.012	0.93	23.20	1.67	57.62
	30–60	0.213	0.033	0.014	1.068	9.47	1.52	84.25
Natural grass (CK)	0–30	0.38	0.033	0.021	1.323	7.38	2.91	84.25
	30–60	0.507	0.042	0.016	1.463	5.31	1.10	76.35

Source: Yang, Y.X. et al., *J. Soil Water Conserv.*, 20(3), 70–73, 2006.

management methods can effectively improve the nutrient content in eroded soil (Table 14.6). The supply of available nutrients such as nitrogen greatly improved the conditions. Hydrolyzed nitrogen content increased by 62.10%, 71.53%, and 68.19%, respectively, in treatments I, II, and III. The available phosphorus in treatment I with pigeon pea was increased by 45.71%. This indicates that the nutrient supply was sufficient. However, the levels of available phosphorus in treatment II with stylo and in treatment III with elephant grass were lower than that of the control, which might be the result of crop removal and soil environment factors. Total potassium content in almost all intercropping systems was lower than in the control, due to the balance of total potassium in the soil.

Soil organic matter decreased as the soil depth increased, suggesting that decomposed surface litter and organic fertilizer supplements did not easily leach down into deeper soil. As a result of root system development in the 30- to 60-cm layer with strong nitrogen fixation capacity, plenty of root exudation, and less nutrient removal, the nitrogen content in the deep soil layer of treatment I was relatively high. In treatments II and III, the nitrogen contents were lower due to the large biomass being removed and the shallow root systems. The relatively higher nitrogen content of the deep soil layer in the control (CK) was caused by serious nutrient loss at the surface. The leaching rate of total potassium in the upper soil layer was higher than the cumulative rate in the lower layer, and the leaching rate of total potassium was reduced in the lower soil layer; as a result, the total potassium content in the upper layer was lower than in the lower layer. The available phosphorus increased significantly in the tillage layer. On the whole, compared to the treatments with elephant grass or stylo, the soil planted with pigeon pea had superior fertility properties, whether or not fertilizer urea was supplied.

14.3.1.3 Changes in the Biological Control Index

Whereas the physical and chemical indices refer to property changes of the eroded soil, the biological indices involved in this study were focused on changes in soil biological features that could be used to assess the effects of different vegetation management methods. As Table 14.7

Table 14.7   Effects of Different Crop Planting Systems

Treatment	Land Surface Coverage (%)	Biomass (t ha^{-1})	Increase of Biomass (%)	Increase of Organic Matter (%)	Economic Output/ Input	Economic Input Return Ratio (%)
Tamarind + pigeon pea	85	64	98.3	43.28	13.5:1	54
Tamarind + stylo	100	56	98.1	42.42	2.5:1	10
Tamarind + elephant grass	100	225.9	99.5	8.22	7:1	28
Natural grass (CK)	45	1.08	0	0	0.25:1	1

Source: Yang, Y.X. et al., J. Soil Water Conserv., 20(3), 70–73, 2006.

shows, the land surface coverage was significantly improved by 47.67%, 55%, and 55% by means of different intercropping systems with tamarind trees as compared to control. This coverage slowed direct rainwater splashing, thereby protecting the land surface while reducing water erosion-caused soil loss. Organic matter increased significantly, particularly in the systems planted with pigeon pea or stylo. On the other hand, relatively low organic matter was observed in the area planted with elephant grass, as a result of heavy cut grass removal due to the rapid growth, high biomass, and large harvest of the grass. In addition, as a result of high land surface coverage and low ground temperature in the treatment with elephant grass, the litter decomposition rate was slow, and nutrients could not be supplied in time; thus, the system exhibited relatively low nutrient content and high biomass due to its biological features. Based on statistics of the production data, the ratio of economic output to cost input, as well as the economic benefit/return ratio, was significantly different among the treatments. The ratio of output to input in treatment I, treatment II, and treatment III was 13.5:1, 2.5:1, and 7:1, respectively, all significantly higher than in the CK (0.25:1). In other words, when the input is the same, the economic benefit produced by treatments I, II, and III was 54, 10, and 28 times that produced by the CK. Labor cost for management was the largest portion of the total cost in this study, and the benefit was generated through restoration of the eroded red soil. In general, the best benefit was derived from the treatment I system, intercropping tamarind with pigeon pea. Although the economic benefit in systems with elephant grass was also high, the organic matter reduction due to the rapid growth and high biomass and the large soil fertility depletion are not favorable for the sustainable development of soil fertility; thus, an increase in fertilizer application is required for that intercropping system (Yang et al., 2006).

14.3.1.4 Analysis of Economic Benefits

The economic yields of each model were investigated, and the average annual income for 2 years was estimated according to local market prices. Pigeon peas are both a forage and a cash crop. In this study, 20% of the pigeon peas were sold as seeds at the price of 10 yuan kg^{-1}, and the rest were used as forage. Stylo, green manure, elephant grass, and pigeon pea leaves were used for feeding goats, cattle, chicken, and fish. Households used the leftover stylo to feed pigs in the form of stylo powder due to its good palatability to livestock and high nutritional value, which could greatly improve the use efficiency and economic value of the forage (Yang et al., 2006). Results indicated that when implemented with agroecological management the average annual income per unit area in each of the four mixed models ranged from 4.79 to 48 times that of the control system (tamarind–natural grass) (Table 14.8).

Table 14.8 Economic Benefits of Different Eco-Agriculture Models

Model	*Tamarindus indica* Yield (t)	*Tamarindus indica* Income (×10⁴ yuan)	*Cajanus cajan* Fruit (t)	*Cajanus cajan* Forage (t)	*Cajanus cajan* Income (×10⁴ yuan)	Forage Grass Yield (t)	Forage Grass Income (×10⁴ yuan)	Avg. Annual Income (×10⁴ yuan)
Tamarind + pigeon pea	15.70	3.93	4.7	4.4	2.79	—	—	6.72
Tamarind + pigeon pea + stylo	0.53	0.13	0.9	1.0	0.56	92.8	0.61	1.30
Tamarind + pigeon pea + green manure	0.57	0.14	1.3	1.2	0.74	42.8	0.40	1.28
Tamarind + elephant grass	0.68	0.17	—	—	—	190.4	0.50	0.67
Tamarind + natural grass	0.45	0.09	—	—	—	10.3	0.05	0.14

Source: Yang, Y.X. et al., *J. Soil Water Conserv.*, 20(3), 70–73, 2006.

14.3.2 Assessment of the Water Preservation Forest–Grass Eco-Agriculture Model in the Rain-Fed System

14.3.2.1 *Water Conservation Effects of the Mixed-Stand Forest–Grass Model*

14.3.2.1.1 *Rainfall Interception by the Forest Canopy*

Planting *Leucaena leucocephala*, mixed with species such as *Albizia kalokora*, *Acacia farnesiana*, *Jatropha curcas*, *Bombax ceiba*, or *Delonix regia* that are adapted to the local environment, was the major method used to restore and improve the gully areas. The forest had canopy coverage of 85%, with an average tree height of 6.14 m, an average diameter at breast height (DBH) of 4.54 cm, and thickness of ground litter of 1.6 to 4.2 cm. The soil was slightly acid torrid red soil. In the region, rainfall was intercepted first by the canopy and litter in the forest ecosystems. As inferred from Figure 14.4, with an increase in precipitation the intercepted flow in the forest canopy in the gully vegetation restoration area increased, but the interception rate decreased with a clear

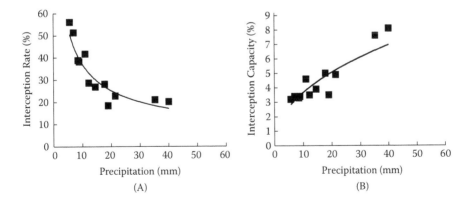

Figure 14.4   (A) Relationship of precipitation and interception rate. (B) Relationship of precipitation and interception capacity. (From Fang, H.D. et al., *Territory Nat. Resour. Study*, 3, 58–59, 2005.)

correlation. The maximum intercepted flow in the forest canopy was 8.10 mm. The average intercepted flow was 4.53 mm. The maximum interception rate was 56.10%, and the average interception rate was 32.63%. When rainfall was above 20 mm, the interception rate decreased rapidly with the increase of precipitation. Regression analysis revealed a significant correlation in this study:

$$P_i = 132.69P^{-0.554} \ (r^2 = 0.8279^{**}, n = 12)$$

where P_i is the intercepted flow (%), P is the amount of total precipitation (mm), and r^2 is the correlation coefficient (Fang et al., 2005).

14.3.2.1.2 *Water-Holding Capacity and Intercepting Effects of the Forest Litter Layer*

The natural water content is the litter water-holding capacity under natural conditions, which can reflect the moisture conditions in the forest during each season. The average monthly natural water content of the forest litter was 25.27%, with a maximum of 56.19% in August and a minimum of 7.80% in January (Table 14.9) (Fang et al., 2005). Rainfall mainly occurs between May and September, and the peak of natural water content occurs between June and October. The high natural water content, appropriate forest moisture and temperature conditions, together with a high degree of litter decomposition, can maintain good litter composition structure and moisture in the forest. The study showed that water absorption of forest litter rose rapidly within 0 to 0.5 hr and then the rate of increase slowed as the soaking time rose and reached a maximum after soaking 4 to 8 hr. Within 0.5 hr, forest litter absorbed 1.2 kg m^{-2} rainfall, which equates to 1.2 mm of precipitation. Essentially, litter could intercept 1.2 mm of rainfall within 0.5 hr. This reflects the ability of the litter layer to intercept rainfall. As Table 14.9 shows, the monthly saturation water-holding rate varied from 66.94 to 182.86%, with the minimum in May and maximum in August. The maximum water-holding capacity was 8.14 to 20.33 t ha^{-1}, with an average of 12.52 t ha^{-1}, which equates to 8.14 to 20.33 mm of water, which was twice its own weight. The litter layer could intercept an average of 12.52 mm of rainfall at most. Overall, the interception of rainfall by the litter layer is an important part of the three rainfall interception mechanisms. Particularly in the rainy season, litter can store a large amount of water and prevent soil erosion by rainfall, thus reducing ground surface runoff and suppressing soil water evaporation.

14.3.2.1.3 *Soil Features and Water-Storage Capacity in the Forest*

Vegetation types vary due to different effects of various species composition on soil physical properties. As Table 14.10 shows, the average capillary porosity of the *Leucaena leucocephala* forest in the 0- to 20-cm, 20- to 40-cm, and 40- to 60-cm soil layers was 28.42%, 29.48%, and 29.72%, respectively. These values were higher than that of the bare land by 3.89%, 1.45%, and 2.34%, respectively. (The values for bare land were 24.53%, 28.03%, and 26.93%, respectively.) In the 0- to 20-cm, 20- to 40-cm, and 40- to 60-cm soil layers of *L. leucocephala* forest, the average non-capillary porosity was 8.56%, 7.38%, and 6.38%, respectively (4.03%, 4.09%, and 3.12% higher than that in the bare land). The maximum water storage capacity of *L. leucocephala* forest was 73.96 mm, 73.71 mm, and 72.20 mm in the 0- to 20-cm, 20- to 40-cm, and 40- to 60-cm soil layers, respectively, with increases of 15.84 mm, 11.07 mm, and 10.40 mm compared to that of the bare land. Among them, capillary water storage was increased by 7.78 mm, 2.89 mm, and 5.58 mm, respectively, and the non-capillary water storage was raised by 8.06 mm, 8.17 mm, and 6.24 mm, respectively. The increase of maximum water storage mainly arose from the increase in non-capillary water storage. In the 0- to 60-cm soil layer, the maximum water storage of *L. leucocephala* forest reached 219.8 mm, 37.29 mm higher than that of the bare land (Fang et al., 2005).

Table 14.9 Litter Water-Holding Features in Vegetation Restoration Forest

Month	Litter Production (t ha^{-1})	Litter Thickness (mm)	Natural Weight (kg m^{-2})	Weight After Absorption (kg m^{-2})	Dry Weight (kg m^{-2})	Water Absorption (kg m^{-2})	Natural Water Content (%)	Maximum Water-Holding Capacity (t ha^{-1})	Saturation Water-Holding Rate (%)
1	0.84	19.82	0.83	1.85	0.77	1.02	7.80	10.21	132.47
2	0.53	21.13	1.04	2.17	0.95	1.13	9.47	11.32	118.94
3	0.33	24.55	1.01	2.01	086	1.00	17.44	10.02	116.28
4	0.21	31.20	1.16	2.09	0.96	0.93	21.83	9.35	96.88
5	0.28	28.82	1.44	2.25	1.21	0.81	19.01	8.14	66.94
6	0.36	27.41	1.67	2.93	1.23	1.70	35.77	17.04	138.21
7	0.41	45.19	1.67	2.89	1.18	1.71	41.53	17.10	144.58
8	0.38	38.22	1.64	2.97	1.05	2.03	56.19	20.33	182.86
9	0.43	35.01	1.25	2.55	0.94	1.30	32.98	13.07	138.30
10	0.67	24.76	1.61	2.99	1.21	1.38	33.88	13.87	114.05
11	1.12	24.85	0.75	1.62	0.63	0.87	19.05	8.75	138.10
12	1.44	20.14	1.85	1.95	0.80	1.10	8.24	11.06	137.50
Average	0.59	28.42	1.33	2.36	0.98	1.25	25.27	12.52	127.09

Source: Fang, H.D. et al., *Territory Nat. Resour. Study*, 3, 58–59, 2005.

Table 14.10   Soil Physical Properties and Water Storage Capacity in *Leucaena* Forests and Bare Soil in Rain-Fed Area of Dry–Hot Valleys

Vegetation Type	Soil Depth (cm)	Soil Bulk Density (g cm⁻³)	Capillary Porosity (%)	Non-Capillary Porosity (%)	Total Porosity (%)	Capillary Water Storage (mm)	Non-Capillary Water Storage (mm)	Maximum Water Storage (mm)
Leucaena forest	0–20	1.67	28.42	8.56	36.98	56.84	17.12	73.96
	20–40	1.67	29.48	7.38	36.85	58.95	14.75	73.71
	40–60	1.69	29.72	6.38	36.10	59.44	12.76	72.20
Bare soil control	0–20	1.88	24.53	4.53	29.06	49.06	9.06	58.12
	20–40	1.82	28.03	3.29	31.32	56.06	6.58	62.64
	40–60	1.85	26.93	3.26	30.19	53.86	6.52	61.80

Source: Fang, H.D. et al., *Territory Nat. Resour. Study*, 3, 58–59, 2005.

14.3.2.1.4 Analysis of Surface Runoff and Soil Erosion Control

As shown in Table 14.11, annual soil erosion was 87.68 t km⁻² a⁻¹ on the bare land and only 4.62 t km⁻² a⁻¹ in the vegetation restoration area. The surface runoff modulus (GRM) reduction rate and soil erosion modulus (SEM) reduction rate were 85.38% and 94.73%, respectively. The maximum GRM and SEM in the restoration area were 2.05 m³ km⁻² a⁻¹ and 532 t km⁻² a⁻¹, respectively; on the bare land, the maximum GRM and SEM were 10.81 m³ km⁻² a⁻¹ and 8768 t km⁻² a⁻¹, respectively. In the restoration area, the minimum erosive precipitation on steep slopes was 8.6 mm, and on bare land it was 1.8 mm. The soil erosion grade shifted from severe to medium; thus, the restoration treatment was rather effective (Fang et al., 2005).

14.3.2.1.5 Rainfall Allocation by Restoration Forests

As Table 14.12 shows, the total rainfall in the vegetation restoration area was 613.80 mm, with 200.28 mm intercepted by the canopy, 150.24 mm by litter, 43.41 mm by surface runoff, and 219.87 mm by soil. These vertical levels accounted for 32.63%, 24.48%, 7.07%, and 35.82% of the total rainfall, respectively. In the control area without forest, surface runoff was 413.24 mm and soil interception was 182.56 mm, accounting for 70.24% and 29.74% of the rainfall, respectively. Compared to the control area without forest, surface runoff was 63.19% lower and soil interception was 6.08% higher in the planted forest. Most of the rainfall in the non-forest area was lost in the form of surface runoff, and the soil interception only accounted for a small proportion of total rainfall. In the planted forest, however, the surface runoff accounted for only a small part of total rainfall, and most of the rainfall was intercepted by the canopy, litter, and soil. Therefore, planted forest not only effectively reduces the production of surface runoff but also effectively improves soil erosion resistance (Fang et al., 2005).

Table 14.11   Comparison of Soil Erosion Between Vegetation Restoration Areas and Control Area

Area	Slope Gradient (°)	Surface Runoff Modulus (m³ km⁻² a⁻¹)	Deduction Rate (%)	Soil Erosion Modulus (t km⁻² a⁻¹)	Deduction Rate (%)	Erosive Minimum Precipitation (mm)	Soil Erosion Grade
Vegetation restoration	29.3	1.06 × 10⁴	90.19	368	95.80	14.2	Slight
	18.4	1.63 × 10⁴	84.92	486	94.46	10.6	Slight
	38.7	2.05 × 10⁴	81.03	532	93.93	8.6	Medium
Control	31.2	10.81 × 10⁴	—	8768	—	1.8	Ultra-light

Source: Fang, H.D. et al., *Territory Nat. Resour. Study*, 3, 58–59, 2005.

Table 14.12   Rainfall Allocation by Artificial Forests in the Vegetation Restoration Areas

Item	Artificial Forest (mm)	Proportion (%)	Control Area without Forest (mm)	Proportion (%)
Rainfall	613.80	—	613.8	—
Canopy interception	200.28	32.63	—	—
Litter interception	150.24	24.48	—	—
Surface runoff	43.41	7.07	431.24	70.26
Soil interception	219.87	35.82	182.56	29.74

Source: Fang, H.D. et al., *Territory Nat. Resour. Study*, 3, 58–59, 2005.

14.3.2.2 *Assessment of Comprehensive Benefits of the Typical Gully-Controlled Model Integrated with* Leucaena leucocephala (Acacia)

14.3.2.2.1 *Plant Community Structural Changes*

Biodiversity structure changes of the restoration models in highly and ultra-highly degraded land were investigated. Two management models were included: (1) planted *Leucaena leucocephala* forest restoration model and (2) enclosed natural recovery model.

14.3.2.2.1 *Community Structure of Planted* Leucaena leucocephala *Forest*

Leucaena leucocephala with 5 m × 4 m spacing was planted in the gullies of the Xiaokuashan watershed in 1993; it included other species already present in the restoration ecosystems. After the planting of *L. leucocephala*, no further management measures were conducted, and *L. leucocephala* grew together with these other species. After 14 years, most of the other species were eliminated or impacted by competition. *L. leucocephala* itself propagated and renewed continuously due to its capacity for rapid growth. The community at the time of this study consisted of the arboreal layer, with *L. leucocephala* as the dominant species (plant height 260 to 1200 cm and basal diameter 5.0 to 18.7 cm); the shrub layer (plant height 70 to 250 cm and basal diameter 0.6 to 5.0 cm); and the grass layer (plant height 0 to 60 cm and ground diameter 0.2 to 0.5 cm). The density of *L. leucocephala* in the arboreal layer reached 2325 stems ha^{-1} with an importance value of 88.06%; the plant density of the shrub layer was 3.16×10^4 stems ha^{-1} with an importance value of 80.18%; and the plant density of the grass layer reached 7.21×10^5 stems ha^{-1} with an importance value of 78.96%. Other species were only found sparsely distributed in some quadrats.

14.3.2.2.2 *Community Structure of the Enclosed Natural Recovery Area*

Enclosing an area for its own natural recovery is an important measure of ecological restoration in a DHV to avoid ecosystem damage by human activities. In particular, grazing may destroy the diversity of an ecosystem, because grazing is a continuous disturbance that can intensively damage or alter natural vegetation. Under grazing pressure, those species with adapted traits will survive, but non-adapted species will be eliminated, thus altering the community.

There were 49 species present in the enclosed natural recovery area. The four arboreal species were *Albizia kalokora*, *Bombax ceiba*, *Eucalyptus camaldulensis*, and *Leucaena leucocephala*. The 11 shrub species were *Dodanaea augustifola*, *Barleria cristata*, *Ziziphus mauritian*, *Flemingia macrophylla*, *Sida acuta*, *Rumex hastutas*, *Gingko biloba* var. *parviflora*, *Acacia farnesiana*, *Osteomeles schwerinae*, *Bauhinia faberi*, and *Phyllanthus emblica*. The 38 grass species included *Heteropogon contortus*, *Bothriochloa pertusa*, *Dichanthium annulatum*, *Aristida adscensioni*, *Eragrostis pilosa*, *Dacrylocienium aegyptiacum*, *Arundinella* spp.,

Arundinella setosa, Eulaliopsis binata, Brachiaria eruciformis, Brachiaria villo, Cymbopogon goeringil, Cymbopogon distan, Adropogon chinensi, Tragus berteronianus, Arthraxon prionodu, Schizachyrium brevifolium, Crotalaria medicaginea, Crotalaria calycina, Zornia gibbosa, Desmodium microphyllum, Indigoferia linifolia, Vigna aconilifolius, Desmodium heterocarpon, Aiylosia scarabaeoides, Shuieria vestiia, Blainvillea camella, Conyza canadenis, Laggera ptorodonta, Pentanema indicum, Phyllanihus urinaria, Striga asiiatica, Tribulus terrestris, Selaginella lebordei, Selaginella puvinata, Fimbristylis orata, Leptodermis tomentella, and *Isoden eriocalyx.*

The open accessible area had 26 plant species: two tree species (*Albizia kalokora* and *Eucalyptus camaldulensis*); six shrub species (*Dodanaea augustifola, Ziziphus mauritian, Flemingia macrophylla, Sida acuta, Gingko biloba* var. *parviflora*, and *Heteropogon contortus*); and 18 grass species (*Bothriochloa pertusa, Dichanthium annulatum, Aristida adscensioni, Eragrostis pilosa, Dacrylocienium aegyptiacum, Arundinella* ssp., *Arundinella setosa, Eulaliopsis binata, Arthraxon prionodus, Schizachyrium brevifolium, Zornia gibbosa, Vigna aconilifolius, Desmodium heterocarpon, Blain villea camella, Conyza canadenis, Laggera ptorodonta, Tribulus terrestris*, and *Selaginella lebordei*).

There were 50% fewer species in the open area than in the enclosed nature recovery area. In the managed and unmanaged areas of the DHV degraded ecosystems, the dominant species also varied, which could be representative of the importance value of each species. *Albizia kalokora* was the dominant arboreal species in both enclosed and open areas, with importance values of 65.99% and 87.44%, respectively. But, in the open area, the density of *A. kalokora* was as low as 75 stems ha^{-1}, which was only 28.09% of that in the enclosed area. The reason why *A. kalokora* can become a dominant tree in DHVs, especially in areas with frequent human activities, lies in its tolerance of poor soil. Our investigation found *A. kalokora* growing in the edges of cracks in some quadrats in the highly degraded ecosystems with frequent human activities, where almost no arbor species could be found. *Dodanaea augustifola* was the dominant shrub species, with importance values in the enclosed and open areas of 77.44% and 78.22%, respectively. The density of *D. augustifola* in the open area was 75 stems ha^{-1}, which was only 10% of that in the enclosed area. For grass species, the importance value varied greatly. In the enclosed area, the dominant species were gramineous *Heteropogon contortus, Bothriochloa pertusa, Dichanthium annulatum*, and *Aristida adscensionis*, with importance values ranging from 10.44 to 16.48%. There was no significantly dominant species. Most of the grass species were evenly distributed, which keeps the system in a steady state; however, in the open area, *Heteropogon contortus* and *Bothriochloa pertusa* were the obvious dominant species, with importance values of 43.52% and 38.62%, respectively. Other grass species were sparsely distributed in the open area. The seed yield of *H. contortus* is usually high, and the seed number can reach 5000 grains m^{-2} in typical savannas in Australia (Meng et al., 2007). These abundant seeds hold great potential for the spread and reproduction of *H. contortus*. Furthermore, the seeds of *H. contortus* are twisted into knots, which easily cause them to lodge during rainy season, during which the seeds absorb plenty of water and germinate quickly. These features have ensured that *H. contortus* became the dominant grass species in DHVs. *Bothriochloa pertusa* grows in clumps, and their underground clumping rooting systems are able to absorb more water for growth and reproduction. They reproduce mainly via aboveground buds and stolons. The stolon length averages 50 to 100 cm, which is beneficial for obtaining water from a depression and ensures that it becomes the dominant grass species in DHVs. From the perspective of plant water physiology, these two grass species have the characteristics of drought resistance. An analysis of the diversity indices indicated that the average Shannon–Wiener index and the average Simpson index of the ten quadrants in the enclosed natural recovery area were 0.388 and 0.517 and in the open area were only 0.319 and 0.477, respectively. Species diversity in the enclosed area was higher than that in the open area.

Table 14.13 Evaluation of Economic Benefits of Different Types of Vegetation in Gully Areas

Vegetation Type	Income (10⁵ yuan km⁻²)	Animal Raised (goat unit km⁻²)
Acacia, eucalyptus, tamarind	1.12	402
Acacia, some *Dodanaea augustifola*	0.74	1097
Acacia, some eucalyptus, *Acacia farnesiana*	1.79	287
Acacia, few *Oxalis corniculata*	0.99	128
Acacia, *Heteropogon contortus*, *Andrographis laxiflora*, *Vigna aconilifolius*	0.81	113
Acacia, *Digitaria cruciata*, *Rottboellia exaltata*, *Andrographis laxiflora*	2.79	0

14.3.2.2.2 Improvement of Soil Physical and Chemical Properties

The average reduction of soil bulk density in the planted *Leucaena leucocephala* forest areas varied from 104 to 118%, the total porosity generally increased by 106 to 104%, and the content of particles with a diameter of 0.1 to 0.01 mm was 102 to 276% of that in the control areas. The natural water content in the planted *L. leucocephala* forests was 5.63 to 15.45%, which is 137 to 219% of that in the control areas (4.09 to 7.05%), suggesting that the soil physical and chemical properties were efficiently improved. The decrease in the proportion of >2-mm particles indicated that water soil erosion had been alleviated in the management areas. The improved water and soil conservation capacity can greatly control soil nutrient loss in the degraded sloped lands. It was reported that *L. leucocephala*, as a leguminous tree species, had great potential for nitrogen fixation. Restoration of the litter layer and herbs strengthened the soil percolation effects. The organic matter, total nitrogen, hydrolyzable nitrogen, and available potassium content all increased by 1.11 to 2.03, 1.04 to 2.12, 1.24 to 2.75, and 1.52 to 2.17 times those in the control areas, respectively.

14.3.2.2.3 Economic Benefits

Economic efficiency was analyzed for *Leucaena leucocephala* with diameters above 5 cm as timber and for seed collection (diameters within 2 to 5 cm are considered firewood), and all firewood and seeds were sold at market prices. As tested by the Institute of Biochemical Technology, Yunnan University, Kunming, the young leaves of *L. leucocephala* contain 1.79% mimosine, which was low enough for it to be used as forage. The *L. leucocephala* leaves of plants less than 2 m in height were considered fodder when calculating the economic return of the restoration areas. Significant economic efficiency has been achieved in the restoration areas, where the income per unit area was 28 times that in the control area, reaching the expected goal (Table 14.13). Therefore, restoration management has demonstrated ecological benefits, some economic benefits, and social benefits. It has demonstrated the ability for local peasants to reduce poverty and improve their economic standing.

14.3.2.2.4 Ecological Benefits

- *Water conservation efficiency*—The value of available moisture was evaluated using the shadow price substitution method (Chakravorty et al., 1995). The average annual precipitation of this area was 619 mm, with a canopy rain interception of 32.7%. It can be calculated by the following formula:

$$V_{wa} = P \times A \times (1 - I_i) \times 1000/100 \tag{14.2}$$

where V_{wa} is efficiency of water conservation of the soil (m³ ha⁻¹ yr⁻¹), P is annual precipitation (mm yr⁻¹), A is the land area (ha²), and I_i is the precipitation interception rate of the canopy. The efficiency of water conservation (V_{wa}) was 416.6 m³ ha⁻¹ yr⁻¹; thus, the water conservation value of

Table 14.14 Comparison of Soil Erosion between Vegetation Restoration Areas and Bare Land Control Areas

Area	Surface Runoff Modulus (m^3 km^{-2} a^{-1})	Deduction Rate (%)	Soil Erosion Modulus (t km^{-2} a^{-1})	Deduction Rate (%)	Erosive Minimum Precipitation (mm)	Erosion Grade
Restoration	1.58×10^4	85.30	462	94.73	12.7	Slight
Bare land	10.81×10^4	—	8768	—	1.8	Ultra-high

L. leucocephala forest in a gully was calculated as 416.6 m^3 ha^{-1} yr^{-1} × 0.67 yuan m^{-3} = 279.11 yuan ha^{-1} yr^{-1} (where 0.67 yuan m^{-3} was the required yearly cost of inputs to construct 1 m^3 of reservoir according to the fixed price in 1990).

• *Water and soil conservation efficiency*—Due to the effects of the root systems, litter layer, increase in soil porosity, improvement of ventilation conditions and water permeability, enhancement of soil infiltration capacity, and canopy interception capacity, surface runoff was reduced. The reduction rate of the soil erosion modulus was 94.73% (Table 14.14).

• *Value of CO_2 fixation and O_2 emission*—The effect of CO_2 fixation and O_2 emission in forest ecosystems was evaluated according to the organic matter productivity of Chinese terrestrial ecosystems, with the equations for photosynthesis shown as follows:

$$CO_2 + 12H_2O = C_6H_{12}O_6 + 6O_2$$

When 1 g dry matter is generated by the consumption of 1.62 g CO_2, 1.2 g of O_2 are released simultaneously. The average yearly dry matter per hectare of *L. leucocephala* forest in the Jinsha River DHV is 7.83 t, and the amount of fixed CO_2 is 121.23 t. The eco-economic value of solid C was 8626.07 yuan ha^{-1} yr^{-1}, based on Chinese afforestation costs of 260.90 yuan t^{-1}. The amount of O_2 released was 24.5 t, and its eco-economic value was 9800.00 yuan ha^{-1} yr^{-1}, inferred from the industrial cost of emitted O_2 of 400 yuan t^{-1}.

14.3.2.2.5 Improvement in the Functions of Water and Soil Preservation

Based on analysis of a 10-year survey on soil erosion grades of the Xiaokuashan watershed, the area of soil erosion decreased by 13.3% (from 108.09 per ha to 93.69 per ha, Table 14.15) from 1995 to 2005. The areas of slightly, medium, and highly degraded land were reduced by 22.03%, 21.15%, and 34.15%, respectively. Reduction of highly degraded land was the greatest, and part of that land has been converted to medium or slightly degraded land, whereas most of the slightly degraded land has been converted to non-degraded land. In addition, the yearly soil erosion amount decreased by 15.42% (58.71×10^4 t) within 10 years, during which the amount of slightly, medium, and highly degraded land

Table 14.15 Soil Erosion Changes in Reforestation Areas of the Xiaokuashan Watershed (1995–2005)

Soil Erosion Grade	Area of Soil Erosion						Soil Erosion Amount		
	Area (ha)		Percent of Total Area (%)		Percent of Eroded Area (%)		Soil Erosion Modulus (t km^{-2} a^{-1})	Annual Erosion Amount (10^4 t)	
	1995	2005	1995	2005	1995	2005		1995	2005
Slight	56.84	50.32	28.1	21.91	52.6	53.71	2000	113.58	100.64
Medium	31.64	28.44	15.7	12.38	29.3	30.36	4000	126.56	113.76
High	18.91	14.22	9.4	6.19	17.5	15.19	7000	132.37	99.54
Ultra-high	0.69	0.71	0.03	0.31	0.6	0.76	10,000	6.90	7.10
Total	108.09	93.69	47.1	40.82	100.0	100.00	—	379.61	321.04

decreased by 11.39%, 9.67%, and 24.80%, respectively, indicating that the efforts played a vital role in water soil preservation in the ecosystems. However, the area of ultra-highly degraded land is still increasing, mainly because the control of such land is rather difficult, and biological measures cannot achieve ideal performance; thus, a combination of engineering measures such as check dams and contour ridges should be applied in the future management of the ultra-highly degraded land.

14.3.3 Assessment of the Economic Benefits of the Irrigation System on Arid Sloped Land of the DHV Using an Integrated Management Model with Special, High-Quality Fruits

14.3.3.1 Economic Benefits of an Integrated Model in Cooperation with Farmers (Case I: Research Group Cooperating with Enterprises)

- *Description*—This was an integrated eco-agriculture model for use in degraded ecosystems. Specifically, the model made full use of local natural resources such as light, heat, water, and land and local social resources such as labor, animal power, and capital to construct multistoried plant and animal husbandry ecosystems and to form mutually beneficial relationships among multiple organisms by using agronomic techniques. The goal was to make good use of space; to reduce input and cost by using mutually beneficial relationships; to form the special space and time structures of agroecosystems; to conduct multiple cropping systems and continue to renew soil fertility through intercropping, relay cropping, and rotation; and to improve matter cycling and make full use of energy through food-chain arrangements and multiple-step uses of energy. The agroecosystem of fruit–crop–pig–chicken–biogas tanks is a typical model (Luo, 2009).
- *Management method*—As the main technical supporter, our research group provided fertilizer for the farm owner, Yang Dewen, who was in charge of daily operations.
- *Irrigation condition*—The water conservation facilities on the demonstration farm were designed so that water could be supplied in accordance with the requirement of crops. Consequently, water and fertilizer levels were adequate for the entire cropping system, plants grew well, and economic benefits were very good.
- *Efficiency evaluation*—As indicated in Table 14.16, longan grew strong and the growth of the 2- and 3-year-old longan trees was particularly high, mainly because in the second and third year more organic fertilizer was applied and a greater number of management methods were applied to the soil, fertilizer, and water (e.g., reshaping, pruning). The 3-year-old longan trees already had flowers and fruits. For the 4-year-old longan trees, the crowns increased gradually. In addition to normal water and fertility conditions, plenty of light and air were also required. In the orchards, pure longan was planted with a spacing of 5 m × 6 m. With strong growth and good fruiting features, the yields and quality were high. In the fourth year, the fruit yield and the output/input ratio were 5.43 kg per tree and 2.61, respectively; 7.59 kg per tree and 2.77 in the fifth year; and 15.29 kg per tree and 5.56 in the sixth year. In a mixture of longan and mango, the mango trees did not grow well due to the high density and poor ventilation conditions. Thus, we suggested cutting the poor-quality mango trees that were too close to the longan trees. In the pure mango orchard, plant growth differed significantly for different soil types. The mango trees growing in the vertisol exhibited poor growth and low economic yields, so it is necessary to improve soil quality to promote plant growth or to adopt other crop species with good adaptations to replace the existing mango trees. Pure mango trees planted in soil with a better texture had moderate or good growth, but rational plantation density is also required. As Table 14.16 shows, the high-density mixture of longan and mango resulted in poor ventilation and light penetration and a yield decrease of 4 to 11 kg per tree. But, overall, the early output values of mangos were higher than those for longan, because at early stages mango management is simpler, requiring less labor and water; however, at the later stages the longan trees began fruiting and required intensive management, especially winter shoot processing, which consumed much labor. In addition, the 5-year-old mango trees were in full fruiting period, whereas the 5-year-old longan trees were just in the early fruiting period; thus, the potential of longan production was greater. Significant economic benefits were obtained by developing a plantation of the tropical fruit trees longan and mango.

Table 14.16 Input and Output Survey for Yang Dewen Demonstration Farm

Plantation Type and Density	Variety	Tree Age (yr)	Ground Diameter (cm)	Tree Height (cm)	Canopy (Crown Breadth) (cm)	Yield per Tree (kg)	Yield (kg ha⁻¹)	Output (yuan ha⁻¹)	Input (yuan ha⁻¹)	Output/ Input Ratio
Pure longan (5 m × 6 m)	Chuliang	1	5.1	80.2	42.8 × 38.8	0	0	0	5328.0	0
		2	12.1	123.2	133.0 × 113.6	0	0	0	3665.0	0
		3	19.5	191.0	239.8 × 247.2	1.95	649.4	3896.1	4895.1	0.79
		4	26.4	254.6	315.0 × 312.8	5.43	1810.8	10,864.8	5028.3	2.16
		5	30.1	312.0	394.6 × 422.5	7.59	2527.5	15,165.0	5461.2	2.77
		6	46.1	384.5	410.2 × 376.6	15.2	5061.6	30,369.6	5461.2	5.56
Longan + mango (5 m × 4 m)	Chuliang + Sannian mango	4	21.1	216.6	242.8 × 231.0	2.28	1138.9	6833.16	5028.3	1.35
		5	38.7	227.8	225.2 × 2176	28.0	13,986.0	41,958.0	6293.7	6.67
Pure mango (5 m × 4 m)	Sannian mango	5	52.2	349.8	373.6 × 372.6	32.8	16,383.6	49,150.8	6293.7	7.80
		5	52.5	382.2	437.6 × 446.1	39.6	19,780.2	59,340.6	6293.7	9.42

Note: First year, longan input was at the price of 15.0 yuan/tree (including the cost of seedlings, planning, digging ponds, base fertilizers, and labor). Second year, longan input was at the price of 11.0 yuan/tree (including the cost of organic and non-organic fertilizer, pesticide, and labor). Third year, longan input was at the price of 14.7 yuan/tree (including the cost of seedlings, planning, digging ponds, base fertilizers, second year labor input conversion, and depreciation of irrigation facility). Fourth year, longan input was at the price of 15.1 yuan/tree (including the cost of seedlings, planning, digging ponds, base fertilizers, and labor). Fifth year, longan input was at the price of 16.4 yuan/tree (including the cost of seedlings, planning, digging ponds, base fertilizers, and labor). Sixth year, longan input was at the price of 16.4 yuan/tree (including the cost of seedlings, planning, digging ponds, base fertilizers, and labor); the input price of 5-year old mango was 12.6 yuan/tree (including organic and non-organic fertilizer, pesticide, labor, first year and second year input conversion, and early input conversion of water facilities).

14.3.3.2 Economic Benefits of the Eco-Agriculture Model in Cooperation with Youth S&T Farm in Yuanmou County (Case II: Research Group Cooperating with Enterprises)

- *Model*—This was a degraded tableland utilization model. The selection of crop species reflected the use of tropical and subtropical fruit tree resources that were adapted to the adjustment of the local economic structure. The main fruit species included longan, litchi, Taiwan jujube, and grapes. Through the development of a fruit seedling breeding base of these trees species, land use efficiency was greatly enhanced and the development of planting tropical fruits was promoted in the surrounding areas. The model made use of the degraded tableland and arid sloped land by means of different plantation types, various plantation densities, and cultural management techniques. Meanwhile, we developed a cropping and animal husbandry model with pigs, goats, chickens, and fish and extended the food chains to realize better eco-economic functions.
- *Management*—As the main technical supporter, our research group provided fertilizer. Daily operations were mainly managed by the farm owner.
- *Irrigation conditions*—The adequate water resources and good water conservation facilities within the ecosystems can meet the water needs of crops, but, due to the high cost of water inputs in the dry season from February to May, the main management method is no-tillage in order to save water. More off-season production was carried out in winter time.
- *Efficiency evaluation*—The major crops are fruit trees with characteristics typical of DHVs. The study was conducted on rational structures, functions, and benefits from the perspective of plantation types and plantation models. The model selection and construction began with the breeding and screening of varieties. Table 14.17 indicates that the constructed model was successful and the benefits were good. The three longan varieties exhibited high adaptability to DHVs and high productivity potential; of these, Shi Xia was the best, Ling Long next, and Chu Liang the worst, for its low fruiting rate and premature senility. Significant differences were also observed between the two major crops of Taiwan green jujube varieties, of which the prolificacy and growth of Gao Lang No. 1 was better than that of Da Shi Jie. The two major crop varieties of litchi did not fruit and the difference in growth was not significant. Bai Tang Ying exhibited thick and strong trunks, while Fei Zi Xiao presented strong shoot branching abilities and a better canopy than that of the Bai Tang Ying. The growth and yield of the same variety varied under different site conditions; for example, the growth and yield of longan were worse on gentle sloped land than on the tableland. The main reason was that to some extent water and soil erosion still occurred on the gentle sloped land; the soil could not hold water and fertilizer as well as the tableland, thus influencing the growth of fruit trees. As Table 14.17 shows, in the fourth year after construction of the orchard the main crops were in the early fruiting period, so their yields and output values were rather low. Within an investigated area of 8.32 ha, the total income was 28.09×10^4 yuan per year, while the yield of green jujube intercropped with longan accounted for 34.24% of the total income. Therefore, reasonably integrated management in the early orchard construction period can greatly improve the economic benefits of the agroecosystem. The income for animal husbandry in the model was 0.8×10^4 yuan per year.

14.3.3.3 Economic Benefits of the Eco-Agriculture Model Constructed at the Experimental Base of the Tropical and Subtropical Cash Crops Research Institute, Yunnan Academy of Agricultural Sciences, Yuanmou (Case III: Constructed by the Institute)

- *Economic benefits*—The experimental base of the irrigated eco-agriculture model was constructed at the Institute with a total area of 76.1 ha, of which 57.16 ha gave economic returns. The main models were tamarind (*Tamarindus indica*)–green jujube, longan–fruit or vegetable (green jujube, wax apple, papaya, guava, star fruit, or vegetable), and mango–forage grass–goat or pig. The models of tamarind–green jujube and longan–papaya exhibited the highest yielding capacities with output values of 13.2 to 18×10^4 yuan ha^{-1}. The economic returns of the longan–vegetable and mango–forage grass–goat models ranged from 6.0 to 9.0×10^4 yuan ha^{-1}. The economic benefits of the main crops are shown in Table 14.18.

Table 14.17 Investigation of the Growth and Yield of Main Crops in the Fourth Year After Establishment of Orchard

Model	Plantation (m²)	Area (ha)	Main Crop Variety	Basal Diameter (cm)	Tree Height (cm)	Crown Width (cm²)	Yield per Tree (kg)	Yield (kg ha⁻¹)	Price (yuan kg⁻¹)	Output Value (×10⁴ yuan)
Longan + cassava	6 × 5	1.33	Shi Xia	27.7	229.6	316.0 × 335.4	6.5	2164.5	5.84	1.68
	1 × 1	1.27	—	—	—	—	8.99	90,000.0	0.30	3.43
Longan + grape	6 × 5	0.54	Ling Long	—	248.6	368.0 × 427.5	4.5	1498.5	5.75	0.46
	2 × 1.5	0.13	—	—	—	—	0	0	—	0
Pure longan	6 × 5	1.73	Chu Liang	21.3	172.4	205.6 × 231.8	3.8	1265.4	5.65	1.24
Longan + green jujube (platform)	2 × 3	1.67	Gang Lang No. 1	—	347.4	—	12.5	20,812.5	2.50	8.69
	6 × 5	0.67	Shi Xia	18.6	168.4	179.8 × 184.2	6.2	2064.5	5.84	0.81
Longan + green jujube (gentle slope land)	1.5 × 6	0.40	Da Shui Jie	—	189.8	212.6 × 218.8	18.4	9337.9	2.50	0.93
	6 × 5	0.33	Shi Xia	18.5	178.2	212.6 × 218.8	5.5	1831.5	5.84	0.35
Litchi + nursery garden	5 × 4	0.13	Bai Tang Ying	13.1	150.2	116.4 × 121.2	0	0	—	0
	15,000 longan seedlings								5.0	7.5
Litchi + nursery garden	5 × 4	0.13	Fei Zi Xiao	11.1	118.0	140.6 × 133.0	0	0	—	0
	10,000 longan seedlings								3.0	3.0
Total	—	8.32	—	—	—	—	—	—	—	28.09

Table 14.18 Economic Benefits of Main Crops of Eco-Agriculture Models in the Pilot Area of the Institute

Crop	1st Year Post-Construction					2nd Year Post-Construction					3rd Year Post-Construction				
	Yield per Tree (kg)	No. of Trees	Area (ha)	Price (yuan kg⁻¹)	Total Income (×10⁴ yuan)	Yield per Tree (kg)	No. of Trees	Area (ha)	Price (yuan kg⁻¹)	Total Income (×10⁴ yuan)	Yield per Tree (kg)	No. of Trees	Area (ha)	Price (yuan kg⁻¹)	Total Income (×10⁴ yuan)
Taiwan green jujube	1.5	100	3.33	2.00	1.50	22	100	3.33	2.00	22.00	50.0	100	3.33	2.00	50.00
Wax fruit	20.0	30	0.66	10.00	6.00	40	30	0.66	10.00	12.00	60.0	30	0.66	10.00	18.00
Papaya	20.0	100	0.66	3.00	6.00	80	100	0.66	3.00	24.00	100.0	100	0.66	3.00	30.00
Guava	12.0	100	0.66	3.00	3.60	40	100	0.66	3.00	12.00	50.0	100	0.66	3.00	15.00
Star fruit	3.0	60	0.66	5.00	0.90	20	60	0.66	5.00	6.00	50.0	60	0.66	5.00	15.00
Longan	—	—	10.00	—	—	—	—	—	—	—	2.5	30	10.0	5.00	5.62
Mango	—	—	0.37	—	—	—	—	—	—	—	1.8	30	0.37	3.00	2.59
Tamarind	—	—	0.80	—	—	—	—	—	—	—	0.5	22	0.80	3.00	0.039
Intercropping crops	—	—	40.00	250.00	15.00	—	—	40.00	250.00	15.00	—	—	40.00	250.00	15.00
Forage grass	41.6 ha; 200 goats				9.00	41.6 ha; 200 goats				9.00	41.6 ha; 200 goats				9.00
Total income						463.14 (×10⁴ yuan)									

Table 14.19   Ecological Benefits of Different Planting Models in Institute Pilot Area

Factor	Monoculture Longan	Longan–Green Jujube	Longan–*Pelargonium graveolens*
Trunk diameter of longan (basal diameter) (cm)	10.61	11.54	11.94
Tree height of longan (m)	3.4	3.76	3.52
Crown area of longan (m²)	14.2	14.3	16.83
Crown density (%)	62.61	87.34	90.12
Leaf area index	3.34	6.57	5.87
Leaf water content (%)	71.18	71.89	72.03
Biomass of litters (t ha^{-2})	7.423	24.1	69.5
Air temperature (°C)	24.0–29.5	23.4–26.9	23.6–27.8
Surface temperature (°C)			
5-cm depth	33.3	27.64	24.6
10-cm depth	27.33	24.58	23.4
15-cm depth	23.7	21.61	21.5
20-cm depth	23.3	21.24	21.4

- *Ecological benefits*—Table 14.19 indicates that the ecological benefits of the integrated models were greater than those of monoculture, specifically in the larger stand with biomass aboveground, higher crown density, higher leaf area index, higher water content, larger litter biomass, and a higher capacity for decreasing air temperatures and ground temperatures. The increase in crown density of the integrated planting models enhanced the increase of basal diameter and tree height, thus more dry matter accumulated.
- *Improvement of soil physical and chemical properties*—Table 14.20 indicates that the integrated planting models had a better effect on soil improvement, with lower soil bulk density and higher water content and heat capacity. In addition, the compound models also enhanced erosion resistance and water-holding capacity and improved the soil available nutrient content and the fertility, as well.

Table 14.20   Comparison of Soil Physical and Chemical Properties under Different Planting Models

Factor	Depth (cm)	Monoculture Longan	Longan–Green Jujube	Longan–Geranium
Soil water content (%)	0–30	4.49	5.64	7.23
Water stability coefficient (K)	0–20	0.44	0.51	0.59
	20–40	0.18	0.32	0.42
Soil bulk density (g cm^{-3})	0–20	1.56	1.40	1.38
	20–40	1.74	1.62	1.57
Organic matter content (%)	0–20	0.81	0.89	0.97
	20–40	0.64	0.68	0.84
Available N (mg kg^{-1})	0–20	37.80	44.60	59.36
	20–40	33.60	51.20	64.19
Available P (mg kg^{-1})	0–20	12.21	8.20	9.46
	20–40	16.31	10.31	15.90
Available K (mg kg^{-1})	0–20	81.80	72.93	77.53
	20–40	49.10	94.17	92.11

Table 14.21   Economic Benefits of Yuanmou Green Jujube Cropping Demonstration Area

Year	Yuanma Town Yield (kg ha⁻¹)	Yuanma Town Price (yuan kg⁻¹)	Laocheng Township Yield (kg ha⁻¹)	Laocheng Township Price (yuan kg⁻¹)	Huangguayuan Township Yield (kg ha⁻¹)	Huangguayuan Township Price (yuan kg⁻¹)	Jiangbian Township Yield (kg ha⁻¹)	Jiangbian Township Price (yuan kg⁻¹)	Nengyu Township Yield (kg ha⁻¹)	Nengyu Township Price (yuan kg⁻¹)
2006	53,100	1.6	51,000	1.5	46,140	1.5	47,820	1.4	54,645	1.6
2007	53,175	2.0	56,850	1.9	50,490	1.9	47,340	1.8	56,025	2.0
2008	54,600	2.3	52,680	2.2	52,605	2.2	47,400	2.1	55,665	2.3
Output (yuan ha⁻¹)	105,630		100,137		93,569		92,870		100,224	
Input (yuan ha⁻¹)	21,690		22,500		23,400		24,000		19,800	

Note: The areas investigated in Yuanma Town, Laocheng Township, Huangguayuan Township, Jiangbian Township, and Nengyu Township were 9.33 ha, 12 ha, 5.33 ha, 2.67 ha, and 4.33 ha, respectively. The yield per ha is an average.

14.3.3.4 Economic Benefits of Yuanmou Green Jujube Cropping Demonstration Area (Case IV: Cooperation with Households)

By comparing the yields and output values of longan and green jujube among the five areas of Yuanmou County, the economic benefits of planting green jujube had a range of 6.9 to 8.4 × 10⁴ yuan ha⁻¹ (Table 14.21), but household awareness of the integrated management models remains low, so there is still great potential to develop this model.

14.4 TECHNIQUES FOR THE CONSTRUCTION OF ECO-AGRICULTURE MODELS IN DHVS

14.4.1 Technology System for the Construction of Eco-Agriculture Models in DHVs

From the perspective of ecosystem composition and the characteristics of DHVs, the technology package includes biological and non-biological restoration technologies. Biological restoration technology mainly includes the combination of forestry, cropping, and eco-cycling engineering methods with practical experience gained from eco-forest and economic-forest construction in order to guide the vegetation restoration in different degraded ecosystems in DHVs. The non-biological restoration technology system mainly includes techniques of soil restoration and both water conservation and utilization. Because the major limiting factor in degraded ecosystem restoration is soil, including soil fertility and soil water, only a restored and improved soil ecosystem can increase biomass and productivity in order to receive economic and ecological benefits derived from ecosystem restoration (Figure 14.5) (Yang and Zhang, 1999).

14.4.2 Key Techniques for the Construction of Eco-Agriculture Models on Sloped Lands of DHVs

14.4.2.1 Vertical Planting and Breeding (Tridimensional Planting and Breeding)

These techniques are used to enhance solar energy use efficiency and land productivity and to improve material production by utilizing the characteristics of different niches in each layer of community, symbiotic relationships, multiple uses of natural resources, and making full use of space. Consequently, a multistory and multisequential community structure should be constructed.

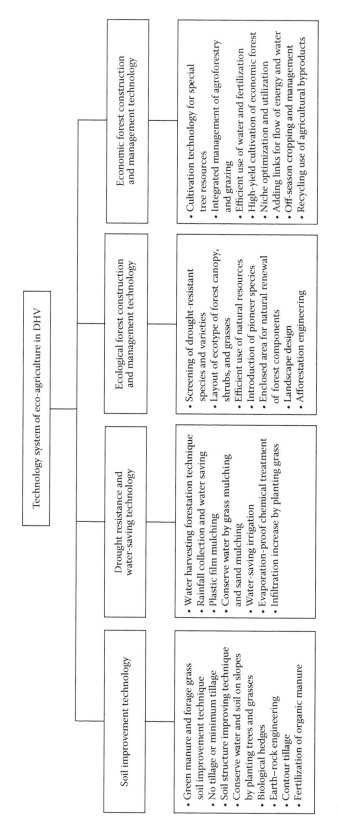

Figure 14.5 Technology system for vegetation restoration of dry–hot valleys.

14.4.2.1.1 Construction of a Multistory Agriculture Community

Rotation, intercropping, and relay intercropping are important techniques that make full use of the light and heat in DHVs. But, in the dry season from February to May, no crops or fewer crops should be produced because of the serious scarcity of water. In the rainy season, including the winter period, the production of off-season crops should be considered in order to take advantage of regional special species and to gain higher output values. Agroforest intercropping is another community construction technique for making full use of light and heat. This technique reflects the effective combination of long-term and short-term benefits. The forests can be firewood forests, economic forests, or mixed stand forests, but if the afforestation density is too high then intercropping efficiency will be affected. The selection of crop species should be mainly based on local natural conditions and market prices (Yang and Zhang, 1999).

14.4.2.1.2 Construction of Compound Agriculture Communities

This technique involves constructing compound communities of agriculture, forestry, and animal husbandry or of agriculture, forestry, and fisheries by means of plantations of crops, forest trees, fruit trees, forage grasses, and green manure. Such mosaic space structures of farmland communities serve to build mutually beneficial interactions between populations by applying the ecological principles of interspecific complementarity and commensal theory, which makes full use of the resources of organisms in agricultural ecosystems to gain maximum sustainable yield and to improve the efficiency of material cycling, energy flow, and land productivity (Yang and Zhang, 1999).

14.4.2.1.3 Vertical Zone Arrangement of Vegetation Communities

This technique develops vertical vegetation communities based on the characteristics of a vertical hilly climate. In high-altitude areas, it is appropriate to develop coniferous forests and firewood forests, whereas in mid-altitude areas various integrated cropping models are suitable and in low-altitude areas tropical cash crops are suitable.

14.4.2.2 Techniques for Recycling Organic Matter

These techniques make multilevel use of agricultural organic waste by converting it into biological energy that can be directly utilized by human beings through feeding animals or microbes, and they realize multistep use of the energy supply by organisms through applying the food chain theory of ecology. No biological energy can be converted 100% from one trophic level of the food chain to the next, but through the construction of eco-agriculture the abandoned food or fecal excretions from each trophic level can be used as food sources for other organisms, which can be further utilized and converted into biological energy. Thus, the biological energy conversion rate is enhanced, and the production and value of biological energy are increased, as well (Yang and Zhang, 1999).

14.4.2.3 Techniques for Energy Exploitation

14.4.2.3.1 Biogas Fermentation and Utilization

Construction and promotion of biogas are of great significance to improving rural energy supplies and fertilizers, to purifying the environment, and to protecting the health of farmers. Techniques include the construction of biogas tanks and utilization of residues for biogas fermentation.

14.4.2.3.2 Solar Energy Utilization

Solar energy in the Jinsha River DHV is abundant, so the direct use of solar energy offers the best way to save energy and increase the income of farmers. Techniques include solar ovens, plastic-film mulching, plastic-film sheds, and solar energy receivers.

14.4.2.4 Techniques for Eco-Environment Control

14.4.2.4.1 Sustainable Use of Land

Soil fertility is the main factor influencing the use of solar energy and other energy resources for biological fixation and conversion, so maintaining land fertility is a basic principle of agricultural production. On arid sloped land, planting leguminous forage plants and developing animal husbandry are ways to employ eco-agriculture engineering by combining cropping and grazing, which can enhance the stability and sustainability of agroecosystems and improve the stress resistance of production systems.

14.4.2.4.2 Control of Water and Soil Loss in Combination with Engineering and Biological Measures

Water and soil conservation should be made the first priority in order to achieve ideal restoration results. The combination of restoration and utilization should be considered. Engineering and biological measures should be coupled as well, with the latter being the primary managed measures. Slope control, land control, and gully control should be conducted at the same time, but the last two should be the focus. The management process should consider soil conservation to be the core while also considering the integration of hills, water surfaces, and farmland-to-road as a whole. A combination of forest and grass plantations should be a priority of biological measures in the DHV area. Mixed plant communities comprised of arboreal species, shrubs, and grass species should be implemented, along with the integrated development of agriculture, forestry, animal husbandry, fisheries, and related processing industries. For land use, every single space should be utilized to take full advantage of solar energy resources (Yang and Zhang, 1999).

14.4.2.5 Techniques for Introducing New Varieties and Filling Niches

Filling niches involves the integration of biological and ecological engineering. Taking advantage of excellent germplasm resources, screening new varieties using traditional and biological technology, and arranging them to their suitable niche can help to dramatically increase ecosystem productivity. Some crop varieties have aged and degenerated in recent years in the Jinsha DHV area. They are no longer adapting themselves to their original niches. Thus, only replacing them with suitable species and varieties to fill all available niches can enhance ecosystem productivity. Screening trials in a pilot area for species and varieties suitable to specific niches should be conducted on the basis of plant adaptations and niche environments (Yang and Zhang, 1999).

14.4.2.6 Fertilization Techniques

To increase fertility, especially organic matter content, the following measures should be implemented: (1) planting leguminous forage to increase the content of nitrogen in the soil, with the forage being used for feeding animals and all animal waste being returned to the farmland; (2) planting green manure and plowing it into the soil to increase soil organic fertilizer content; and (3) returning stalks and other crop residue associated with organic fertilizers to the farmland to promote the conversion and utilization of organic matter. The timing of the use of these organic resources is critical and should be adjusted according to plant growth stages, seasonal rainfall, and cropping system (Yang and Zhang, 1999).

14.4.2.7 Techniques for the Control of Diseases, Pests, and Weeds

In the process of eco-agriculture construction, no or fewer chemicals should be used to control diseases, insect pests, and weeds. Integrated management should be implemented through the use of systematic management methods according to the ecological principles of mutualism and competition theory. However, it is sometimes not practical to carry out only biological control methods to control diseases, insects, and weeds in the beginning. Thus, biological methods can be integrated with chemical methods initially and then the chemical input can be gradually eliminated through the use of the following non-chemical methods (Yang and Zhang, 1999):

1. Using high-density leguminous forage plants to suppress the growth of weeds
2. Shifting the time and space of crop plantings to prevent or reduce pests
3. Utilizing rotation, intercropping, and relay intercropping to reduce the chance of pest and disease damage
4. Taking advantage of the natural enemies of pests to regulate pest populations

14.4 CONCLUSIONS

This chapter summarizes the results of six research projects, including national key research projects during the ninth 5-year economic development planning period (1996–2000), tenth 5-year planning period (2001–2005), eleventh 5-year planning period (2006–2010), and twelfth 5-year planning period (2011–2015). The results confirm that, in spite of the unfavorable climate and soil conditions, dry–hot valleys can be restored and well utilized through the integrated use of various system configurations and adaptable technological packages. During eco-agricultural development in DHV regions, it is important to evaluate the available water resources. For rain-fed areas, supplemental irrigation areas, and irrigation areas, our experience in the eco-agricultural development of these three types of DHVs could be useful to be applied both locally and elsewhere.

REFERENCES

Chakravorty, U., Hochman, E., and Zilberman, D. 1995. A spatial model of optimal water conveyance. *J. Environ. Econ. Manage.*, 29: 25–41.

Chen, L.D., Wang, J., and Fu, B.J. 2001. Strategy on sustainable development of eco-fragile area of xerothermic valley in Southwest China. *Chin. Soft Sci. (Reg. Econ.)*, 6: 95–99.

Fang, H.D., Ji, Z.H., Yang, Y.X., Bai, D.Z., and Liao, C.F. 2005. Assessment on the ecological economic value of vegetation restoration in Jinsha River hot-dry valley: taking Yuanmou as an example. *Territory Nat. Resour. Study*, 3: 58–59.

Ji, Z.H. 2009. *Theory and Practice of Eco-Agriculture in Arid-Hot Valley*. Kunming: Yunnan Scientific and Technical Press, pp. 130–138.

Ji, Z.H., Yang, Y.X., Liao, C.F., Fang, H.D., and Bai, D.Z. 2005. Construction of an eco-agricultural model of stereoscopic plantation and breeding on the degraded sloping land in the arid hot valley of Yuanmou. *J. Southwest Agric. Univ. (Social Sci.)*, 3(3): 1–4.

Ji, Z.H., Pan, Z.X., Sha, Y.C., Fang, H.D., Liao, C.F., Bai, D.Z., and Yang, Y.X. 2006. Model construction of ecological restoration in arid hot valley of Jinsha River. *J. Agro-Environ. Sci.*, 25(Suppl.): 716–720.

Ji, Z.H., Fang, H.D., Yang, Y.X., Pan, Z.X., and Sha, Y.C. 2009. Assessment of system functions after vegetation restoration of the degraded ecosystem in arid-hot valleys of Jinsha River: a case study on small watershed of Yuanmou. *Ecol. Environ. Sci.*, 18(4): 1383–1389.

Luo, S.M. 2009. Fundamental classification of eco-agricultural models. *Chin. J. Eco-Agric.*, 17(3): 405–409.

Meng, T.T., Ni, J., and Wang, G.H. 2007. Plant functional traits, environments, and ecosystem functioning. *Chin. J. Plant Ecol.*, 31(1): 150–165.

Pan, R.C. and Dong, Y.D. 1984. *Plant Physiology*. High Education Press: Beijing, p. 107.

Yang, W.X. and Zhang, Q.M. 1999. *Eco-Agricultural Engineer Technology*. Chinese Agricultural Science and Technology Press: Beijing.

Yang, Y.X., Ji, Z.H., Sha, Y.C., Pan, Z.X., Fang, H.D., Bai, D.Z., and Liao, C.F. 2006. Study on soil and water conservation benefit of models of eco-agriculture on dry slope land in Yuanmou dry hot valley. *J. Soil Water Conserv.*, 20(3): 70–73.

Zhang, R.Z. 1992. *Arid-Hot Valley in Hengduan Mountain*. Science Press: Beijing, pp. 16–30.

Zhang, X.B., Yang, Z., and Zhang, J.P. 2003. Lithologic types on hill slopes and revegetation zoning in the Yuanmou hot and dry valley. *Sci. Silvae Sin.*, 39(4): 16–22.

CHAPTER 15

Eco-Agricultural Practices of State Farms in Heilongjiang Province

Wang Hongyan, Wang Daqing, Dai Lin, Ying Nie, An Menglong, Xu Maomao, and Cao Can

CONTENTS

The state farms in Heilongjiang Province have three characteristics: intensive agriculture, large-scale monoculture production, and high mechanization. The practice of eco-agriculture in the state farms significantly influences the path of agricultural development in all of China. This chapter introduces some critical issues and typical experiences for the state farms in Heilongjiang Province.

15.1 INTRODUCTION TO ECO-AGRICULTURE IN STATE FARMS OF HEILONGJIANG PROVINCE

Heilongjiang Province is a grain production province in China with high yields. Figure 15.1 shows the main distribution of animal husbandry and crop production in the province. The state farms of Heilongjiang Province are located in the northeast in the Xiao Xingan Mountains, scattered over a wide area of Sanjiang Plain, and in Songnen Plain, one of the four famous black soil belts in the world (Figure 15.1). A 5.62-km² area within the state farms has 2.88 million ha of cultivated land, 0.93 million ha of forest land, 0.35 million ha of grassland, and 0.25 million ha of

Figure 15.1 Distribution of agriculture, forestry, animal husbandry, and agriculture production in Heilongjiang Province.

Figure 15.2 Distribution of Heilongjiang State Farms. (From Wang, H.B., *Farm Econ. Manage.*, 8, 54–56, 2013.)

bodies of water. This area is also a state-level ecological demonstration region featuring 113 farms, 866 state-owned or state-holding enterprises, and 152 non-state-owned enterprises within the state farms. Twelve cities with a total population of 1.734 million, of which 983,000 are employees, are scattered throughout Heilongjiang Province (Figure 15.2).

The state farms of Heilongjiang Province have become an important strategic base for commodity grain and food production, as they offer the largest scale, greatest degree of modernization, and most comprehensive production capacity in China. The entire reclamation area has produced 285.32 billion kg of grains and distributed 222.61 billion kg of commodity grain within China so far; thus, the state farms are contributing greatly to the country's grain, food, and ecological security.

The overall grain production capacity of the state farms can reach up to 22.5 billion kg yr^{-1}. The state farms can provide over 20 billion kg of commodity grain a year, which meets the annual needs of an urban population of 120 million. Black soil and meadow soil comprise 50% of the cultivated land. The black soil layer reaches 30 to 50 cm deep, with a soil organic matter content between 3 and 5% on average. The 46 rivers that cross the region have 1000 km^2 of catchment area that provides 9.75 billion m^3 of water resources. There are 19 national and provincial wetland conservation areas and 155 national and provincial wildlife reserved zones—extremely important national protected biodiversity centers—covering a total area of 753,000 ha, which accounts for 13.49% of the land cover. The mechanization rate in this reclamation area has reached 96%. The rate of science and technology contributions to agriculture has reached 67%. The average grain production capacity can be as much as 40 tons per farmer working on the state farms.

15.2 REVIEW OF ECOLOGICAL CONSTRUCTION
IN HEILONGJIANG STATE FARMS

15.2.1 Brief Review

Throughout history, the Great Northern Wilderness was famous for its humid climate, lush forests, swamps, and wetlands throughout the vast prairie—a lonely, vast wild wasteland. Old folk descriptions of this magic and rich land are vivid: "the black soil is so rich that when holding it you can see oil drops bubbling out"; "even a chopstick inserted into the soil can survive and sprout"; and "sticks could easily hit a deer, ladles could easily scoop up a big fish, and pheasants could occasionally fly into your rice cooker." Unfortunately, destruction caused by the large number of immigrants beginning during the Japanese occupation period and continuing into the 1970s has reduced the forest and wetland areas. In the late 1970s, the forest coverage rate dropped below 10%. Wind erosion, water erosion, desertification, and natural disasters have threatened agricultural production for years. Wind erosion and desertification have occurred on over 40% of the farmland; seeds are blown away and production is reduced. Ecological environmental deterioration has threatened the survival and development of the Great Northern Wilderness.

In the 1980s, the state farms initiated ecological construction, primarily through afforestation. By 2000, forestry in the reclamation area reached the goal set for afforestation. After 2001, a policy of "giving up farming for forest restoration" was implemented in some early reclaimed and degraded hilly land. These regions were in poor condition, with thin soil and low fertility, and were subjected to water and wind erosion. According to the census at that time, this type of cultivated land covered up to 183,000 ha. If people continued to cultivate this area, not only would the vicious cycle continue but it would also endanger the surrounding areas. For this reason, the Heilongjiang Land Reclamation Bureau decided to return the desertified, low-yielding arable land back to forest.

15.2.2 Developmental Path of Ecological Construction

In 1996, the development of eco-agriculture led to construction of an ecological demonstration area in the state farms. In 1998, the eco-demonstration on the 291st farm received recognition and approval from the National Environmental Protection Bureau. In 1999, construction of a comprehensive eco-demonstration area was carried out by the local land reclamation bureau within the framework proposed by the Heilongjian provincial government. In 2002, the Bao Quanling administration area was approved as the state-level ecological demonstration area. By 2006, seven other administration regions and land reclamation bureaus also obtained approval from the Environmental Protection Bureau for state-level ecological demonstration regions. Today, the entire reclamation area has become a state-level ecological demonstration area and is the largest regional state-level ecological demonstration area in China.

Since 2005, the ecological demonstration stage has formally moved on to full ecological construction in all land reclamation areas. A special management group for the ecological construction was set up within the Bureau of State Farms. Similar management structures were also set up in all of the state farms in the region. The ecological construction plan for the reclamation area was implemented beginning in 2006. Leaders in top positions had to sign responsibility contracts at the beginning of each year for the tasks and important goals to be achieved. They were required to report the results and to be evaluated at the end of each year. Ecological construction in the reclamation area proceeded steadily and systematically.

In 2007, the 17th National Party Congress deployed a strategy of constructing what was termed an "ecological civilization." The need for harmony between humans and nature was pointed out for the first time in China. The goal of building a resource-conserving and environmentally friendly society was written into the party constitution. In 2008, a work plan for accelerating ecological construction was issued by the Land Reclamation Bureau. In the same year, ecological and environmental protection indicators were included in the evaluation system together with economic and social development goals for farm leaders. A critical environmental protection standard for the evaluation of leaders was also implemented in 2008.

The report of the 18th National Party Congress in 2013 proposed an "ecological civilization" strategic action plan. Since then, administration leaders have tried their best to implement this strategy actively and effectively. Ecological construction at both Bao Quanling Farm and Sanjiang Farm passed provincial assessment in 2013. In 2014, the state farms began to operate according to the standards of the National Ecological Civilization Demonstration Region and played a leading role in ecological development for the entire Heilongjiang Province.

15.2.3 Current Status of Ecological Construction

After years of effort, the state farms have completed the first stage of the ecological construction. The basic ecological civilization policies, regulations, and legislation in China have been strictly implemented. These policies and regulations are presented in agricultural law, planning law, land management law, water law, environmental protection law, atmospheric pollution prevention law, soil and water conservation law, environmental impact assessment law, forest law, grassland law, and regulations on ecological environmental protection and wetland management. The economic and administrative policies of state farms are also subject to pollution discharge control, public participation, ecological and environmental evaluation, and mechanisms for rewards and punishments.

The goal of state farms is to pay equal attention to both economic development and ecological protection. Protecting natural conservation areas and protecting prime farmland are clearly defined as the bottom line. There are 753,000 ha of natural preserved area and 2.73 million ha of basic prime farmland that fall under this "protected bottom line" policy. An ecological compensation mechanism has been established; market mechanisms that could be used for ecological compensation were explored in order to establish a fair and effective compensation mechanism. Ecological compensation payments from the financial budget were increased, and economic development that addresses resource conservation and encourages environmentally friendly industries will help realize effective use of energy and materials.

15.2.4 Eight Ecological Projects Implemented in State Farms

Construction of the ecological demonstration region in Heilongjiang state farms began in 2000. The eight ecological projects were proposed and implemented by putting more than 6 million yuan into special project funding and 3.0 billion yuan into the construction work. A system of accountability was set up for the leaders in the top positions of the organization in order to implement these eight projects: (1) ecological agricultural development, (2) ecological restoration in forests, (3) ecological animal husbandry and grassland restoration, (4) degraded land restoration, (5) wetland biodiversity conservation and ecological restoration, (6) industrial pollution control, (7) ecological urban construction, and (8) development of rural clean energy. Four key functional zones—priority development, key development, development-restricted, and development-prohibited—were divided clearly within the area managed by the state farms. Wetland was protected. Part of the rain-fed fields were converted to paddy fields. Soil conservation measures were taken to increase the productivity of low-yielding land. Conservation tillage methods were also adopted by installing advanced agricultural machinery. Since 2003, 107 demonstration zones for modern mechanized farming have been set up. Devices for satellite global positioning, automatic navigation, precision seeding, and variable fertilization were installed in these machines. Deep plowing, shallow cultivation, leveling, seeding, mixing, and suppression can be accomplished in one operation. Such developments not only have increased production efficiency but have also protected farmland more effectively. The farmland is turning into field gardens.

Table 15.1   Ecological Indicators for Heilongjiang State Farms (2012)

Category	Indicator	Value
Economic development	Gross domestic product (GDP) used for environmental protection (%)	2.2
	Protected land area (%)	14.5
	Enterprises passing clean production exam (%)	30.0
Ecological protection and construction in agriculture	Forest and grass coverage rate (%)	18.4
	Discharge of SO_2 (%)	1.6
	Discharge of COD (%)	3.0
	Chemical fertilizer application rate (N+P+K kg hm^{-2})	156.4
	Agricultural plastic-film recovery rate (%)	96.7
Reuse, recycle, and reduce use of resources	Reuse and recycling rate of livestock waste (%)	92.0
	Green food and organic food production area (%)	69.4
	Green area per capita in urban areas (m^2)	18.9
	Sewage treatment rate in urban areas (%)	60.0
	Properly treated garbage (%)	80.0
	Energy consumption (tons of standard coal per 1,000,000 yuan GDP)	68.7
	Water consumption (m^3 per 1000 yuan GDP)	102.0
	Reuse and recycling rate of straw (%)	94.3
	Central heating rate (%)	69.1
Social progress	Level of urbanization (%)	85.0
	Satisfaction rate of public regarding the environment (%)	95.0

15.3 ACHIEVEMENT OF ECO-AGRICULTURE CONSTRUCTION

Through implementation of eco-agriculture and the eight key ecological projects, the economy of the state farms in Heilongjiang Province improved rapidly. The social and ecological environments have benefited each other. Green development, cyclical development, and low carbon development became characteristic of the state farms.

15.3.1 General Results

The statistics for 2012 (Table 15.1) show that the water-saving irrigation area reached 243,000 ha, and green food and organic food production areas reached 1.993 million ha, which accounted for 44.5% of the total green food and organic food certification area in Heilongjiang Province. There were 63 national green food production bases. The regional forest coverage rate reached 18.4%. Vegetation-covered areas grew to 41 m^2 per capita, even as the urbanization rate reached 85%. The development of animal husbandry was conducted in a green and healthy way. Light industrial production increased by 17.51 billion yuan, with an annual growth rate of 43.8%. Income from ecological tourism was 2.28 billion yuan. Locations such as Hailin, Qixing, and Xingkai Lake were identified as popular tourist towns in Heilongjiang Province. There are 21 nature reserves with a total area of 515,000 ha, accounting for 9.1% of the total area. Over 60% of the farms improved their protection of drinking water resources.

In general, the data from 2012 for the Heilongjiang state farms showed satisfactory results with regard to ecology and environmental construction. The agricultural waste reuse and recycling rates were high, and the chemical fertilization rate was lower than the high rates set in European countries and the United States. The percentage of green food and organic food production in one region was the highest in China. The continued development of science and technology, education, culture, medicine and hygiene, social insurance, and social well-being is improving steadily.

15.3.2 Eco-Agriculture Achievement at Bao Quanling Farm

Bao Quanling Farm is a good example of eco-agriculture among the Heilongjiang state farms. An investment of 300 million yuan in a natural forest protection project at this farm resulted in a 15.6% forest coverage rate after replanting trees over 110,000 ha and returning 20,000 ha of farmland back to forest. The investment of 30 million yuan for grassland protection resulted in 24,200 ha of improved grassland. Over 74,000 ha of farmland in desert areas and 2800 ha of farmland in salinization zones have been reformed and improved, and 90.1% of the soil and water erosion area has been controlled. Clean production audits were carried out for industries such as dairy production, paper making, and oil production and helped to speed up pollution control in these traditional industries. Also, 92% of discharged industrial wastewater and 98.3% of urban solid waste treatment achieved levels set by national standards (Qin, 2014). The ecological environment at Bao Quanling Farm provides a unique experience. During the development of eco-agriculture at this farm, pollution-free techniques were implemented within all 300,000 ha of farmland, including soil testing for rational fertilization, the use of organic fertilizer, and abandonment of highly toxic or high-residue pesticides. The green food and organic food production area reached 152,000 ha. Green food products such as Baroque Sauce and Bezique rice have become famous both at home and abroad. Bezique rice has been successfully exported to the United States, United Kingdom, Canada, and other countries (Qin, 2014).

15.4 TECHNICAL APPROACHES USED IN ECO-AGRICULTURE IN STATE FARMS

Eco-agricultural practices have been effectively implemented in the state farms of Heilongjiang Province. Eco-agriculture has increased soil fertility, improved environmental quality, and protected biodiversity. The ecological situation has greatly improved at the farms while the high level of food production has been maintained.

15.4.1 Returning Farmland to Forest and Farmland Shelterbelt Construction

15.4.1.1 Scale of Forest Restoration and Shelterbelt Construction

At least 122,000 ha of farmland have been returned back to forest. Reforestation areas have reached 56,300 ha, and key protective forest areas cover 11,000 ha. There are now 895,000 ha of forest land, among which 553,000 ha are planted or restoration forest. Total wood stock has reached 55 million m³. Over 40,000 windbreak networks formed by 70,000 tree belts are protecting more than 2 million ha of cultivated farmland. The tree planting area in the shelterbelt system reached 140,000 ha in 2012, accounting for 5% of the arable land. Regional forest coverage has increased to 18.4%.

15.4.1.2 Benefits and Effects

Under the protection of the shelterbelt system, average grain and bean yields increased 15.2%, wind speed decreased 48%, temperature increased in a range of 0.8 to 3.6%, evaporation decreased 7.5%, soil moisture increased between 3.8 and 4.3%, absolute humidity increased between 2.3 and 4.9%, and precipitation increased about 5.1% (Guo, 2007).

15.4.1.3 Practices in the Construction of Protective Forest

A good planning and support policy is vital in the beginning of protective forest system construction. Developing an integrated forest economy promises higher economic returns from the forest, in addition to ecological benefits. A well-organized forest management system is essential for long-term forest development. The plan to build protective forest in Heilongjiang Province focused on improving poor farmland and edge areas and on developing a more effective shelterbelt system. The poor sloped land, land that had undergone desertification, and eroded land were targeted for transition back to forest. The remaining bare hills and separated small parcels of wasteland were listed as edge areas for reforestation. The broken shelterbelts were repaired. Towns and villages were projected to be surrounded by forest, shrubs, and grass. Both sides of roads and irrigation channels were slated to become green corridors with a mix of needleleaf and broadleaf trees and shrubs.

The development of edible fungus, Chinese medicinal herbs, tree seedlings, special poultry production, and economic forest products using advanced technology can take advantage of forest resources and promote a strong forest economy. At present, the state farms include 6700 ha of economic forest, 2500 ha of seedling nurseries, 7100 ha of medicinal herb production fields, and 4500 ha of wild vegetable production fields. The annual edible mushroom production is over 2000 tons. The output of tree seedlings reaches 250 million. The number of artificially raised wild animals has reached 196,000 (Guo, 2009).

Forest management system reform divided forest areas owned by the state farms into small patches of woodland to be operated by family members for a long period of time, 20 to 70 years. Developing family forest farms, implementing an accountability system in forest management, and exploring property rights reform for the state-owned forests were all encouraged. There are now 15,000 family forest farms and 133,300 ha of forest under responsibility contracts. Forest management systems at the state farm level, farm level, and team level, including firefighting teams, forest patrol teams, and pest control teams, are now well organized. Money for the Forest Ecological Compensation Foundation is raised annually at a rate of 15 yuan per ha of farmland for the construction of forest. These practices will be further strengthened during future forest development at the Heilongjiang state farms.

15.4.2 Soil and Water Erosion Control

According to a 2006 survey, soil and water erosion occurred in 73 of the Heilongjiang state farms, which accounted for 16.8% of the total land area. Among that area, 824,900 ha were subjected to water erosion and 83,100 ha were subjected to wind erosion. Soil erosion often occurred in cultivated land. The area of light erosion was 282,900 ha, the area of moderate erosion was 377,100 ha, and the area of heavy erosion was 282,800 ha. The 4915 erosion gullies had a total length of 2301.6 km, and gully density reached 0.17 km km^{-2} (Jiang, 2006). Measures to prevent soil and water loss included vegetation recovery measures, agricultural measures, and engineering measures.

15.4.2.1 Vegetation Recovery Measures

Vegetation recovery measures included managed tree and grass planting and enclosing hills for natural vegetation recovery. Increasing the rate of vegetation coverage can reduce surface runoff and wind speed, thus reducing erosion and protecting farmland. The revegetation efforts were conducted in gullies, on ridges, on dams, on slopes, on bare hills, and in poor fields. In eroded gullies, engineering methods were used in combination with vegetation methods. Revegetation was carried out in different parts of each gully (upper part, sides, sloped part, and bottom part), according to the specific situation. Vegetation on field ridges or on a dam in front of a gully could provide protection for the field or gully. Windbreak systems and shelterbelt systems that protect against wind erosion are more important in the western state farms due to the more serious wind erosion risks there. Vegetation covers for slope protection were planted horizontally in the lower part of the slope through a mix of trees, shrubs, and grass. Poor farmland on steep slopes was required to be returned back to forest or grassland. Natural restoration methods were carried out by enclosing poor forest or shrub areas.

15.4.2.2 Agricultural Measures

Agricultural measures include contour plowing, organic fertilizer application, shallow tilling, subsoil plowing, straw mulching, and field ridge building. These measures could help to stop runoff and improve water-holding capacity, thus reducing runoff and topsoil erosion in fields. Changing the cropping row direction from an up-down direction to a horizontal direction on farmland with 1.5 to 3° slopes could increase interception of surface runoff, increase soil water storage capacity, and increase crop yield. This measure was used for farmland with moderate slopes but without a high-density and well-established shelterbelt system. Hence, the row change operation was easy to implement.

Building a soil ridge barrier on the up-down direction field furrow was another method used to reduce surface runoff. During the last cultivation operation, an earth-retaining soil ridge barrier was constructed in the furrows of the sloping land. This small change in terrain on sloped land can help to capture and retain surface runoff. Western state farms with well-developed shelterbelt systems that made it difficult to change crop row directions adopted this measure. Constructing horizontal field ridges on farmland with 3 to 5° slopes, together with contour tillage, can effectively prevent slope runoff and seedling loss. Conducting subsoil tillage along cropping rows proved effective in breaking plow layers, improving soil structure, increasing soil porosity, improving soil storage capacity of water, and reducing surface erosion. Subsoil tillage methods were used for field crops such as soybeans, corn, potatoes, and beets in almost all state farms. Applying more organic fertilizer helped to improve fertility, increased topsoil thickness, improved soil moisture, and enhanced soil erosion resistance. This method was used in poor soil with low fertility, especially by the smaller farms.

15.4.2.3 Engineering Measures

The engineering methods used included construction of intercepting ditches between upper slopes and crop fields, intercepting ridges on the eroded gully heads, and check dams on the lower part of an eroded gully. The intercepting ditch prevented the invasion of runoff water from the upper sloped land into crop fields. The intercepting ridge was used in the eroded gully with a small amount of upper runoff water from a small watershed or from a big watershed with good vegetation cover. It can prevent further erosion from developing at the head of a gully. Check dams were usually built using soil, stone, concrete, willow knitting fence, woven bags with soil inside, or wood. Willow and grass were planted on the check dams. In the northern state farms, stone dams and concrete dams were adopted because of the abundant stone in the mountain area. In the moderately sloped land found in western sites, the soil, willow, and woven bag methods were used more often. Water storage ponds behind the check dams were used to facilitate field irrigation.

15.4.3 Eco-Agriculture Technology Applied in State Farms

The major eco-agricultural technology used in state farms included methods that could help to reduce the use of chemical fertilizers and pesticides in crop production and to improve livestock production.

Figure 15.3   Straw being returned to soil after harvest on Tangyuan Farm.

15.4.3.1 Comprehensive Utilization of Crop Straw

The comprehensive utilization rate of straw in the Heilongjiang state farms has reached 91%, providing 50% of the nutrition needs of crop production. Through long-term continuous straw cycling practices, soil organic matter content increased 0.02 to 0.04% each year. Straw is also used for biogas and electricity generation. Animal waste is mixed with straw for biogas production. The production of rice hulls in the state farms has reached 490,000 tons. Six power-generation plants using rice hulls have been set up. Figure 15.3 shows returning the stubble at the Tangyuan farm after corn harvest.

15.4.3.2 Improved Fertilizer Efficiency

Methods of testing soil and plant nutrition to guide fertilization practices were widely used in the Heilongjiang state farms. Together with returning straw and animal waste after biogas generation and the use of mechanized fertilizer application, fertilizer application rates decreased and the efficiency of chemical fertilizers increased. Table 15.2 shows that the average fertilizer application level in the state farms in 2010 was lower than the averages in China, the United States, United Kingdom, Germany, Japan, and South Korea (Wang, 2013). Although this number is higher for the state farms than in Heilongjiang Province as a whole, the average fertilizer application rate in terms of unit grain output in the state farms was lower than in Heilongjiang Province and was only about one quarter of the average fertilizer level in China.

15.4.3.3 Pesticide Control Measures

Since 1990, integrated pest management techniques have been actively promoted in the Heilongjiang state farms. Pest control methods emphasized more accurate pest forecasting, the wide use of resistant varieties, suitable fertilization and irrigation methods, more biological measures, and the use of advanced spraying machines. The application of methamidophos, methyl parathion, phosphoric amine, and monocrotophos was strictly prohibited, and the application of parathion was also gradually ended. Instead, the pesticides used originated from plants, such as pyrethroids

Table 15.2 Fertilizer Application Rate and Grain Production Efficiency in Selected Regions and Countries from 2007 to 2010

Location	Fertilizer Application Rate (ton ha⁻¹)				Grain Production Efficiency (kg kg⁻¹)			
	2007	2008	2009	2010	2007	2008	2009	2010
State Farms[a]	150.8	157.5	165.0	172.5	30.4	27.8	26.5	26.6
Heilongjiang[b]	147.0	149.3	143.3	150.8	44.2	42.8	45.7	42.9
China[c]	333.0	335.3	340.5	346.5	101.8	99.1	101.8	101.9
United States[d]	294.8	180.0	186.8	204.0	60.2	36.3	34.8	39.9
United Kingdom[d]	352.5	300.0	362.0	376.5	80.0	51.5	65.1	70.9
Germany[d]	237.0	207.8	235.5	282.8	54.6	38.2	43.6	56.6
France[d]	341.3	213.0	219.8	218.3	65.1	38.0	38.8	40.6
Italy[d]	313.5	198.8	184.5	171.0	63.2	50.7	52.2	44.6
Canada[d]	150.8	105.0	93.0	136.5	73.8	43.3	40.1	58.4
Japan[d]	559.5	449.3	392.3	438.0	121.6	97.1	86.4	96.7
Korea[d]	556.5	572.3	552.8	398.3	112.9	101.5	84.5	76.6

Source: Wang, H.B., *Farm Econ. Manage.*, 8, 54–56, 2013.

[a] Data obtained from *Statistical Yearbook of Heilongjiang State Farms*, 2010.
[b] Data obtained from *Heilongjiang Statistical Yearbook*, 2010.
[c] Data obtained from *Statistical Yearbook of China*, 2010.
[d] Data obtained from *United Nations Statistical Yearbook*, 2010.

in the form of bromine hydrogen chrysanthemum ester, fluorine cypermethrin, and cypermethrin; low toxic organic phosphorus in the form of marla sulfur, phoxim, chlorpyrifos, acephate, and bromine phosphorus; ammonia carbamate pesticides and biological pesticides in the form of *Bacillus thuringiensis*, abamectin, and liuyang; and matrine, all of which were recommended to be used carefully according to the guides. Currently, pesticides used in the state farms amount to 3.69 kg ha⁻¹, compared to an average of 15.23 kg ha⁻¹ in China. These measures have effectively protected the environment and food safety.

15.4.3.4 Integrated Use of Water Surfaces

Bawujiu Farm has over 100 ha of water surfaces, such as lakes and ponds, and has adopted methods that integrate various uses of this resource. As an example, the farm raises crabs underwater, ducks and geese on the surface of the water, and pigs on the dike. The annual income of this farm from such integrated use of its water surfaces has risen to over 1 million yuan. Another type of water surface is rice paddy fields. Small grids of rice paddies were merged into a larger grid, which was not only beneficial for machinery operation but also created more wetlands area for wildlife. A reduction in the use of fertilizers and pesticides, an increase in the forest belt system, and the large area of water surface have attracted migratory birds, including swans, in the spring (Figure 15.4).

15.4.3.5 New Method for Livestock Production

The microbial fermentation method was adopted at the breeding center for Australian Holstein cattle at the Rongjun farm. Through the introduction of microbial fermentation processing technology and the use of fermentation bedding technology, excrement (e.g., cow dung, urine) can be decomposed into water and carbon dioxide by rapid oxidation, which produces a large amount of heat to warm up the cowsheds and keeps them cleaner. No water cleaning is needed, and zero water pollution discharges can be achieved. Animal waste also becomes high-quality organic fertilizer through this process.

Figure 15.4 Migrating swans at Huangye Farm wetland.

15.4.4 Construction of a Circular Agricultural Economic Cycling System

In recent years, the Heilongjiang state farms have explored the practice of developing a more circular economic system for agriculture. This so-called circular agroeconomy is based on the concept of recycle–reduce–reuse, with the intention of promoting the recycling and reuse of agricultural by-products and agricultural wastes such as crop straw and animal manure. Such an approach helps push agroecosystems toward more sustainable design and management in order to replace the linear input–output industrial model. This cyclical economy approach has been supported by implementation of the Circular Economic Promotion Law in 2009 in China. The positive results of this approach and practice in Heilongjiang Province and elsewhere in China should soon be available

15.4.4.1 Utilization of Renewable Energy and Biogas

The state farms have actively promoted 10 new energy and energy-saving projects, including rice-hull-generated power, biomass energy, solar energy, wind energy, and other clean energies. From 2006 to 2010, 30 renewable energy projects were undertaken. Among them, 15 large biogas generation projects were completed. With a total tank volume of 10,450 m³, they produce 2.86 million m³ of biogas annually. The 350-kW generators are able to produce 1.64 million kWh of electricity, and 9510 households are using biogas for cooking and heating. There are also several projects for straw gasification and establishment of a centralized gas supply. An experimental project for straw biogas generation is designed to provide biogas for 2300 households. These projects promote the use of renewable energy resources and the recycling of materials within farming systems (Figure 15.5).

15.4.4.2 Progress in Biomass Power Generation and Energy-Saving Projects

The Heilongjiang state farms generate large quantities of crop straw and by-products from forest trees. This biomass can be used as an energy resource to generate power by gasification methods. An investment of 20 million yuan has yielded 46 rice hull power generation plants with the ability to

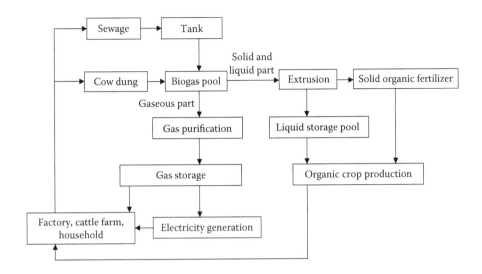

Figure 15.5   Reclamation area of Heilongjiang State Farms under the circular agriculture model.

generate 200 kWh of power. They can generate 552×10^5 kWh of power annually using 165.6×10^3 tons of rice hulls. These generation plants utilize all of the rice hull resources, reduce environmental pollution and production costs, improve economic returns, and improve social welfare.

Oil, rice, and dairy production in the state farms has achieved significant results by focusing on saving energy. For example, the Beidahuang Rice Industry Limited Company invested more than 20 million yuan to build a rice-hull heating plant and a biomass power generation plant to replace coal-based power. By using 150,000 tons of rice hulls each year, 15 million kW of power can be generated and 165,000 m^3 of area can be heated. This approach has reduced emissions of CO_2 and other greenhouse gases by over 200,000 tons each year. The company has also created benefit-returning mechanisms for farm workers, rice varieties suitable for processing, energy-saving and pollution reduction facilities, and research and development for new products. The Wandashan Dairy Company has modified its boiler system so it can use rice hulls, which saves 3000 tons of standard coal each year. Other power-saving renovations by this company include the use of solar energy for the preheating process and better insulation of cowsheds, helping to save 7 million kWh of electricity each year.

15.4.4.3 Recycling System

A recycling system has been established in the state farms. At Hailin Farm and others, the use of agricultural wastes is key to their eco-agricultural practices.

15.4.4.3.1 Integrative Use of Rice Straw and Rice Husks

- *Biogas method*—Recycling systems that move through such steps as pig (cow)–biogas–fruit (vegetable) or pig–biogas–cooking and heating fuel have been established. At present, 96.7% of the households at Hailin Farm are using biogas. Families with three people can save about 450 yuan per year. In 2008, Hailin Farm began to use biogas slurry for organic vegetable, grape, and rice production. At present, there are 150 ha^2 of organic rice, 8 ha^2 of organic grapes, and 50 greenhouses for organic vegetable production on this farm (Dong, 2008).
- *Generating electricity*—Using rice hulls and rice straw to generate electricity can generate low-cost power with a price of 0.3 yuan kWh^{-1}. Because the ash from burning the rice husks and straw is rich in silica, it can be used in rubber production and thermoplastic elastomer (TEP) production,

in addition to being used as high-quality fertilizer. A win–win situation for both the economy and the environment has been achieved. Biogas-generated power for industrial usage reduces production costs. In 2007, Hailin Farm invested 170,000 yuan to install a 75-kW biogas generator. At present, the generator is running normally and can save 10,000 yuan in electricity fees each year (Dong, 2008).

15.4.4.3.2 Use of Animal Waste

At the state farms, animal waste is used as organic fertilizer for crop production, and crop residues are used as animal feed for animal production. This is an important way to link animal production with crop production. Crop straw is also returned directly to the fields. The remaining crop straw and animal wastes are used for energy production. Animal wastes used for energy generation amount to 320,000 tons per year, and the amount of crop straw used for energy generation is 800,000 tons per year.

15.5 ECO-AGRICULTURE DEVELOPMENT ON HAILIN FARM

Hailin Farm is located in the southeast of Heilongjiang Province within Hailin City. The farm was founded in 1954, under the administration of the Mudanjiang Management Bureau, Heilongjiang Land Reclamation Bureau. There are 8733 ha of cultivated land within a total 17,533 ha of land area with a population of 7300. There are 3 agricultural management districts, 6 joint-stock enterprises, and 13 dairy farms. Hailin Farm is considered to be a medium-sized state farm and combines crop, animal, and processing enterprises. The largest breeding farm for Australian dairy cattle in Heilongjiang Province, Hailin Farm has 15,000 head of cattle.

The rapid development of animal husbandry on Hailin Farm produced a lot of manure and waste. The animal waste not only occupied a lot of land but also caused serious land, water, and air pollution. The environmental quality of the farm deteriorated, which seriously affected local farm production and the living conditions of farm workers. To solve this problem, farm leaders conducted an investigation in 2003. They visited biogas developments in Zunhua City in Hebei Province, Shunyi and Daxing County in Beijing, and Yanji County in Jilin Province. After that, a strategic plan was made to construct an environmentally friendly, energy-saving, and high-economic-benefit biogas system by comprehensively using local biomass resources.

The Hailin Farm biogas project is located on the southern part of the farm and covers an area of about 10,000 m². The main parts of the biogas project include raw material pretreatment, biogas digesters, biogas purification, gas storage, gas distribution, gas utilization, separation of residue and slurry, and utilization of solid residue and liquid slurry. This biogas system can process 24,000 m³ of cattle waste and generate 432,000 m³ of biogas each year, meeting the needs of 1000 households and saving 519 tons of standard coal each year. In addition, organic fertilizers in the form of biogas residue and biogas slurry reached 22,000 tons and 1050 tons each year, respectively (Guo and Dong, 2015).

In 2005, construction began on another biogas project located in the northwest part of the farm, covering an area of 15,000 m³. The total investment was budgeted at 4.58 million yuan. The volume of the biogas generation tank is 1900 m³, and it can process 80 m³ of cattle waste and generate 1200 m³ of biogas daily. It provides gas for 1000 households (Figure 15.6)

A large-scale biogas project utilizing an up-flow anaerobic solids reactor (USR) for fermentation has also been set up at Hailin Farm. The raw material comes from 3000 head of dairy cattle. It produces 2000 m³ of biogas daily, of which 500 m³ are used by 9000 households and the remaining 1500 m³ for power generation. The generated power is used by a glycoside production factory and to light public places such as streets and squares. The biogas residue is used for vegetable production inside the farm or is sold as organic fertilizer to the neighboring farmers at a price of 60 yuan/ton.

Figure 15.6 Hailin Farm biogas system.

Biogas slurry is mainly used in 200 ha of paddy rice production. In the off-season, biogas slurry is stored for the next cropping season. Only a small part of the surplus must be treated by aerobic measures before discharge (Guo and Dong, 2015).

In 2007, Hailin Farm further reformed and improved its biogas systems. The biogas system is now running at its full capacity. The reform and improvement of biogas systems included (1) increasing the temperature of raw material from a stevia processing factory, (2) increasing the temperature of the biogas tank using a solar heating device, (3) increasing the temperature in the input part of the biogas tank, and (4) building 200 greenhouses for organic vegetable production by using solid and liquid waste from the biogas tank. In the same year, Hailin Farm invested 3.5 million yuan to build a new biogas station in the Second Management Region. This new gas station can provide 219,000 m³ of biogas, 400 tons of solid residue, and 7000 tons of slurry. Biogas is used as fuel for residents in their daily lives and for generating power and central heating in this region. The solid and liquid wastes are used as organic fertilizer for organic rice and organic vegetable production. An attempt is underway to use rice straw as raw material for biogas production. The surplus gas generated from this second biogas station is transferred to the network of the first biogas station in order to benefit more people.

Recently, Hailin Farm has linked animal husbandry, crop production, and daily energy consumption of biogas (Figure 15.7). An economic analysis of the Hailin Farm eco-agricultural system shows that the total investment in biogas system construction was 4.5 million yuan and operation costs have been about 822,400 yuan each year, including a 10-year depreciation fee of 450,000 yuan per year. Annual income from the system is 834,900 yuan: (1) 219,000 yuan generated from 500 m³ of cooking fuel with a price of 1.2 yuan/m³, which is less than the market price of 2.5 yuan/m³; and (2) 615,900 yuan from electricity generated by 1500 m³ of biogas with an efficiency of 1.5 kWh of electricity from each cubic meter of biogas at the price of 0.75 yuan kWh⁻¹. The annual net income is 12,500 yuan, which does not include income from the use of biogas residue and biogas slurry. The price of gas calculated here is only half the market price. Because the quality of discharged water has been improved, the pollution emission fee charged by the government has been reduced. Because greenhouse gas emissions are reduced, some income can be generated if some biogas is sold on the greenhouse gas market. In general, eco-agricultural construction has achieved significant social, economic, ecological, and environmental effects (Guo and Dong, 2015).

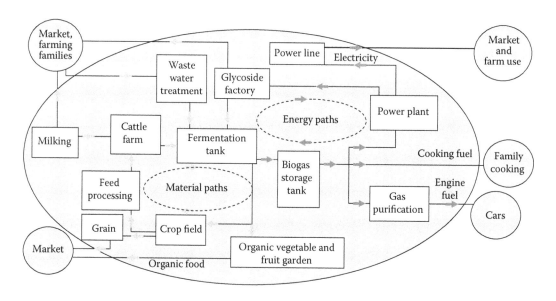

Figure 15.7 Framework of eco-agriculture on Hailin Farm.

REFERENCES

Dong, Q.Y. 2008. Great potential for recycling economy in state farms: investigation and implication of biogas development in Hailin Farm. *State Farm Econ. Manage.*, 3: 53–54.

Guo, B.S. 2007. Evaluation of the ecological benefits of the farmland's shelter belt in Heilongjiang land-reclamation area. *Forest Byprod. Speciality China*, 3: 31–32.

Guo, B.S. 2009. Practice and exploration in the construction of ecological civilization in the spirit of the Great Northern Wilderness State Farms. *Farm Econ. Manage.*, 3: 24–26.

Guo, X. and Dong, Q.Y. 2015. To develop circular economy and to develop low carbon agriculture: investigation report on circular economy based on animal husbandry in Hailin Farm, Heilongjiang Reclamation Region. *Agric. Modern.*, 426(1): 6–7.

Jiang, S.Q. 2006. Present situation and control measures of soil and water loss in State Farm of Heilongjiang Province. *Modern. Agric.*, 11: 26–28.

Qin, J.C. 2014. Exploration and practice of construction of ecological management in Baoquanling Branch. *Farm Econ. Manage.*, 5: 25–29.

Wang, H.B. 2013. To see the development of green agriculture in Heilongjiang reclamation area from the amount of chemical fertilizer. *Farm Econ. Manage.*, 8: 54–56.

CHAPTER 16

Developing Agroecology Practice in Baiquan County

Wang Shuqing

CONTENTS

Baiquan County, located in the eastern part of Qiqihar City Municipality, is in the transition zone between the Xiao Xing'an Mountains and Songnen Plain in the midwestern part of Heilongjiang Province in northeast China (Figure 16.1). There are 650,000 people in the county, among whom 515,000 are rural residents. Baiquan County is located within one of the four black soil regions in the world. Agriculture in Baiquan County is primarily conducted on the Songnen Plain. The main staple crops include corn, soybean, and potato. The main economic crops are sugar beet, sunflower, coarse cereals, and mixed beans. As a traditionally important agricultural county, Baiquan has had a glorious history. During the initial stage of reclamation in this region, the black soil was so fertile that it was said that "good yields can be obtained from these lands without any fertilization." People even vividly suggested that "you could hunt a deer easily just by throwing out a stick or harvest fish just by scooping up a cup of water from a river. Wild birds could even occasionally fly into your rice bowl." Baiquan County used to be a major grain-producing county renowned throughout the country; however, due to unreasonable cultivation methods and short-sighted behaviors, the soil and water erosion area accounted for less than 60% of the 3506-km^2 total area in Baiquan by the

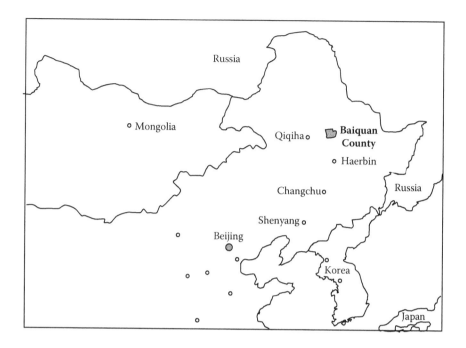

Figure 16.1 Location of Baiquan County in northeast China.

early 1980s (Zhou and Zhang, 2014). The 160,000 ha of farmland on sloped areas accounted for 70% of the 270,000 ha of arable land. Each year, 100 million m³ of water from rainfall and 14 million tons of soil were lost from this sloped land. The eroded water and soil removed 119,000 tons of nitrogen (N), phosphorus (P), and potassium (K). Each year, 6 mm of topsoil were washed away by water erosion and 4 mm of topsoil were blown away by wind erosion. The thickness of the black soil layer decreased from about 1 meter at the initial stage of reclamation down to only 20 to 30 cm after the 1980s. The soil organic matter content decreased from 8% to only 3 to 4% (Wang, 1995). The amount of soil eroded in the entire county each year was equivalent to the total topsoil in one village. Severe cases of desertification and basification occurred. There were 27,000 erosion gullies occupying 5070 ha of arable land within the county, which was equivalent to the arable land of a medium-sized township. If the ecological environment of agriculture were to continue to deteriorate like this, there would be no arable land within 200 years in Baiquan County. In fact, the deterioration of the ecological environment has led to economic poverty. Once a major grain-producing county, Baiquan became a poverty-stricken county surviving on government aid (Wang et al., 1996).

I grew up in Baiquan County. Since 1984, I served in leadership positions as deputy magistrate, deputy secretary of the CPC Baiquan County Committee, county magistrate, and secretary of the CPC Baiquan County Committee. During this time I paid close attention to the importance of promoting eco-agriculture development. When I served as a member of the CPC Qiqihar Standing Committee and then as deputy mayor of Qiqihar City from 2004 to 2007, I was able to apply the experience with eco-agriculture development gained in Baiquan to the entire city, which led to improvements in the ecological situation. I still serve as an ecological development consultant for the Qiqihar government and am actively working on relevant consulting and public education.

Eco-agriculture development has been occurring in Baiquan since the 1980s. After 30 years of unremitting efforts, the overall ecological environment has fundamentally changed in the county. In 1992, the county became the first one in China to develop artificial or introduced forests on more than a million acres in the plain regions and was honored by the provincial government. The county has also won many awards, including being designated as a National Advanced County in Water and Soil Conservation and a National Advanced County in Eco-Agriculture Construction, in addition to being included among the National Hundred Counties in Forestation and being named as an outstanding international green industry demonstration zone by the United Nations Industrial Development Organization (UNIDO). In recent years, Baiquan County has been utilizing ecological development to strengthen its agriculture, industry, and social services. The goal is to build an ecological, prosperous, civilized, harmonious, and beautiful Baiquan County. The effort so far has laid a solid foundation for the realization of a flourishing society. Great achievement has been made in both economic and social development in Baiquan County.

16.1 AN OVERVIEW OF AGROECOLOGY PRACTICE IN BAIQUAN

Before 1981, agricultural land reclamation destroyed forest and grassland and resulted in the ecological destruction of Baiquan. In the early years of the new China, the reclamation areas occupied 66.2% of the total land of Baiquan County. This number reached 74.8% in 1956 and 77.1% in 1980. Only 3700 ha of natural grassland remained. All of the unprocessed agricultural products were sold in the form of raw materials. Every year, 100,000 tons of soybeans and 160,000 pieces of leather were produced and directly sold to other places.

In 1979, multisite eco-agriculture demonstrations were conducted in Sandao Town, Baiquan Town, Jianguo Village, and Dazhong Village in Baiquan County. In 1986, an evaluation of the painful experience of severe soil erosion caused by reclamation activity, the poor economic returns, and the impact on these villages led the party committee and government of the county to make the decision to implement a strategy of eco-agriculture based on sustainable principles. These leaders realized that it was necessary to follow natural and economic laws and to adhere to the principles of holism, coordination, circulation, and regeneration in order to change the situation from one of confrontation to cooperation between humans and nature. It is necessary to respect nature in order to live in harmony with it and to apply socioeconomic and technical measures. Natural resources should be utilized but protected. An eco-economic system linking energy flow and material cycles between living biological communities and the non-living environment can provide ecological services and support sustainable economic development (Shi, 1995).

The agricultural development and ecological changes in Baiquan can be divided into six stages:

1. In the 1950s, grain production was the key activity and forests were destroyed during reclamation.
2. In the 1960s, farmers in villages were organized as production teams. The concept of "eight important aspects for crop production"—water, fertilizer, soil, seed, density, protection, machinery, and management—was proposed. The yield level in northeast China was required to reach the level in the Yellow River basin according to the National Agricultural Development Outline.
3. In the 1970s, attempts were made to change the situation of "three running-offs"—water runoff, fertilizer runoff, and soil runoff—to the "three conservations" of water conservation, fertilization conservation, and soil conservation. At that time, consideration was only given to reversing the bad situation, and our understanding of ecology was superficial.
4. In the 1980s, after implementation of the reform and opening up policies in China, reforestation was vigorously carried out and gullies and slopes were controlled. The integrative management of small watersheds was initiated. The course of development was changing, and ecological and ecosystem theory was used to guide the direction and the strategy of further development.
5. In the 1990s, further improvement of the environment was made in recovering the denuded mountains and cleaning the dirty rivers. An eco-economy and circular economies were put on the agenda.
6. After the year 2000, a scientific outlook on development was established and the construction of an ecological civilization related to society, economy, politics and culture was carried out.

The strategy of eco-agriculture development in Baiquan County began with identifying and analyzing the ecological and agriculture situation and then assembling an overall strategy for eco-agriculture development in Baiquan that combined biological, environmental, and social factors. Eco-agriculture patterns and economic structures that could adapt to various ecological situations and social development levels were proposed, and the resulting eco-agriculture development strategy was implemented in several stages. Qualitative, quantitative, and positioning goals were set up for each stage. In this way, the strategy became a series of practical actions.

With the application of the principles of ecology and systems engineering, we coordinated economic development and environmental processes. Under the guidance of agroecologists, we divided the county into four zones: hilly areas, semi-hilly areas, plains areas, and low-lying waterlogged areas along rivers. In accordance with the principle of "reforestation in mountainous regions, fishing in aquatic areas, animal production in valley regions, crop production on flat land, and industry located in the joining points," six eco-economic development patterns were established in Baiquan County: (1) integrated management of forest, grassland, orchard, animal production, and grain production; (2) system that cycled among animal, fish, and rice production; (3) maximizing the use of land resources in farmyards through the integration of grain fruit and animal production; (4) holistic management of sloped land, water bodies, farmland, forestland, and roadways in a watershed as a system; (5) integrated management of industry, agriculture, trade, and scientific research as a whole; and (6) eco-agriculture development for resource saving and environmental protection.

16.2 TEN MAJOR ACTIONS FOR AGROECOLOGY PRACTICE IN BAIQUAN

Ten eco-economic projects were prioritized for action to be taken: (1) water and soil conservation as well as drought and waterlogging control; (2) farmland shelterbelt development; (3) increased use of agricultural machinery and optimization of farming system; (4) improving cropping systems by filling ecological niches; (5) ecological animal husbandry; (6) eco-industry development; (7) eco-aquaculture; (8) integrative development of rural energy systems; (9) eco-economy development in farmers' courtyards; and (10) protection and conservation of ecology and environment. Implementation of these projects is described below (Wang and Su, 1995; Ma et al., 2005).

16.2.1 Integrative Control of Water and Soil Erosion

In order to stop erosion and avoid flooding disasters, efforts were focused on the recovery of eroded gullies, sloped land, and eroded watersheds to establish a solid foundation for eco-agriculture development. Baiquan is a county with serious water and soil erosion in the black soil area of the Songnen Plain. Over the years, the focus has been on controlling water and soil erosion. Residents were motivated to reverse the serious gully erosion. In 1975, the first terrace and first reservoir were built in the Tongshuang watershed located in Xinsheng Town. The first parcel of water conservation forest was also planted in this eroded gully. Integrative management of sloped lands, water, farmland, forest, and roadways within this watershed was carried out. Later this watershed became a model for other medium and small watersheds. During construction of the terrace and reservoir, a warning banner was posted that stated "For how many more years can our mother land stand for this serious runoff situation?" to raise awareness of the problem and to help highlight the serious crisis the county was facing. It was hoped that the ignorance and barbarism that existed at the time could be replaced with an ecological civilization.

The development of eco-agriculture in the small watersheds in Baiquan County went through four development stages: (1) current tillage and cropping practices were changed to contour farming; (2) terraces and infrastructure (e.g., road systems, irrigation and drainage systems) were added to the watershed; (3) an integrative forest, water, farmland, roadway, and slope management plan was developed for the watersheds; and (4) a period that can best be described as "to plant pines at the top of a hill like a cap, to grow leguminous vetch on terrace ridges like a belt, to grow grass instead of crops on slopes as land cover like a blanket, to build ponds in the bottom of a valley for fish farming, to raise ducks behind an erosion control dam, to grow rice in farmland outside the dam with irrigation, to plant fruit trees on the lower part of the hill, to build a shelterbelt grid on flat plains, to set up factories near towns, and to develop our economy in comprehensive ways" (Guo, 2001).

Biological, engineering, and agronomic measures were combined to manage erosion in gullies and slopes and to set up three lines of control to reduce water and soil erosion. The first line of control was slope protection, which was achieved by planting pines at the top of mountains and then digging interception ditches at the borders of forest and farmland to channel water running down along the slope. The second line of control was farmland engineering, which was achieved by planting shelterbelts, building terraces, and using contour cultivation to store water and preserve soil moisture through infiltration. The third line of control was valley engineering, which was achieved by building cascades to prevent erosion at the gully head, building check dams at gully bottoms, lowering slopes and planting willow trees along both sides of a gully, and preserving forest areas at the end of a gully to stop any remaining potential soil erosion. Terraces were built with the contour direction set according to the mountain terrain, the width of a field was set according to soil properties, and the distance between fields was set according to the steepness of the slope (Figures 16.2 and 16.3).

The development pattern of the Houjiagou watershed in Baiquan Township could be described as "forest on mountain; crops under mountain; vetch in between; geese behind check dam; rice outside check dam; chickens, pigs, and ducks on dam; fish inside reservoir behind dam; fruits, vegetables, and trees around house; and biogas used by each family." After years of continuous efforts, 21,000 erosion gullies and 162,000 ha of soil erosion area were brought under control in Baiquan, which had accounted for 80% of the eroded area of the entire county. Within these watersheds, 182 eco-economic zones were established that not only integrate resource protection and economic development but also combine green food production and tourism development.

Zhang Zhanxue, a farmer in the San Dao Town of Baiquan County, made use of the land, dam, and water to simulate the famous mulberry dike–fish pond circular system in the Pearl River Delta on 24 ha of waste gullies and slopes. He applied the concept of agroforestry by intercropping trees with various crops and grass on sloped areas. He planted willow trees and built a check dam at the gully

Figure 16.2 (See color insert.) Author Wang Shuqing standing before a newly developed agroecosystem with terraces and shelterbelt in Baiquan County.

Figure 16.3   (See color insert.) Well-developed agroecosystem with terraces, shelterbelt, and pond in Baiquan County. Author Wang Shuqing is standing in front.

head. He raised fish and geese behind the dam and grew rice outside of the dam. At the same time, he set up a cattle farm and a pig farm to consume crop wastes. Animal wastes were returned to the fields as organic fertilizer. He also set up a small food processing workshop. This optimized the application of eco-agriculture, made good use of land resources, and established an effective circular system by extending the food chain. His net income exceeded 200,000 yuan for many years, even without accounting for the "green bank savings" in forest cover. He was named a national model worker in 1994 by the State Department and All China Federation of Trade Unions (Ye, 1998).

Zhongxin Village, Sandao Town, Baiquan County, planted pines at the top of mountains and grew crops on contour terraces on the sloped land. In all, 280 ha of terraces, 12 cascades, and 42 check dams were built, 28 gullies were reformed, and 423 ha of forest were planted. The coverage rate of natural vegetation reached 32% in Zhongxin Village. Eco-agriculture increased the grain yield from 1312 kg ha^{-1} in the early 1980s to 5587 kg ha^{-1} in recent years. The annual income per capita rose from around 1000 yuan to around 6000 yuan. The annual income for harvesting the branches of willow (wicker) alone reached 165,000 yuan in this village. Since 1985, 8000 ha of waste ravines, slopes, and erosion gullies have been reformed in the county, and 3700 ha of the drought-tolerant legume *Lespedeza* have been planted for slope stabilization. Six willow weaving plants were established. Today, Baiquan County has 40,000 ha of terraces, 27,000 ha of horizontal buffering zones, and 40,000 ha of horizontal ridges. The diversified agriculture includes soybean, corn, sugar beet, trees, fruits, medicinal herbs, grass, and livestock.

16.2.2 Sustainable Use of Water Resources

Baiquan is located in the upland farming region of western Heilongjiang Province. There are no large rivers, and the groundwater level is quite deep. A scarcity of water resources has been a major limiting factor restricting development of agriculture and a rural economy in the area. A history of "nine spring droughts in ten years" (i.e., 90% probability of a drought situation in spring) gave rise to the saying that "the soil is so dry that the only thing we can do is to cry facing the sky." Due to serious soil and water erosion, 100 million cubic meters of valuable surface water were wasted every year in Baiquan. The continued development of Baiquan depends on its water resources. Whether it runs away or is preserved, the same water can produce entirely different scenarios.

Accordingly, a project known as "100 reservoirs, 1000 ponds, and 10,000 wells" was implemented at the foot of the Xiao Xing'an Mountains in the 1980s to help Baiquan to overcome its water scarcity. A priority was put on maximizing the area's water resources by combining a blocking method to keep water from running directly downward and a diversion method to channel water to suitable locations for storage and usage. Rainfall was carefully preserved, surface water used effectively, and groundwater pumped up while balancing the amount within each watershed and river system in Baiquan. To improve water use efficiency, water was saved during periods of surplus and then used during dry periods. Economic zones were developed according to the availability of water resources. In this way, a bad situation has been turned into a good one. Over the last 20 years, 10.5 million m^3 of soil have been reallocated by hydrological engineering projects. By 2001, the total number of engineering reservoirs and ponds reached 138 and 1352, respectively, and the water-holding capacity approached 240 million m^3. Irrigation areas for spring crops reached 130,000 ha and paddy fields for rice increased by 13,000 ha. The water stored in reservoirs and ponds, surrounded by green mountains and hills, have secured the county's economic development and provided bountiful harvests of rice and fish in the county (Xu et al., 2012).

To further save water resources, five water-saving patterns were developed:

1. In plains areas, pumping wells were rationally distributed as a network. A sprinkle irrigation system was developed over a large area to replace surface flooding irrigation.
2. In low, rolling hill areas, reservoirs and ponds were used to block running water. Deep pumping wells acted as a backup. Sprinkle irrigation was used in the integrative crop, animal, and fish production system.
3. In mountainous areas, pumping wells were built at the base of the mountains, and reservoirs and ponds were built on the middle part of mountains. Water was pumped up for watering orchards and gardens using sprinkle irrigation systems.
4. In lowland areas of a valley or along a river, the water storage, drainage, irrigation, and lifting techniques were combined.
5. In suburban areas, microdrip irrigation methods were adopted for intensive farming systems producing high-value crops.

After water-saving irrigation became established in Baiquan, farm irrigation water quotas were greatly reduced. The field irrigation quota was reduced from 3000 m^3 ha^{-1} for surface irrigation to only 1200 m^3 ha^{-1} for utilization of water-saving techniques, saving 1800 m^3 ha^{-1} of water (60%). According to data obtained at test stations, the utilization coefficient of irrigation water was increased from 33% to 86.9%. Water-saving irrigation techniques used on 75,000 ha of farmland saved a total of 130 million m^3 of groundwater each year in the county and effectively solved the problem of scarce surface water and dropping groundwater levels. In severe drought years, drought farmland accounted for as much as 85.1% of the arable land. Only crops grown in fields with water-saving irrigation flourished and provided a good harvest. During the severe drought year of 2000, each person owned an average of 0.13 ha of high-yielding fields with water-saving irrigation capacity. Crop production increased more than 40% that year instead of decreasing. Because the pressure of water scarcity has been removed, it has been possible to develop 4600 ha of fish ponds and 6700 ha of paddy fields, in addition to raising around 20,000 geese and 10,000 ducks in the county. Sustainable water resource management has also included converging and diverging methods. Heavy water flow can be diverged by terraces, reservoirs, irrigation channels, and cascades, thus avoiding potential flooding disasters. Separate, small water resources can be converged through conduction, interception, storage, and lifting, also eliminating the threat of drought disasters (Sun, 2011).

16.2.3 Development of Protective Forest Systems and an Integrative Forest Economy

Baiquan is one of the key counties within the "Three North" shelterbelt development project in China (Li, 1994). After 30 years of unremitting efforts, planted forest areas have reached 82,000 ha. The percentage of forest cover increased from 3.7% in 1978 to 22.1% in 1997. Living wood stock reached 5.49 million m^3, which was worth 5.4 billion yuan. The situation of "bare field, bare road, and bare village" without surrounding trees and woodlands has been fundamentally changed. A protective forest system has been established that is composed of trees, shrubs, and grasses and forms a network of shelterbelts and corridors. Residents of Baiquan have planted 2010 ha of economic forest, 28,400 ha of firewood forest, 12,000 ha of timber forest, 2700 ha of *Lespedeza*, and 36,300 ha of protective forest. The shelterbelt system is comprised of 10,629 grids, each 500 m by 500 m. The influence of the microclimate effects of the shelterbelts has been described by farmers as "like a cap on rainy days and like a wall on windy days." Wind erosion on the farmland has not occurred again within the 20 years after the forest network was established. In addition to the engineered reservoir, the vegetation cover constitutes a second reservoir, a biological reservoir.

It was also important to develop diversified communities within the forests. Depending on their different demands for light and nutrients, various tree, shrub, and grass species were configured to improve their resource use efficiency. For example, woody and herbal plants, leguminous and non-leguminous plants, light-demanding and shade-tolerant plants, and deep-rooted and shallow-rooted plants can all be grown within a community to provide mutual benefits for each other. Diversified forest communities formed by different plant species of different ages provide the good shade and high-humidity environment demanded by many medicinal plants, non-cultivated fruit and vegetable species, mushrooms, and wild animals. Animal husbandry can also be developed under these diversified plant communities, and the energy captured by plants can be further converted and utilized. Diversified forest communities offer a strong buffering capacity and self-regulation ability against adverse external impacts. Gradually, soil and water erosion have been greatly reduced. The nutrient return rate has increased and the ecological environment improved. Grain, timber, and fodder production has also grown.

Baiquan County is located on the edge of a desert. In order to suppress the occasional raging sandstorms in the plains area, six programs based on diversifying the plant community to prevent and control desertification were put into place:

1. *Manor approach*—Each manor, or homestead, with around 7 ha of farmland, an energy-saving house, and a water-lifting well is surrounded by a well-developed shelterbelt grid. A sprinkle irrigation system is installed in the field for watering crops. The former sand dunes become irrigated lands and the wastelands become fields with high and stable yields. The alkaline lands become oases.
2. *Plant barriers*—To prevent sandstorms from striking villages, shelter systems were developed with smaller grids and higher plant densities. The main tree species in the shelter system, *Pinus sylvestris* var. *mongolica*, is grown with other shrub and grass species. Such an approach can stabilize sand dunes and protect villages effectively.
3. *Irrigation systems*—Water pipelines were installed within each grid of the shelterbelt system to provide water for sprinkler irrigation.
4. *Development of forest products*—Based on their resource superiority, special varieties of *Hippophae rhamnoides* without thorns and with large fruits were selected to serve as the shelterbelt species. The fruit of *H. rhamnoides* is used to make jam, and the economic returns from the jam industry further aids in the fight against desertification.
5. *Landscape design*—All land resources used by a village are included in the landscape planning. Mountains and hills, water and irrigation systems, fields and villages, crops and livestock, forests and grasslands, roads and electricity are all considered. These elements are coordinated and handled as necessary for desertification control.

6. *Conservation*—Since 1989, a grassland enclosure measure for conservation and recovery of grasslands has been implemented. Farmers are subsidized to build fences, and local industries are developed to improve the life of local people who are no longer cultivating the fragile grasslands. A total
of 7100 ha of grassland have been fenced and effectively improved, further curbing the desertification trend of the former grasslands.

From 1998 on, forestry development in Baiquan entered a new phase of improving the structure
and increasing economic returns. Baiquan, with its 350 ha of newly expanded seedling nurseries,
has become an important seedling provider in northern China. Over 5270 ha of plantation forests
have been planted with evergreen tree species. In 2002 alone, for example, 158 ha of seedling nurseries and 1930 ha of *Pinus sylvestris* reforestation areas utilized 2.3 million tree seedlings. Baiquan
is developing planted forests that include economic species and *P. sylvestris* instead of being dominated by the more water-demanding poplars planted during previous reforestation projects. The
complex forest network mixes trees with bushes and conifer species with broadleaf trees.

For the poplar forests, an approach known as "reforestation by stump" has been adopted. After
clear cutting a mature poplar forest, the remaining stumps are covered with soil. Weeding is also
carried out. After new shoots emerge, they are thinned through a process of selecting only the strong
shoots and cutting any lateral and redundant shoots. Selling the shoots from thinning alone can yield
3000 yuan ha^{-1} which can cover the cost of reforestation. In the 39,000 ha of poplar forest in Baiquan,
this return has helped the county avoid investing a total of 140 million yuan for forestation efforts.
Poplar wood board production capacity has reached 160,000 m^3 a year, returning a benefit of as much
as 80 million yuan for farmers and providing over 35 million yuan in tax income for the government.
Because of the careful balance of forest growth and cutting, the wood board industries and poplar
forests are both developing quickly and seem to be sustainable (Zhang et al., 2006).

16.2.4 Integrating Animal Production with Crop Production

In the early 1980s, crop production in Baiquan accounted for 87% of the total agricultural output
while animal production was only 7.1% according to annual economic reports for the county. This
low level of animal husbandry not only constrained income levels but also became the limiting
factor for crop production due to the lack of organic matter in the soil. However, one year in the
late 1980s, 31 households in the Minle village of Limin fed 75 cattle and earned 97,260 yuan in
revenue. That same year, 39 m^3 ha^{-1} cow dung manure was applied to the farmland and crop income
reached 65,540 yuan. The revenue from non-agricultural work was 20,000 yuan. The total income
of the village reached 167,800 yuan, with per-capita income of 3054 yuan. The revenue from animal
husbandry accounted for 57% of the total income. Another example is a farmer in the Shangsheng
village of Jinbu. He worked by contract on 9 ha of farmland and grew 3 ha of forage crop. He also
owned a small plant for grinding rice and animal feed. In 1989, he raised 100 pigs and sold 70 of
them. Over 3 years, he sold a total of 201 pigs. One year he collected 416 m^3 of manure and applied it
to his soybean field as organic fertilizer at a rate of 45 m^3 ha^{-1}. His soybean field production reached
3420 kg ha^{-1} that year. He became famous in the county for earning a good return from both animal
and crop production. Another example is a farmer in the town of Dazhong who planted alfalfa on
her contracted land and also intercropped corn and the leguminous forage crop *Melilotus alba* Desr.
in her managed field. She raised two milk cows, two beef cattle, and two horses. One year she sold
6086 kg of grain and animal products, and her total agricultural income was 18,106 yuan. The livestock income alone was 13,800 yuan, 76.6% of her total income. These returns were much greater
than other farmers in Baiquan experienced.

These examples and others attracted the attention of county government. They clearly demonstrated a virtuous agroecology cycle involving more forage grass, more livestock, more organic fertilizer, more fertile soil, more crops harvested, more animal products, and more income. Agriculture

without animal husbandry is baseless agriculture. Efforts were concentrated on encouraging large-scale animal farming wherever possible. The animal husbandry of Baiquan is shifting from traditional ways to more modern production systems. Large-scale, standardized animal farming has increased sharply, replacing small-scale animal farms which had replaced farmyard production; also, special animal raising zones have replaced rearing animals within villages. The market-oriented animal industry in Baiquan is now managed mainly by companies located within special zones, associations organized by individual farmers, or villages specializing in animal husbandry.

During development of the animal industry in Baiquan, the concept of microecology feed (MEF) began to be promoted. The production of MEF involves an explosive process caused by a sudden drop of air pressure to loosen the internal structure of crop tissue and a fermentation process caused by suitable microorganisms. Whole corn stalks can be used in such a way to produce high-quality feed that offers not only rich nutrition but also large amounts of beneficial microorganisms and disease-fighting factors for animals. The use of such feed could eliminate the need for antibiotics and hormones, thus improving the quality of animal products. The animal by-products do not contain the residues of toxic and hazardous substances, and the meat products meet national green food standards, even organic food standards. The use of MEF allows crop residues to replace grass as animal feed, reduces grazing pressure on grasslands, and preserves vegetation cover.

A farmer in Fuqiang town who has been recognized as a national model worker makes wine and uses local corn as a raw material. He uses the dregs from the winemaking and stalks from his corn production to fatten cattle. This is a good example of increasing product value through processing, feeding, and recycling. One year, this farmer's number of fattened cattle grew to 1200 and his net income was 500,000 yuan. The five farmers working for him all used cow dung as fertilizer to cultivate their total 14 ha of farmland without any chemical fertilizer. Soil properties were improved and consecutive bumper harvests were achieved (Wang and Tian, 2015). Another farmer in Baiquan set up an ecological pig-raising farm. After collecting information on green food marketing and related techniques, he prepared his green feed for pigs by using abundant pollution-free and toxic-free maize. The annual output of pigs earning the green food label reached 12,000. The pigs were all sold to markets in large cities such as Shanghai, Harbin, and Daqing at a 4.8 yuan kg^{-1} higher price than non-green-food pigs. Another example is a farmer in the Changrong village of Furong who intercropped corn with *Melilotus*, an herbal leguminous crop, by using a 2:1 land ratio on his 0.37 ha of farmland. The yield of corn reached 13,500 kg ha^{-1} on the 0.25-ha cornfield, and the fresh forage from *Melilotus* reached 8000 kg ha^{-1} on the 0.12-ha *Melilotus* field. He fed two cows with fresh forage from *Melilotus*. The milk production increased to 2.5 kg per day.

In this way, the corn production belt is being converted into a milk and meat production belt, food processing belt, and generally a high-income belt. In the year 2000, 185,000 beef cattle were produced in Baiquan. That year, the output value of the developing cattle economy reached 255 million yuan. Baiquan County was identified as the "Hometown of Beef Cattle" by the Committee of Chinese Hometowns for Special Products. By 2001, the output value of animal husbandry in Baiquan reached 470 million yuan, which accounted for 41.1% of total agricultural output. To reduce the threat of frequent disasters and low agricultural yields, it is necessary to coordinate the ecological structure of crop, forestry, and animal husbandry and to establish a sustainable eco-economic foundation that can be complementary and reinforcing (Xu, 2007).

16.2.5 Crop Production Using Resource-Saving and High-Yielding Techniques

Baiquan County's science and technology promotion program focuses on saving resources, reducing pollution, producing high yields, and developing new techniques. Various techniques in crop production, such as plastic mulch for corn production, deep subsoiling and precision fertilization for soybean production, standardized operation techniques for wheat production, and low density and upland seedling techniques for rice production, have all been promoted in Baiquan. These have helped to dramatically increase crop yields and production efficiency. Because standardized methods for wheat and other major crops were developed specifically for the farmland situation in Baiquan, the adoption of these new methods of cultivation could help farmers to save inputs and increase crop yield. When a standardized method was applied to 23,000 ha of wheat in Baiquan, the yield reached 5300 kg ha^{-1}, and income increased 2200 yuan ha^{-1}. This application alone provided 15.78 million yuan more income to the county. Standardized planting of a 4500-ha area of soybean produced a yield that reached 2798 kg ha^{-1} and income increased 3000 yuan ha^{-1}, providing 19.87 million yuan more in income. A rice production technique for upland seedlings and low-density planting in a cool region was applied to 1900 ha of rice fields, increasing the rice yield to 6750 kg ha^{-1}. When the technique of transplanting sugar beet seedlings in paper containers was extended to 2000 ha, the sugar beet yield reached 45 ton ha^{-1}, more than double the yield for direct seeding.

The use of plastic mulch for corn production can increase the accumulated temperature in soil and prolong the growing season of corn. Plastic mulch improves water use efficiency by reducing evaporation. It can also reduce weed competition and increase corn planting density. When the plastic-mulch technology was applied to 7330 ha of corn, the average yield reached 12,000 kg ha^{-1}, much greater than the 4650 kg ha^{-1} yield achieved by direct seeding. The corn yield was increased by 144%. When a soil-less seedling method for corn production was applied to 20,000 ha, seedlings germinated and grew in a greenhouse up to the one-leaf stage before being transplanted (Zhao and Fu, 2012).

16.2.6 Multistep Processing and Utilization of Agricultural Products

Multistep processing systems for agricultural products were a consideration during the development of eco-agriculture in Baiquan. Such systems allowed superior production yields to be translated into superior market commodities. The added value of agricultural products increases with each step of processing. At least ten processing systems have been developed:

1. Soybeans to soybean products, including vegetable oil, instant bean flour, bean skins for cooking, and bean cakes for animal feed
2. Corn to corn products, including wine and alcohol, starch, and vinegar
3. Four-in-one system of pigs–biogas–vegetables within an energy-saving greenhouse, such that pig manure is digested in biogas tanks, biogas slurry is used as organic fertilizer, vegetable leaf waste is used as pig feed, and biogas is used as fuel
4. Potatoes to potato products, including crude starch, common flour, modified starches, potato chips, and French fries
5. Animals to animal products, including meat, eggs, leather, and wool
6. Underground bentonite to building materials and products
7. Wood to wood products, including composite board, fine woodboard, chipboard, furniture, knit processing products, and edible fungi
8. Sugar beets to sugar and alcohol (100,000 tons of sugar beet residues after processing can be used to feed 30,000 cattle)
9. Flax to flax products, including long or short fibers, medium-density hemp board, and cloth
10. Mineral water to products such as soda water

Higher corn production yields in the Baiquan village of Changrong were obtained through the use of mulching methods. In addition to harvesting the corn, the corn husks can be used as weaving material and corn stalks used for mushroom production. In Sandao, rice straw has been used as feed for cattle and for straw bag weaving. The straw bag weaving plant collects 275 tons of straw each year; production has reached 560,000 yuan yr^{-1}, and the plant employs dozens of people. The virtuous cycle between resources and products has been realized.

16.2.7 Crop Production Based on Adaptations of Organisms and Their Environments

A survey of the natural resources of Baiquan County provided a scientific basis for agroecological zonation and production zonation. There are considerable soil fertility differences throughout Baiquan. Farming system design should consider not only light utilization by the aboveground parts of plants but also the nutrition conditions below ground. The factors of environment, crop density, and characters of crop varieties should all be coordinated. In hilly areas where soil is poor, mixed cropping systems with low density are used. This can help reduce the problem of low soil fertility leading to a waste of solar energy. Intercropping *Melilotus* with corn using the mulching method and intercropping sugar beet with corn using the soil-less seedling method are widely utilized in hilly areas. In plains areas where soil fertility is in the medium range, sorghum, wheat, and sugar beets are the main crops wherever soil salinity is at a low level.

Agricultural production in Baiquan faces several environmental challenges, such as a short frost-free season, spring droughts, and summer flooding. It is necessary to choose suitable crop species and varieties that are adapted to these environmental stresses. Proper timing and spacing of the cropping community are essential for increasing the efficiency of resource use and to obtain maximum yields. Early autumn frosts and floods occur regularly along the river in Sanhe, Baiquan. For this area, a wheat variety (e.g., Longfu) with a short growing season is ideal. Soil moisture can be used during the early seeding stage, and the characteristic early maturing and harvesting can avoid the flooding in summer. Such an approach has changed a passive, unfortunate situation into a beneficial one. Another approach that makes good use of the flooding season is the cultivation of rice. In Baiquan, the town of Sandao was a pure upland cropping region before the year 2000. The grain yield was less than 3700 kg ha^{-1}, and the per-capita annual income was 2000 yuan. From then on, though, rice production has utilized upland seedling and low-density cultivation methods such that the available water resources have been utilized more effectively. The grain yield is close to 9000 kg ha^{-1} and the annual income per capita has approached 8000.

To increase biological diversity, suitable crops must be chosen and the environment must be properly managed. Greenhouse nurseries and protective cultivation techniques have been used to create suitable environments for various crops, thus allowing high-value varieties that require longer growing seasons but produce high yields (e.g., peanut, sweet potato, broccoli, Chinese medicinal herbs such as *Glycyrrhiza uralensis*, *Saposhnikovia divaricata*, *Gentiana scabra*, and *Oenothera biennis*) to be produced in Baiquan. Intensive horticultural cultivation has been encouraged in vacant farmyards to maximize the area's land resources as much as possible. Currently, 40,000 farmyards are being used in this way, and total land use in these farmyards has reached 3700 ha, providing a total annual income of 180 million yuan income for the farmers.

16.2.8 Agricultural Mechanization

It is necessary to explore how to combine machinery and land resources in order to better coordinate farm labor, production scale, and production techniques. The formation of cooperatives has helped to realize significant changes:

1. Machinery operations have changed from single operations such as cultivation or harvest to whole-season operations including sowing, irrigation, and spraying.
2. Property owned cooperatively has changed from being machinery only to also including land resources.
3. Machinery owned by cooperatives includes a variety of types.
4. Management of cooperatives has changed from addressing only agricultural machinery operation to also overseeing the use of land resources.

5. Government supervision and management of cooperatives has changed from being composed of only one department (e.g., agriculture bureau, agricultural machinery bureau) to encompassing multiple departments (e.g., bureaus of agriculture, agricultural machinery, water management, land resources, finance and commerce).

Currently, 273 farmer cooperatives have been set up. The land resources managed under these cooperatives have reached 150,000 ha. The 21 agricultural machinery cooperatives that have been established have equipment valued at more than 10 million yuan. The total power of large agricultural machinery has reached 381,900 kW, and 88.1% of field operations are now performed by machines.

Machine cultivation has changed from shallow surface tillage to deep harrowing. After deep harrowing, alternative row plowing, surface leveling, and raised ridge operations are utilized. Deep harrowing and loosening operations are essential to adjust the relationships among soil water, soil fertility, humidity, and temperature; to allow the soil to absorb more water; and to increase soil temperatures under the drought, frosty, and low-soil-moisture conditions in early spring in Baiquan County. When a tractor makes ridges, soil temperatures can increase. The temperature on the ridge surface has been shown to increase 6 to 20% (increased 1 to 3°C) as compared to the temperature on level surfaces. The physical, chemical, and biological traits in soil all improve, and the root systems of crops can expand deeper and more widely. After deep plowing, the volume weight of surface soil can be reduced 0.1 to 0.2 g cm^{-3}, porosity can increase 3 to 8%, and moisture content can increase 2 to 4% (Zhou et al., 2000). The drought resistance of the soil can be improved by capturing rainfall in the fall for use in the spring during drought conditions. Well-managed farmland soil is referred to as the "third reservoir" in Baiquan, the other two being the engineering reservoir and the biological reservoir. Soil fertility is carefully protected, actively improved, and reasonably utilized in Baiquan.

16.2.9 Developing Renewable Energy in the Countryside

Energy shortages in rural areas are an important issue that must be addressed in the development of eco-agriculture. A shortage of cooking fuel resulted in the destruction of vegetation by cutting trees, which caused more serious water and soil erosion. Another way to deal with the fuel shortage was to burn crop residue, straw, and stalks, which interrupted the material cycling path and caused a shortage of organic fertilizer for the fields and a shortage of feed for animals. An adequate supply of straw for fuel, feed, or fertilizer became a major problem. Diversifying energy resources and improving energy efficiency were both considered to solve rural energy issues while adhering to a policy based on specific local conditions, exploring multiple and complementary energy resources, encouraging integrative energy utilization, and improving energy efficiency.

In order to save energy, energy-saving electricity transformers were installed and the use of energy-saving bed heaters, coal-saving stoves, and transformed brick kilns was promoted. Such an approach helped to purify the air and reduced coal consumption by 200,000 tons each year, worth 80 million yuan. The straw and root stubble of crops were crushed and returned to 80,000 ha of farmland, effectively increasing organic matter in the soil (Zhang et al., 1996).

In an effort to explore more energy resources, fuel wood forests were expanded, the use of solar energy grew, and biomass energy was further developed. In addition to solar greenhouses and solar-heated pig houses (refer back to the four-in-one system of greenhouses–pigs–biogas–vegetables, mentioned earlier), the use of biomass energy was expanded as a result of Jiangsu Xintiandi Environmental Science and Technology Company investing in Baiquan and setting up the Heilongjiang Shengyan New Energy Development Company, which developed a biomass energy project requiring a total investment of 2.5 billion yuan. It produced 400,000 tons of compressed straw briquettes, a type of solid biomass fuel. This company also helped to reform 100,000 biomass stoves in households, to build a biomass-based heat and energy plant with a 24,000-kW capacity, and to set up a 100,000-ton capacity plant to produce methanol, ethanol, and other short-chain alcohols.

Over 82,000 hectares of planted forest can produce 2.1 million m^3 of woody biomass in the mature and over-mature forests in Baiquan. Also, there are 27,000 ha of fuel wood forest, and 1,193,000 tons of crop straw can be harvested from farmland. These resources are enough to ensure the annual production of 400,000 tons of solid fuel briquettes. This biomass fuel can replace 266,600 tons of standard coal, thus reducing 7800 tons of dust emissions, 1578.6 tons of sulfur dioxide emissions, and 152,500 tons of carbon dioxide emissions. Meanwhile, the fuel can provide revenue for farmers of 92 million yuan, which amounts to 939 yuan per household. This biomass fuel project created 3500 jobs and has provided 54 million yuan in annual salaries. Further development of this project will enable Baiquan to more widely use biomass energy as a source of heat, electricity, gas, and alcohol. It is expected that eventually all energy consumed in Baiquan will be biomass energy (Wang, 1995).

16.2.10 Improving Urban Ecological Environments

The principles guiding the improvement of the ecological environment in Baiquan include (1) people first, (2) combining natural and human factors to form a comfortable environment, and (3) coordinating infrastructure, industry, and tourism. The county has focused on landscape design and developed a round city shelterbelt system, round city park system, round city drainage system, and round city road system in Baiquan Town. The networks of forest, water channels, and roads are combined together. The average vegetation area for each resident in Baiquan Town has reached 40 m^2. Currently, the natural landscape and cultural landscape are interdependent and mutually beneficial. Forests are hoods on the mountain tops. Buildings are wound around the slopes. Factories are concentrated on one side. This landscape looks like a town within a forest, but also like a forest within a town. Buildings can be found among the green vegetation, and the people are living among flowers. An eco-city next to a river in North China with a cold climate has arisen in Baiquan.

16.3 STRENGTHENING THE AGROECOLOGY PRACTICE MANAGEMENT SYSTEM

The current management system supporting agroecology practice is far from perfect in China. Effective prevention measures were not established early on, and neither responsibility nor compensation for unfortunate consequences has been established. Some prominent problems include the following (Wang, 2000):

1. A legal system related to eco-agriculture has not been established. Funding in support of eco-agriculture is not adequate. Eco-compensation levels are low and not sustainable. This situation actually encourages adverse short-term agricultural behaviors such as relying heavily on chemical fertilizers to increase yield and on pesticides to control pests, seriously impairing the basis for sustainable agriculture development.
2. A comprehensive decision-making mechanism for environmental protection and economic development has not been well developed. A mechanism for public participation in eco-agriculture development has not been established.
3. The enforcement cost of environmental law is very high; however, the penalties for violating environmental law are usually small. A monitoring and management mechanism is not well established.
4. An evaluation system for the performance of government leaders in many locations places too much emphasis on economic development indicators and pays little attention to eco-environmental indicators. Relying on the concept of "gross domestic product (GDP) first" to guide social and economical development has not fundamentally changed and has resulted in intense economic development characterized by high resource consumption, low utilization rates, and serious environmental pollution.
5. The understanding of eco-agricultural development is far from sufficient. The significance of a well-developed eco-agriculture management system for environmental protection, effective utilization of resources, and sustainable development has not been fully appreciated.

In order to ensure the smooth progress of eco-agriculture, policies and regulations must be improved:

1. Resource consumption, potential environmental damage, and ecological services required should be evaluated with regard to economic and social development, together with relevant standards, assessment methods, and incentive mechanisms that reflect the requirements of eco-agriculture.
2. Protective mechanisms for agricultural resources and the environment should be established, and farmland protection systems, management systems for water resources, and environmental protection systems should be strengthened.
3. A combination of assessment and guidance for government actions should be enforced. An initial evaluation, mid-term supervision, and follow-up inspection, tracking, and rectification should be conducted.
4. Farmers should be fully mobilized for the construction of eco-agriculture.

Only in this way can we follow the public will to make our plans, rely on the capacity of our people to set up actions, and rely on public wisdom to improve supervision and management measures. The following actions have strengthened the management system for eco-agriculture in Baiquan:

1. Assessment standards for leaders have been modified to reflect more than simply the GDP. Target standards, assessment methods, and incentives reflecting the requirements of eco-agriculture have been established. The appraisal system is the baton helping to change the concept and behavior of people. Production levels are no longer the only standard for assessment. Long-term development potential and the quality of development have become more important. Deng Xiaoping said, "It doesn't matter whether it is a black cat or a white cat. The cat that can catch a mouse is a good cat." The "cat" destroying the eco-environment is definitely not a "good" cat. The collectively owned mountains cannot become the property of only a few persons. If arable land is no longer able to be cultivated, the leaders in charge should no longer be able to be cultivated (i.e., promoted). If the earth is wounded, the leaders in charge must also be wounded in some way. More than 80 people have been promoted due to their outstanding achievements in the construction of eco-agriculture over the last 30 years at the county, town, or village levels in Baiquan; 10 people have been subject to disciplinary or criminal penalties.
2. An eco-agriculture office was established in Baiquan, and management responsibilities were carefully allotted, together with administrative resources and financial resources. This helped to effectively solve prominent problems affecting resources, the environment, and ecological security.
3. Legislation regarding and law enforcement of eco-agriculture have been strengthened. Judicial procedures are used to bring charges against persons who pollute the environment and damage the county's ecological foundations. This has helped Baiquan County to realize social fairness and justice and promote the sustainable development of ecosystem, economy, and society. The county has further developed economic policies and regulations for enhancing ecological services. The minimum ecological threshold for a company to set up in Baiquan County was increased, and the minimum ecological standards for criminal charges to be brought against a company when violated have been raised. A lifelong accountability system for ecological environmental damage has been established and a penalty system implemented.
4. The county has tried to improve the role of markets in eco-agriculture development. Pesticides and fertilizers are being used less to protect the land from non-point-source pollution. An ecological compensation system was established to encourage a reduction in the use of pesticides and fertilizers. Such resource-saving policies have put pressure on the area's industries to transform and upgrade their structure, hence changing the course of economic development.
5. A 100-year development plan for Baiquan County has been put into place and clearly defines the ecological space boundaries for residential areas, industrial areas, urban built-up areas, rural residential areas, basic farmland, farmland shelterbelts, water bodies, and wetland areas. Regulations on the various uses of land were established. Based on the principle that the mountains, water, forests, fields, and lakes are all linked together as one community, land use management systems for eco-agriculture throughout the county have also been established. Any changes in arable land are

strictly controlled, as are changes in other ecological land uses such as natural grasslands, woodlands, river channels, reservoirs, wetlands, and riverbanks. Changes from ecological land use to constructed land use are strictly controlled to ensure that the ecological space in the county is not reduced. The plan includes monitoring and early warning systems for any changes in the carrying capacity of the county's resources and environment. These can prevent the serious and irreversible consequences caused by excessive development. Further industrialization, urbanization, information revolution, and agricultural modernization in Baiquan County will be realized only within the framework of an ecological civilization.

16.4 AGROECOLOGY ACHIEVEMENTS

History proves that the development of eco-agriculture is the path toward enriching Baiquan County. Runoff from sloping fields has been reduced by 78% and sediment loss by 88%. The organic matter in the soil has improved by 0.51%. The air humidity has increased by 10 to 14%, and the wind speed has dropped 58%. Wind erosion has been greatly reduced such that no wind stripping has occurred recently. Soil erosion has been reduced from an average of about 4000 t km^{-2} yr^{-1} to the current 1650 t km^{-2} yr^{-1}. Over 67,000 hectares of planted forest in Baiquan are absorbing 24 million tons of carbon dioxide and releasing oxygen at a rate of about 17 million tons each year. The GDP of the county reached 5.5 billion yuan in 2012, and per-capita income increased to 6500 yuan. The grain output increased from 1350 kg ha^{-1} in 1980 to 6750 kg ha^{-1} in 2012. Total grain output was stable at 1 billion kg for many years and exceeded 1.5 billion kg in 2012. In 2013, grain output reached 2 billion kg and per-capita income reached 8000 yuan.

Currently, the 2.06 million m^3 of growing stock of mature forests in the county have the potential to bring in revenues of 1 billion yuan. After repaying all of the village-level debts, the remaining wood stock is equivalent to 3 million yuan being deposited in each village. The top ten eco-economic industries have become pillars of the county's finances. The ecological, economic, and social benefits have all increased.

When a great flood occurred in the northeast in 1998, potential losses in Baiquan were reduced by 750 million yuan thanks to the development of eco-agriculture. In the year 2000, areas surrounding Baiquan endured a great drought while Baiquan experienced only a partial drought (Hang et al., 2011). A documentary entitled "Away from Disaster" told the story of eco-agriculture in Baiquan and was shown at the World Summit on Sustainable Development in 2002. In 2003, a great drought occurred again in Qiqihar, but a good harvest was reaped in Baiquan and the county's revenue remained high. In 2013, an article entitled "Forestation in the Three North Areas" recorded the experience of eco-agriculture development in Baiquan (Li et al., 2013).

In 1992, Baiquan became the first county with a million acres of planted forest on plains areas and was given an award by the provincial government. Former Vice Premier Tian Jiyun wrote the inscription "greening motherland, transforming landscape" for Baiquan. Since then, Baiquan has been awarded the National Advanced Unit for Soil and Water Conservation, National Advanced Unit for Forestation, Heilongjiang Province Governor Special Award for development of eco-agriculture, first prize for International Ecological Engineering at the International Ecological Engineering Conference, and Third Global Award. The county has been recognized as a National Top Ten Pacesetter in Forestation, National Advanced County in Ecological Agriculture Development, and National Advanced County in Soil and Water Conservation Project. Baiquan has been identified by UNIDO as an International Green Industry Demonstration Zone, and in 2012 Baiquan was named the National Ecological Civilization County in Soil and Water Conservation. The construction of eco-agriculture in Baiquan has passed the long-term tests provided by reality, history, and its people.

ACKNOWLEDGMENT

Thank you to Professor Luo Shiming from South China Agricultural University for his help with translating and editing the original draft.

REFERENCES

Guo, S.T. 2001. Hope of Chinese eco-agriculture, inspiration of eco-agricultural construction in Baiquan County of Heilongjiang Province. *Chin. J. Eco-Agric.*, 9(3): 105–107.

Hang, Z.J., Chen, S.G., and Zhong, X.H. 2011. Discussion on ecological agriculture construction and sustainable development. *Heilongjiang Environ. J.*, 35(1): 3–5.

Li, C.J., Liu, S.Y., Le, K.Y., Bai, R.X., and Han, B. 2013. *Forestation in the Three North Areas*, http://news. xinhuanet.com/politics/2013-09/25/c_117508134.htm.

Li, J.S. 1994. The state of development of the protective forest systems in "Three Norths" and some countermeasures. *World For. Res.*, 1: 64–69.

Ma, L.H., Zahng, Y.L., Zhang, W.Z., and Meng, F.X. 2005. Soil and water conservation and ecoagriculture development in Baiquan. *Heilongjiang Environ. J.*, 29(2): 56.

Shi, S. 1995. It wakes us up: a brief introduction about the achievement, future and significance of ecoagriculture development in Baiquan County. *Agric. Environ. Dev.*, 43(1): 9–11.

Sun, A.L. 2011. Study on Water Technique of Maize for Alternative Furrow Irrigation in Semi-Arid Region of Western Heilongjiang Province, master's thesis, Northeast China Agricultural University, Harbin.

Wang, K.J. and Tian, D.F. 2015. An old model worker, a newly created business. *Qiqihar Daily*, May 8.

Wang, S.Q. 1995. The strategy and practices of ecoagriculture development in Baiquan County. *Eco-Agric. Res.*, 3(4): 76–78.

Wang, S.Q. 2000. The achievement of Baiquan County in eco-agriculture development. *Local Econ.*, 10: 53–54.

Wang, S.Q. and Su, J.C. 1995. The proposing and implementing ecoagriculture development strategy in Baiquan County. *Agric. Environ. Dev.*, 43(1): 5–8.

Wang, S.Q., Wen, J.R., and Wang, G.Q. 1996. To construct eco-agriculture for soil and water conservation and to promote county economic in Baiquan. *China Soil Water Conserv.*, 5(3): 9–12.

Xu, J.H., Song, J.F., and Sun, X.W. 2012. Review on the measures for water runoff in watershed located in high latitude black soil region. *Appl. Tech. Soil Water Conserv.*, 2012(1): 37–38.

Xu, N. 2007. From eco-agriculture development in Baiquan County to ecological development in Heilongjiang Province. *Heilongjiang Grain*, 5: 14–18.

Ye, C.H. 1998. The leading person in management and use of "waste land." *Water Conserv. Irrigation*, 10: 20.

Zhang, R.W., Zhang, Y.Z., Ji, W.Y., and Zhang, T. 1996. Reasonable use of resources is an effective way for sustainable development of agriculture. *Agric. Environ. Dev.*, 47(1): 36–39.

Zhang, W.L., Yu, M., and Yang, X.G. 2006. On the "three north" protective forestry construction and its development trend in Baiquan County. *Protect. For. Sci. Technol.*, 74(5): 67–68.

Zhao, X.H. and Fu, J.Y. 2012. Cultivation technique for soilless seedling from greenhouse. *Agro-Technol. Serv.*, 29(10): 1093.

Zhou, N. and Zhang, C.S. 2014. Effects and experience of ecological civilization in soil and water conservation in Baiquan County. *Soil Water Conserv. China*, 1: 55–57.

Zhou, S.G., Li, C.H., Chang, S.M., Lian, Y.X., and Liu, K. 2000. Effects of ridge culture on summer maize ecological environment and growth development. *J. Henan Agric. Univ.*, 34(3): 206–209.

PART **IV**

Perspective

PART IV

Perspective

.

CHAPTER 17

Development and Prospect of China's Eco-Agriculture—Agroecology Practice

Wu Wenliang, Li Ji, Wang Jian, Zhao Guishen, Du Zhangliu, and Liang Long

CONTENTS

17.1 EMERGENCE AND PROSPECTS FOR CHINA'S AGROECOLOGY PRACTICE

By adapting to their actual conditions, many countries in the world have set up their own strategies for developing ecologically based agriculture, and these deserve our attention. In China, these strategies have been guided by a "scientific approach to development" and adhere to the sustainable development strategy supported by the central government (Zhang, 2004). Agriculture in China is also following the idea of developing a modern agroecology practice referred to as eco-agriculture, which features "comprehensive management of agricultural issues, transformation of major sectors, reuse and recycling, input control, clean production and safe output" to promote the ecological transformation of conventional agriculture (Li and Lai, 1994; Li, 2003, 2008; Luo, 2007). The term "eco-agriculture" used in China is essentially the same as the term "agroecology practice," which is used in the United States and some other Western countries, and "environmentally friendly agriculture," a term used in eastern Asia countries.

17.1.1 Agroecology Practice Is an Inevitable Choice for China's Agriculture

In China, the development of agroecology practice (or eco-agriculture) has undergone three stages (Li et al., 2005; Luo, 2010): (1) beginning and exploration stage, (2) testing and development stage, and (3) diversified and integrated development stage. The first stage began in the late 1970s and continued to the early 1990s. During this stage, the concepts of eco-agriculture were introduced within China's academia. The basic principles of eco-agriculture were primarily described, and many village-level eco-agriculture practices and explorations were launched nationwide. The second stage began in the mid-1990s and lasted until the beginning of the 21st century. The major achievements of this period were the clarification and popularization of the basic concepts of eco-agriculture; numerous research papers were published, which helped to initially formulate the theoretical foundation of agroecology practice with Chinese characteristics, or eco-agriculture. Meanwhile, supported by a national eco-agriculture development project, a number of county-level eco-agriculture pilot projects were established. The first decade of the 21st century marked a steady development of eco-agriculture in China, represented by biogas projects, formula fertilization, clean production, biodiversity conservation, and other key projects. Meanwhile, eco-agriculture industrialization is growing quickly. Since 2010, China has entered a stage of rapid industrialization and urbanization, and such a social transformation has resulted in huge problems and challenges for China's agricultural modernization. As a vulnerable and fundamental industry, agriculture is confronted with multiple risks from society, nature, and the market, while the ecological transformation of agriculture is facing an even more complicated historical situation. Despite this, the trend of ecological transformation is unstoppable. Ecological protection and development are essential for fast-growing but extremely unbalanced countries such as China.

Before the 1950s, China's agriculture was known as a self-sufficient traditional agriculture. In the 1960s, within the context of industrialization, the transition toward conventional (or industrialized) agriculture began. During the 1970s and 1980s, China's agriculture focused on modernization and was characterized by high inputs of fossil energy and other agro-inputs, as well as the pursuit of high economic efficiency and high product output (Zhao et al., 2007). Although this process guaranteed a national supply of agricultural products, it also resulted in ecological damage, environmental pollution, safety problems with agricultural products, excessive consumption of resources, and other issues. Some local and national environmental problems were exacerbated, including increased pollution of local environments caused by excessive application of fertilizers and pesticides, overexploitation of water resources due to the significant increase in agriculture irrigation, dropping groundwater levels, excessive land reclamation and conversion to agriculture, and downgraded carrying capacity of the environment caused by overgrazing. All of these problems posed a substantial threat to the sustainable development of agriculture and thus threatened healthy and sustainable socioeconomic development. Consequently, these problems gained the attention of scientists and extension workers in the field of agriculture, resources, environment, and ecology as well as managers of related industries and policymakers. They began to rethink the future of agricultural development in China.

Currently, the transition from conventional agriculture to eco-agriculture is based on the trend of China's economic growth and characteristics of agricultural development. The transition is an inevitable choice for the future, thus it is imperative to establish a set of agroecology practices that incorporate proper management and protection of the ecological environment as well as conservation and efficient use of resources for agricultural production and rural economic development. In an "integrative, harmonious, recycling, and regenerative" agriculture, there is a need to utilize both traditional and modern agricultural wisdom and to emphasize structural optimization among crop, forestry, animal husbandry, ancillary production, and fisheries within an ecosystem. The larger goal of sustainable agriculture development should also be combined with the goals of developing business and alleviating poverty in order to establish a comprehensive management system that will be able to coordinate different levels, different professions, and different industrial sectors.

17.1.2 Theory and Practice in the Ecological Transformation of International Agriculture

Since the 1960s, many countries around the world have investigated ecologically based agriculture. The Western concept for agroecology practice is closely aligned with organic agriculture at its earliest stages. The origin of agroecology practice can be traced back to Europe from the 1920s to the 1940s, when organic farms appeared in Switzerland and the United Kingdom. In the 1960s, many of the farms in Europe turned to ecological farming. In the 1970s, other countries, including Japan in Asia and the United States and Canada in North America, began to explore ecological agriculture, thus spurring on the beginning of an ecological transition in agriculture. By the 1990s, developed countries around the world began to move segments of their agriculture toward agroecology practice, eco-agriculture, or environmentally friendly agriculture.

Most developed countries have put forward some programs that support the transformation from conventional agriculture to environmentally friendly agriculture. In many ways, these could be defined as modern agroecology practice, or modern eco-agriculture, but China has yet to establish harmony between agriculture and the environment. In fact, agriculture has served multiple functions in terms of product service, environment service, and cultural service since the early days of Chinese civilization. Modern industrialized agriculture looks only at high productivity, which will surely come at the cost of ecological integrity, which in turn will harm the sustainability of productivity in the long run. Overlooking the landscape and cultural heritage functions of agriculture will push agriculture and rural areas to the edge of recession, striking a heavy blow to the cultural and social basis of the country. The question is whether it is a key time for China to move in ecologically sound directions. What are the risks that come along with such a transformation? South Korea and Japan achieved a per-capita gross domestic product (GDP) of over US$3000 in 1987 and in 1973, respectively. They initiated the transformation toward environmentally friendly agriculture in 1994 and 1992, respectively. They were both influenced by the "Den Bosch Declaration on Sustainable Agriculture and Rural Development" presented at the 1991 International Conference on Agriculture and Environment.

China's per-capita GDP exceeded US$3000 in 2008 and reached US$6767 in 2013. Because the global situation is changing rapidly and the 25th anniversary of the Den Bosch Declaration is approaching, it is now suitable for us in China to take the critical step toward eco-agriculture. In terms of the production function, eco-agriculture does not necessarily decrease food production. Major economies such as the United States, European Union, Australia, and Japan have not reduced food production due to the implementation of environmental policies. In fact, the promotion of eco-agriculture has allowed these countries to make full use of world trade liberalization to obtain food products at a lower cost while avoiding many trade barriers imposed by the World Trade Organization (WTO) through their environmental policies or green box policies, thus achieving a smooth transition to a more sustainable agriculture. Lessons to be learned from the successful transformation going on in developed economies will help us design a policy framework for agricultural development over the next several decades. Meanwhile, pilot programs involving related industries, technologies, and policies should be launched so as to develop feasible pathways and solutions in China.

All of the terms used for ecologically based agriculture—agroecology practice, eco-agriculture, or organic agriculture in the United States; bio-agriculture, eco-agriculture, biodynamic agriculture, or low-input agriculture in Europe; and natural farming or alternative agriculture in Japan and South Korea—reflect many similarities and approaches that can all be generalized as agroecology practice, or eco-agriculture. A primary purpose of alternative agriculture in developed countries is to provide solutions for resource and environmental problems caused by conventional industrial agriculture. In this regard, developed countries attach great importance to building protective agricultural ecosystems. They design and manage agroecosystems that function on the basis of returning high levels of organic matter to the land and using crop rotations. These approaches have encouraged incorporating legumes, green manure, and straw; the use of biological prevention and control methods for the control of insects, diseases, and weeds; prohibiting or limiting the use of synthetic chemicals, such as pesticides, fertilizers, and hormones; and reducing energy consumption and costs.

17.1.3 Major Approaches of Modern Eco-Agriculture: Industrialized Eco-Agriculture

As industrialized eco-agriculture has begun to develop quickly, many scholars have conducted comparative research on what this term means. Eco-agriculture usually has operated as an independent industry sector in accordance with marketing rules and requirements. Through efficient organization and industrialized operation, and supported by modern science and technology, eco-agriculture could become more market oriented and farmer based by working with numerous processing or cooperative economic units; by targeting higher profit through the intensification of production factors such as capital, technology, land, and labor; and by creating various links before, during, and after the production process of eco-agriculture. With ecological conservation as the precondition, and marketing as the direction, industrialized eco-agriculture offers comparative advantages in terms of ecology, location, and products. It also focuses on the production of high-quality, safe, and pollution-free products to build a network that integrates production, processing, and marketing so as to enhance overall productivity and achieve sustainable development of rural economies while still being eco-friendly. An industrialized eco-agriculture system can help to improve the value of products and the comprehensive competitiveness of the industry, thus achieving a much better balance among ecological, economic, and social benefits in agriculture.

17.1.4 Characteristics of Eco-Agriculture

17.1.4.1 Basic Concept of Eco-Agriculture in China

Chinese eco-agriculture pays great attention to building up an agricultural production system by reviewing and adopting the successful experiences of different agricultural production systems where the application of modern science and technology is linked with ecosystem engineering approaches. It is based on the principles of ecology, eco-economics, biology, and material recycling under the guidelines of economic and environmental co-development. This approach combines agricultural production and rural economic development with improvement and protection of the ecological environment, natural resource nurturing, and their efficient utilization. It requires the integrated development of food production and cash crop production. It balances production among crops, forestry, animal husbandry, and fisheries, as well as among primary agriculture and secondary and tertiary industries. Chinese eco-agriculture aims at promoting the sustainable development of agriculture, rural economy, and society and sticks to "integration, harmony, recycling, and regeneration" as its basic principles. Chinese eco-agriculture coordinates these relationships through constructed ecological engineering designs so as to form a "virtuous circle" encompassing ecology and economy that promotes sustainable development of the agriculture industry. It emphasizes the optimization of production scale and embraces micro-business and the alleviation of poverty among farmers as part of its strategic goal of development. It establishes an integrated management mechanism to coordinate among departments of different levels, different professions, and different sectors.

Compared with conventional agriculture, eco-agriculture is a socialized agricultural practice that features wide application of modern science and technology, as well as use of the production materials and scientific management methods provided by modern industries. However, it focuses on the development and recycling of natural resources, pays attention to ecological effects, and gives full play to the ecological functions of agriculture. Modern eco-agriculture is a new type of development mode based on both traditional agriculture and conventional agriculture. It integrates production, operation, distribution, and consumption through eco-labeled products and balances ecological, environmental, economic, and social benefits. In a word, it is the eco-friendly and environmentally friendly version of modern agriculture. In China, eco-agriculture allows what is called "reasonable" use of fertilizers and pesticides that are forbidden in eco-agriculture and organic agriculture in other countries. China has set up a healthy food program that is unique in the world and includes pollution-free agricultural products, green food, and organic food.

17.1.4.2 Main Features of Chinese Eco-Agriculture

First, Chinese eco-agriculture is coordinated with the development of food security strategies for the rest of Chinese agriculture. The development of modern eco-agriculture strives to be an effective path to the sustainable development of agriculture that gives full play to the macro-function of agroecosystems and adjusts and optimizes agricultural structure in comprehensive planning. Adhering to the principle of "integration, harmony, recycling, and regeneration" and from the starting point of mega-agriculture, it strives to realize the balanced development among agriculture, forestry, animal husbandry, and fishery industries and the coordinated development among the first, second, and tertiary industries in rural areas so as to promote the transformation of China's agricultural development mode. On the other hand, by respecting industrial layout, circulation, and regeneration, there is a greater chance of eco-agriculture being able to guarantee a solid foundation for food production and therefore ensure national food security.

Second, Chinese eco-agriculture has distinct regional characteristics. Faced with the vast territory of China and differences in local natural conditions, resources, and economic and social development throughout the country, the development of eco-agriculture fully absorbs the essence of traditional agriculture. It also makes the most of modern science and technology to equip agricultural production with a variety of ecological modes, ecological engineering, and rich types of technology so that regions can draw upon each other's strengths and give full play to their own regional advantages and so that various industries can develop in response to social needs and local conditions.

Third, eco-agriculture is a multifunctional industry. The development of modern eco-agriculture ensures agricultural output while at the same time reducing inputs and promotes the development of environmentally friendly and sustainable agriculture. It realizes added economic value through material recycling, multilevel and comprehensive utilization of energy, and serialized processing of agroproducts. It practices the recycling of waste materials to reduce agricultural costs and improve efficiency. It creates employment opportunities within the agricultural industry for surplus labor in rural areas and maintains the enthusiasm of farmers to engage in agriculture.

Fourth, eco-agriculture ensures long-term stability in agriculture development. Eco-agriculture protects and improves the ecological environment, prevents and controls pollution, maintains the ecological balance, and improves the security of agricultural products. It reorients the regular development path of agriculture and rural economy into a sustainable development path and closely combines environmental construction with economic development. While meeting people's growing demand for agricultural products, eco-agriculture also improves the stability and sustainability of ecological systems and reinforces the momentum of agricultural development.

Fifth, eco-agriculture requires policy support. In the environment of a market economy that puts economic benefit first, any weakness in the agricultural industry will be fully reflected in eco-agriculture. Therefore, to develop eco-agriculture, policies must be put in place and the institutional support and eco-compensation mechanisms need to be established in advance.

17.1.5 Reasons for Including Eco-Agriculture in China's National Agricultural Strategy

Chinese eco-agriculture does not totally exclude the use of modern synthetic chemicals. It places more emphasis on the overall design of agro-ecosystems, thereby reducing resource consumption and environmental pollution while improving agricultural productivity and ecosystem security. Modern eco-agriculture is based on simulating the structure and functions of the natural ecosystem. It applies modern scientific methods and technological achievements to the construction of efficient, stable, and safe modern agricultural systems. It is a new development mode of agriculture that uses the best from both traditional agriculture and conventional modern agriculture while integrating production, operation, distribution, and consumption through eco-labeled products and balancing ecological, environmental, economic, and social benefits. In a word, it is the eco-friendly and environmentally friendly version of modern agriculture.

The core of "eco-civilization" is the transition from confrontation between humans and nature to the coordinated development of humans and nature. Because of the fundamental status of agriculture in the national economy and the role of agriculture in ecology, production, and people's lives, agriculture is fundamental to eco-civilization construction throughout the nation. Therefore, eco-agriculture is not only the main component of agriculture and rural eco-civilization construction but also the core of eco-civilization construction of the entire nation.

China's agricultural development is always constrained by resource-related, environmental, and ecological factors. If initiatives are not taken to soften these ecological and environmental constraints for sustainable agriculture development in China, the country will face more and more severe crises of agricultural resources and environment in the coming 10 years. It can be expected that, even if things can stand unchanged for a while, agricultural development will eventually be forced to move in an eco-friendly direction, but only in a much more painful and costly way. As China entered the Twelfth Five-Year Plan for National Economic and Social Development period from 2011 to 2015, eco-agriculture development faced new strategic opportunities. The awareness of citizens about the ecological environment and food safety is greater than ever before and has become the most important public relations aspect of eco-agriculture development. Agricultural operations are more market oriented, and business entities based on agricultural enterprises, family farms, and farmer cooperatives are becoming the main force of agricultural production. These factors serve as an important impetus for the development of eco-agriculture. Scientific research on eco-agriculture, on related agricultural biological diversity and circular agriculture, on the essence of traditional Chinese agriculture, on ecological compensation, and on ecology-related laws and policy has reached new heights. The Eighteenth Communist Party of China National Congress recognized the need to combine eco-civilization construction with all aspects of political, economic, cultural, and social construction as our nation's strategy. Facing such a new opportunity, promoting the constructive development of modern ecological agriculture is not only necessary but also feasible (Liu, 2013). The opportunity for modern eco-agriculture development as a national strategy in China has matured.

17.2 STRATEGIC DESIGN OF AGROECOLOGY PRACTICE

The practice of agroecology in China, or the development of modern eco-agriculture, can be characterized as "one major goal, two transformation pathways, three development stages, four strategic areas, and five transformation measures." The "one major goal" of modern eco-agriculture is to build a resource-conserving and environmentally friendly agricultural system. The concept of "two transformation pathways" refers to the pathway toward constantly increasing agricultural productivity to meet the needs of society and the pathway toward caring about protecting and preserving agricultural resources, environment, and ecology. The "three development stages" refer to the development stages of the near term, middle term, and long term. In the near term, the quality and safety of agricultural products should be tackled and solved first. In the middle term, attention should be paid to the ecological transformation of leading sectors in the agricultural industry to establish sound ecological mechanisms and promote the recycling of waste into usable resources. As a long-term goal, management of the entire system should be ecologically sound and should encompass agricultural input, agricultural production, product transportation, marketing, and consumption. Eventually, organic, green, and healthy agroecology products should dominate the market. The "four strategic areas" that should be recognized in China include (1) eco-sensitive areas that require production control and structural optimization, (2) eco-rich areas that require resource protection and diversified production, (3) advantaged main producing areas that require coordination of farming and animal husbandry, and (4) coastal and urban areas that are suitable for exportation and urban service-oriented agriculture. Agroecology development patterns should be adapted to their natural and social environment. Finally, the "five measures" for strategic transformation for agriculture are (1) establishing pilot zones of modern eco-agriculture in typical areas, (2) building a system of laws and regulations for modern eco-agriculture, (3) improving the market transaction platform and monitoring system of modern eco-agriculture, (4) building a system of technology innovation for modern eco-agriculture, and (5) setting up a special development fund for modern eco-agriculture.

17.2.1 Comprehensive Management of an Agroecology Foundation

Conventional intensive agriculture is now faced with acute resource and ecological problems such as heavy metal pollution, groundwater depletion, non-point-source pollution, severe plant diseases, insect pests, and contamination of agroproducts. To effectively control this degradation, a portfolio of strong measures in terms of technology, economy, administration, and regulations should be adopted in a targeted manner. Such measures will go a long way toward comprehensive management of major problems in severely polluted areas. No effort should be spared to maintain the bottom line for the safety of agricultural resources, environment, and ecology so as to lay a solid foundation for ecological transformation and the healthy, coordinated, and sustainable development of the agriculture industry.

17.2.2 Transformation of Major Production Sectors

Apart from comprehensive management of major ecological and environmental problems, the key to eco-agriculture lies in the major production sectors of the agricultural industry. Major ecological transformation is needed in the conventional agricultural sectors, including the cropping, fish breeding, and animal raising sectors, in order to strengthen their ecological functions. Through optimizing their special structure and production processes, China can conserve land, water, fertilizers, seeds, and energy. Conservation tillage, precise seeding, rational fertilization, biological pesticide use, water-saving irrigation, rain-fed farming, ecological breeding, and comprehensive use of straw and other conservation technologies should all be considered to promote ecological transformation in the major sectors of the agriculture industry and to promote its sustainable development.

17.2.3 Recycle and Reuse

A weakness of conventional agriculture is the separation of various sectors resulting in resources such as straw and animal excrement being discarded and polluting the environment. Modern agro-ecology will recycle and reuse agricultural wastes and rigorously develop the recycling of biomass resources through organic fertilizer, edible fungi, biomass fuel, biomass fodder, etc. The links among sectors will be strengthened to add high-value by-products, to increase resource efficiency, and to better protect the eco-environment.

17.2.4 Input Control

Quality screening and safe use of inputs in planting and breeding are the basics of eco-agriculture. The production, circulation, and use of banned agrochemicals should be strictly prohibited through law enforcement, while the use of safe and green agricultural inputs should be encouraged and process controls imposed.

17.2.5 Clean Production

Development of a set of technical specifications for clean production to ensure the safety of the eco-environment and products should be considered during the ecological transformation of conventional agriculture in China.

17.2.6 Output Safety

A major defect in China's eco-agriculture has been the lack of a direct link to product safety. Modern eco-agriculture must ensure the quality and safety of agroproducts and must keep up with the standards of green food or organic food in China.

17.3 STRATEGIC GOALS AND STAGES OF ECO-AGRICULTURE DEVELOPMENT

17.3.1 Strategic Goals

The primary goal of China's modern eco-agriculture is to establish a resource-conserving and environmentally friendly agricultural system that will ensure continuous increases in agricultural productivity and protect the country's agricultural resources, environment, and ecology (Cheng et al., 1997; Wang, 2001; Huang et al., 2011). It has been estimated that it could take China 20 years or so to complete the ecological transformation of its agriculture while maintaining the quality and safety of its agricultural resources, environment, and agroproducts. Such an effort will facilitate efficient, coordinated, sustainable, and healthy development of this industry (Chen and Ma, 2014). In the future, organic and green eco-agriculture products will occupy 60% of the market (Yin et al., 2015). The primary goal consists of three key points. First, the time span will be around 20 years. Second, the idea of "making eco-agriculture the mainstream of China's agricultural industry" reflects the importance of the primary goal, as well as the difficulty in achieving this goal. Third, China's eco-agriculture will make direct contributions to agro-environment, food safety, efficient and healthy agriculture development, and construction of a new rural area.

These goals would seem attainable, as China has already gained some experience in developing eco-agriculture, and agroecology is a global trend that has attracted the attention of the Chinese government and even the global community. These goals are proposed based on the foundations and trends of eco-agriculture at home and abroad. Eco-agriculture and organic agriculture, after two decades of development, still represent a small proportion of agriculture in most other countries, varying from 1 to 10% area-wise (Chen et al., 2015). The development of eco-agriculture in China has been relatively rapid but more is needed. Areas devoted to eco-agriculture now account for more than 10% of the total farm production area in China (Cao and Wang, 2006). Within two decades, China should have moved forward in setting up legal safeguards, technical support, and the organizational and management mechanisms necessary to grow its agroecology program (Cao, 2013).

17.3.2 Agroecology Development Stages

The development of China's modern agroecology should be a three-stage endeavor.

17.3.2.1 Near Term (2015 to 2020)

The target of this stage will be the comprehensive treatment of urgent problems with agricultural resources, environment, and ecology. A portfolio of strong measures in terms of technology, economy, administration, and policies should be adopted to resolve problems with agricultural resources, environment, and ecology and to curb degradation. In the first 3 years, a "mandatory restrictive plan on the negative list (red list)" should be carried out, which includes periodic treatment and avoidance of heavy metal pollution, monitoring the safety and application of agricultural inputs (e.g., imposing limits on the use of major fertilizers), imposing mandatory restrictions on extracting groundwater, adopting water-conserving technologies and management measures, eradicating highly toxic pesticides, prohibiting the burning of straw, mandatory waste treatment and recycling in medium- and large-sized farms, implementing complementary reward and sanction measures according to the performance assessment of relevant departments at all levels, and building demonstrations to successfully deal with major ecological issues. In the latter 3 years, a "green incentive action plan (green list)" should be implemented. A portfolio of strong measures regarding technology, economy, administration, and policies should be proposed and adopted to encourage and support the development and application of eco-friendly technologies, offer subsidies to green-labeled products and products with "organic conversion" certification, support ecosystem conservation, encourage eco-friendly use of land and water and use of green means of production, and build eco-friendly demonstration examples. Based on these 6 years of efforts, the goal for the share of organic and green eco-agriculture products can be set to about 20%.

17.3.2.2 Medium Term (2021 to 2025)

This is the stage for ecological transformation of major agricultural sectors, which will recycle and reuse resources and establish eco-agriculture systems. On the basis of comprehensive measures, accumulated experience, and demonstration efforts of the previous 6 years in the short term, leading sectors in the agriculture industry will be transformed ecologically by means of legislative mechanisms setting up green budget and eco-compensation policies. China can achieve ecological transformation of its agriculture as a whole through eco-agriculture regulation and control ensured by the implementation of relevant regulations and laws. Major projects such as establishing socioeconomic mechanisms for agroecology and pilot projects for the ecological transformation of agriculture should be launched in this period. In the meantime, a system for eco-compensation and successful demonstrations should be built up. The share of organic and green eco-agriculture products can be expected to reach 40%.

17.3.2.3 Long Term (2026 to 2035)

This will be the stage for all-around construction and optimization of an eco-agriculture system. It anticipates that input safety control, process safety control, product safety control, and ecology safety control systems will work at high efficiency in this stage. A comprehensive, modern eco-agriculture system will be established to ensure the safe, efficient, coordinated, healthy, and sustainable development of eco-agriculture and to make continued breakthroughs in the ecological transformation of agriculture. On the basis of previous efforts, large-scale ecological transformation of agriculture will be conducted in an efficient manner. Specifically, action plans including land fallowing for ecological restoration and protection and sustainable exploitation of agricultural biodiversity will be implemented to build up a production system and an environment safeguarding system that rely more on ecological resources. The level and scale of ecological transformation will increase and guarantee the increased agricultural productivity and safety of resources, environment, and ecology. The share of organic and green eco-agriculture products is expected to be above 60% by the end of this stage.

17.4 STRATEGIC MEASURES IN THE TRANSITION TOWARD AGROECOLOGY

After three decades of reform and opening, China's agriculture has basically transformed from a decentralized and self-subsistence type to a specialized and market-oriented industry. In decades to come, agriculture should eliminate the resource depletion and environmental pollution associated with conventional farming and move toward resource-conserving and environmentally friendly eco-agriculture. Modern eco-agriculture requires compliance with the so-called "four coordinations" (coordination of production, people's lives, and ecology; coordination of the cyclic advancement among resources, products, and reused resources; coordination of resource conservation and clean production; and coordination of industrial upgrading and added value of products) and the "three combinations" (combination of effective supply of major agroproducts and prevention and control of environment pollution; combination of ideas in agricultural development with those in the whole process control of environment pollution; and combination of prevention and control of urban, industrial, and rural pollution). Specific measures are as follows.

17.4.1 Building Eco-Agriculture Demonstration Zones in Typical Areas

The State Council has mandated governments at all levels to divide areas according to their specific major natural and social functions. Land should be developed in strict accordance with its function. Based on function-specific planning and the eco-environment features of different areas, China will coordinate the ecology, people's lives, and production by establishing national pilot eco-agriculture zones in typical areas and in provincial pilot zones. Based on local eco-environmental challenges, more detailed function-specific zoning and landscape planning will be carried out at county and township levels. In this way, major challenges with the resources, ecology, and environment of local agricultural development, as well as the potentials and opportunities, can be fully analyzed. Major targets of modern eco-agriculture development can be identified while establishing eco-agriculture pilot zones. For leading eco-agriculture enterprises, small and medium enterprises, eco-villages, eco-gardens, eco-agriculture-specialized cooperatives, and eco-households in various areas will be set up. Pilot zones integrating modern eco-agriculture industry, technologies, and eco-compensation and offering capital, technical, and policy support should be established for the accumulation of experiences.

17.4.1.1 Rigorously Promoting the Application of Green Inputs

The use of toxic chemicals should be strictly prohibited and any illegal production or use of banned agrochemicals should be penalized through agricultural law enforcement as a way to further regulate and promote the production and application of green inputs. Straw, animal excrement, and by-products should be recycled to develop compost, biogas, and other biomass products. China needs to keep expanding the scale and scope of such clean production in rural areas. With villages as basic units, clean production integrates and promotes water-, fertilizer-, and energy-saving technologies; establishes eco-projects to intercept nitrogen and phosphorus in farmland; and builds waste recycling facilities for straw, excrement, household garbage, and wastewater according to local conditions. These will promote the reuse of rural waste resources. China needs to renew its ideas and approaches to solving pollution within the production process and people's daily lives, while embedding environmental protection into the goal of increasing food output, agricultural productivity, and farmers' income. Measures for saving water and reusing wastes should be implemented; these could include clean production methods and waste-recycling technologies, such as water-saving and pesticide-saving technologies and eco-healthy breeding; increasing the efficiency of fertilizers and pesticides; reducing wastage rates; lowering external inputs and therefore pollution discharge; and implementing key measures to treat products and pollution in some areas. China will strictly control the transfer of industrial wastewater, waste gas, and solid waste to rural areas; prohibit the discharge or dumping of these wastes in agricultural production areas; prohibit the direct use of urban garbage or mud as fertilizers; and prohibit the stacking, storage, or treatment of solid waste in producing areas. If solid waste is to be stacked, stored, or treated near producing areas, effective measures must be taken to prevent potential pollution of the producing areas. Monitoring and control of heavy metal pollution sources should be strengthened. Pollution control standards in urban and rural areas should be gradually aligned to realize synchronized treatment in both areas.

17.4.1.2 Improving Eco-Agriculture Landscapes in Different Areas

More detailed function-specific zoning and landscape planning should be carried out at county and township levels, including the following:

1. Identify eco-protection areas, eco-sensitive areas, and pro-production areas in agricultural or nearby regions.
2. Coordinate the land use of secondary industry and transportation in core agricultural areas so as to avoid potential external pollution sources from upwind or upstream directions.

3. Sort out important balance relationships and their layout in a certain area. In planning, there is a need to calculate and predict several "balances": supply–demand balance of agroproducts over the long term; supply–demand balance of water resources; balance between the self-cleaning capacity of water bodies and the projected discharge of pollutants (e.g., nutrients, COD); balance between permitted number of livestock and actual excrement-absorbing capacity of the land; balance between organic matter consumption by farmland and its generation; and the temporal and special layout of the source, pool, and flow of such energies and substances.

4. Optimize the spatial arrangement of planting and breeding sectors. Different plantations should be embedded within one another in an orderly manner, and planting and breeding sectors should be combined organically.

5. Add aesthetic value to the landscape planning of living areas in villages, taking into consideration not only the ecology and environment but also the convenience of people's lives, thus combining traditional folk culture with the beauty of nature.

6. Carry out vegetation planning, which is crucial to wind speed adjustment, water filtration, habitats of natural enemies, and protection of species. Vegetation planning includes planning the forest network for farmland, filtration and buffering vegetation systems along irrigation and drainage systems, vegetation systems to protect natural enemies habituating on ridges, systems of scenic forests in villages, vegetation systems at sources of water in mountainous villages, and forest corridors that connect species in different forests.

17.4.1.3 Building a Recycling System of Eco-Agriculture

The agricultural recycling system is centered on the utilization of crop straw and residues, livestock waste, agro-rocessing wastes, and urban and rural organic refuse. The utilization methods can be physical (e.g., straw crushing machinery, compression equipment, carbonization equipment), chemical (e.g., straw decomposing agents, pH conditioning agents), biological (e.g., microbes, edible fungi, earthworms, insects), or a combination of these (e.g., biogas, compost, artificial wetlands, ecological ditches). Through such recycling processes, more diversified products can be made, and the excess energy and organic matter can be recycled back into agricultural production. The recycling systems can happen within a single production unit or across multiple, even regional production units by creating professional producers of recycled products. As the old wisdom goes, "Where the blood flows unimpeded, there shall be no pain and illness"; hence, the recycling process will fundamentally address agriculture-induced non-point-source pollution and other environmental and food security issues, reduce inputs needed, and increase resource use efficiency.

17.4.1.4 Promoting the Sustained Use of Biodiversity Resources

Using the essence of natural ecosystems and traditional agricultural practices, an agricultural pattern that makes the greatest use of biodiversity under intensified and large-scale conditions should be established: (1) promoting multistory cropping based on differentiated growth periods in perennial woody plantations (rubber, fruits, tea, and cash-crop trees); (2) establishing a system of rotational cropping, intercropping, and relay cropping that both uses and nurtures the farm land; and (3) enhancing pest prevention by improving biodiversity based on gene resistance, food chain restrictions, biochemical attraction and repulsion, etc.

17.4.2 Establishing and Improving the System of Laws and Regulations Related to Modern Eco-Agriculture

China has made much progress in establishing a legal system for agricultural production and environment and resource protection over the years. The legal system for resource and environment protection has been preliminarily put in place. Nearly 20 laws on environmental protection have been promulgated and amended (Wei and Zeng, 2014). Over 30 administrative regulations have been drawn up by the State Council. The support of ecological protection and construction and the requirements and methods for ecological compensation have been articulated in many rules, regulations, and policies. In general, however, a strong and sound legal system for eco-agriculture has yet to be established. There is no clear-cut identification of stakeholders' rights and obligations, or specifics on the content, methods, and standards of such support. The related legislation lags behind the development of ecological protection and construction, with some laws and regulations being incompatible with institutional changes and economic growth needs and some legal support for emerging ecological issues and protection methods being ineffective. Also, some laws of significance fall short of effective regulation of ecological protection and compensation.

Learning from the experience of developed countries such as the United States and Japan and the European Union, improvements in the legal system of agroecology practice in China are suggested. A law on eco-agriculture in the People's Republic of China needs to be promulgated in a timely manner to serve as the legal bedrock of eco-agriculture development. Based on near-term plans and mid- to long-term objectives already made in the National Program of Eco-Agriculture Development, progressive objectives, targets, and measures should be incorporated into the national five-year economic development plan as a guide for practice.

A modern eco-agriculture compensation mechanism and its policy structure should be established. Eco-agriculture and organic agriculture are characterized by high investments, low direct benefit returns, and significant environmental externalities, all of which warrant effective fiscal, financial, and taxation support at both national and local government levels. More public finance should be earmarked for agriculture and eco-agriculture. In particular, the proportion of investment in the agricultural infrastructure and ecological projects should be increased, and more emphasis should be placed on agricultural technology research and development, extension, and application. Taxation as a lever controlling socioeconomic life must be given full play in the form of tax credits for proceeds gained from eco-agriculture investment. Public subsidies should be increased and skewed to higher science technology input, infrastructure construction, and rural and ecological input. More policy-based financial support should be given to eco-agriculture and organic agriculture.

In view of the long-term development of eco-agriculture, greater policy support is necessary. Production equipment for ecological and organic agricultural products should be listed in the national subsidy directory. National demonstration areas, parks, bases, flagship enterprises, and cooperatives of green and organic agriculture should be established, and policy support should be given to those showing good demonstration effects. Great assistance should be given to marketing green and organic agricultural products, government-sponsored marketing events should be organized, and a broader market for green and organic producers explored. Also, effective economic compensation should be given to the production bases based on the sizes and phases of transition or stability of production.

17.4.3 Improving the Market Trading Platform and Supervising System for Modern Eco-Agriculture

The certification of eco-agriculture products in China is a threefold system involving pollution-free agricultural products, green food, and organic food, which are certified by specific agencies under the Ministry of Agriculture and the Ministry of Environmental Protection. Despite its positive effects, this certification system suffers from ambiguity among the three types of products, leading to confusion among producers and consumers. In reality, much effort is put into the certification process while much less is done with regard to supervision. The absence of governance over the certification bodies may result in lax enforcement and unreasonable charges by the certification bodies. For this reason, green and organic product certification must be put into better order. The supervisory power of the certification and accreditation administration of China should be enhanced. A market exit mechanism should be adopted, the authority of these bodies should be enhanced, and the cost of becoming certified should be reduced through government and market regulation. A multifold supervisory system including certification bodies, local administrative management, and self-discipline of farmers should be established. The concepts and standards regarding eco-agriculture products must be further specified. Pollution-free agricultural products and green food can be referred to as agricultural products with reduced chemical input or as eco-agriculture products.

17.4.4 Improving the Technology Innovation System for Eco-Agriculture

Multiple technology systems used in eco-agriculture face the challenge of holistic innovations. Given the fact that modern agriculture is large scale, enterprise dominated, and market driven, the technological systems packages used in eco-agriculture require holistic innovations that are locally adapted, suitable for large-scale application, and standardized for easy implementation. Better eco-agriculture model system designs as well as supporting technologies should be realized by learning from and drawing upon the experience accumulated during the tenth and eleventh five-year economic development plan periods and the latest research results, both domestic and foreign. Technologies can come from proven practices in the past; from modern agriculture technologies such as new varieties, new fertilizers, new pesticides, and new equipment; from new materials, information technologies, and biological technologies; and from the wisdom acquired by farmers with regard to rotational cropping, intercropping, relay cropping, and disease and pest prevention, as well as traditional farming successes such as combining fisheries and rice paddies, duck breeding and rice paddies, or dikes and ponds. Training programs customized to different localities based on the types of ecological resources and production setups are needed. The trainees should be representatives from agricultural authorities at provincial, municipal, and county levels, in addition to those in charge of pioneering agriculture enterprises, large professional farmers, farmer cooperatives, and family farms. By grasping the essence of measures and management of eco-agriculture, a talent pool of operational, managerial, and innovative expertise for eco-agriculture can be established.

17.4.5 Establishing a Special Fund in China for Modern Eco-Agriculture

As the first step, a special fund for comprehensive demonstration zones of modern eco-agriculture should be created as a buttress for national endeavors in developing and demonstrating eco-agriculture. Successful experiences can then be promoted aggressively. After systemwide progress has been made in national and provincial demonstration areas, a special fund from the public budget dedicated to modern eco-agriculture should then be created. It will be an integral part of the national fiscal budget and green budget in particular and should be increased each year in an effort to guarantee China's strategic transformation to modern eco-agriculture and agroecology.

REFERENCES

Cao, J.J. and Wang, X.Z. 2006. On the situation of modern eco-agriculture between China and foreign countries. *Ecol. Econ.*, 6: 108–111.

Cao, Z.P. 2013. Future orientation of ecological agriculture. *Chin. J. Eco-Agric.*, 21: 29–38.

Chen, D.W. and Ma, L.Q. 2014. Selection of ecological agriculture development mode in China—based on the comparative perspective of developed countries. *J. Anhui Agric. Sci.*, 42: 8395–8396.

Chen, X., Yu, L.W., Kang, Y.X., and Chen, W.Z. 2015. Experience and enlightenment of the development of ecological agriculture in foreign countries. *Tianjin Agric. Sci.*, 21: 90–93.

Cheng, X., Zeng, X.G., and Wang, E.F. 1997. *The Introduction of Sustainable Agriculture.* Beijing: China Agriculture Press.

Huang, G.Q., Zhao, Q.G., Gong, S.L., and Shi, Q.H. 2011. Overview of eco-agriculture with high efficiency. *J. Agric.*, 4: 23–33.

Li, W.H. 2003. *Ecological Agriculture: Theory and Practice of China Sustainable Agriculture.* Chemical Industry Press: Beijing.

Li, W.H. 2008. *The Problems of Ecological Agricultural and Comprehensive Governance.* China Agriculture Press: Beijing.

Li, W.H. and Lai, S.D. 1994. *Agriculture and Forestry Compound Management China.* Science Press: Beijing.

Li, W.H., Min, Q.W., and Zhang, R.W. 2005. *Ecological Agriculture Technology and Model.* Chemical Industry Press: Beijing.

Liu, S.H. 2013. Retrospective on the construction of socialist ecological civilization: a review of the "five in one" overall goal of the socialism path with Chinese characteristics. *J. Econ. Shanghai School*, 4: 36–40.

Luo, S.M. 2007. To discover the secret of traditional agriculture and serve the modern ecoagriculture. *Geogr. Res.*, 26: 9–15.

Luo, S.M. 2010. On the technical package for eco-agriculture. *Chin. J. Eco-Agric.*, 18: 453–457.

Wang, Z.Q. 2001. *Chinese Ecological Agriculture and Agriculture Sustainable Development.* Beijing Press: Beijing.

Wei, Q. and Zeng, Z. 2014. Comparative analysis of China–Japan agroecological environment policy. *World Agric.*, 2: 76–78.

Yin, C.B., Cheng, L.L., Yang, X.M., and Zhao, J.W. 2015. Path decision of agriculture sustainable development based on eco-civilization. *Chin. J. Agric. Resour. Reg. Plan.*, 36: 15–21.

Zhang, X.L. 2004. Policy introduction: restructuring China's social policy system: an examination of the impact of the scientific approach to development on social policy formation and implementation. *Social Sci. China*, 4: 99–101.

Zhao, Q.G., Huang, G.Q., and Qian, H.Y. 2007. Ecological agriculture and food safety. *Acta Pedol. Sin.*, 44: 1127–1134.

Index